FOUNDATIONS OF
AGRICULTURAL EDUCATION

FOUNDATIONS OF

AGRICULTURAL EDUCATION

Fourth Edition

B. Allen Talbert, Barry Croom,
Sarah E. LaRose, Rosco Vaughn, and Jasper S. Lee

Purdue University Press • West Lafayette, Indiana

Cataloging-in-Publication Data is on file with the Library of Congress.
978-1-61249-752-5 (print)
978-1-61249-753-2 (epub)
978-1-61249-754-9 (epdf)

Cover credit: Image 898449496 by Kinwun/iStock/Getty Images Plus via Getty Images;
Image 931566626 by torwai/iStock/Getty Images Plus via Getty Images; Image 1201407573 by
Ridofranz/iStock/Getty Images Plus via Getty Images

About the Authors

Dr. B. Allen Talbert is a professor of agricultural education at Purdue University. He currently teaches courses in school-based agricultural education program planning, SAE/FFA, and student teaching. His research focuses on recruitment and retention of students from underrepresented minority groups. His engagement work is focused on professional development of agriculture teachers and service to FFA on all levels.

Dr. Barry Croom is a professor at the University of Georgia and professor emeritus at North Carolina State University. He has more than thirty-five years of experience in agricultural education. Croom began his professional career as a high school agricultural education teacher. While a high school teacher, he was selected by the National FFA Organization to develop and present in-service workshops to teachers across the United States. Croom maintains a research program that focuses on effective teaching, career and technical education policy, and diversity in agricultural and extension education.

Dr. Sarah E. LaRose is an assistant professor of agricultural education at Purdue University. She began her career in agricultural education as a high school teacher and FFA advisor in Woodbury, Connecticut, where she developed curriculum on local food production and extensively used agricultural teaching laboratory spaces to deliver instruction. Her research seeks to cultivate the development of agricultural educators who actively create student-centered, inclusive programs so that all students can experience the transformative benefits of agricultural education.

Dr. Rosco Vaughn is a professor and agricultural teacher educator at California State University, Fresno. He was a coauthor of previous editions of this title.

Dr. Jasper S. Lee is a retired agricultural educator. He served as a faculty member at Virginia Tech and Mississippi State University, after which he worked full-time as an author and publisher. He was a coauthor of previous editions of this title.

Preface

Foundations of Agricultural Education was previously published by Pearson Education and Professional Educators Publications. This edition with Purdue University Press continues to expand on the major goal of the book: to introduce future agricultural educators to their profession and support professional development of those now in the profession.

The second edition updated and expanded on the content of the first edition. The third edition enhanced the useful features of the first two editions. Now, this fourth edition updates each chapter using current literature, legislation, and initiatives. In recent years, digital learning, life–work balance, and DEI (diversity, equity, and inclusion) discussions have been prominent in school-based agricultural education. Therefore, this edition adds a new chapter on digital learning, includes a new chapter with a life–work balance focus, and highlights DEI efforts throughout the textbook. The authors were determined to prepare a relevant book for the agricultural education profession and are pleased to be able to bring this book to you.

The audience for this book includes college students in agricultural teacher education programs, agriculture teachers, state supervisors, teacher educators, and others interested in agricultural education. The overall purpose is to provide a foundational resource, one that broadly covers each element necessary to be a teacher of agricultural education. The book is appropriate for introductory as well as advanced classes in agricultural teacher education. Incumbent teachers will also find information useful to them as they go about their roles as professionals in agricultural education.

Agriculture and education are both fast paced and ever changing. Agricultural education is a blend of each of these. This book focuses on current content, terminology, practices, and theory while giving historical and philosophical foundations to agricultural education. Opening scenarios, examples, and terms have been used that will help to keep the book current.

We strongly feel the secondary agricultural education model of classroom/laboratory instruction, supervised agricultural experience, and FFA has withstood the test of time. This model, when properly followed, will result in enhanced student learning, a better prepared agricultural workforce, more competent community leaders, and an agriculturally literate society.

Delve into *Foundations of Agricultural Education*. Review the contents and thumb through the chapters. You will note a user-friendly organization, images that reflect successful local program practices, and a unified professional approach. You will also note wide regional representation across the United States. This reflects the backgrounds of the authors as well as nationwide practices. Your review of this book should pique your interest and promote your professional enthusiasm for greater study. And do the authors a favor: commit to an energetic and productive career as a teacher of agricultural education.

Acknowledgments

We acknowledge the important roles of our families in this work. The extra time and effort required often took us away from family duties. We appreciate their understanding and hope they view this product as worthy.

We are grateful to many individuals who have directly or indirectly contributed to the production of this book. Some were high school teachers, and others were teacher educators who guided us in our own professional development. Others who should be acknowledged are our current professional associates, who daily enhance our knowledge and help mold our professional practice.

We appreciate the universities where we are teacher educators for allowing and supporting the development of this book. Purdue University; California State University, Fresno; and University of Georgia are acknowledged in this regard. We are also grateful to other universities where we have studied or had close personal contact. These include Virginia Tech, Mississippi State University, New Mexico State University, University of Illinois, University of Connecticut, University of Florida, The Ohio State University, Texas Tech University, Clemson University, The Pennsylvania State University, Texas A&M University, Oregon State University, Cornell University, North Carolina State University, and University of Minnesota.

We wish to acknowledge the organizations affiliated with agricultural education, including the National Association of Agricultural Educators, the American Association for Agricultural Education, and the National Association of Supervisors of Agricultural Education. The staff of the National FFA Organization is also acknowledged for direct and indirect support of the authors.

Several schools are acknowledged for extra efforts in supporting the book with images. These include agriculture teachers at Lyman Hall High School (specifically Kathryn Dal Zin, Rachel Holden, and Emily Picard), Ledyard High School (specifically Devon O'Keefe), and Nonnewaug High School (specifically Thomas DiMarco, Eric Birkenberger, Marisa Bedron, and Kathleen Gorman) in Connecticut; Switzerland County, North Decatur, Eastern Hancock High School, Franklin Community High School, and Manual High School in Indiana; Eastern Randolph High School, Chase High School, and Chatham Central High School in North Carolina; Florin High School, Casa Roble High School, Ponderosa High School, Lemoore High School, West Central High School, Madera High School, Kingsburg High School, Elk Grove High School, and Clovis High School in California; Montrose High School in Colorado; Molalla High School, Canby Union High School, and North Clackamas High School in Oregon; Tallulah Falls School, Oconee County High School, Madison County High School, and Franklin County High School in Georgia; Sandra Day O'Connor High School in Texas; and Millsaps Vocational Education Center, Starkville Academy, and East Mississippi Community College in Mississippi. The assistance of students at Piedmont College, Georgia, as technical models is gratefully acknowledged. Education Images, c/o Dr. Jasper Lee, is acknowledged for its assistance as a source of many images used in this book.

We would also like to thank the reviewers of previous editions for their thoughtful comments and suggestions. They are Thomas Dobbins, Clemson University; Gary Briers, Texas A&M University; Kristin Stair, New Mexico State University; and Gregory Miller, Iowa State University.

Formerly published by Professional Educators Publications (PEP), Inc., of Illinois and Pearson Education, Upper Saddle River, New Jersey, the book is now published by Purdue University Press, West Lafayette, Indiana.

The change in publishers brings exciting new opportunities. A thank-you is extended to all individuals at Purdue University Press who made this book possible.

Contents

PART 2 Program Development and Management

PART 3   Instruction in Agricultural Education

18
Adult and Postsecondary Education, 334

19
Evaluating Learning, 349

20
Meeting the Needs of Diverse Students, 367

21
Using Laboratories, 383

PART 4 Supervised Agricultural Experience, FFA, and Community Resources

22
Supervised Agricultural Experience, 413

23
FFA, 431

24

PART 5   Career Stages in Agricultural Education

25

Part 1

Introduction to the Agricultural Education Professions

1

A Career in Agricultural Education

Agricultural education offers a number of important, challenging, and rewarding professional opportunities. You most likely have carefully investigated the possibilities for you as an agricultural educator. Some possibilities were likely obvious to you; others might not have been quite so obvious.

Teaching is the first opportunity many agricultural education graduates consider. It allows you to utilize the education you have received in teacher preparation and enjoy unique success as a teacher of agriculture. You may also want to consider other areas of agricultural education such as a state leader or university agricultural teacher educator. In addition, you may consider numerous opportunities in education (such as school administration or career counseling) and the agricultural industry (such as human resource director or manager of staff development). Regardless, success in agricultural education will require setting goals and putting forth the needed effort to achieve your goals.

You are now likely enrolled in university-level classes that will prepare you to become an agricultural educator. You are probably pursuing a college degree that meets teacher credentialing requirements. You may be in the process of changing your thought perspective from that of a student to that of a teacher. You are likely thinking about professional matters related to being an excellent teacher. This chapter and others in this book are intended to help you develop into a highly competent, professional agricultural educator.

TERMS

administration	profession
agricultural education	professional
agricultural education program	professional code of ethics
Carnegie unit	professional growth plan
continuing education	reciprocity
counseling	salary schedule
Council for the Accreditation of	stakeholder
Educator Preparation	standard teaching license
credentialing	teach
creed	teacher
demeanor	teaching
formal education	teaching license
LEAP	transition-to-teaching program
nonformal education	workplace specialist teaching license

OBJECTIVES

This chapter addresses the National Quality Program Standards for Agriculture, Food, and Natural Resources Education (National Council for Agricultural Education, 2016), specifically Standard 6: Certified Agriculture Teachers and Professional Growth. It has the following objectives:

1. Describe the meaning and importance of teaching.
2. Explain teaching as a profession, including professionalism and the role of ethics.
3. Discuss the meaning and scope of agricultural education.
4. Discuss credentialing requirements.
5. Describe practices in gaining an initial position.
6. Describe compensation and other benefits of teaching.
7. Identify relationships of agricultural education to education in the United States.
8. Identify the roles and responsibilities of agriculture teachers.
9. Relate teaching agriculture to the school community.

FIGURE 1.1 A preservice agriculture teacher is participating in a curriculum workshop.

TEACHING AGRICULTURE

People have different perspectives on the meaning of teaching. Some aspects are acknowledged by all as essential to defining teaching. Foremost is that the role of teaching is to promote learning. It is used in the education and training of individuals so they have necessary knowledge, skills, and values that promote a gainful and successful life.

Its Meaning

Teaching is defined in terms of two other words: *teach* and *teacher*.

To **teach** is to guide or cause others to learn something. It also means to instruct others using various strategies, such as examples and experiences. The process is neither simple nor easy. To teach involves far more than contact with learners. It includes planning and evaluating the educational process as well as providing agricultural education program leadership.

Anyone who teaches is called a **teacher**. We may also think of a teacher as an individual whose professional occupation is to teach. Teachers are often further identified by the subject they teach, such as agriculture, horticulture, or veterinary science. An agriculture teacher is a teacher of agriculture, including closely related subjects.

Teaching is the art and science of directing the learning process (Lee, 2001). This is an active role that involves many functions in helping others learn. Teaching can occur in a wide range of environments, such as in the pasture of a ranch, at a bench in a greenhouse, or in a school classroom. Teaching environments are discussed in detail later in this book.

Educators sometimes limit the definition of *teaching* to "instruction that occurs in an educational institution." That definition does not fit agricultural education very well. Think of the teaching that occurs in the agricultural industry outside the walls of a school building!

FIGURE 1.2 Agriculture teachers often use realia (real things or likenesses of real things) in teaching. (COURTESY OF EDUCATION IMAGES.)

Educators also sometimes limit the definition of *teaching* to "instruction provided by a credentialed, certified, or highly qualified individual." Every state has certification requirements. Some local school districts also have requirements beyond those of the state. In general, credentialing requires that a teacher be a graduate of a college teacher education program and meet other requirements a state may impose, such as minimum scores on standardized tests.

Teaching is a profession of more than 50 million people throughout the world. The field of education, however, includes more than teachers. It also includes administrators, curriculum specialists, counselors, and others at elementary, middle school, high school, and post–high school (college) levels. But not all teachers are in education! Some teachers are employed by government agencies, agricultural businesses, and nonprofit organizations.

The roles in teaching agricultural education are many and diverse. Regardless, the overall aim is to direct the learning process of students—children and adults—efficiently and effectively toward educational objectives and goals.

Its Importance

The well-being of a nation requires an educated citizenry. The ability to read, write, use mathematics, and perform other basic skills is essential for people to lead productive and satisfying lives. Education promotes the appreciation of diversity and increases acceptance of individual differences in people. An educated population can lead to increased productivity and a higher standard of living.

Following are some important reasons for teaching, with an emphasis on agricultural education:

- Promote overall development of youth and adults for productive and successful lives.
- Assure that people have sufficient agricultural literacy to be good consumers and citizens.
- Promote agricultural practices to assure sufficient food and fiber supply to meet the needs of an increasing human population.
- Promote safe food handling and preparation practices.
- Develop knowledge and skills needed to enter and advance in an agricultural or related career.
- Provide preparation for students to pursue education in agricultural areas beyond high school.
- Promote the use of sustainable resource practices in agriculture, horticulture, forestry, and other areas.
- Provide related education to promote achievement in science, mathematics, technology, and other academic areas.
- Help individuals enjoy a higher standard of living.
- Promote an appreciation for the environment and its protection.
- Promote wise use of natural resources to achieve human goals.

Where Teaching Occurs

Teaching occurs in formal and nonformal settings. *Formal education* is training or education that is provided in an orderly, logical, planned, and systematic manner and often associated with school attendance. Formal settings are those in schools where structured teaching is used to assure essential standards are met. *Nonformal education* is typically offered outside

FIGURE 1.3 Laboratories with animals, plants, and other facilities are often used in teaching agriculture.
(COURTESY OF LEDYARD AGRI-SCIENCE & TECHNOLOGY PROGRAM.)

of schools and in settings other than those that involve the implementation of structured standards and fulfillment of diploma or degree requirements.

In agricultural education, formal education is provided through public and private schools at the primary, middle school, secondary, postsecondary, and college and university levels. Formal agricultural education on the middle and high school levels is often referred to as school-based agricultural education. This instruction and classes may help students meet graduation requirements.

Nonformal agricultural education may be provided by agencies, associations, and entrepreneurial businesses. A major provider of nonformal agricultural education is the Cooperative Extension System. A collaborative effort of the federal, state, and local governments, the teaching is by local agents, specialists, or other qualified individuals. The National Institute of Food and Agriculture (NIFA) of the U.S. Department of Agriculture administers programs at the federal level. At the local level, extension agents or educators are hired to organize and provide instruction in areas of agriculture, health and human sciences, 4-H youth, as well as a few other areas related to community development and the environment. Nonformal education may meet certification needs (such as pesticide applicator requirements) and continuing education requirements but does not usually offer credit toward a high school diploma or college degree.

TEACHING AS A PROFESSION

A *profession* is an occupation that requires a long, specific program of preparation and uses a code of ethics to guide the conduct of individuals in the profession. Testing may be used as an additional approach to assure that only individuals with high capability enter a profession. Most professions also have some degree of self-regulation and control over credentialing, though this area has sometimes drawn criticism as lacking in the education profession.

Professional Status

A *professional* is an individual who has acquired those traits ascribed to the profession practiced by the individual. This includes the appropriate education, training, and competence in carrying out the duties of the profession. You might have had teachers who demonstrated that they were professionals in all regards. Unfortunately, you might have had a few teachers who did not demonstrate good professionalism.

Following are a few generally held expectations for agriculture teachers to become professionals and continue to enjoy high professional status.

Preparation   The agriculture teacher has the credentials to be an outstanding professional. This includes having the education and experience to be an effective teacher and the desire to stay current in the profession. Most agriculture teachers have a college degree in agricultural teacher education plus additional formal and informal education. Some individuals become agriculture teachers following alternative routes of preparation such as having a degree in an area of agriculture followed by teacher education preparation. Certification as a teacher is typically provided by a state agency with the legal authority to do so. Agricultural teacher educators at a university or college with a teacher education program will be able to provide all details related to preparation.

Participation   A professional agriculture teacher takes an active role in the agricultural education profession and in education. The teacher is a member of professional organizations and participates in professional activities to assure continued growth and development. A professional speaks positively about the profession and promotes its well-being. Agriculture teachers also interact professionally with other teachers in the school systems where they are employed.

Continuing education   Staying up-to-date is an important responsibility of a professional. *Continuing education* is education after an individual has completed a general level of formal education. College courses, workshops, field days, reading, and other approaches are used to keep current. These may contribute to the requirements of an advanced degree or a higher salary level, or they may be for general knowledge and skill development. Many states have in-service activities for agriculture teachers.

Demeanor   *Demeanor* is the outward manner of an individual. It includes the way in which a person relates to others, as well as the manner in which he or she dresses and grooms, the skill with which the person uses language, and the general state of organization and cleanliness in his or her environment. These attributes can be improved to increase professionalism.

Citizen responsibilities   Teachers should follow regulations, pay taxes, vote, meet financial obligations, and maintain honesty and integrity in all regards. Teachers who fail to do so discredit themselves as well as their fellow professional teachers.

Community standards   Though such standards are largely unwritten, agriculture teachers should meet acceptable moral and personal standards of the communities in which they live and teach agriculture. Teachers are typically held to higher standards than nonteaching citizens of the community. Most communities do not approve of teachers using recreational drugs or similar substances, socializing with students in inappropriate ways, and violating laws established for the good of the overall population.

Code of Ethics and Creeds

A *professional code of ethics* is a statement of ideals, principles, and standards related to the conduct of individuals within a profession. Codes of ethics establish rigorous ethical and moral obligations of individuals in a profession.

The education profession in the United States does not have a universally adopted code of ethics. Some state education agencies and local school districts have statements of principles and ideals for teachers. Associations related to the education profession may also have statements resembling codes of ethics. A code of ethics for the education profession was established by the National Education Association (NEA) and adopted by its representative assembly in 1975. This code of ethics has a preamble followed by two principles. Principle I focuses on commitment to the student. Principle II focuses on commitment to the profession of education. A limitation is that the NEA is not represented in all school districts and has limited visibility among some teachers. Lawrence Baines (2010) stated that the NEA Code of Ethics is the closest facsimile to a national code of ethics for teachers. Some states and local school districts copy the NEA Code of Ethics verbatim and adopt it for their needs. (This code of ethics for the education professional may be viewed at the NEA website: www.nea.org/resource-library/code-ethics-educators.)

Creeds may serve as guides for behavior of individuals in a profession. A *creed* is a statement of general beliefs of individuals in a profession or organization. The National Association of Agricultural Educators (NAAE) has developed the Ag Teacher's Creed (see Figure 1.4).

MY CREED

I am an agricultural educator by choice and not by chance.

I believe in American agriculture; I dedicate my life to its development and the advancement of its people.

I will strive to set before my students by my deeds and actions the highest standards of citizenship for the community, state and nation.

I will endeavor to develop professionally through study, travel and exploration.

I will not knowingly wrong my fellow teachers. I will defend them as far as honesty will permit.

I will work for the advancement of agricultural education and I will defend it in my community, state and nation.

I realize that I am a part of the school system. I will work in harmony with school authorities and other teachers of the school.

My love for youth will spur me to impart something from my life that will help make for each of my students a full and happy future.

naae

Updated 3/5/2011

FIGURE 1.4 Ag Teacher's Creed developed by the National Association of Agricultural Educators.
(COURTESY OF NATIONAL ASSOCIATION OF AGRICULTURAL EDUCATORS.)

CREDENTIALING REQUIREMENTS

There are more than 13,000 agriculture teachers in more than 8,500 schools in the United States (Foster et al., 2020). Each year around 1,400 new teachers are needed to replace retiring teachers and teachers leaving the profession, as well as to fill openings for additional teachers and for teachers in new programs. In a typical year, there is a 7% turnover of agriculture teachers in the country. How are these new teachers credentialed to teach agricultural education? This question will be answered in the following sections.

The responsibility of teaching agricultural education demands that the teacher be qualified to teach. One method of ensuring that teachers are qualified is credentialing. *Credentialing* is the process of determining and certifying that a person can perform to required standards and meets competencies necessary to be a teacher. The most typical route for credentialing is through a teacher education program at a university. Upon meeting the requirements of credentialing, a person receives a teaching license.

A *teaching license* is a certificate issued by a legally authorized state office documenting that a person is qualified to teach and the conditions under which that person may teach. Conditions typically include the subject area(s) the person can teach, the grade level(s) at which the person can teach, and the period covered by the license. Each state has its own unique credentialing requirements and licenses, although some components tend to be common among all states. For example, in agricultural education, a license may include career and technical education, agriscience, grades 5–12, and expires in five years as the conditions.

To become a licensed teacher, a person must demonstrate competence in content and pedagogical knowledge, dispositions, and performances. Potential teachers may demonstrate their competence by successful completion of college teacher education programs on either the baccalaureate or the master's degree level or through alternative means. Each of these is discussed next.

A *standard teaching license* is a license issued upon successful completion of an accredited teacher education program and any other requirements of the state. An individual successfully completing a teacher education program may receive a baccalaureate or master's degree or a postbaccalaureate certificate, depending on the state and the university. For an agricultural education teaching license, other requirements may include passing national and/or state examinations, submitting a teaching portfolio, having no felonies or applicable misdemeanors on a criminal history check, verifying occupational work experience in agriculture, and documenting successful completion of required hours/weeks of student teaching internship. States may have other requirements such as a minor or degree in a content area, certification in cardiopulmonary resuscitation (CPR), completion of modules on suicide prevention or child abuse and neglect recognition, or dyslexia awareness training.

There may be more than one level of standard license. For example, a state may issue a beginning teacher a probationary license, sometimes also called an initial, induction, provisional, preliminary, or beginning license. A probationary license may be valid for one to five years, during which time the beginning teacher receives mentoring, is observed by experienced educators, and is evaluated by an administrator. These observations and evaluations, using approved forms, become part of the documentation to obtain the next-level license. Also, the probationary teacher may be required to submit a portfolio before obtaining the next-level license.

When held by an experienced teacher, this standard license is typically renewable every four or more years. Names for this license include professional, professional educator, proficient practitioner, collegiate professional, and others denoting competency and professionalism. Some states may have higher-level licenses for individuals with advanced degrees, National Board certification, or other indicators of advanced accomplishment. The advantages of higher-level licenses may include a higher base salary or a longer renewal period.

Although some practicing teachers may hold "life licenses" that do not require renewal, most agriculture teachers must renew their teaching licenses every four to 10 years. Renewal requirements are in place to ensure that teachers stay current in their field for content, pedagogy, and technology. Most states allow teachers to apply graduate credits earned from college courses toward license renewal. Many salary schedules have steps that teachers can move up as they complete certain numbers of graduate credits, such as baccalaureate plus 15 credits, master's degree, master's plus 15, and master's plus 30. Other renewal options include *professional growth plans*, which are set goals for professional development with points awarded for attendance and participation at workshops, conferences, and seminars and for participation in other in-service activities. Some states allow even greater flexibility in documenting professional growth and improvement that leads to license renewal.

Alternative Licensure

Alternative licensure allows individuals to be licensed without having completed traditional programs of teacher education preparation. Several approaches are presented. One or more of these may be used in your state.

Workplace Specialist

Many states, in addition to standard licenses, provide a separate route for teacher licensure for career and technical education subjects. A *workplace specialist teaching license* is a license for an individual with substantial occupational experience in the field they wish to teach. Examples of such fields are automotive mechanics, cosmetology, electronics, and building trades. Many times the best people for teaching these subjects will not have baccalaureate degrees and almost always will not have degrees from teacher education programs in their subjects. For this reason, states provide an alternative route for these people to obtain teaching licenses.

Teachers licensed through the workplace specialist route typically have three or more years of work experience in their field and are recognized as technical and content experts. Once employed, they receive intensive instruction in teaching methodology, classroom management, and other topics necessary for effective teaching. They are assigned mentors and are regularly observed. Just as with standard licenses, occupational specialist teaching licenses may have different levels and require coursework or continuing renewal credits in order to be renewed.

As a career and technical education subject, agricultural education in many states also allows for licensing of workplace specialist teachers. Common areas include horticulture, landscaping, animal and plant sciences, and natural resources. Depending on the state, these teachers may be restricted to teaching only the agricultural education courses that match their occupational specialization. In many cases, an agriculture teacher who holds a workplace specialist license has a baccalaureate degree or higher in agriculture.

Transition to Teaching

Another route to licensure is a system for people who have had careers outside of education and then want to be teachers. The programs go by different names but can generically be called transition-to-teaching programs. A *transition-to-teaching program* is a credentialing approach for people who hold content degrees and have pursued careers in the content field. Examples include laboratory researchers who want to become science teachers, retired military trainers who want to enter teaching, and agribusiness specialists who want to become agriculture teachers. A transition-to-teaching program typically consists of a limited number of education courses taught in an intensive manner, combined with an internship similar to student teaching. Standardized testing and criminal check requirements must still be met to obtain a teaching license. These programs may also have grade point average requirements and/or occupational experience requirements.

Emergency Certification

Some states allow teachers to teach outside of their field or even allow unlicensed people to teach on an emergency basis. For example, a licensed agriculture teacher may teach one biology class for a school year because a biology teacher left one week before classes started in the fall. Or, a licensed science teacher may teach a mathematics course. In other instances, people without teaching degrees or experience are hired because of a shortage of licensed teachers. Emergency certification is not desirable as a solution to the shortage of qualified teachers.

Distance-Based Certification Programs

LEAP, or Licensure in Education for Agricultural Professionals, is a distance-based certification program in agricultural education. The program is designed for people who already hold baccalaureate degrees in agriculture and meet specified minimum qualifications.

The LEAP program appeals to people such as career changers who are unable to return to college full time. The required courses are delivered online so students have the flexibility to complete the courses at times and places convenient to them. LEAP is administered by the Agricultural and Human Sciences Department at North Carolina State University. LEAP is accredited by the Council for the Accreditation of Educator Preparation (CAEP).

The *Council for the Accreditation of Educator Preparation* is an organization that accredits college and university teacher education programs. It uses a performance-based system and verifies that certain standards have been met. Completing a teacher education program at a CAEP-accredited institution has certain advantages, such as reciprocity among some states and institutions.

Successful completers receive teaching licenses from the state of North Carolina. They are then eligible to teach in almost all 50 states through a process called reciprocity. *Reciprocity* is the recognition by one state of the validity of a teaching license issued by another state. This allows teachers to take teaching jobs more freely in other states. Reciprocity does not relieve a teacher from state-specific requirements. For example, to continue teaching in the new state, the teacher may have to pass a state or national competency test or take additional courses.

AG*IDEA is an alliance of universities providing online courses and degrees. Students select a home university, which provides a base for advising, enrolling in courses, and obtaining a degree. An advantage of AG*IDEA is that all courses cost the same tuition regardless of which institution originates the teaching. The Agricultural Education program within AG*IDEA can lead to a career as an agriculture teacher.

AGRICULTURAL EDUCATION: MEANING AND SCOPE

Agricultural education is a program of instruction in and about agriculture and related subjects such as natural resources and biotechnology. This means that agriculture is both a content to be learned and a context in which learning can occur (Roberts & Ball, 2009). It is most commonly offered in the secondary schools of the United States, though some elementary and middle schools and some postsecondary institutes/community colleges also offer such instruction. Subjects included are those of less than the four-year college level in plant and animal production, supporting biological and physical sciences, horticulture, natural resources and environmental technology, mechanics, and forestry. Some areas of food processing and distribution are included.

Agricultural education has been carried out, in one form or another, for centuries. Much of the early emphasis on school-based agricultural education at the federal level began in 1917 with passage of the Smith-Hughes Act, which promoted vocational education. Additional federal laws have been enacted since inception. These laws have included agricultural education as a part of vocational education. In recent years, vocational education has become known as career and technical education. Federal support continues to be provided in collaboration with states and local school districts. For many years, agricultural education in the secondary schools was referred to as "vocational agriculture."

In recent years, the term "school-based agricultural education" (SBAE) has gained relatively widespread use. This is to distinguish it from agricultural education that is not based in local schools. However, most individuals continue to use the term *agricultural education*.

The National Association of State Directors of Career Technical Education Consortium has identified 16 career clusters (see Chapter 8). One of these clusters is Agriculture, Food, and Natural Resources (AFNR). The AFNR cluster includes careers and instructional areas associated with the production, processing, marketing, distribution, financing, and development of agricultural commodities and resources, including food, fiber, wood products, natural resources, and other plant and animal products and resources.

The Agriculture, Food, and Natural Resources cluster structures the instructional content for agricultural education into eight pathways (seven pathways were initially specified but an eighth, biotechnology systems, was added). The pathways and brief descriptions of their content are as follows:

FIGURE 1.5 A teacher is demonstrating the use of a computer-controlled system for a greenhouse.

Plant systems—the study of plants and cultural practices in plant production; areas include agronomy, horticulture, forestry, turf, viticulture, and soils

Animal systems—the study of animals and their production and well-being; areas include large and small animals, wildlife animals, research animals, animal health, and others

Biotechnology systems—the study and use of data and applied science techniques in the solution of problems associated with living organisms; areas include plant and animal biotechnology, microbiology and molecular biology, genetic engineering, and environmental applications

Power, structural, and technical systems—the study and use of equipment, engines and motors, fuels, and precision technology; areas include power, structures, controls, hydraulics, electronics, pneumatics, and the like

Natural resources systems—the study of the management practices for soil, water, wildlife, forests, and air; areas include habitat conservation, mining, fisheries, soil conservation, and the like

Environmental service systems—the study and use of technology and instruments in environmental quality; areas include water and air quality, solid waste management, pollution prevention, hazardous waste management, sanitation, and the like

Agribusiness systems—the study and use of economic and business principles related to economic systems, management, marketing, and finance; areas include agricultural sales and service, entrepreneurship, agricultural management, and the like

Food products and processing systems—the study and application of science and technology to assure quality and wholesome food products; areas include processing, preserving, packaging, food safety, quality assurance, regulations, and distributing food

The content standards for each of these pathways have been identified and published for use by agricultural educators and other individuals (see Figure 1.6). These standards focus on the content or subject matter of agriculture instruction in grades 9 through 14 (high

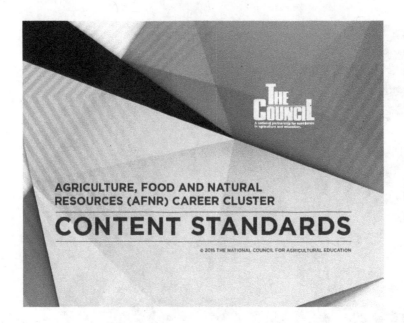

FIGURE 1.6 Content standards help guide the direction and focus of an agriculture program.

school and postsecondary school). In addition to the eight pathways, content standards have also been prepared for an area known as "life knowledge and cluster skills." The content standards are valuable in planning programs of study and developing overall curriculum guides by agricultural educators and others. Individuals who develop instructional materials make considerable use of these standards. (The *National Agriculture, Food and Natural Resources [AFNR] Career Cluster Content Standards* are available for downloading or printing at https://thecouncil.ffa.org/afnr/.)

AGRICULTURAL EDUCATION IN AMERICAN EDUCATION

Agricultural education is very much a part of American education. It is offered in public and, sometimes, private secondary schools just as other required core curriculum and elective courses are offered. State-level curriculum guides, standards, and testing initiatives are typically provided for all accredited public secondary schools.

Education in the United States

Education in the United States is primarily offered in public schools and is universally available to all children. Child education is compulsory, with the ages for compulsory education ranging from ages 5–8 until ages 14–18, depending on state law. While the vast majority of children attend public schools, some attend private and parochial schools or may be home schooled (about 10% attend private schools and 7% are home schooled). School governance has been viewed as a location function. Most school policies are set locally by a board of education, though control and support for education comes from three levels: federal, state, and local governments.

In the early years of the United States, education was viewed as a function of local and state governments. The federal government was not involved in education. This began to change in the early 1900s with passage of the National Vocational Education Act of 1917, also known as the Smith-Hughes Act. This act provided funds to the states and local schools to establish vocational education instruction. Agricultural education was part of vocational education, though agriculture classes had been offered in a number of schools for many prior years. Since 1917, several federal initiatives have supported local school education.

The No Child Left Behind Act of 2001 (NCLB) was a major effort at federal influence over state and local education. From its beginning NCLB was controversial among state and local education officials. Overall, NCLB was intended to support standards-based education with measurable goals to improve education outcomes. However, many educators felt that NCLB established unrealistic goals for local schools and teachers and was too punitive when these goals were not met. NCLB, with its far-reaching provisions, replaced the Elementary Secondary Education Act (ESEA) of 1965. In 2015, the Every Student Succeeds Act (ESSA) replaced NCLB. ESSA was intended to ensure every student regardless of circumstances had the opportunity for a quality education. ESSA was designed to provide states more flexibility in their curriculum and in how they documented accountability for meeting goals.

The National Center for Education Statistics (NCES) (https://nces.ed.gov) of the U.S. Department of Education gathers and reports information about education in the United States. NCES (2019) reported there were 98,158 elementary and secondary schools in the United States. Of these, 23,814 were secondary schools (middle and high school). In addition, there were 1,528 two-year colleges (sometimes referred to as technical schools). The number of schools has greatly decreased in the past century due to school consolidation and increased size of school enrollments.

Schools throughout the United States are increasingly offering and requiring similar courses for graduation. This is much different from a century ago when local school districts had greater control over curriculum and other education matters. High school courses are typically measured as Carnegie units. A *Carnegie unit* is a standard for measuring school subjects by length of class participation, with 120 sixty-minute hours being the minimum requirement for schools to grant 1 unit of credit. One Carnegie unit is equivalent to one academic year of study. Course scheduling and period lengths vary widely. Many high schools have 180-day years.

The minimum courses in mandatory subjects in U.S. high schools are the following (one Carnegie unit is typically earned with each course):

Science—three units or years minimum of science courses; normally biology, chemistry, and physics are required

Mathematics—four years minimum of mathematics courses; courses include algebra, geometry, precalculus, statistics, and, in some schools, calculus

English—four years minimum; including composition, literature, and oral language

Social studies—three years minimum, including history, government, and economics

Physical education—none to two years minimum

Agricultural education courses (and all other career courses) are typically viewed as electives. This means that instruction and enrollment by students in agriculture is not required but is optional in most schools. In order to gain enrollment in agricultural education, the courses offered, the curriculum outlines followed, and the teaching approaches used must be attractive to students and provide highly useful learning opportunities to students. Students must feel that enrollment in agricultural education is beneficial to their educational and lifelong pursuits.

Agricultural Education Enrollment and Schools

Agricultural education (at the secondary school level, formerly known as vocational agriculture) is primarily offered at the high school level in the secondary schools of the United States. Some 15,500 public high schools (grades 9–12) enroll 15.2 million students. Of these schools, a few over 8,000, or 52%, have agricultural education offerings. With an estimated 1.3 million student enrollments in agricultural education, only 8.5% of the total

FIGURE 1.7
Agriculture facilities vary by local program emphasis. The Sound School in Connecticut has an aquaculture focus.

high school student population is enrolled in agricultural education. These data reflect a huge opportunity for agricultural education to increase enrollment and serve additional numbers of students.

In agricultural education, programs are offered in elementary, middle, and high schools as well as postsecondary institutions. The focus of this textbook is on instruction in middle and high schools. At the time of the most recent national study on agricultural education (Lee, 2009), the largest proportion of agriculture teachers (74%) were teaching in comprehensive high schools. Only slightly over 12% were teaching in middle schools and 6% were teaching in career and technical centers (formerly known as vocational-technical education centers). Nearly 2.5% were teaching in magnet or theme schools and academies. The remainder (about 5%) of teachers were teaching in correctional facilities, exceptional education schools, and related kinds of institutions.

The largest number of programs and student enrollments has been and continues to be at the high school level and in comprehensive high schools. Since this 2009 study was conducted there has been a considerable increase in middle school programs. Teachers may teach high school and middle school classes or be full-time middle school agriculture teachers.

GAINING AN INITIAL POSITION

Once you have a license, you are ready to seek a position. Several important procedures are presented here that will help you to be successful in gaining an initial position.

Locating Available Positions

The first step in gaining an agricultural education teaching position is to locate schools that are seeking to hire agricultural education teachers. Finding out about openings is sometimes a challenge. You want to be sure that an opening exists before making a job application. Some individuals view it as unprofessional to apply for a position before it has been declared open.

Following are several strategies in locating open positions.

Consult with teacher educators   Most teacher education programs maintain lists of vacancies. A program may post vacancy announcements on the teacher education website. Teacher education programs may also send regular position announcements to students' email addresses. Always seek the help of your academic advisor with position openings and applications.

Consult with state supervisory staff   State staff members in agricultural education are often the first to hear about position openings. Because they visit local schools, they usually know quite a bit about local school situations in terms of support for agricultural education and the nature of the programs that have been in place.

School district websites   Most school districts maintain employment sections on their websites. Among other information, position openings in a district are announced on its site. Regular checking of the website is needed, depending on how often it is updated.

Professional organizations   The state agriculture teachers' organization may maintain a list of teaching vacancies. The information may be posted on a website, be available in a newsletter, or be obtained by contacting an officer.

Other organizations   The state department of education may maintain a list of teaching vacancies. The information is typically posted on a website. Also, the National Association of Agricultural Educators (NAAE) maintains an agricultural educator teaching positions website.

Family and acquaintances   People who work in a school or school district often hear about position openings. Teacher education students who took agriculture classes in high school often ask their former teachers about openings. School employees share the information with others, including family members. You will want to be sure that news about an opening is a fact, not a rumor, before you apply. Check out what you hear by calling the personnel director of the school system to see if an opening actually exists.

Searching for the Right Position

How does a newly licensed agriculture teacher decide what positions to interview for and which job offer to accept? Following are some considerations to help guide the selection process.

Program Emphasis

It is important that beginning teachers find programs that are a good fit with their expertise and personalities. An agricultural education program that has a strong emphasis area, such as agriscience or horticulture, needs an agriculture teacher who has the expertise to teach in that area. A mismatch, without remediation, in this consideration can lead to frustration and ineffective teaching. Appropriate remediation can include mentoring, additional coursework, and experiences such as internships and on-the-job training.

Program Characteristics and Resources

It is also important that beginning teachers decide whether they work better in single- or multiteacher programs. Do you prefer to work by yourself or with someone else? Teachers who have had positive working experiences in multiteacher programs tend to gain greater job satisfaction and potentially have greater teaching effectiveness.

Another consideration is the resources and support that will be available for the program. These are important factors. A facility that is less than state of the art can often be improved. The same is true for inadequate instructional materials, equipment, and other resources.

Here are a few questions to answer in assessing program characteristics and resources:

- What is the history of the agricultural education program? How many changes in agriculture teachers have occurred in the past five years? How many years has the program been at the school?
- Does the program receive a budget each year? How are funds made available and expended?
- What do the facilities look like? Are they in good repair and appropriate for the program emphasis?
- Are the appropriate instructional materials available, or will the administration commit to obtaining needed materials?
- Are the proper equipment and tools available for the program emphasis?
- Is there evidence that the administration supports the local agricultural education program?
- What are the plans regarding the program? Are adequate fiscal resources provided for the program?

FFA and SAE Characteristics

FFA (see Chapter 23) and supervised agricultural experience (SAE; see Chapter 22) are major components of the agricultural education program. You will want to assess these. Some teachers prefer to take a program that is "down" and build it "up." Other teachers prefer to go into a program that is in good condition.

Here are a few questions to answer:

- What are the expectations regarding FFA and SAE?
- Does the program have a history of success in career and leadership development events?
- What are school and community service-learning expectations?
- Does the program have more of a leadership, career, or personal development emphasis?
- Regarding the SAE program, what are the expectations for home/worksite visits, and what travel support is provided?
- How are travel, per diem, and lodging expenses funded for agriculture teachers when they attend FFA events and activities?
- How are the FFA members' expenses funded?

School Characteristics

The physical and administrative layout of a school is somewhat important in selecting where to teach. Agricultural education programs located in comprehensive high schools, area career centers, and middle schools have different emphases and characteristics. The grade levels taught at a school and the proximity to schools of other grade levels influence the opportunities for the agricultural education program. The location of the agricultural education program within a school and the facilities associated with the program give an indication of the value placed upon the program.

Administrative support is an important consideration. Assess the support given teachers and the agricultural education program in the past. Gain an understanding of the current administrators' support by talking with the principal, career and technical education director, or other administrator.

Another consideration is the other teachers in the school. Included are those in agriculture as well as those in other subject-matter areas.

Here are a few questions to answer:

- Is the overall faculty experienced or inexperienced?
- Do teachers tend to stay at the school for several years, or is there frequent turnover of teachers?
- Do teachers socialize together?
- From what you can observe, such as from walking down the hallways, is there good teacher–student interaction in the classrooms?
- Is active learning occurring?
- Are classrooms well kept and inviting?
- Are students' work and accomplishments displayed within the classrooms and hallways?

Community Characteristics

Although teachers do not usually begin and end their career at the same school, considering community characteristics when selecting a teaching position is still important. Some teachers like small towns, while others like urban or suburban areas.

Here are a few questions to answer:

- Is the rural, suburban, or urban location of the school what you want?
- If you have a geographic restriction, is the community in the part of the state or country where you want to live?
- Does the community have the shopping, cultural, social, and religious amenities you want? If not, are these within driving distance?
- Are job opportunities available for a spouse?
- What is the overall cost of living?
- How available is housing?
- What agricultural structures and businesses are located in the community?

Position Benefits

Although important, salary should not drive your decision of where to teach. A high salary does not improve a bad situation or a mismatch. Salary and fringe benefits should be considered along with other factors when deciding whether you can be successful in a school situation.

Contract length is an indication of the value placed upon the total agricultural education program and the expectations for the agriculture teacher. Salary level compared with that of other school districts is usually an indication of the socioeconomic status of the community rather than the value placed on teaching. Some schools have contracts with extra days added beyond those in a regular teaching contract; others provide a stipend for FFA/SAE work. You will want to know if either or both are part of the master contract or of the agriculture teacher's contract.

Fringe benefits are an important part of the salary package and are often overlooked. Benefits must be included in assessing your overall compensation.

Here are some questions to answer about benefits:

- How is the retirement plan structured?
- What are the medical, dental, and vision care benefits?
- Are other types of insurance available at a group rate?
- What is the sick leave and personal days policy?
- Does the school have funds for reimbursing teachers for graduate coursework, educational travel, or sabbaticals?

Most of these questions should be addressed to the school's personnel department, not asked during the job interview. In many cases, an interviewee receives an employee handbook and other materials that explain benefits as part of the initial interview.

Nonteaching Duties

Another consideration is the nonteaching duties required of teachers. Some schools expect teachers to perform duties beyond those in agricultural education.

Here are some questions to answer:

- Are teachers expected to perform hall, bus, or lunchroom duty?
- Are there expectations that teachers attend or work at a certain number of athletic or other events?
- Are new teachers expected to become athletic coaches or extracurricular club sponsors?
- Do teachers have homeroom, study hall, or other nonteaching duties during the school day?
- Is the agriculture teacher expected to be a certified school bus driver? (For an agriculture teacher, who many times holds or expects to get a bus driver's license, it is important to find out if he or she will be called upon as a substitute, field trip, or athletic team bus driver.)

Applications and Interviewing

Most school districts have standard application forms for teaching positions. Some districts accept only online applications. An application should be completed accurately, neatly, and promptly. It should be typed. Beyond the application form, an application packet usually includes a cover letter, college transcripts, letters of recommendation, a résumé, and verification of teaching license. Sometimes an application packet also includes a philosophy statement and other components. The application packet should be professionally organized and submitted in a timely fashion.

For many positions, the interview determines who receives the job offer. A job interview serves two purposes. For the school, it is a chance to verify that what appears on paper in the application packet is reality. Is the interviewee knowledgeable and articulate? Will this person be a good fit for the agricultural education program, the school, and the community? Is the interviewee energetic, creative, and passionate about educating young people? For interviewees, the interview is a chance to determine whether they can be successful in this school and whether it is a good fit.

Interviewees should dress professionally for the interview, arrive early, and be prepared. They should have a firm handshake, look interviewers in the eye, and speak with energy and confidence. Nervous energy should be channeled into enthusiastic facial expressions and speech rather than exhibited as annoying mannerisms and fidgeting. This is helped by eating, sleeping, and exercising properly during the weeks before the interview. It is also helped by going through mock interviews and practicing answers to probable interview questions.

Interview questions may cover a wide variety of topics. In general, interviewers want to discover how the interviewee will work with students, parents, and colleagues. They want to find out what instructional and classroom management techniques the interviewee will employ. They also want to determine why the interviewee is interested in the particular position and what makes this person better than the other candidates for the position. The interview is also a good time for assessing the interviewee's oral communication skills. Avoid poor grammar, incomplete sentences, and bias in your speech.

Some interviewers like to create situations to get a feel for how the interviewee will respond with special needs students or in certain circumstances. Others try to determine the interviewee's philosophical orientation—that is, how the person thinks about and approaches the teaching situation. Still others have a practical orientation and want to determine what the interviewee knows and is able to do.

At the conclusion of the interview, the interviewee should thank the interviewers for their time. Within 24 hours email a thank-you note, then within one to three days mail a thank-you note. Typically, the interviewers should explain the timeline for making a decision and give instructions for whom the interviewee should contact if further questions arise. The interviewee should avoid being pushy or demanding. If a teacher is selected for a position, approval by the board of the school administrative unit is required before a contract can be offered. It is appropriate to share with family and friends news of a job offer and acceptance; however, it is not official until the board votes approval.

COMPENSATION AND OTHER BENEFITS OF TEACHING

Teacher compensation is typically based on years of experience and educational level. These are combined into a salary schedule. A *salary schedule* is a list or table that shows the salary at each experience level and educational level. Increments are added for each year of teaching experience, though some schedules may not go beyond 15 or 20 years. Compensation may be based on teacher performance using a system commonly known as merit pay. Some merit pay systems do not use a salary schedule.

Compensation

Teacher compensation is composed of base salary, extended days, extra duties, and benefits. Merit pay can be given as a one-year bonus or included in the base salary. Most teacher contracts are for a 180- to 190-day school year. The year includes 180 or more instructional days, one or more in-service days, one or more preparation days, and other times, such as evening parent–teacher conferences. Typically, every teacher receives this same base contract. For the 2018–2019 school year, the U.S. average beginning teacher base salary was $44,150, and the average teacher base salary was $57,950 (NCES, 2019).

Agriculture teachers often receive higher salaries because of job duties and extended contracts. Experience and advanced degrees also increase the salaries. Salaries for experienced agriculture teachers in the $70,000- to $100,000-a-year range are not uncommon around the United States. Teachers who have completed National Board certification may receive several thousand dollars a year more.

Some school districts may use collective bargaining. Representatives of the teachers' association negotiate salary and benefits with school administrators. After negotiations are concluded, a master contract is in force for a set time period. You may ask for a copy of the master contract to help you in making a position decision.

Extended-Day Contracts

Not all states treat extended days for agriculture teachers the same. Some states require that every agriculture teacher be employed on a 12-month contract. This type of contract provides for vacation days and holidays and boosts teacher pay greatly over that of the base salary.

A different type of yearlong contract is the 240-day contract. The agriculture teacher works 240 days and is off the remaining days of the year. This is the equivalent of receiving 21 vacation days and holidays (365 days in a year minus 104 Saturdays and Sundays leaves 261 days eligible for working). The agriculture teacher must ask how evening and weekend work as well as overnight trips count toward days worked.

Other extended-day contracts are written for any number of days between 180 and 240. Some states allow agriculture teachers to direct supervised experience or teach courses (including for adults and young adults) in the summer. These also increase pay over that of the base salary.

Most master teacher contracts also include a schedule for paying teachers for extra duties. These duties include serving as athletic coaches, department heads, class sponsors, or directors or advisors of band, chorus, and certain other teams and clubs. This extra pay is in recognition of teachers working with the respective students and activities outside of the normal school day. In some cases, the agriculture teacher receives extra-duty pay for FFA, SAE, or other agricultural education duties. Extra-duty pay is calculated in several ways. Most of the systems place all extra duties into categories and pay for every duty within a category at the same rate. In some systems the category rates are set, in others they are adjusted for years of experience, and in still others the pay is included in calculating a teacher's salary and salary increases.

Benefits can add 25% or more of base salary to the total compensation package. Most teachers are a part of the Social Security/Medicare system. For these teachers, the employer pays 7.65% of their salaries into the system. This is one component of the retirement benefit for teachers. Another component is the school- or state-sponsored retirement program. In most cases, a set percentage of the teacher's salary is paid into the retirement program each pay period. A final component of the retirement benefit is optional retirement and savings programs, such as 401(k), 403(b), and other such plans. Contributions to these may be entirely from the employee's pay or matched by the employer. Teachers typically receive medical insurance and, possibly, dental and vision care insurance. The number of insurance plans and combinations are too numerous to discuss in this section.

Other benefits are available depending on the school system and the state. Schools are typically large employers within a community. Because of this they are able to negotiate for reduced group rates for insurance, such as disability, term life, and long-term care. These are optional for teachers to purchase, depending on their own needs and circumstances. Some school systems reward their teachers for pursuing graduate degrees and other certifications. These rewards may be in the form of tuition reimbursements, higher steps on the salary schedule, and educational travel funds. Other school systems have paid or unpaid sabbatical programs that allow teachers to take a year to pursue graduate degrees or other educational opportunities and return to their jobs after the sabbaticals.

SUCCESSFUL AGRICULTURE TEACHERS

Agriculture teachers have a number of roles and responsibilities that go far beyond meeting classes each day. They deliver programs of agricultural education, which include instruction, directing students in supervised experience, and advising the student organization—FFA.

This is true in whatever area of agricultural education they may be teaching—horticulture, wildlife, food science, agricultural mechanics, animal care, veterinary assisting, or other area.

Roles of Agriculture Teachers

Following are some major roles and responsibilities of agricultural educators.

Being a school team member   An agriculture teacher is a member of a team of teachers, counselors, administrators, and others who go about fulfilling the mission of a school. A teacher should participate and support the school. Attending school functions not related to the agricultural education program is always a positive step. Helping other teachers with activities and events is another step in developing as a team player in a school. Agriculture teachers should support the work of other teachers and programs in a local school and not isolate themselves.

Planning and developing a program   *Agricultural education programs* have several major components. Nearly all have classroom and laboratory instruction, supervised experience, and student organization (FFA) components. Some have adult and young adult components and may include school farms, canneries, forestry labs, and similar kinds of facilities. Programs are planned to meet local needs as well as implement state-established standards and guidelines.

Preparing to teach classes   Instructional planning includes making the necessary preparation for classes so time and other valuable resources are used efficiently. It may involve developing lesson plans, organizing demonstrations, and preparing hands-on laboratory activities to achieve instructional objectives. Thorough planning is essential to provide students with the maximum time for learning.

Delivering instruction   Instructional time is used to guide student learning in classroom and laboratory activities. Resources must be available and efficiently used. Students must have relevant materials if they are to achieve the maximum. These include textbooks appropriate to the grade level and with content of suitable depth. The materials should have high appeal to students. All activities should focus on learning and mastering the established objectives. Lack of such focus leads to off-task student behavior.

Evaluating student progress   Assessing the progress students are making toward achieving the instructional objectives is a regular part of the teaching process. Information gained from such assessment can be used to reteach as needed or to move to another instructional area.

Advising student organizations   The major student organization in agricultural education is FFA. Agriculture teachers are also FFA advisors. In a school with one agriculture teacher, that teacher is the sole advisor. Teachers may divide the advisor role in a school with more than one agriculture teacher.

Supervising student experiences   Agriculture teachers are responsible for promoting supervised experience and for helping students plan, carry out, and evaluate activities. This includes teaching record keeping as well as providing on-site supervisory visits.

Managing resources   Agriculture teachers typically have a wide range of resources to manage. Resources are budgets, facilities, land, equipment, instructional materials, and

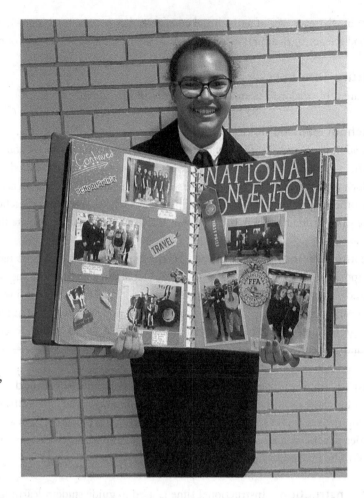

FIGURE 1.8 As the FFA advisor, an agriculture teacher is responsible for supervising activities such as travel to the National FFA Convention. (COURTESY OF LEDYARD AGRI-SCIENCE & TECHNOLOGY PROGRAM.)

other items, such as a pickup truck or van. These duties must be carried out efficiently and effectively.

Relating to publics Good relationships are needed with the publics important in agricultural education, including students, parents, prospective students, school personnel, local agriculture officials and businesses, and others. A good personality and the ability to meet and greet students and parents will be useful. A friendly, happy, and outgoing manner will serve a teacher well. Skills in planning good public relations, such as with newspaper and broadcast journalists, are also needed.

Lifestyle Agriculture teachers should follow lifestyles worthy of being modeled by students and others in the school community. Participation in civic clubs, agriculture organizations, and related groups is often beneficial in an agricultural education program. Holding to high moral standards is a typical expectation. Recreational drug use and involvement with illicit activities is never appropriate.

Agriculture Teacher Traits and Abilities

What are the qualities of a good teacher? Every individual has different ideas about what makes a good teacher. People don't always agree on the meaning of good teaching and the traits of individuals who are good teachers. In the 1980s, elaborate testing requirements were established in the states in order to distinguish potentially good teachers from individuals who would not be good teachers. Within a couple of decades it was determined that what a teacher scores on a test has little correlation to his or her effectiveness as a teacher (Baines, 2010). Many states still require passage of some form of standardized test to gain credentials as a teacher. Be sure to get full details from your university advisor on requirements for teacher certification.

In recent years, education agencies have attempted to define or relate good teaching (and teachers) to the scores their students gain on achievement tests, a practice that has been controversial from the start. Many factors are involved in student test scores. Some online testing opportunities have been widely used among the states and local school districts in assessing student achievement. Compromised integrity of the testing process has resulted in cheating scandals created by teachers, school principals, and other individuals. Achievement testing is available in areas of agricultural education, with tests varying from those directly correlated to state instructional guides to testing in areas without regard to instructional guide content.

The Horace Mann: Reach Every Child website (2011) listed 11 traits of good teachers, as follows:

Be unsatisfied—Good teachers are eager to learn new things and are lifelong learners.
High expectations—Good teachers hold high expectations for their students.
Create independence—Good teachers encourage students to seek answers on their own.
Knowledgeable—Good teachers possess a deep knowledge of their subject matter.
Humor—Good teachers have a sense of humor.
Insightful—Good teachers provide quick and accurate assessment of student work.
Flexible—Good teachers utilize the resources of the community.
Diverse—Good teachers use a wide array of strategies to promote student learning.
Unaccepting—Good teachers do not accept false excuses or less than the best work of students.
Unconforming—Good teachers challenge and motivate students with a range of approaches.
A communicator—Good teachers are good communicators.

Research has been carried out in an attempt to identify the specific qualities (competencies, traits, abilities, and characteristics) of good and successful agriculture teachers (Harlin et al., 2007; Roberts et al., 2006; Roberts & Dyer, 2004). Several research reports have been published in the *Journal of Agricultural Education* and the *Journal of Career and Technical Education* identifying qualities of good agriculture teachers. Different methods and sources of research data often result in findings that appear to support similar but varying traits. Reading the reports has resulted in the following traits being identified as those of strong agriculture teachers:

- Plans and manages the agricultural education program (including budgetary areas)
- Follows appropriate state and local education policies and guidelines
- Has knowledge and skill in content area(s) and a love of agriculture
- Plans instruction, both for program and courses and on a daily basis
- Directs student learning activities and is on task in the classroom and laboratory
- Evaluates student learning and uses findings to reteach as necessary

- Advises, encourages, and counsels students appropriately
- Manages the learning environment to motivate students, recognize achievements, and maintain classroom discipline
- Advises student organization through a sound knowledge of FFA and effective student preparation for CDEs/LDEs and other FFA activities
- Directs supervised experiences of students (including student planning, supervision, record keeping, and reporting)
- Relates to publics in the community, including parents, agriculture groups, and other stakeholders
- Markets the program through effective public relations and student recruitment
- Promotes and follows professionalism
- Communicates well with others
- Cares for students
- Follows appropriate ethics and moral standards

These teacher qualities are discussed in more detail elsewhere in this book. Of course, local expectations may vary and teachers should know and follow these as they go about leading agricultural education programs.

THE SCHOOL COMMUNITY

Agricultural education is a part of a larger school community. It is not carried out in isolation from other classes, teachers, and students. A school community performs best when all members work together as a team to attain student achievement goals.

Community Elements

A school community at the secondary level has several elements. No one function or area is carried out in isolation from the others.

Administration

The board of education has the legal authority to take action and use resources in the operation of schools. Some of this legal responsibility may be delegated to hired personnel. Other duties, such as hiring teachers, cannot be delegated in most states. Overall, boards of education employ administrators to run the schools.

Administration is the management of the procedures used to implement policies in an educational organization, such as a high school or career center. Individuals who carry out administration are called administrators, with a superintendent being the top administrator of a school district and a principal the top administrator of an attendance center (school). Several assistants may help superintendents and principals in their work.

Overall, administrators are to be instructional leaders. Some individuals describe good administrators as cheerleaders for students and teachers. Administrators organize and facilitate teaching and learning processes. They also relate to the public; handle student discipline problems; employ, supervise, appraise, and discipline employees; and oversee all other areas, including purchasing, transportation, food service, and building maintenance.

Teaching

In a school community, the teachers are responsible for providing the instruction and implementing the curriculum. They plan and deliver learning activities to promote student mastery of the objectives established for the school and the classes offered. They appraise student achievement and provide reports on progress.

A larger school may have a cluster of teachers in a similar area of content, with a lead teacher, often called a department head. This approach is an extension of administration. Department heads often continue in the classroom but with a reduced teaching responsibility.

An agriculture teacher should support all teaching areas and roles in a school. Instructional content may be planned so that teachers in academic areas and those in career areas integrate instructional content. This means that a teacher in one area includes relevant content from other areas. An example would be a horticulture teacher articulating instruction with a biology teacher. Resources are sometimes shared, such as biology lab equipment being used in horticulture instruction and horticulture facilities being used in biology instruction.

Counseling

All schools have services available to counsel students and assist teachers. *Counseling* is the process of appropriately trained individuals helping persons with problems find satisfactory solutions.

Counseling may deal with personal problems, educational problems, and vocational or career problems and with the general advisement of students. Most teachers appreciate the role of counseling and refer students for help.

A school may have one or more counselors. Some schools have individuals in counseling-related positions, such as psychologists. Testing experts (psychometrists) may also be in larger schools or serve several schools within a district.

Extracurricular Staff

A school often has many activities for students outside the required curriculum. These are known as extracurricular activities. Staff members are hired and/or available to help with these activities. Supporting these activities helps promote the agricultural education program. (*Note:* Some activities or groups are considered cocurricular. This means that they are a part of the instructional program carried out in classes. Science clubs, FFA, and history clubs are cocurricular or intracurricular examples.)

Common extracurricular activities include the following:

Music—Chorus, marching band, orchestra, and other music programs often involve directors and other individuals. Music programs may have flag performers and precision dance groups.

Drama and debate—Drama clubs select, organize, prepare for, and give dramatic presentations, or plays. Some schools offer debate or speaking groups. These groups allow students to develop the ability to think on their feet and deliver presentations.

Dance and gymnastics—Schools may offer dance and gymnastics activities. Teachers in these areas may collaborate with other school groups. For example, a dance teacher may assist the band with a precision dance group.

Athletics—High schools often have several competitive athletic teams, such as soccer, tennis, swimming, and volleyball. Cheerleading may be included in the athletics area.

Student government—A high school usually has a form of student government that allows students to develop leadership skills and promote an improved school environment. The organization is often called the student council. Students are typically elected through democratic voting.

Honorary and other curriculum-related and special-interest organizations—High schools typically have many organizations for student participation. The honorary organizations, such as Beta Club and Golden Key, are important in promoting academic achievement. Organizations related to the curriculum are also common, such as science clubs and speech clubs.

Special Services

Instructional technologists, media specialists, special education teachers, and language or speech specialists may be included in this element. In recent years, the number of instructional technologists and their importance has markedly increased. Often referred to as IT specialists, instructional technologists provide valuable support in the use of computer-based approaches in education and accountability. The agriculture teacher should maintain good relationships with all these. Good relationships lead to assistance when it is needed.

Support Personnel

Support personnel fill a range of positions that help a school operate efficiently.

Office staff includes secretaries, administrative assistants, attendance clerks, registrars, purchasing agents, payroll clerks, and others employed in a school office or at the district level. They interface between students, parents, teachers, administrators, and other school and community individuals. Positive interactions of the agriculture teacher with these individuals promote a quality school community and help when problems arise in agriculture. These staff members are good resources for the agriculture teacher.

The custodial staff is responsible for cleaning classrooms, hallways, restrooms, and other facilities and for assuring that these areas are in good order for students and teachers. The work often occurs after the end of the regular school day or early in the morning before a new school day begins. Custodial service should be available in the agriculture facility. However, the agriculture teacher and students are usually responsible for cleaning the agricultural laboratory.

The food service staff is responsible for operating the school dining service (cafeteria). These workers plan meals, requisition food, prepare and serve food, and clean eating and cooking utensils. Meals are intended to provide adequate nutrition at a reasonable cost. These individuals may also prepare and serve meals for special functions, such as banquets, field trips, and receptions.

The transportation staff includes bus drivers, schedulers, maintenance personnel, and others needed to assure that safe and efficient transportation is provided. This includes transportation on daily routes as well as for field trips and events. The agriculture teacher should follow all procedures in scheduling and using transportation.

All school facilities require routine maintenance, such as replacing a light bulb or unstopping a drain. Scheduled maintenance, such as painting, reroofing, and servicing heating, ventilation, and air-conditioning systems, is also included. The grounds staff picks up trash, mows, establishes and maintains landscapes, and removes snow from sidewalks, among other duties.

Security and safety officers promote a safe learning environment that is free of hazards. These individuals are sometimes associated with local law enforcement offices.

Stakeholder Elements

A *stakeholder* is an individual or a group that has a stake or strong interest in an enterprise or program. Schools have a number of stakeholders. Parents are interested in the educational success of their children. Taxpayers are concerned about the efficient use of tax funds

FIGURE 1.9 Good rapport with security personnel is needed in maintaining security of facilities, including farms and other laboratories. (COURTESY OF EDUCATION IMAGES.)

appropriated for public schools. Businesses want schools to prepare future employees who will be good workers. Some stakeholders have specific interests in agricultural education.

Family Members

Family members, particularly parents and guardians, have a great deal of interest in the quality of the education their children are receiving. Parents/guardians want their children to be well prepared for seeking additional education beyond high school and/or for entering occupations. They want their children to gain high scores on standardized tests for graduation, college admissions, and other purposes.

Family members are often very willing to assist with school activities. Sometimes they may help as chaperones for field trips. Other times they may serve as resource persons in class or provide supplies and materials for activities. They may also provide sites for supervised experience, help judge local FFA competitions, and help in other ways.

Relating well to family members helps build support for the school and agricultural education. Taking the initiative to speak to family members and showing sincere interest helps them to feel comfortable with school personnel.

Taxpayers

Taxpayers want accountability in use of their tax money. Large property owners may pay more taxes than small property owners and be in a position to be more demanding. Good performance by a school demonstrates accountability. Excellence in the agricultural education program also shows strong evidence of accountability with funds it has received. Maintaining good public relations promotes a positive feeling among taxpayers.

Businesses

Businesses in the local community have a great deal of interest in the local school. The agricultural businesses can be particularly useful to the agricultural instructional program and to specific agriculture classes. They may readily offer their facilities for field trips, lend or donate equipment for educational purposes, and serve as supervised experience training sites.

Business owners and employees may have children enrolled in the local school. This provides increased incentive for supporting efforts of the school.

Other Educators

Other educators may be viewed as stakeholders. This includes administrators, counselors, and teachers. Good, positive relationships with academic subject teachers promote the value of agricultural education in helping students gain knowledge and skills in these subjects. The integration of science, mathematics, and other areas into agricultural education has become increasingly important. Careful planning and implementation are needed.

REVIEWING SUMMARY

To teach is to guide others to learn. The process is not always easy and involves active, on-task approaches in teaching. Teaching is the process of directing the learning experiences of others. It is far more than meeting with students in a classroom. It involves planning and evaluating learning—managing the learning environment so students are on task in mastering knowledge and skills. Consideration is given to the situations of the learners and the resources available for their use.

Agricultural education is education in and about agriculture and related subjects. It involves a cluster of careers and eight pathways. These pathways help organize instructional programs and the instruction offered.

Agricultural education is offered in more than 8,500 secondary schools by more than 13,000 teachers. An estimated 1.3 million students are enrolled. Enrollments have been increasing more rapidly than the number of teachers and the funding available to support the instruction.

Teaching is important for many reasons. In agricultural education, teaching is viewed as a means of assuring adequate and safe supplies of food and fiber for the well-being of the people in our nation. Sustainable approaches are advocated to assure the ability to provide indefinitely for the needs of people.

Agricultural education teachers have many roles and responsibilities. Above all, they are in charge of a program of agricultural education. That program includes classroom and laboratory instruction, student organization (FFA), and supervised experience components. In some places, programs may have other components, such as adult and young adult education, school farms, canneries and food processing facilities, and land laboratories.

Teaching agricultural education is a profession. It involves meeting certain high standards of educational accomplishment, using various means of screening in only the best-qualified individuals, and following a code of ethics and a creed. Professional agricultural educators embrace continuing education and the importance of projecting an appropriate image in a community.

Agricultural education teachers are part of an overall school community. They do not go about their jobs in isolation. Being supportive of the entire school program helps build support for agricultural education. School personnel, as well as community stakeholders, appreciate an agriculture teacher who embraces all aspects of the school community, including extracurricular activities.

QUESTIONS FOR REVIEW AND DISCUSSION

1. What is teaching? How does the word *teaching* relate to *teach* and *teacher*?
2. Why is teaching important? List at least five reasons for teaching that emphasize agricultural education.
3. What are the two settings for education? Distinguish between the two.
4. What agency or service is the leading provider of nonformal agricultural education?
5. What is a profession?
6. What are the expectations for agriculture teachers to have and enjoy professional status?
7. How are codes of ethics used in professions?
8. What is a creed? What organization established the Ag Teacher's Creed? What impresses you the most about the Ag Teacher's Creed?
9. What are the expectations of an agricultural education teacher in becoming a professional and continuing to enjoy professional status?
10. What is agricultural education?
11. What are the pathways in the Agriculture, Food, and Natural Resources (AFNR) career cluster?
12. What are content standards? How are these standards useful? Why are they needed?
13. What have been the national trends in enrollment, teacher load, and funding for agricultural education?
14. How is agricultural education a part of American education?
15. How has school governance and control of education in the United States changed over the past hundred years?
16. What are two federal laws that have brought greater federal involvement into local education?
17. What is a Carnegie unit?
18. What are the minimum mandatory subject requirements in most U.S. high schools? Is agriculture one of these? Why?
19. What does "being a school team member" mean?
20. Name and briefly explain 10 roles and responsibilities of an agricultural education teacher.
21. Why is continuing education important?
22. What is demeanor? How is it related to a teacher's image in a community?
23. What are the major elements in a school community? Briefly explain each.
24. Who are the major stakeholders in a school community? Briefly explain each.

ACTIVITIES

1. Investigate the professional code of ethics for educators. Use the website of the National Education Association (www.nea.org/resource-library/code-ethics-educators) for your investigation. Print out the code of ethics and incorporate it into your professional portfolio. Carefully read the code of ethics and, in 100 words or less, summarize its meaning to a teacher.
2. Take a field trip to a modern high school agricultural education program. Observe the students and faculty, as well as the facilities, instructional resources, and other elements of the program. Take digital photos, provided you have permission. Prepare a report that describes what you observed and offer an evaluation of your observations.
3. Interview a secondary agriculture teacher and/or administrator of a school that offers secondary agricultural education. Prepare a list of questions ahead of time. Ask about

the nature of the work, the trends that are occurring, the biggest challenges faced in agricultural education, the teacher's approaches that are most effective, curriculum/courses offered, instructional resources available, and enrollment. Prepare a report and present it in class.

4. Explore agricultural education teacher openings in your state. Develop a mock application packet and have your teacher educator or state staff personnel review it.

REFERENCES

Baines, L. (2010). *The teachers we need vs. the teachers we have*. Rowman and Littlefield.

Foster, D. D., Lawver, R. G., & Smith, A. R. (2020). *National agricultural education supply and demand study, 2019 executive summary*. http://aaaeonline.org/Teacher-Supply-and-Demand

Harlin, J. F., Roberts, T. G., Dooley, K., & Murphrey, T. (2007). Knowledge, skills, and abilities for agricultural science teachers: A focus group approach. *Journal of Agricultural Education, 48*(1), 86–96. https://doi.org/10.5032/jae.2007.01086

Horace Mann: Reach Every Child. (2011). *Top 11 traits of a good teacher*. Retrieved November 18, 2011, from http://www.reacheverychild.com/feature/traits.html

Lee, J. S. (2001). *Teaching agriscience*. Pearson Prentice Hall Interstate.

Lee, J. S. (2009). *Report of a National Study of Agricultural Education in the United States*. Lee and Associates.

National Council for Agricultural Education. (2016). *National quality program standards for agriculture, food, and natural resources education: A tool for secondary (grades 9–12) programs*. https://thecouncil.ffa.org/

NCES. (2019). *Digest of educational statistics*. https://nces.ed.gov/programs/digest/d19/tables/dt19_211.20.asp

Roberts, T. G., & Ball, A. L. (2009). Secondary agricultural science as content and context for teaching. *Journal of Agricultural Education, 50*(1), 81–91. https://doi.org/10.5032/jae.2009.01081

Roberts, T. G., Dooley, K., Harlin, J., & Murphrey, T. (2006). Competencies and traits of successful agricultural science teachers. *Journal of Career and Technical Education, 22*(2). https://doi.org/10.21061/jcte.v22i2.429

Roberts, T. G., & Dyer, J. (2004). Characteristics of effective agriculture teachers. *Journal of Agricultural Education, 45*(4), 82–95. https://doi.org/10.5032/jae.2004.04082

2

Philosophical Foundations of Agricultural Education

The Thursday night Philosophy of CTE class meeting had just ended. Charles Taylor, the agriculture teacher at Lakewood High School and part-time graduate student, packed up his notebook and materials in preparation for the drive home. The course instructor, Dr. Barbara Wilson, walked over to his seat. "Do you have a minute, Charles?"

"Sure, Dr. Wilson," said Charles, and the two settled into a couple of desks as the other students departed the class.

Dr. Wilson got right to the point. "I cannot help but notice that you have contributed very little in our classroom discussion. I was wondering what was going on. Your assignment submissions are good, but you just don't seem to be engaged in the class."

"I'm sorry, Doc," said Charles. "But I am just not into this philosophy stuff. I know that I need to take this course for my master's degree, but as you know, I'm an agriculture teacher working full time, and I just don't see much use for philosophy in my work."

Dr. Wilson thought for a moment, then said, "I appreciate your sharing this with me. And I'd like you to do me a favor. When you get back to your classes tomorrow, see if you can find evidence of philosophical thought in conversations with your students and colleagues. You don't have to tell them what you're doing; listen to what they're saying. Then we will talk again after class next week. What do you think? Are you up for the challenge?"

TERMS

aesthetics	metaphysics
Aristotle	Charles Sanders Peirce
axiology	Johann Heinrich Pestalozzi
Johann Amos Comenius	philosophy
constructivism	Plato
John Dewey	politics
epistemology	pragmatism
ethics	progressive education
idealism	realism
William James	Socrates
logic	

OBJECTIVES

This chapter addresses the National Quality Program Standards for Agriculture, Food, and Natural Resources Education (National Council for Agricultural Education, 2016), specifically Standard 7: Program Planning and Evaluation. It has the following objectives:

1. Describe the meaning and importance of philosophical foundations in agricultural education.
2. Identify and explain appropriate philosophical foundations.
3. Relate philosophical foundations to educational practice.
4. Discuss current philosophical concepts in agricultural education.

"Well, sure thing, Doc, but why?" said Charles.

Dr. Wilson replied, "I know how busy teachers can be, and sometimes it is hard to see the concepts we discuss in this class at work in the real world. I just want you to listen to what your students and colleagues are saying."

The following week after the class ended, Charles came down to the front of the classroom where Dr. Wilson was putting away her lecture notes.

"Well, Dr. Wilson, you are right," said Charles.

"How so?" asked Dr. Wilson.

"I was in the teacher's workroom making copies of a test and overheard the principal talking to our English teachers about possible budget cuts next year. Two teachers were seated at the conference table discussing how to deal fairly with a student who has been disrupting their classes."

Charles continued, "On my way back to class, I saw a poster in the hallway encouraging students to treat other students with respect. In the math and science hallway, I ran into our assistant principal, who was speaking to some students about how to keep the hallways litter-free."

"And in my classroom, I taught a unit on leadership and personal development, reminding students to set goals and to create a personal development plan. And at lunch, I noticed that the art department had set out watercolor paintings by students for other students to view."

"That's wonderful," said Dr. Wilson, "Philosophy drives everything we do in education, wouldn't you agree, Charles?"

"I surely see it now. In just a few hours, I saw almost every field of philosophy at work in my school, politics, aesthetics, ethics, metaphysics, axiology . . ."

". . . and logic," Dr. Wilson interjected. "From the hallway litter discussion."

Charles thought for a minute. "Oh yeah, that's right!"

Dr. Wilson said, "You may recall from our very first class meeting that we discussed that the basic tenet of philosophy is that it is a search for truth. Where better to do that than a school?"

PHILOSOPHY, THE SEARCH FOR TRUTH

The search for truth is as old as humankind. Humans have always had an unquenchable desire for meaning and truth. The story of this search is written across the ages. Around 580 BCE, the Greek philosopher Thales posited that all matter could be reduced to the quintessential element water. Watson and Crick's discovery of the DNA molecule as the smallest package of life and the discovery of subatomic particles certainly weakened Thales's argument. Still, humankind's insistence on knowing the truth has not diminished. Why do you suppose this is? Perhaps it is, as Leonardo da Vinci called it, "the noblest pleasure, the joy of understanding" (Durant, 1961, p. 3).

Our quest for understanding takes us to questions about such things as the best methods for thinking, teaching, and learning. We concern ourselves with questions about the ideal structure of society and the purpose of government. We devote time to discerning real things that have substance and things that do not. We all have experienced fear in one way or another, but philosophy requires that we examine our fears to determine if they are real or unfounded. This examination of our thoughts is a philosophical question rooted in *metaphysics*.

But philosophy can address other things. The search for the ideal form or beauty and the search for exemplary conduct of humans in society are mindful pursuits grounded in philosophy.

It is difficult to grasp the importance of philosophy because it seems, from one perspective, to be merely an exercise in thinking without anything productive to show for it. Philosophy often deals with questions that have not yet reached scientific study. As Will Durant stated, "Science is analytical description, philosophy is synthetic interpretation" (Durant, 1961, p. 2). Philosophy examines problems not yet reachable by science, such as the questions of good and evil, human freedoms, and life and death, as well as those questions that involve the development of the human mind—the purview of the educator.

Philosophy is a discipline that attempts to provide a general understanding of reality and interpret the meaning of what we observe. It seeks to draw meaning from truth and knowledge. People are naturally curious about what they experience in the world around them. Overall, philosophy describes the efforts of people to understand some of why we do what we do. While there is much difference of opinion on major philosophical questions, we can share the fundamental beliefs we hold and cite those of well-known philosophers, along with the commonly accepted principles of the profession.

For teachers, philosophy can seem like a frivolous pursuit and wholly unnecessary when they face a whole classroom of students for the first time. Teachers, especially novice teachers, worry about how they will teach their students. They focus on how to impart knowledge to students and how to teach the lessons best. But this worry about teaching methodology is a philosophical dilemma. The method by which we transfer knowledge is called *epistemology*. It is the philosophical study of the nature and origin of knowledge. Through the ages, Plato, Descartes, Kant, Aristotle, and other philosophers concerned themselves with studying expertise and how it is acquired.

But this isn't the whole story. Teachers use philosophical tenets, albeit tacitly, in the ordinary course of their duties. Teachers constantly deal with the problems that students bring to the classroom. For example, disruptions in learning can occur almost every day. Teachers have to manage the classroom environment fairly and consistently, using their personal and professional *ethics* as their guide. Ethics is the philosophical theme that attempts to explain the difference between right and wrong. It is the moral compass the teacher uses to lead in the learning environment.

Teachers, through their unions and professional organizations, press for more funding and resources for education. Teachers must constantly allocate scarce resources in the best possible way. This allocation process is rooted in the philosophical theme of *politics*, defined as the best and wisest use of resources in a society.

By teaching students how to calculate the amount of fertilizer to use in a given situation, agriculture teachers apply the philosophical theme of *logic*. Logic, another philosophical theme, seeks to find the best solution possible through an ideal form of reasoning.

Through the FFA, teachers provide leadership opportunities to help students understand their human value (the philosophical theme of *axiology*) and their place in this world (metaphysics). In designing projects in agricultural labs or growing plants for sale to the public through the school's greenhouse, teachers apply concepts associated with beauty and functionality. Beautiful plants are a valuable sales item in school greenhouses, and shop projects must be somewhat aesthetically pleasing to be appreciated. This focus on that which has beauty draws upon the philosophical theme of *aesthetics*.

Practically every day in school, teachers apply all seven themes of philosophical thought (see Box 2.1) without even knowing it. The search for answers, and the search for truth, is a philosophical endeavor and a core function of teaching and learning.

> **BOX 2.1** *The Seven Themes of Philosophy*
>
> 1. **Ethics**—The search for that which is correct.
> 2. **Politics**—The search for the best way for society to continue.
> 3. **Aesthetics**—The search for beauty.
> 4. **Metaphysics**—An attempt to explain the nature of being.
> 5. **Logic**—The search for the ideal form of reasoning.
> 6. **Epistemology**—The study of knowledge.
> 7. **Axiology**—The search for an understanding of the value of things.

THE EARLY PHILOSOPHERS

Mention the word philosophy to some individuals, and they will immediately conjure up a mental picture of ancient scholars in robes lounging about the columns and steps of a stone edifice. A discussion of Socrates, Plato, and Aristotle today would probably be based upon some desire to be amusing in the presence of colleagues instead of an attempt to cite a reference in a scholarly argument. However, one should not have any significant discussion about philosophy without considering the influence of the great thinkers of Athens, who arrived on the scene around 470 BCE.

Socrates (469–339 BCE)

Socrates is perhaps the best known of the early Greek philosophers, although he left no manuscripts for our study. Most of what we know about Socrates comes from the writings of his students, particularly Plato. Perhaps he is best known as the creator of the "Socratic method," or dialectic method, of teaching. The teacher posed a series of questions to help the learner arrive at the more profound truth and understanding of an idea. The general result of such questioning was to prove that the first answer to the first question posed was wrong. Socrates believed that people couldn't know all the answers to life's questions, which often infuriated those people around him. Not only did Socrates's questions reveal a deeper understanding of ideas, but they also had the unpleasant result of revealing one's ignorance. The questions asked by Socrates eventually embarrassed the Athenian leaders, who arranged for his trial on a charge of corrupting the youth of Athens and blasphemy against the gods (Nails, 2020). His trial and execution were vividly portrayed in Plato's *Crito*, *Apology*, and *Phaedo*.

Plato (429–347 BCE)

As a pupil of Socrates, *Plato* carried on the tradition of Socrates's dialectic method through his writings and teaching. Plato's *Republic*, which attempted to define the nature of the ideal state, is his most famous work. Plato established the first known college of higher education in his home, the Academy, where he taught his concept of *idealism*—the essence of existence was the idea. In its purest and ideal form, the idea for something is changeless and eternal (Kraut, 2017). The material or physical structures of an idea are merely decaying copies of the original idea. For instance, the concept of a chair is perfect and indestructible because it exists outside of the physical world. The physical form of a chair can be destroyed easily. Regardless of how we design or build a chair, its physical form will eventually end up in the local landfill. Plato suggests that it is impossible to kill an idea regardless of the many

FIGURE 2.1 *The Death of Socrates* by Jacques-Louis David. (COURTESY OF THE METROPOLITAN MUSEUM OF ART, NEW YORK.)

forms in the physical world. As Plato puts it, "Everything in the world is always becoming something else, but nothing is ever permanently the same" (Magee, 2001).

Aristotle (384–322 BCE)

Around 335 BCE, the world was introduced to Plato's pupil, *Aristotle*. If Plato was the father of a system of education, then Aristotle was the father of scientists. As the founder of the Lyceum, a school in ancient Athens, he taught such luminaries as Alexander the Great and influenced generations of scientists for more than 2,000 years. Aristotle differed from Plato in that he believed in *realism*—that the truth of an idea rested in its form. For instance, a pile of rubber, plastic, metal, and circuitry might be the essential parts of an automobile. Until someone assembles these parts in the correct order and fashion, the car does not exist. Aristotle's realist philosophy urged individuals to seek the cause of a concept, for it is the grounds by which we find the true meaning of the concept. The work of Aristotle yielded what scientists know today as the scientific method and what scores of agriculture teachers recognize as the problem-solving method of instruction.

These three early philosophers, Socrates, Plato, and Aristotle, developed the foundation for education. However, many other philosophers contributed to the philosophy of education along the way.

Johann Heinrich Pestalozzi (1746–1827)

Johann Heinrich Pestalozzi was a Swiss educator and educational reformer who gave us the principle of informal education and the need to make children feel welcome and comfortable in the learning environment. He also gave us the concept of hands-on learning in the sciences. Pestalozzi wrote, "My idea of elementary education. It consists in developing, according to the natural law the various powers of the child—the moral, intellectual, and physical

FIGURE 2.2 Johann Heinrich Pestalozzi. (COURTESY OF THE REAL ACADEMIA DE BELLAS ARTES DE SAN FERNANDO, MADRID.)

powers, with the subordination necessary to their true equilibrium. This equilibrium alone produces a tranquil, happy, and useful life" (in De Guimps, 1889, p. 250). Pestalozzi believed that children must be active to learn. Students should be encouraged to express themselves in speech and written form. Teachers should design learning activities to progress from easy to complex tasks. Students need time and instruction to master a concept before moving to the next idea (De Guimps, 1889). Finally, Pestalozzi wrote at length about the need for parents to be the foundational support for a child's education (Pestalozzi, 1898). Similarly, agriculture educators involve parents in the learning process, potentially through SAE.

Johann Amos Comenius (1592–1670)

Comenius was a Moravian philosopher, educator, and theologian. He is a leading figure in the early development of modern education. Comenius believed that education should be universal and available to all. Students would be best served by studying the sciences and the arts, thus refining the concept of a holistic education program.

Johann Amos Comenius (1592–1670) advocated for modern education, lifelong learning, and extracurricular activities for children. Comenius proposed that teaching should promote a student's natural tendency to learn and that learning should proceed from easy subjects to more difficult ones. Comenius was one of the first to encourage education for career preparation. As a survivor of the Thirty Years' War, he was much attuned to human rights issues and fought for the education of all people, including women. Comenius proposed that schools develop educational programs based upon the student's life stage. Teachers should teach subjects in which they have the requisite content knowledge, and plan lessons based on the social, emotional, and cognitive stages in which the learner presently exists. Learning activities should be sequenced to allow the learner to apprehend critical concepts most effectively. Learning should proceed in stages, from easy to complex, and consider the maturity of the learner.

Further, teachers should plan and prepare in advance of the learning experience. The modern-day planning period in schools may well have been an original idea derived from Comenius. The curriculum should be developed in such a manner as to consider the cognitive development stages of youth. That is, curriculum specialists plan the scope and sequence of the curriculum using the cognitive development stages as their guide.

Comenius was born before his time, as is evidenced by his thoughts on modern education. Comenius believed that classrooms should have realia and reusable learning objects for use by students. Teachers should stock classrooms with visual aids such as geographical maps, historical timelines, and pictures related to the subject matter. Most importantly, Comenius held that students' interests and existing knowledge should drive instruction. Further, Comenius proposed that critical thinking and problem-solving are effective propellants in education. He wrote, "The proper education of the young does not consist in stuffing their heads with a mass of words, sentences, and ideas dragged together out of various authors, but in opening their understanding to the outer world, so that a living stream may flow from their minds" (Comenius, 1907, p. 147).

John Locke (1632–1704)

Locke was a British philosopher and professor at Oxford University with specializations in the Greek language and rhetoric. He published his seminal work, *An Essay Concerning Human Understanding,* in 1690.

An Essay Concerning Human Understanding was one of the first significant works to propose empiricism as the mode of human understanding; that is, humans derive knowledge through their senses and experiences (Uzgalis, 2020). Locke's theory stands in contrast to other philosophers, Aristotle chief among them, that knowledge comes from reason and logic. And it stood in opposition to the Church of England, which proposed that knowledge comes from God.

Locke followed the essay with *Some Thoughts Concerning Education* in 1693. Locke believed that humans were born with minds ready to apprehend knowledge through experience. Locke proposed that the mind, at birth, is a blank slate (Locke, 1889). In Locke's view, children learn best through practice and observation rather than memorization and recitation. Practice forms indelible habits (Locke, 1889). Learning should be exciting and related to the student's interests, and teachers should design lessons that engage and maintain student attention.

THE EMERGENCE OF PRAGMATIC THOUGHT

The philosophy of pragmatism centers on the concept of "knowing by doing." Knowledge is a practical activity, and questions about meaning and purpose fit within the context of pragmatic thought (Magee, 2001). It is a relatively young philosophy when compared with the older philosophies of idealism and realism.

This method is proper when applied to the agricultural sciences. We measure the value of ideas through their practical application. For example, a new metal welding technique may be assessed by whether or not the welds resist breakage. If the welds consistently hold under pressure, then the new welding method has value. Proponents of this line of philosophical reasoning were Charles Sanders Peirce, William James, and John Dewey.

Charles Sanders Peirce (1839–1914)

Although several other philosophers and scientists had worked on concepts relating to pragmatic thought, the first well-known proponent of this new philosophy was ***Charles Sanders Peirce***. Peirce served as a scientist with the U.S. Coastal Survey upon graduation from Harvard University and used a scientific approach to his study of philosophy. In his

FIGURE 2.3 Charles Sanders Peirce.
(COURTESY OF THE NEW YORK PUBLIC LIBRARY, NEW YORK.)

1878 *Popular Science Monthly* article, "How to Make Our Ideas Clear" (Peirce, 1878), he described the pragmatic philosophical approach.

Peirce believed that true meaning evolves through a three-stage process. The first stage of true understanding begins at a rudimentary level as the "first grade of clarity." Peirce defined this stage as having an unconscious or unreflective knowledge of a concept. For instance, most people do not think about air as they perform their daily activities. However, their lack of reflective thought about air does not mean that they are unaware of its existence, nor does it indicate an aversion to the concept. Most people simply breathe normally and give air no conscious thought at all. They take for granted that air will exist in their offices, in their cars on the way home from work, and in their homes at night.

Peirce defined the second stage of clarity as being aware of the characteristics of the concept such that one can explain it to others. For instance, the air that humans breathe is approximately 21% oxygen and 79% nitrogen. Humans inhale a mixture containing 21% oxygen and exhale a mix that includes roughly 16% oxygen. Understanding this second level of clarity puts a person in a position to understand the final stage.

Peirce describes the third stage of clarity as one's ability to apply what one knows about a concept to other situations involving that concept. If one knows the concept of air, the third stage of clarity might be to distinguish between poor air quality and good air quality. Anyone who has ever crossed the street behind a city bus knows that air laced with diesel fumes is neither healthy nor desirable and thus they have achieved pragmatism's third stage of clarity.

Unfortunately, Peirce's research papers and manuscripts were not well preserved (Burch, 2021). It would be left to William James to introduce pragmatism to the world.

William James (1842–1910)

William James entered Harvard Medical School with some degree of reluctance about his study. The field of medicine could not satiate his intellectual hunger, so he engaged himself in philosophy, physiology, and psychology studies. Because of his friendship with Charles Sanders Peirce, he lectured on pragmatic theory in 1898 but departed from Peirce's definition of the concept. James asserted that *pragmatism* is about searching for truth in contradiction to Peirce's assertion that pragmatism is about finding the meaning of a concept. James believed that the test of an idea is in its truth or workability. Does the idea withstand criticism? As a result, James's position on pragmatism has often been described as "Whatever works is true." This position is not precisely how James defined pragmatism. The workability of an idea is usually independent of its practicality. Still, the work of Peirce and James set the stage for John Dewey's work.

John Dewey (1859–1952)

Plato's philosophy on education conceptualized the purpose of education as discovering individuals' inherent strengths and training them for mastery in those strengths. In mastering intellectual powers, individuals are most beneficial to society. It was upon this foundation that *John Dewey* based his work. As a pragmatist, Dewey believed that the importance and value of vocational education stemmed from the ability of the individual to "learn by doing." Dewey thought that inquiry was an orderly and scientific process. The learner experienced a felt need, created a hypothesis or problem statement that accurately defined the problem, developed a series of potential solutions, and attempted to solve the problem using these solutions. The result of the process identified those solutions that worked and those that did not. Once the problem was solved, the learner could move forward to solve other issues.

Dewey's problem-solving model serves a dual role in education—it encourages students to be creative in solving problems while using relevant theory to help students master competencies in their field of endeavor. Dewey believed that a student's interest in a subject was a powerful motivating force to help that student learn the subject matter more thoroughly. This is a fundamental component of *constructivism*. Constructivism is the theory that individuals construct what they know and understand from previous learning. Constructivists believe that learners make their knowledge. Learning is an active process whereby learners produce knowledge based upon their framework of beliefs. Because each learner possesses a unique set of ideas, the knowledge constructed is also unique. Since the learner also establishes how new knowledge is built, the amount and value of social interaction with teachers and others vary accordingly.

Dewey further proposed that social interaction was essential to the learning process. In *The Child and the Curriculum*, he suggested "what the best and wisest parent wants for his

FIGURE 2.4 John Dewey. (COURTESY OF THE LIBRARY OF CONGRESS PRINTS AND PHOTOGRAPHS DIVISION, WASHINGTON, D.C.)

child, that must the community want for all of its children" (Dewey, 2001, p. 5). Individuals require a social setting to learn, and teachers in the social setting mediate the construction of knowledge in learners. The teacher structures the learning environment so students can be actively involved with learning essential concepts. For example, instead of showing photographs of landscape plants for identification purposes, the teacher can allow the students to explore the plants in a greenhouse and identify them through guided self-study.

The philosophical ideas of Peirce, James, and Dewey are the foundation upon which career and technical education is constructed. Understanding and appreciating the intellectual meanderings of learned scholars is one thing, but putting educational theory into practice is quite another. In the early twentieth century, the industrial revolution was indeed revolutionizing the way America did business. Henry Ford's assembly-line production method for automobiles in 1913 soon spread to other industries. The entrance of America into World War I hastened the expansion of industry as America geared up for the war effort. The need for workers in this new industrial age created the need for vocational education.

Dewey founded the Laboratory School in Chicago in 1896 to test his theories about teaching and learning, and he served as its director until 1904. In time, Dewey's ideas became known as progressivism, a philosophy based on the Montessori method of instruction.

Progressive education focuses on students' extraordinary and unique experiences, interests, and abilities as the driving force behind instruction. Progressive educators design lessons based on students' interests to provoke authentic curiosity and engage the learners. Education should begin with the existing knowledge of the learner and expand outward (Dewey, 1998, 2001). The best educators, according to Dewey, recognize and capitalize on the unique attributes of every child.

THE 16 THEOREMS APPLIED TO AGRICULTURAL EDUCATION TODAY

By the early 1920s, the Vocational Education Act of 1917 was entirely in effect, and vocational education began to grow exponentially in the United States. In 1925, Charles Prosser and Charles Allen developed 16 theorems to explain the aims and purposes of vocation education (Prosser & Allen, 1925). These theorems defined the meaning and purpose of vocational education and, consequently, agricultural education. However, Prosser and Allen introduced these ideas more than 90 years ago during the industrial revolution. How do these 16 theorems relate to agricultural education today? Are they still relevant? For agricultural education, Prosser and Allen's work can be condensed into the following six categories.

Authentic Instruction

If it is to be practical and valuable, agricultural education must mirror the current state of technology in the agricultural industry. Students must use the same types of equipment, perform the same tasks, and be exposed to the same risks as someone currently employed in an agricultural profession. It is wasteful and impractical to teach students outdated skills or skills not essential to their chosen fields of endeavor. Instruction should go beyond general training in agricultural occupations and provide training specific to jobs in the local community.

Authentic Self-Evaluation

The agricultural education curriculum in a school should provide courses that capitalize on student interests and abilities. Students who are interested in the subject matter and see it as an essential tool for helping them become established in the careers of their choice will be motivated to learn. Agricultural education is not a good choice for every student, and only those interested in pursuing careers in agriculture or related areas should enroll.

Higher-Order Thinking Skills

The agriculture profession requires highly skilled workers. Agricultural education should consistently provide learning experiences that encourage higher-order thinking. Teachers should prepare learning experiences that cause students to analyze and critically appraise information and to be creative. Students should be able to connect knowledge acquired through cognitive means and psychomotor skills related to that knowledge. For instance, a student must understand the principles of electricity before safely using a shielded metal arc welding machine.

Psychomotor skills learned in the agriculture classroom must be directly related to current occupational skills and authentic. The learning experience should be the same as one might expect to find in the industry. Furthermore, students need to practice psychomotor skills to the point of mastery to establish good work habits. Students unprepared for the minimum level expected for gainful employment in the industry are not well served by the agricultural education program.

Agricultural Education Curriculum

The market demand should drive the agricultural education curriculum in agricultural occupations. The instructional program should provide learning experiences that prepare students for an entry point into agricultural jobs in the community, even if the skills needed by the agricultural industry in that local economy are less efficient and less technical than in other communities. For instance, if the local economy needs workers to operate farm

FIGURE 2.5 Students practicing higher-order thinking skills.

FIGURE 2.6 Veterinary science is a relatively new subject in the agricultural education curriculum and reflects progress in developing relevant and rigorous learning experiences for students. (COURTESY OF EDUCATION IMAGES.)

machinery, then instruction should prepare students for jobs of that type. The curriculum should prepare students accordingly if the local community requires workers highly skilled in biotechnology subjects.

The Agriculture Teacher

Agricultural education in the local school community will be only as successful as the skills and abilities of the agriculture teacher will allow. The teacher is essential to the success or failure of the program and must be highly qualified, well trained, and enthusiastic about the profession of teaching. Teachers must not only master the art and practice of teaching, but they must also stay current in the technical content of the job. Teachers must have professional development plans that allow them to stay abreast of recent developments in agriculture. Even the best teachers become ineffective when the technical content of their lessons becomes outdated.

FFA and Agricultural Education

Agricultural education must also consider the whole student in the development of the curriculum. Students must have experiences that allow them to grow intellectually, physically, and socially. FFA is needed to provide experiences in teamwork, leadership, cooperation, conflict resolution, management, and interpersonal communications. The agricultural industry requires workers who can cooperate with other individuals in carrying out the goals of an organization. Those individuals who are deficient in social skills are at a disadvantage.

PHILOSOPHY OF AGRICULTURAL EDUCATION TODAY

By the mid-1980s, there was a tremendous groundswell of support for education reform in America. Government agencies with responsibility for public education were spurred into action to create a world-class education system. In *Understanding Agriculture: New Directions for Education*, the National Research Council (National Research Council, 1988) recognized the need for agricultural subject matter at all levels of education. The National Research Council's report significantly changed the traditional focus of agricultural education.

For the most part, agricultural education was meant for those students who planned to enter the farming profession. The Vocational Education Act of 1963 broadened the scope of agricultural education to include training in nonfarm agricultural occupations. Students were now receiving instruction in agricultural sales and service, horticulture and natural resources, and other nonfarm trades. This law was instrumental in shaping new curriculum areas in agricultural education.

Understanding Agriculture recommended broadening the scope of agricultural education even further by concluding that agricultural literacy was an essential ingredient of instruction. Agriculture became a universal subject, and agricultural education assumed responsibility for agricultural literacy among all students. The agriculture industry, and the agricultural education programs it served and that serve it, continues to evolve as our understanding of emerging issues grows.

REVIEWING SUMMARY

The original purpose of agricultural education was to prepare youth for careers in agriculture. That core purpose has not changed, but certain factors influence how agricultural education is delivered and what agricultural education does. Global markets and international trade are fundamental to the success of American agriculture. American farmers are now buying and selling directly on the international market. This broader view of agriculture also serves to broadens the knowledge of teachers and students regarding the complexity of the agriculture, food, and natural resources industry. This highlights the need for all citizens to understand the complex role of agriculture in society and the economy. Students in agricultural education courses want and need technical agricultural knowledge, but everyone needs to have a basic understanding of agriculture.

QUESTIONS FOR REVIEW AND DISCUSSION

1. Identify the impact of each of the following individuals on the emergence and practice of vocational education in the United States.
   a. Aristotle
   b. Charles Sanders Peirce
   c. Johann Amos Comenius
   d. Johann Heinrich Pestalozzi
   e. John Dewey
   f. Plato
   g. Socrates
   h. William James
2. What are the basic tenets of each philosophy listed below?
   a. Constructivism
   b. Idealism
   c. Pragmatism
   d. Progressivism
   e. Realism

3. What are the three stages of clarity in pragmatism? Give an example of each stage.
4. What was the significant impact of the National Research Council in its report entitled *Understanding Agriculture: New Directions for Education*?
5. What was the central purpose of the initiative entitled Reinventing Agricultural Education for the Year 2020?
6. Prosser and Allen developed 16 theorems of vocational education. Which theorem is or has been most influential in agricultural education? Why?

ACTIVITIES

1. Complete a brief research project on the differences and similarities between progressivism and pragmatism. How are these two philosophies interrelated?
2. What is your philosophy of agricultural education? Write a statement that presents your understanding of agricultural education and methods of program delivery.

REFERENCES

Burch, R. (2021, Spring). Charles Sanders Peirce. In E. N. Zalta (Ed.), *The Stanford encyclopedia of philosophy*. Metaphysics Research Lab, Stanford University.

Comenius, J. A. (1907). *The great didactic* (M. W. Keatinge, Ed.). Adam and Charles Black. (Original work published 1657).

De Guimps, R. (1889). *Pestalozzi: His aim and work* (M. C. Crombie, Trans.). C. W. Bardeen.

Dewey, J. (1998). *Experience and education*. Kappa Delta Pi. (Original work published 1938).

Dewey, J. (2001). *The school and society; & The child and the curriculum*. Dover Publications. (Originally published in 1899 and 1902).

Durant, W. (1961). *The story of philosophy*. Simon & Schuster.

Kraut, R. (2017, Fall). Plato. In E. N. Zalta (Ed.), *The Stanford encyclopedia of philosophy*. Metaphysics Research Lab, Stanford University.

Locke, J. (1889). *Some thoughts concerning education* (2nd ed.). Cambridge University Press.

Magee, B. (2001). *Story of philosophy* (J. Metcalf, Ed.; 1st ed.). DK.

Nails, D. (2020, Spring). Socrates. In E. N. Zalta (Ed.), *The Stanford encyclopedia of philosophy*. Metaphysics Research Lab, Stanford University.

National Council for Agricultural Education. (2016). *National quality program standards for agriculture, food, and natural resources education*. https://thecouncil.ffa.org/

National Research Council. (1988). *Understanding agriculture: New directions for education*. National Academies Press.

Peirce, C. S. (1878, January). How to make our ideas clear. *Popular Science Monthly*, *12*, 286–302.

Pestalozzi, J. H. (1898). *Letters on early education: Addressed to J. P. Greaves, Esq.* (C. W. Bardeen, Trans.). C. W. Bardeen.

Prosser, C. A., & Allen, C. R. (1925). *Vocational education in a democracy*. Century.

Uzgalis, W. (2020, Spring). John Locke. In E. N. Zalta (Ed.), *The Stanford encyclopedia of philosophy*. Metaphysics Research Lab, Stanford University. https://plato.stanford.edu/archives/spr2020/entries/locke/

3

History and Development of Agricultural Education

"Mr. Hidalgo, why do we have to learn these dates and events in FFA history? All this stuff happened so far back in the past that it just doesn't seem important. It's just a collection of dates and people's names."

Mr. Hidalgo started to reply with his usual answer that it is nice to know a little about our FFA history. But he stopped, and after a moment, decided on another approach.

"You are exactly right, Annie," he replied.

"You have been asked to memorize a bunch of facts and dates. But we have not discussed why they are important," Mr. Hidalgo continued. "Okay, everyone, let's look at this list of dates in the official *FFA Manual*. Specifically, look at the year 1965. What happened that year?"

David spoke up. "That was the year that the FFA and NFA merged."

Mr. Hidalgo replied, "That's right, but what is important about that event?"

No one raised their hands to answer. "Annie, why do you think it is important to know that the NFA, the organization for African American farm youth, merged with the FFA in 1965?"

Annie replied, "I don't know why they merged. Maybe they wanted to be together. The two clubs, that is."

Mr. Hidalgo said, "That's an excellent start to the answer, but tell me more. Why do you suppose the NFA existed in the first place?" Again, silence from the class. Finally, Ray spoke up. "Mr. Hidalgo, all this happened so long ago. It does not seem important."

TERMS

Simeon De Witt	Charles Prosser
Hatch Act	Hoke Smith
Dudley Hughes	Smith-Hughes Act
Justin Morrill	Smith-Lever Act
Morrill Act of 1862	A. C. True
Morrill Act of 1890	Jonathan Baldwin Turner
Carroll Page	Vocational Education Act of 1963
Carl Perkins	

OBJECTIVES

This chapter addresses the National Quality Program Standards for Agriculture, Food, and Natural Resources Education (National Council for Agricultural Education, 2016), specifically Standard 1A: Program Design and Instruction—Curriculum and Program Design. It has the following objectives:

1. Describe situations in the United States that contributed to the development of agricultural education.
2. Describe the creation of land-grant colleges.
3. Discuss legislation that created and expanded secondary school agricultural education and the roles of key individuals in gaining the legislation and guiding program implementation.

Mr. Hidalgo replied, "I hear you, Ray, but it is important to know how we arrived where we are at this point in our American society." Mr. Hidalgo went further. "By looking at how society has changed over time, we get a better understanding of how we can best deal with the problems of society today."

Mr. Hidalgo picked up the *FFA Manual*.

"Studying history positions us to see patterns that help us determine where we are going in society, and perhaps help us solve problems."

Mr. Hidalgo looked at the list of dates and events in the *FFA Manual*.

"Why did we have a separate organization for Black students and a separate organization for White students?"

Several hands went up in response, and for the next several minutes, there was a rich discussion of how education has changed over the years to become more inclusive. After the bell rang and the students had filed out of the classroom, Mr. Hidalgo paused for a moment to reflect on the discussion. There is a need to spend more time discussing historical events because they have implications for addressing today's problems.

THE HISTORY OF AGRICULTURAL EDUCATION

The history of agricultural education in the secondary schools of the United States is closely aligned with that of career and technical education. Common legislation and similar missions have held the two together over the years. Agricultural education was created under the overall umbrella of what is now a part of career and technical education today.

Several early leaders emerged to form and shape vocational education. Justin Morrill, Jonathan Baldwin Turner, Hoke Smith, Dudley Hughes, and Charles Prosser are leaders recognized for establishing agricultural education in the United States. Prosser had many insights and proposed an overall direction for the programs. He is well known for his statements on the purpose and mission of vocational education. His writings reflect a strong sense of understanding about promoting improved levels of living through gainful employment.

The purpose of vocational education is to help a person secure a job, train him so that he can hold it after he gets it, and assist him in advancing to a better job. —CHARLES PROSSER

The purpose of studying history is to utilize the events of the past to understand the present more clearly. This chapter provides the groundwork that, along with the chapter on philosophical foundations, brings perspective to the work of modern agricultural educators.

PRE-COLUMBIAN PERIOD

Native Americans were farming in the Americas as early as 5000 BCE, and as such, they were the first farmers in what became the continental United States. Native Americans domesticated several vegetables and by 1000 CE had developed a sophisticated system of agriculture based on vegetables and grains. The methods by which Native Americans and Alaskan Natives transferred knowledge about agricultural practices from one generation to the next was a form of agricultural education. Tribal elders taught youth how to cultivate plants, and this knowledge was handed down as oral tradition over hundreds of years. Native Americans

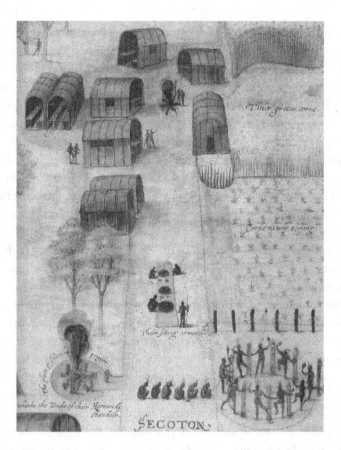

FIGURE 3.1 John White (c. 1540–c. 1593) was an artist and mapmaker who recorded early interactions between settlers and Native Americans. John White's drawings provide an insight into early agricultural practices. This 1585 drawing shows maize cultivation by Native Americans. (COURTESY OF THE BRITISH MUSEUM, LONDON.)

taught early European colonists how to plant and grow maize, pumpkins, squash, beans, and tobacco (Cochrane, 1979).

COLONIAL AND EARLY NATIONAL ERA

Farming was the chief economic engine of the new English colonies in America. Nearly every colonial family raised their food and sold or bartered agricultural products. Like Native Americans, colonial farmers used a method of trial and error with crop and livestock production. Knowledge about agriculture was transmitted orally through the generations. In the colonial period, farmers began to rely on farmer's almanacs for knowledge about agricultural practices. These almanacs provided advice based on astrological phenomena and guided American agricultural education in its infancy (Hurt, 1994).

Farming by the astrological signs as proposed by the farmer's almanacs may have been sufficient for subsistence farming. However, even the earliest European colonists knew the value of farm products and sought to barter or trade them with Native Americans. The early European settlers in America may have started farming first to meet the family's needs, but that was one of many intermediate goals to use agriculture production to improve the quality of life for the settler's family. Subsistence agriculture placed beef, grains, and vegetables on the dinner table. Still, the barter and sale of agricultural products bought salt, gunpowder, nails, fabrics and linen, farm implements, and oxen for improved agricultural production. Most efforts to colonize the North American continent were motivated by economic forces. The settlers who established the Jamestown colony in Jamestown, Virginia, were entrepreneurs seeking to replicate the discovery of wealth experienced by Spain in its colonization of South America. In many ways, they were successful. Colonist John Rolfe learned how to cultivate tobacco from Native Americans (Kell, 1966). The dried leaves of this plant had both medicinal and recreational uses, as demonstrated by the natives. Rolfe's production of tobacco made him very rich. Before long, Rolfe's fellow yeomen were growing tobacco, which significantly added to increasing trade between the New World and Europe. Trading companies in Europe financially supported the growth and development of American agriculture in these early years.

As the sophistication of the colonial American farmer grew with each successful growing season, there arose a need for more substantial training in agricultural methods. Agricultural societies promoted and developed an interest in productive farming methods. These societies were primarily a European idea. The colonies in America took their cue from agricultural societies in Europe. The Edinburgh Society of Improvers in the Knowledge of Agriculture was organized in Scotland in 1723. The Society for the Encouragement of the Arts, Manufactures, and Commerce was established in England in 1754. The first veterinary school was established in France in 1766 (R. H. True, 1935).

In the United States, farmers formed the Philadelphia Society for Promoting Agriculture in 1785. Its purpose was to draw attention to agriculture, increase productivity, and improve rural life. The motto of the society was "Venerate the Plough." The organization published articles in newspapers about agriculture and encouraged experiments that led to greater productivity in all aspects of agriculture. A secondary purpose of the Philadelphia Society was to promote the development of other similar agricultural societies throughout the United States. Furthermore, the society also distributed awards to members for essays produced on subjects of importance to the society. In 1788, for instance, the organization awarded essayists for papers on the best experiments in crop production, animal husbandry, innovative cultivation methods, pest control, food preservation, and soil conservation (R. H. True, 1935).

In 1792, the Massachusetts Society for Promoting Agriculture explored new methods of improving agriculture. Throughout the 1800s, agricultural societies disseminated research in agriculture through publications, newspaper articles, and lectures (A. C. True, 1929). These societies urged legislative action to create formal schools of agriculture. In 1855 the Agricultural College of the State of Michigan was founded. Yale University added three professors of the agricultural arts in 1845. A Farmers High School was created in Pennsylvania by the state legislature in 1854 (Farmers High School of Pennsylvania Incorporated, 1855).

These societies and schools encouraged the development of innovative agricultural practices and fostered the inventiveness of the American people. Eli Whitney's cotton gin demonstrated the powerful processing technology available in the late 1700s. Whitney's gin significantly reduced the hand labor needed to separate seeds from cotton lint and led to increased use of slave labor. Other developments followed, such as the grain reaper by Cyrus McCormick in 1842 and the steel moldboard plow by John Deere in 1837. McCormick's invention was the first significant advance in developing what we refer to today as harvesting equipment. Deere's contribution was important because it effectively tilled heavy soils, opening vast land areas for agricultural production. These inventions increased the potential for agricultural productivity.

THE CIVIL WAR ERA AND AN INDUSTRIAL REVOLUTION

The early 1860s saw the nation embroiled in a great civil war. With almost every man on the front lines or engaged in running the war industries, agriculture in both the North and the South had to adopt new technologies for production. The early 1860s saw a change from hand-powered agriculture to horse-powered agriculture as the first American agricultural revolution got underway (National Center for Agricultural Literacy, 2018). As markets for agricultural exports increased in the 1860s, so did the need for suitable methods of transporting those products to market. The era saw a significant expansion of railroads across the nation.

Economic fluctuations caused by wartime inflation, recession, and recovery fueled the fires of ambition for many Americans. In 1862, the Homestead Act was passed, granting 160 acres of land to those Americans willing to carve out a niche in the West. The western territories developed rapidly under the agrarian interests of farmers and ranchers. The late 1860s were the days of quickly expanding cattle production in the western United States.

The 1860s were a period of rapid industrial growth. A significant new direction was making steel and iron for the railroads and other development. Railroads were built in the western United States, opening up a new era in transportation. The American population was also on the move, shifting westward across North America. John Deere's steel plow allowed farmers to cut through the hard Midwestern soil for planting. Farmer societies and organizations rapidly expanded to promote the causes of agriculture. There were almost 1,000 agricultural organizations in the United States by the mid-1860s (National Center for Agricultural Literacy, 2018).

During this period of westward expansion and wartime industriousness, some leaders with foresight saw the need for a nationwide agricultural, mechanical, and military education system to catalyze the era's rapid social and industrial growth.

Creation of Land-Grant Colleges

Today, the land-grant colleges are significant forces shaping agricultural education and research in the United States. These were established and prospered due to the Land-Grant College Acts of 1862, 1890, and 1994, known as the Morrill Acts.

It is important to note that the concept of using land grants to fund education was not a novel idea in the 1850s and 1860s. New England states had used land grants to support education as early as the 1700s (Carleton, 2002). The Land Ordinance of 1785 provided land allocation to help education, and this land-grant concept was further refined in the Northwest Ordinance of 1787 (Carleton, 2002). It is also important to note that the idea of agricultural colleges was being discussed in the United States as early as 1819 when *Simeon De Witt* (1756–1834) of New York proposed colleges for teaching agricultural subjects and conducting experimental research (Cross, 1999). De Witt may have broached the idea of agricultural colleges for the populace. Still, *Jonathan Baldwin Turner* (1805–1899), a professor of classical literature at Illinois College, brought the concept to American awareness. Turner transformed the idea of agricultural and industrial colleges into an issue worthy of attention by influential people.

Jonathan Baldwin Turner was born and raised on a farm in Massachusetts and educated at Yale University. His classical education at Yale led him to a teaching position at Illinois College. Turner was a fiery orator who espoused his populist views with much vigor and zeal. His passion for the common people stirred the public when the national economy teetered between recession and depression. In the late 1840s, Turner's ideas about how to serve the educational needs of the people crystallized into the concept of making higher education in the practical arts and sciences available at low cost. Turner believed that the federal government should pay for the establishment of universities to provide such education. His argument eventually convinced the Illinois legislature, which sent a resolution to Congress asking for the development of land-grant colleges (Cross, 1999).

Justin Morrill (1810–1898) entered the U.S. House of Representatives in 1855 as the new congressman from Vermont. In the first few months of his first term, Morrill distinguished himself as a proponent of legislation designed to help the American farmer and the working-class

FIGURE 3.2 Justin Smith Morrill. (COURTESY OF THE LIBRARY OF CONGRESS PRINTS AND PHOTOGRAPHS DIVISION, BRADY-HANDY PHOTOGRAPH COLLECTION.)

American. However, as Morrill was to discover, his colleagues from the South did not share his zeal for the methods he was proposing. His resolution to establish a board of agriculture and national agricultural schools failed because of southern opposition to any legislation deemed threatening to states' rights (Cross, 1999).

Not to be deterred, Morrill continued to work on land-grant legislation through his first term in office. On December 17, 1857, Morrill introduced a bill granting lands for agricultural colleges. After a year of parliamentary wrangling, the bill passed the House by a vote of 105 to 100. The Senate passed its version in February 1859, and the bill eventually found its way to the desk of President James Buchanan.

Unfortunately, President Buchanan vetoed the bill. Under pressure from southern Democrats, Buchanan avoided another fight on state's rights with land-grant legislation. Morrill waited until the 1860 elections for perhaps a more favorable audience with the next president of the United States. Abraham Lincoln would support the legislation, but Morrill had once again to navigate the political minefield to get his bill on the president's desk. Southerners opposed to the bill would not take kindly to its return to the congressional agenda in 1861, and the battle for land-grant colleges looked to be a tough one.

In January 1861, Morrill received somewhat unintended but not altogether unexpected support from President Lincoln. With Lincoln's inauguration that year, the southern states began to secede from the Union, starting with South Carolina in December 1860 and ending with Tennessee in June 1861. With the southern states gone from the House and Senate chambers, conditions were favorable for the bill's return. On December 16, 1861, Morrill introduced a revised land-grant bill for consideration by the House of Representatives. After some additional editing by both houses, the bill finally passed the Congress and went to the White House for President Lincoln's signature in July 1862.

FIGURE 3.3 The University of Illinois in Urbana established the Morrow Plots in 1867. They are the oldest continuous experimental plots in the United States. (*OBSERVATORY, DAY, MORROW*, COURTESY OF THE UNIVERSITY OF ILLINOIS AT URBANA-CHAMPAIGN ARCHIVES PHOTOGRAPHIC PRESERVATION PROJECT, HTTPS://UIHISTORIES.LIBRARY .ILLINOIS.EDU/CGI-BIN /VIEWIMG?ID=21876)

The Morrill Act of 1862

On July 2, 1862, President Abraham Lincoln signed the *Morrill Act of 1862* into law. With the stroke of his pen, he created a system of higher education that would free millions of people from the poverty of ignorance and set into action a course of events that would establish the United States as the leader among all nations in agricultural production. The creation of colleges to teach the agricultural, mechanical, and military arts opened new opportunities in the United States. The Morrill Act of 1862 kindled the flames of expansion for agriculture and industry in the United States. The industrial revolution of the early 1900s can trace its roots back to Congressman Justin Morrill's ability to shepherd legislation through the U.S. Congress.

The Morrill Act of 1862 set aside 30,000 acres of land for each member of Congress from those states then incorporated into the United States of America. Upon the sale of public lands, the proceeds were used for the creation in each state of at least one college where the leading subjects would be agriculture and the mechanical and military arts. For the states to receive the proceeds from the sale of public lands, the legislatures had to accept the act's provisions within five years. Furthermore, the act provided free public land to be made available for each new state as it entered the Union.

Only recently have universities begun to acknowledge that these public lands were taken from Native American tribes and communities. The land was obtained by the federal government through various treaties, below-value purchases, and confiscation. Virginia Tech's Special Collections and University Archives has posted an acknowledgment statement from its American Indian & Indigenous Community Center (Virginia Tech, n.d.). Statements such as this one are an important first step, but should not be the only step.

Reconstruction Era

The United States government and agricultural colleges and universities supported the agricultural education movement through short agricultural courses made available to farmers. In 1868, Kansas Agricultural College provided training to farmers on the application of modern agricultural practices. In the summer of 1884, the Alabama State Agricultural College conducted meetings with farmers regarding agricultural problems. Colleges in Massachusetts, Illinois, Iowa, and New Hampshire also adopted similar institutes, and state boards of agriculture also conducted farmers' institutes (A. C. True, 1929).

The *Hatch Act* of 1887 established experiment stations for agricultural research across the United States. These research stations provided and continue to provide for research innovation in the agricultural sciences. But that is only part of the story. The Hatch Act provided for agricultural research and the dissemination of research findings to the public. Furthermore, these research findings were written so that farmers and ranchers could readily understand and apply research findings to their operations. The research station at the University of Minnesota opened a school of agriculture for students who were below college level in 1889 (Moore, 1988). Alabama soon followed with similar schools in 1889.

In 1888, the Office of Experiment Stations in the United States Department of Agriculture recognized the value of farmer's institutes and began collecting data and researching the work of the institutes. By the end of the nineteenth century, agricultural education had expanded outward from farmer's institutes and university short courses into public schools across the nation. In 1906, public school officials in Michigan, Arizona, and Georgia would invite institute speakers to speak to the students on agricultural subjects (A. C. True, 1929).

The Morrill Act of 1890

Justin Morrill was eventually elected to the U.S. Senate and continued to work for legislation that supported the land-grant college movement until he died in 1898. The *Morrill Act of 1890* set aside funds for teacher education in agricultural and mechanical arts. However, the act restricted funds to those colleges that made no distinction based on race or color in student admissions. It was acceptable for states to establish separate colleges for White and African American students under the principle of separate but equal education.

The second Morrill Act provided funds to establish African American agricultural colleges. Many scholars assume that the second Morrill Act created Black land-grant colleges. It did not. The first Morrill Act did not expressly exclude funding for African American institutions. Funding went to Alcorn State University, established in 1871, and Hampton University, established in 1872. However, most land-grant schools did not welcome Black students. Two decades would pass before specific legislation assisted in developing agricultural colleges for Black students (Cross, 1999).

The Morrill Act of 1890 was a little more complex than the first one. To provide instruction in food and agricultural sciences, this act appropriated funds from the sale of public lands, beginning with a single appropriation of $15,000 in 1890, then increasing in annual increments of up to $50,000.

THE BEGINNING OF AGRICULTURAL EDUCATION IN THE SECONDARY SCHOOLS

In 1821, Robert Hallowell Gardiner and his neighbors in Gardiner, Maine, petitioned the Maine legislature for a school for agricultural and mechanical subjects. Gardiner had supplied the land and buildings for the prospective school, so the state of Maine obliged Mr. Gardiner's request for a school on the banks of the Kennebec River (Stevens, 1921). Gardiner's Lyceum was likely the first school of agriculture in the United States, and it combined liberal arts subjects with studies in agriculture and mechanical arts. The school eventually became a high school academy and operated until a fire destroyed the building in 1870. Other small schools like the Gardiner Lyceum operated in that period almost exclusively as community schools. In 1845, Theodore S. Gold established Cream Hill Agricultural School in Cornwall, Connecticut, to educate young men on the virtues of farming and rural life (Gannett & Cornwall Historical Society, 1986). Massachusetts established two similar schools for agricultural education in 1858 (Hamlin, 1962).

For almost three centuries after European settlers arrived, no nationwide curriculum or program in agricultural education existed. Local schools that offered agricultural subjects were left to their own means. The Hatch Act was the first significant unified national effort at less than the college level for agricultural education. The Hatch Act of 1887 established agricultural experiment stations and educated the public about the implications of the research conducted at these experiment stations. In 1889, a successful school was established on the agricultural experiment station near the University of Minnesota to teach agricultural subjects to students below the college level. Other states soon adopted similar arrangements to disseminate information arising out of agricultural research in the experiment stations (Moore, 1988). New York's legislature appointed the Cornell University Agricultural College in 1897 to supervise and establish agricultural education in public schools. The Farm Life Act of 1911 in North Carolina created boarding schools for agriculture and home economics. The state school superintendent and the farm-life school advisory board established the curriculum, including practical farm work (Stimson & Lathrop, 1942).

Eventually, the federal government would recognize the need and importance of agricultural education and create legislation that specifically encouraged states to develop agriculture teacher training programs and fund local agricultural education programs.

An early proponent of agricultural education in public schools was *A. C. True*. As director of the Office of Experiment Stations in the federal government, he wrote extensively on the subject. Director True believed schools in farming communities could provide instruction for rural farm youth. He devoted a significant portion of his time to promoting the cause of agricultural education to citizens, colleges, and the federal government (A. C. True, 1929).

True's work led to increased funding for agricultural education and continued growth in schools below the college level. In the years before the passage of the Smith-Hughes Act, it

FIGURE 3.4 Alfred Charles True. (COURTESY OF THE U.S. GOVERNMENT PRINTING OFFICE, WASHINGTON, D.C.)

FIGURE 3.5 Land-grant universities maintain several experiment stations that focus on particular problems. The experiment station at Montana State University operates an agronomy farm. (COURTESY OF EDUCATION IMAGES.)

is evident agricultural education was well established in schools. By 1917, 30 states had agricultural education in schools (Hamlin, 1962). The task remained of organizing agricultural education to be available to every student who wished to study it.

The Smith-Lever and Smith-Hughes Acts

Efforts to build support for vocational education were unsuccessful in the Republican-controlled Congress in the early 1900s. During the presidency of Theodore Roosevelt, the United States became a strong world power. The era of progressivism yielded significant improvements in food safety, farming, and conservation. Roosevelt modernized the federal government and began efforts to clean up illegal business trusts and the economy.

Theodore Roosevelt held to his pledge not to run again for the presidency in 1908, and the nomination and subsequent election went to William Howard Taft. Taft could carry out many of Roosevelt's initiatives, but he could not rally the people behind the Republican platform during the 1910 congressional elections. In 1910, the Democrats gained control of Congress for the first time in almost two decades (Morison, 1994). Momentum began to build for federal legislation on several fronts. In 1912, the election of President Woodrow Wilson resulted in increased federal legislation that would have far-reaching effects across the nation into the farms and towns of America.

For southern farmers, federal assistance could not come too soon. By 1910, more than 80% of all Southerners lived in rural communities, and many were in poverty. Unfortunately, half of all farms in the South were unimproved. Southern farmers lacked the skills and resources necessary to improve farmland and lacked knowledge of the best practices in animal science. In 1910, the average southern farm was valued at half the national average for farms.

The U.S. Congress wasted little time in passing legislation designed to help the American farm family. The *Smith-Lever Act* of 1914 created the Cooperative Extension Service. This law established a partnership between the federal government and the land-grant colleges to extend knowledge about the best practices in agriculture to rural communities. However, this federal law did not address the growing need for national leadership and agricultural education funding in the early 1900s.

Beginnings of Federal Agricultural Education Legislation for Public Schools

Senator *Carroll Page* of Vermont had a passion for vocational education. He believed that it was the duty of the federal government to provide job-specific training for young people. From an economic viewpoint, it made sense to produce skilled workers who could step into the American workforce without going through an apprenticeship period. Vocational education could improve job training for millions of youth, and Senator Page was motivated to sponsor legislation that provided the necessary federal support.

The chief advisor to Senator Page was *Charles Prosser*. Dr. Prosser knew from firsthand experience the condition of education in the years preceding the Smith-Hughes Act. As an elementary school teacher, principal, and school superintendent in New Albany, Indiana, he knew the kind of education needed in the workforce. Because of his strong stance on education, he served as president of the Indiana Teachers Association and later became the deputy superintendent for industrial education (Moore & Gaspard, 1987).

In a few years, Charles Prosser became the secretary of the National Society for the Promotion of Industrial Education. His association with this organization placed him in contact with legislators who could sponsor legislation friendly to vocational education, including Senator Carroll Page, Senator Hoke Smith of Georgia, and others.

The same year that Senator Page came to Washington, Senator Jonathan Dolliver of Iowa and Representative Charles Davis of Minnesota attempted to provide federal legislation for vocational education. The Dolliver-Davis bill languished in Congress as attempts were made to amend it into an acceptable form. The death of Senator Dolliver, a champion for vocational education, in 1910 effectively removed hope of the passage of the bill. Senator Page and Representative William Wilson of Pennsylvania put their own version of the bill up for debate in Congress.

Many in Congress did not like the Page-Wilson bill. The main argument against the bill seemed to be that it was a competing measure to the more popular Smith-Lever bill, already making its way through Congress. Since the Page-Wilson and Smith-Lever bills were similar, it was possible that one bill could swallow the other during the debate (Barlow, 1976). Senator Page and Charles Prosser negotiated several amendments to the Page-Wilson bill to make it more palatable to its opponents. Still, the Smith-Lever bill would eventually win the fight. Senator *Hoke Smith* of Georgia was a sponsor of the Smith-Lever Act creating the Cooperative Extension Service.

As a skilled politician and parliamentarian, Smith successfully blocked the passage of the Page-Wilson bill until his legislation to create the Extension Service had passed both houses. Smith's chief argument against the Page-Wilson bill was that it was poorly written and under-budgeted to meet vocational education needs. Smith believed in vocational education, but he also thought that the Page-Wilson bill was not the correct measure to establish a national vocational education system (Barlow, 1976). In 1914, the Smith-Lever Act establishing agricultural extension passed both houses of Congress and was signed into law by President Wilson.

After Smith-Lever became law, President Wilson established the Commission on National Aid to Vocational Education to study the need for federal aid for vocational education. The commission's membership was stacked in favor of vocational education, so the committee's report was characteristically pro-vocational education. Specifically, the report recommended the Congress of the United States create legislation for vocational education.

FIGURE 3.6 Senator Hoke Smith of Georgia. (COURTESY OF THE LIBRARY OF CONGRESS PRINTS AND PHOTOGRAPHS DIVISION, WASHINGTON, D.C.)

This commission's report established guidelines that later became the framework for the Smith-Hughes Act (Smith, 1914).

With Smith-Lever safely passed, Smith turned his attention to the matter of vocational education. Acting on the recommendations of the Commission on National Aid to Vocational Education, Senator Smith introduced his vocational education bill on December 7, 1915. Representative ***Dudley Hughes*** of Georgia introduced companion legislation in the House of Representatives two days later. The Smith-Hughes bill was born.

Over the next year, Congress would amend the Smith-Hughes bill through committee work and floor debate. In time, the Congress of the United States finally reached a general agreement on the merits of the bill, and it passed both houses of Congress on February 17, 1917. Six days later, President Wilson signed the bill into law. After six years of effort by Charles Prosser, Senator Carroll Page, Senator Hoke Smith, and others, the National Vocational Education Act of 1917, known as the ***Smith-Hughes Act***, was law.

Provisions of the Smith-Hughes Act
Funding for Vocational Education

The Smith-Hughes Act provided funding for training teachers in agricultural education, industrial arts education, and home economics education. Furthermore, the act paid the salaries of teachers in these subjects. The act provided funding for the establishment of teacher education programs in colleges and universities. It funded the hiring of supervisors to manage the expenditure of funds at the school level. These supervisors gave direct assistance to teachers in the teaching of their respective subjects.

State Boards for Vocational Education

Smith-Hughes also created a state board for vocational education in each state receiving funding under the act. There was some flexibility in how the board was set up in each state. In some cases, the responsibilities of the state board for vocational education were absorbed into the work of the state board of education.

For the states to receive and continue to receive funding under Smith-Hughes, a provision of the law required that schools expend the minimum appropriations for the training of teachers before any appropriations for salaries could be released.

Federal Board for Vocational Education

The Smith-Hughes Act also created the Federal Board for Vocational Education. The purpose of this board was to see that the act's provisions were carried out according to the law. President Wilson appointed federal board members on July 17, 1917, which met for the first time on July 21, 1917, with Secretary of Agriculture David Houston as chair. On August 15, 1917, Charles Prosser was appointed director of the board. The first members of the board were

- David Houston, Secretary of Agriculture
- William Redfield, Secretary of Commerce
- William B. Wilson, Secretary of Labor
- P. P. Claxton, U.S. Commissioner of Education
- Charles Greathouse, Agriculture Representative
- Arthur E. Holder, Labor Representative
- James Munroe, Commerce Representative

The federal board was authorized to hire a staff to carry out the administrative tasks of the board. The state boards for vocational education were required to submit annually to the federal board detailed plans of how vocational education funds would be expended.

The Beginning of Agricultural Education for African Americans

Before the twentieth century, Southerners' prevailing mood was skepticism toward strong central governments, and government programs included public education. In the prewar South, schools were the responsibility of the family and, to some extent, the church.

Consequently, a weak system of schooling prevailed, starved to near death by a lack of public tax support (Cash, 1991). After the American Civil War, schools did not fare much better. Public taxes went into rebuilding those parts of the South destroyed by the war. There was inadequate funding for school construction, the hiring of teachers, and the equipping of schools. Southern states, because of segregationist legislation, had established a dual system of schooling. Whites attended schools supported by public taxes, and Blacks attended schools supported by donations and taxes mainly derived from the African American community. African Americans occupied the poor lower working class where what little money could be earned was spent on the immediate subsistence needs of the family, and more often than not, their schools were significantly underfunded.

In 1865, 90% of the southern Black population was functionally illiterate, yet efforts were underway by education supporters to mitigate this problem. During the American Civil War, teachers from the North arrived behind the advancing Union armies in the South. These teachers set up shop in old stores, churches, and homes and began educating those who came to school each day. The funding for these schools was mainly through the Bureau of Refugees, Freedmen, and Abandoned Lands (Freedmen's Bureau) and from donations by the members of the American Missionary Society and other Christian missionary groups. These societies often paid for teacher salaries and helped secure teaching materials for teachers. By the end of the work of the Freedmen's Bureau in 1870, there were approximately 10,000 teachers for Black schools in the South, and most of these were White Northerners (Wright, 1949).

TABLE 3.1

Major Philanthropic Organizations Supporting Agricultural and Industrial Education for African Americans

Organization	Year Established	Amount of Initial Funding	Donor
General Education Board	1903	$1,000,000	John D. Rockefeller
Phelps-Stokes Fund	1911	$1,600,000	Olivia and Caroline Phelps-Stokes
Anna T. Jeanes Foundation	1907	$1,000,000	Anna T. Jeanes
Julius Rosenwald Fund	1917	$3,030,000	Julius Rosenwald
John F. Slater Fund	1882	$1,000,000	John F. Slater

Adapted from Butler (1932).

As it became evident that public funding of African American schools for agricultural and industrial education would not materialize from state legislatures or the federal government, educational leaders began to lobby vigorously for the support of these schools through other means and for different reasons. In 1902, Charles Dabney, educator and scientist, wrote,

> Everything in the South waits upon the general education of the people. Industrial development waits for more captains of industry, superintendents of factories, and skilled workmen. . . . We must educate all of our people, blacks as well as whites, or the South will become a dependent province instead of a coordinate portion of the nation. (Dabney, 1902, p. 208)

Both Dabney and Winston voiced the concerns of many prominent clergy members, political figures, and wealthy industrialists. These individuals began to mobilize financial resources to build schools, hire and train teachers, and provide agricultural and industrial education for African American students. Table 3.1 lists the major philanthropic organizations that supported this effort.

FIGURE 3.7 The Penn School of St. Helena Island in South Carolina was a school for formerly enslaved people during and after the American Civil War. Established in 1862 with the assistance of the federal government and northern Christian missionaries, the school taught basic academic subjects and vocational training.

The Growth and Development of Supervised Agricultural Experience

Supervised agricultural experience in America likely began as youth apprenticeship in the colonial period or earlier. The concept of learning agricultural subjects under the direction of a mentor or skilled practitioner is an old one indeed. Evidence of apprenticeship can be found among Native Americans in the North American archaeological record (Struck, 1958). As the population of the American colonies grew, the economy supported the growth of the apprenticeship. In early American schools, students learned basic skills in reading, mathematics, history, Latin, and Greek (Urban & Wagoner, 2004) and then went home to the farmstead to understand animal husbandry, crop science, and engineering.

In the early 1900s, Rufus W. Stimson, the Smith Agricultural School principal, developed the project method. Students learned the basics of agricultural production methods and applied them to their home farms (Moore, 1988). Emphasizing the benefits of a practical hands-on instructional program, Stimson insisted these projects be completed away from the school campus and on the home farm (Stimson, 1919). Stimson thought school projects were impractical, unrealistic, and involved too many students in one project. Students had no personal ownership of school projects. Stimson's model for farm projects specified projects be realistic and valuable and measured by specific learning goals. The two-pronged curriculum at Stimson's school curriculum included the study of production agriculture and individual project work.

The Beginning of the Integrated Model of Agricultural Education

Agricultural education is typically organized into interrelationships between three significant concepts: classroom and laboratory instruction, supervised agricultural experience, and the FFA youth organization. How did these three concepts become linked together as the model for agricultural education?

Dr. Glen C. Cook wrote the *Handbook on Teaching Vocational Agriculture*, first published in 1938. Cook and others revised the text with the latest revision occurring in 2008. This textbook represented more than seven decades of agriculture teacher education and provided much insight into the priorities of agricultural education in its formative years.

In the 1938 handbook, Cook identified not three but four components of agricultural education. These were classroom instruction, supervised farm practice, farm mechanics, and extracurricular activities. Cook (1938) described supervised farm practice as an integral part of agricultural education but stopped short of making the same judgment about the FFA. While FFA activities were acceptable for agricultural education students, Cook did not limit agricultural education to a sole partnership with the FFA. Extracurricular activities also included 4-H and agricultural clubs in addition to the FFA.

Cook's 1947 *Handbook on Teaching Vocational Agriculture* cleared up the role of FFA in agricultural education. Cook identified the five major phases of agricultural education to include classroom instruction, supervised farming programs, farm mechanics, community food preservation activities, and Future Farmers of America (FFA) activities. Cook described farm mechanics as an essential subcomponent of supervised farming programs. He further identified community food preservation activities as necessary during the World War II era of rationing and food shortages (Cook, 1947).

Later editions of Cook's textbook (Phipps, L. J., 1965, 1980; Phipps, L. J. H., 1966, 1972; Phipps & Cook, 1956; Phipps & Osborne, 1988; Phipps et al., 2008) continued to prescribe the three-component model of agricultural education with one notable change. The component devoted to agricultural youth organizations included the New Farmers of America and Young Farmers in the 1966 and 1972 editions of the Phipps text. References to the New

Farmers of America in the agricultural education model disappeared in editions of the text written after 1965. In the 1988 edition of the handbook (Phipps & Osborne, 1988), references to Young Farmers in the model disappeared. The four instructional components were classroom instruction, supervised experience, laboratory instruction, and vocational student organization.

Cook's textbook explained the agricultural education model but did not reduce the model precisely to the present-day three-component version. In the 1970s, the FFA began a series of teacher development programs designed to improve the quality of agricultural education programs. The 1975 *FFA Advisor's Handbook* was prepared by the National FFA Organization, emphasizing teacher development. It included the three-component classroom and laboratory instruction model, supervised experience, and FFA (National FFA Organization, 1975). Page 7 in the handbook contained the Venn configuration of three overlapping circles portraying these three components. The handbook justified the integral nature of FFA with the instructional program and explained FFA activities require both supervised agricultural experience and instruction to be valid.

The three-component model continues to be the prevalent method for explaining the concept of agricultural education. The question remains, is the model truly of an integral nature? SAE and FFA activities have become optional elements in some agricultural education programs, and FFA activities may not always reflect the goals of the instructional program.

"High school vocational agriculture has been composed of three parts: classroom teaching, supervised practice, and the Future Farmers of America organization. These parts have developed separately, and unfortunately, they remain to a considerable extent discrete" (Hamlin, 1950, p. 224).

FEDERAL LEGISLATION FROM 1918 TO 2018

As the nation grew, so did the need for vocational education. More importantly, the complexity of the American education system developed during the twentieth century. The school curriculum evolved into a comprehensive curriculum with career and technical education subjects alongside traditional academic subjects. The United States Office of Education because the United States Department of Education in 1980. The Vocational Education Act of 1963 superseded the Smith-Hughes Act, and the Federal Board of Vocational Education no longer exists. Much of the federal education legislation enacted in recent years has a more comprehensive purpose. For instance, Congress passed the Elementary and Secondary Education Act (ESEA) in 1965 as part of President Lyndon Johnson's Great Society program. Its purpose was to provide resources so that every child has an opportunity for quality education. All other acts of the same title that followed this 1965 legislation typically amend it or reauthorize funding. The Improving America's Schools Act of 1994 and the No Child Left Behind Act of 2001 are reauthorizations of the original ESEA of 1965. Both of these acts provided funding to schools with agricultural education. The 1994 version of the ESEA increased federal support for technology in education. The 2001 reauthorization of the ESEA increased accountability for teachers and sought to find new methods for improving student performance in all subjects.

The United States government usually provides direction and guidance to states on the improvement of educational programs. The federal government provides only 10% of the funding for public education in the United States, but it is a vital partner in funding education. The federal government's funding share handles shortfalls in local and state funding. For instance, the United States Department of Education (ED) works with the United

States Department of Agriculture to provide the school lunch program and the Department of Health and Human Services to provide the Head Start early childhood program. The ED provides funding where the most good can be derived. Funded mandates are the traditional response of the federal government to the improvement of education, and the most significant period of federal funding for education began in the early 1900s.

Between 1918 and 1963, Congress passed several federal laws to modify or increase funding for vocational education. Senator Walter George of Georgia was the principal sponsor for many of these acts. The "George Acts" amended the Smith-Hughes Act by increasing funding in existing vocational education programs and expanding the reach of federal legislation into new areas of vocational education. Here is a summary of four "George Acts":

George-Reed Act of 1929—Increased federal support for vocational education and gave home economics independent status as a division.

George-Ellzey Act of 1934—Repealed the George-Reed Act. This legislation provided additional funding for vocational education and implemented new distributive education financing.

George-Deen Act of 1936—Increased funding for vocational education in four areas: agriculture, home economics, trade and industrial education, and distributive education.

George-Barden Act of 1946—Significantly increased annual appropriations for vocational education and altered the formula for distributing funds in favor of agricultural education. This act also established area vocational schools for training students in vocational subjects. Because the federal funding provided by this act allowed for vocational guidance and the purchase of equipment, vocational programs grew substantially in the years following its passage.

National Defense Acts

The National Defense Acts are reauthorized periodically, as they provide funding and leadership policy for the defense of the United States. For instance, the 1940 National Defense Act and the 1958 National Defense Education Act provided funding for creating a highly skilled workforce in times of war. These acts also provided training for all workers in war production, including women. The 1958 National Defense Education Act provided financial assistance for students enrolling at colleges and universities to prepare for these highly skilled and technical jobs of the future.

The Vocational Education Act of 1963

By the 1960s, vocational education under the Smith-Hughes Act needed revision to meet the changing needs of the American economy. More than any other law since Smith-Hughes, the *Vocational Education Act of 1963* broadened agricultural education and made great strides in moving the program forward. It notably provided that instruction could be in nonfarming areas of agriculture.

The 1960s was a turbulent period in America's history. Matters related to civil rights and America's involvement in the Vietnam War were lightning rod issues for most Americans. Underneath all this, the face of American agriculture was changing. The Fair Labor Standards Act extended federal minimum wage requirements to employers of most farmworkers, and the United Farm Workers began unionizing California farmworkers. Great strides were made in farm mechanization, and the need for skilled mechanics and technicians increased exponentially. Agricultural commodity groups significantly increased their lobbying efforts in Congress. Agricultural exports made up almost one-fifth of total U.S. exports.

To meet the new challenges of the agricultural industry in the last half of the twentieth century and to meet the growing demand for a highly skilled workforce in all vocational disciplines, the federal government's presence in the classroom needed to change. The Vocational Education Act of 1963 replaced the categorical funding of specific vocational programs, such as agricultural education and trade and industrial education, with a population-driven funding formula to the states. Federal funding under this act was distributed proportionally to the states based on the number of students within a specific age group.

The increased federal funding provided by the Vocational Education Act of 1963 made vocational education more flexible and open to emerging trends. This act, sponsored by Senator *Carl Perkins* (1912–1984) of Kentucky and Representative Wayne Morris of Oregon, opened the door for vocational training in other areas, such as business education. Agricultural education expanded the options for youth to include careers beyond the farm in such areas as marketing, horticulture, agribusiness, and natural resources. The act further provided funding for training youth with physical disabilities, creating work-study programs, and establishing vocational training centers.

The Vocational Education Act of 1963 also changed supervised experience in agriculture to include more than production agriculture. Unfortunately, this expanded variation of supervised experience led some states to abolish or reduce its value as part of the agricultural education model. The act also abolished the supervisory role of the state agricultural education staff and relegated it to a consulting role.

Congress amended the Vocational Education Act of 1963 in 1968 and again in 1976. The 1968 amendments increased federal funding for vocational education and allowed for vocational guidance, career counseling, and the construction of other vocational schools. Fifty percent of the budget in vocational education was to be used for students with disadvantages or disabilities and for employment services to students. The 1976 amendments increased federal funding and established guidelines for eliminating gender discrimination.

The Perkins Acts

The Federal Vocational Education Acts of 1984, 1990, 1996, 1998, 2006, and 2018 bear the name of Carl Perkins. Carl Perkins was a member of the United States Congress from Kentucky who served from 1949 to 1984. He was a strong supporter in Congress for vocational education legislation. These laws attempted to modernize vocational education further and to expand its emphasis on career and technology education. The 1984 act created federal legislation to support the efforts of vocational education in making the United States more competitive economically. The primary focus of the 1984 act was to improve the quality of vocational education for all students and improve the accessibility of vocational education for all students who wanted it. Congress earmarked more than half the funding provided by the first Perkins Act for special student populations—students with disadvantages or disabilities, adults, and single-parent families.

The Carl D. Perkins Vocational Education Act of 1990 increased funding for vocational programs to meet advances in technology. This act also increased funding for curriculum integration efforts between vocational and academic subjects. It sought to move vocational education away from training for specific jobs and toward education in the more general aspects of careers. The act further provided for the establishment of articulation agreements between secondary schools and colleges, and universities.

In 1996, the third Carl D. Perkins Vocational Education Act reauthorized funding for vocational education. Still, the emphasis was even stronger on curriculum integration and articulation agreements between secondary and postsecondary institutions. The 1998 reauthorization of Perkins retained the same focus on integration and required that students enrolled in career and technical education courses receive the same rigorous curriculum as other students.

Congress enacted the latest revision of the Carl D. Perkins Vocational and Technical Education Act in 2018. The new law, known as Perkins V, increased the emphasis on the career achievement of all students, not just those enrolled in career and technical education. It transferred the authority for determining academic performance measures to the states and stakeholders. Perkins V broadened the definition of special needs populations to include homeless students, foster children, and the children of parents serving in the armed services. The new act also provided funding for students in state correctional systems. Perkins V emphasized the focus for educational efforts in rural areas through increased funding (Strengthening Career and Technical Education for the 21st Century Act, 2018).

Morrill Revisited

One of the first federal laws to establish some form of agricultural education specifically suited to supervised experience was the Civilization Fund Act of 1819, which provided funding to teach Native Americans "the mode of agriculture suited to their situation" (Fraser, 2019, p. 47). For some years afterward, the United States government made additional attempts to assimilate Native Americans into Western culture (Stahl, 1979). Still, the most significant legislation to provide agricultural education for Native Americans was the Equity in Educational Land-Grant Status Act of 1994. This act conferred land-grant status on 29 tribal colleges to teach agricultural and mechanical subjects per the original Morrill Act of 1862. This third "Morrill Act" resulted from a successful lobbying effort by the American Indian Higher Education Consortium (AIHEC). In 1992, this group secured the necessary legislation to provide land-grant funding to the tribal colleges. This "1994 Morrill Act" created an endowment fund for Native American education at these 29 colleges and offered matching grants to improve buildings and laboratories. Most of these colleges are two-year technical schools, but at least three offer baccalaureate degrees.

Modern Legislation Influencing Agricultural Education

The following are major legislative acts and initiatives that influenced agricultural education in the United States.

Elementary and Secondary Education Act of 2001

The Elementary and Secondary Education Act of 2001, also known as the No Child Left Behind Act, authorizes most federal funding for elementary and secondary education. The act was to supplement local and state education funding to improve student achievement. The standards imposed by the act are challenging. To receive continued funding, schools must meet selected goals. Schools that fail to meet rigorous standards have sanctions levied against them. There is a concern that the act will harm career and technical education because of the focus on high-stakes testing and accountability. Will schools find it necessary to cut vocational programs to focus resources on helping students meet annual academic goals? This act is a politically charged issue among educators, and only time will tell whether it meets its plans to improve student achievement in American schools.

The American Reinvestment and Recovery Act of 2009

The global financial crisis of 2007–2008 caused considerable damage to the worldwide economy. To put the United States back on the path to recovery, the United States Congress enacted the Reinvestment and Recovery Act of 2009. This act provided more than $100 billion in spending on education, and included funds for students with disabilities, disadvantaged youth, rural youth, and otherwise vulnerable populations. The act also provided training and employment services, especially in critical need vocational areas such as health care and information technology. Funds were also allocated for the training of workers in areas of the economy where significant skill gaps existed. Funding did not specifically go to agricultural education in the states, but the influx of funds indirectly supported agricultural education. Just as a rising tide lifts all boats, the increased federal funding paid for administrative support structures that benefited agricultural education.

Common Core Standards and Agricultural Education

As the American Reinvestment and Recovery Act progressed through Congress, state officials worked toward a common core of educational standards that every child should master. The Council of Chief State School Officers and the National Governors Association prepared these common core state standards.

The purpose of these standards is to ensure that every child has the requisite knowledge and skills at each grade level. The Common Core is research-based, evidence-based, and aligned with the expectations required for college and career readiness. The standards encourage critical thinking and problem-solving skills. The Common Core initiative encourages agricultural educators to collaborate with other disciplines to help students meet the standards. For example, agricultural teachers might assign readings that encourage critical thinking and deeper reading comprehension. Schoolteachers and administrators are working to align and crosswalk Common Core standards throughout the curriculum. The Common Core Standards initiative is an example of how nongovernmental organizations move to affect policy related to agricultural education.

Every Student Succeeds Act of 2015

The Elementary and Secondary Education Act of 2001 (ESEA), also known as the No Child Left Behind Act, was a significant step toward measuring student progress and determining areas of improvement. However, the authoritarian nature of the act made it a difficult fit for the thousands of school districts in the United States, each with its unique characteristics, problems, and concerns. The United States Congress passed the 2015 version of the Every Student Succeeds Act (ESSA), amending the ESEA to allow local school leaders to decide how to account for student achievement. Funding served students with special needs, students in poverty, and minority students. Testing and accountability measures remained in place, but the act provided additional assistance to students so they could succeed regardless of societal stumbling blocks.

ESSA funding helped agricultural education by providing resources for underserved students in agricultural education courses. Funds went to career and technical education in general and agricultural education specifically for program improvement.

REVIEWING SUMMARY

The founders of vocational education are gone, but their legacy lives on. Today, millions of young people have the opportunity to have the careers of their choice because of career and technical education. Millions of Americans have learned valuable skills, acquired technical jobs, raised families, and, as responsible citizens, contributed to the overall well-being of the United States. There can be no better epitaph for these great visionaries.

Agricultural education in the United States came about primarily through the efforts of individuals responding to needs within their local communities. Before the Smith-Hughes Act of 1917, much of the leadership for agricultural education was found within the individual states. It often came from visionaries who worked under the authority provided by federal legislation for experiment stations and land-grant universities. The Vocational Education Act of 1917 (Smith-Hughes Act) provided the organization with federal aid for vocational education until 1963 when a new vocational education act replaced it.

The new focus for agricultural education is in curriculum integration and articulation agreements between secondary and postsecondary institutions. Today, career and technical education is available for any student who wishes to participate, can profit from it, and needs it to meet career goals.

QUESTIONS FOR REVIEW AND DISCUSSION

1. What were some of the chief provisions of the Morrill Acts of 1862, 1890, and 1994?
2. What did the Smith-Hughes Act accomplish?
3. What was the primary purpose of the "George Acts"?
4. What trends do you notice in federal legislation for vocational education in the 1900s?
5. What were significant provisions of the Vocational Education Act of 1963 as related to agricultural education?
6. Briefly identify the following individuals: Justin Morrill, Dudley Hughes, Hoke Smith, Charles Prosser, Carl Perkins, and Jonathan Baldwin Turner.

ACTIVITIES

1. Complete a biographical research project on one of the people mentioned in this chapter. Specifically, how did that person's background, interests, and vocation lead the individual to support career and technical education legislation?
2. Investigate the history of secondary agricultural education in your state. Who were the early leaders? What were the first schools to initiate agriculture classes? What was the nature of the instruction in these early classes?
3. Investigate the structure of the land-grant system in your state. Prepare a report that provides the names and locations of the colleges or universities and discuss how they accomplish the mission of teaching, research, and extension.

REFERENCES

Barlow, M. L. (1976). *The unconquerable Senator Page: The struggle to establish federal legislation for vocational education*. American Vocational Association.

Butler, J. H. (1932). *An historical account of the John F. Slater Fund and the Anna T. Jeans Foundation*. University of California.

Carleton, D. (2002). *Landmark congressional laws on education*. Greenwood Publishing Group.

Cash, W. J. (1991). *The mind of the South*. Vintage Books.

Cochrane, W. W. (1979). *The development of American agriculture: A historical analysis*. University of Minnesota Press.

Cook, G. C. (1938). *Handbook on teaching vocational agriculture*. Interstate Printers & Publishers.

Cook, G. C. (1947). *Handbook on teaching vocational agriculture.* Interstate Printing Company.

Cross, C. F. (1999). *Justin Smith Morrill: Father of the land-grant colleges.* Michigan State University Press. http://site.ebrary.com/id/10514601

Dabney, C. W. (1902). *The public school problem in the South.* U.S. Government Printing Office.

Farmers High School of Pennsylvania Incorporated, Pub. L. No. 46, 50 (1855).

Fraser, J. W. (2019). *The school in the United States: A documentary history.* Routledge.

Gannett, M. R., & Cornwall Historical Society. (1986). *The Cream Hill Agricultural School at West Cornwall, CT.* Cornwall Historical Society.

Hamlin, H. M. (1950). *Agricultural education in community schools.* Interstate Printers & Publishers.

Hamlin, H. M. (1962). *Public school education in agriculture: A guide to policy and policy-making.* Interstate Printers & Publishers.

Hurt, R. D. (1994). *American agriculture: A brief history.* Iowa State University Press.

Kell, K. T. (1966). Folk names for tobacco. *Journal of American Folklore, 79*(314), 590. https://doi.org/10.2307/538224

Moore, G. E. (1988). The involvement of experiment stations in secondary agricultural education, 1887–1917. *Agricultural History,* 164–176.

Moore, G. E., & Gaspard, C. (1987). The quadrumvirate of vocational education. *Journal of Career and Technical Education, 4*(1), 3–17.

Morison, S. E. (1994). *The Oxford history of the American people. Vol. 1, Prehistory to 1789.* Meridian.

National Center for Agricultural Literacy. (2018). *Growing a nation.* https://growinganation.org/

National Council for Agricultural Education. (2016). *National quality program standards for agriculture, food, and natural resources education.* https://thecouncil.ffa.org/

National FFA Organization. (1975). *FFA advisors handbook.*

Phipps, L. J. (1965). *Handbook on agricultural education in public schools.* Interstate Printers & Publishers.

Phipps, L. J. (1980). *Handbook on agricultural education in public schools* (4th ed.). Interstate Printers & Publishers.

Phipps, L. J. H. (1966). *Handbook on agricultural education in public schools* (2nd ed.). Interstate Printers & Publishers.

Phipps, L. J. H. (1972). *Handbook on agricultural education in public schools* (3rd ed.). Interstate Printers & Publishers.

Phipps, L. J., & Cook, G. C. (1956). *Handbook on teaching vocational agriculture* (6th ed.). Interstate Printers & Publishers.

Phipps, L. J., & Osborne, E. W. (1988). *Handbook on agricultural education in public schools* (5th ed.). Interstate Printers & Publishers.

Phipps, L. J., Osborne, E. W., Dyer, J. E., & Ball, A. (2008). *Handbook on agricultural education in public schools* (6th ed.). Thomson Delmar Learning.

Smith, H. (1914). *Vocational education: Report of the Commission on National Aid to Vocational Education together with the hearings held on the subject, made pursuant to the provisions of Public Resolution No. 16, Sixty-third Congress (SJ Res. 5).* U.S. Government Printing Office.

Stahl, W. K. (1979). The U.S. and Native American education: A survey of federal legislation. *Journal of American Indian Education, 18*(3), 28–32.

Stevens, N. E. (1921). *America's first agricultural school.* Science Press.

Stimson, R. W. (1919). *Vocational agricultural education by home projects.* New York: Macmillan. http://archive.org/details/vocationalagricuoostim

Stimson, R. W., & Lathrop, F. W. (1942). *History of agricultural education of less than college grade in the United States: A cooperative project of workers in vocational education in agricultural and in related fields.* Federal Security Agency, U.S. Office of Education, 1942.

Strengthening Career and Technical Education for the 21st Century Act, Pub. L. No. 115–224, 132 STAT. 1564 79 (2018).

Struck, F. T. (1958). *Vocational education for a changing world.* Wiley.

True, A. C. (1929). *A history of agricultural education in the United States, 1785–1925.* U.S. Government Printing Office.

True, R. H. (1935). *Sketch of the history of the Philadelphia Society for Promoting Agriculture.* Philadelphia Society for Promoting Agriculture.

Urban, W. J., & Wagoner, J. L. (2004). *American education: A history.* McGraw-Hill.

Virginia Tech. (n.d.). *Land & labor acknowledgement.* Special Collections and University Archives, University Libraries. https://digitalsc.lib.vt.edu/exhibits/show/indigenous-vt/land-acknowledgement

White, J. (1585). *Village of Secotan in North Carolina. Watercolor by John White in the British Museum, London.*

Wright, S. J. (1949). The development of the Hampton-Tuskegee pattern of higher education. *Phylon (1940–1956), 10*(4), 334. https://doi.org/10.2307/272114

4

Organization and Structure of Agricultural Education

Mariah has accepted a position as the agriculture teacher several states away from the one where she grew up and attended college. She is excited for the opportunity, but realizes she is unfamiliar with how education is implemented on the state and local levels. She has many questions about curriculum, funding, and local school boards. Mariah has decided to contact the state's specialist in agricultural education to prepare herself for this new adventure.

Education in the United States is a responsibility of each state, which may organize state and local educational levels differently from other states. It is important for the agriculture teacher to understand each level of administration and how each impacts the local agricultural education program.

OBJECTIVES

This chapter addresses the National Quality Program Standards for Agriculture, Food, and Natural Resources Education (National Council for Agricultural Education, 2016), specifically Standard 6: Certified Agriculture Teachers and Professional Growth. It has the following objectives:

1. Describe the relationship of agricultural education to local school boards.
2. Explain the levels of local, state, and federal administration.
3. Describe the organization of school-based agricultural education in charter, online, and private schools.
4. Relate local programs to the National FFA Organization.
5. Discuss issues related to local autonomy.

TERMS

autonomy
board of education
curriculum
elective course/program
high school
junior high school
middle school

policy
postsecondary program
principal
procedure
required course
school district
superintendent

FIGURE 4.1 A career center director discusses proposed facility renovations with the head of the agriculture department.

AGRICULTURAL EDUCATION AND LOCAL SCHOOL BOARDS

Agricultural education is provided at the local level through state-approved programs in the nation's schools. Agricultural education programs in different schools include the three major program components—classroom and laboratory instruction, supervised agricultural experience, and student organization (FFA). However, the content and methodology of instruction varies from school to school. The program should be designed to meet the needs of the local community.

State administration for agricultural education is provided through various state agencies and institutions. The most common lead agency for agricultural education programs is a state department of education or a state department of agriculture. Federal administration is provided through the Office of Career, Technical, and Adult Education (OCTAE), U.S. Department of Education.

Today's agricultural education programs are focused on the science, business, and technology of the plants, animals, and natural resources systems. More than 1 million students participate in agricultural education programs offered in grades 7 through adult throughout the 50 states and three of the U.S. territories.

School Districts

A *school district* is a geographical area under the supervision of a given school board. It may include several schools or attendance centers, or it may have only one school. Sizes of school districts vary considerably. A district may be congruent with the limits of a city or county or some other defined geographical area.

FIGURE 4.2 A school district typically provides for student transportation.

A central administrative office may be responsible for coordinating functions among the schools. The superintendent and several assistant administrators, as well as support staff, are usually part of this central office.

A *board of education* is a group that has the legal authority to take action and use resources in the operation of a school district. Boards of education are provided for in the laws of a state and are sometimes known as local school boards. Members of a board may be elected or appointed. The superintendent reports directly to the board. All other employees of the board report through the superintendent. Usually a board delegates matters only to the superintendent. The superintendent may further delegate to staff members. Together, the board and the superintendent may establish guidelines for operation of the schools.

A local school board is a public body, and its meetings are open to the public. Teachers, students, parents, and others may attend school board meetings to keep abreast of issues discussed by the board. If a teacher or another individual wishes to present an item to the board, they must follow the board's policies for having the item placed on the board meeting agenda. It is wise for a teacher to inform the school principal if intending to present information at a school board meeting.

A good agricultural education advisory committee can often be utilized to interface with the school board through both formal and informal communications channels. Advisory groups have no administrative or legislative authority and cannot establish policy. Their function is to increase understanding between the school and the community. A strong agriculture program has a local advisory committee to provide community input into the content and design of the local agriculture program. Good teachers working with a good local advisory committee are able to design and deliver high-quality programs that meet the needs of the local students and community. It is at the local level that teachers have the most opportunity for using their initiative and creativity in shaping the agricultural educational program to meet the needs of their students, school, and local community.

Policies and Procedures

A *policy* is a general guiding principle. Many policies are based on laws. For example, a state law may require a local board of education to have a policy on executive sessions. The board must then establish the policy on appropriate subject matter, timing, and who may participate. Policies are made by individuals or groups that have the authority to do so, such as school boards. Policies are established to guide the management, procedures, and decision making especially of government bodies, such as school boards. In policy making, a local board may seek teacher and citizen input.

A *procedure* is the way a policy is to be implemented or carried out. Some procedures are routine or standard operating procedures. An example is purchasing supplies for a horticulture program. School districts have definite purchasing procedures that should be followed.

State approval is usually required to establish and maintain a local agricultural education program, but the local program operates at the pleasure of the local board of education. Local school board members represent the community in establishing policies for the operation of the local school system.

Local school boards must follow state and federal guidelines to ensure their schools are eligible to receive state and federal funds and to ensure their graduates are qualified to enter the workforce and/or advance to higher education programs. While much of a school's curriculum is mandated by state laws and regulations, there remains considerable local control of educational programs and school policies. For example, the local school board is responsible for approving and overseeing the school's budget and for hiring and terminating employment of administrators, teachers, and other school employees. The local board also establishes and/or approves most of the operational policies and procedures that students, teachers, and other school employees must follow.

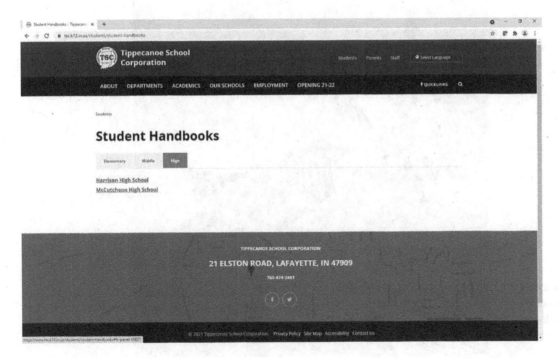

FIGURE 4.3
School districts publish student handbooks with information and expectations in line with school board policies.

Programs and Curriculums

The local board of education is responsible for assuring that all programs and curriculums meet the standards established by the state board of education.

The programs or courses in a school district are approved by the local board of education. The professional staff of the district is responsible for assuring the programs or courses are properly implemented so that students receive the appropriate education.

The *curriculum*, to many, is the list of courses offered in a school. Curriculum is sometimes defined more specifically as the learning experiences a student has under the direction of school personnel. Some disparity develops between what may be specified in the outline of a course and the learning experiences actually provided.

At the secondary level, some courses are required, and others are elective. *Required courses* are those needed to meet core requirements for graduation or admission to an institution of higher education. The programs or courses offered become the local curriculum. These offerings are ultimately determined by the local community, as expressed through local school board policy. *Elective courses/programs* are those that are not mandated by state policies or standards and that students are not required to take to graduate from high school. These courses and programs, which include agricultural education, are either offered or not offered according to the will of the local board of education.

FIGURE 4.4 A high school principal discusses student learning activities with a horticulture teacher at the school's land lab.

Administrative Staff

A school district may have several administrative positions. The role of individuals in these positions is to see that quality education is provided in an efficient manner.

A *superintendent* is the one who has executive oversight and charge. The local school board hires a superintendent as the chief executive for the school system, and the superintendent, in consultation with the board, provides for the preparation and administration of the school system budget and the hiring of other school administrators, including school principals. A state usually has policies requiring certain qualifications or levels of preparation for an individual to become a superintendent.

A *principal* is a person who has controlling authority or is in a leading position. The school principal is in charge of a local school site and responsible for leading and managing a particular school. Principals evaluate their school faculty and staff and make recommendations regarding hiring and firing employees at their school. They also administer policies and manage the budget for their school. The principal reports to the school system superintendent.

LOCAL, STATE, AND FEDERAL ORGANIZATION

Local, state, and federal levels are organized so educational programs can be carried out. Local districts differ in their organization but have many similarities. The same is true of the states.

Local District Organization

Agricultural education at the local level may be offered in middle schools, high schools, and career technical education centers and in special settings, such as academies and magnet schools. One or more of each may be found in a school district, depending on the size and student population of the district. Each site where students go to school may be known as an attendance center. A rural school district with sparse population may have one attendance center where all grades attend.

Agricultural education is offered at the middle school level, where the programs focus on exploring the broad subject of agriculture. A *middle school* is a school that commonly includes grades 6 through 8. Instruction at this level is focused on helping students understand the plants, animals, and natural resources systems and helping them make informed choices about their future (Jones et al., 2020). The first year of agricultural education is usually introductory in nature and covers FFA, supervised agricultural experience programs, an overview of the agricultural industry, and agricultural careers. The introductory course may be provided at the middle school, junior high school, or high school. In addition, some basic leadership and agricultural career skills are often taught in introductory courses. The middle school instruction should be articulated with that of the high school the students will attend after completing middle school.

A *junior high school* is a school that includes grades 7 and 8 and sometimes grade 9. Agriculture instruction may be offered at the junior high school level. The curriculum is articulated with that of the high school the students will attend after completing junior high.

A *high school* is a school that includes grades 9 or 10 through 12. In some states, agricultural education is first offered at the junior high school level and/or high school level. Most high school courses are designed to develop specific skills in agriscience, agribusiness, technology, leadership, and human relations. Agricultural education is also offered beyond the high school level.

High school programs may be articulated with postsecondary programs. A *postsecondary program* is one that is sequenced following the high school level, such as a program at a

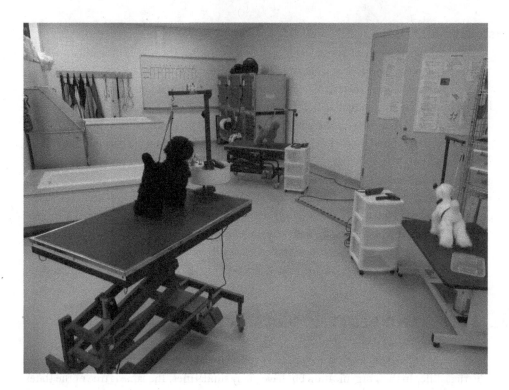

FIGURE 4.5 A secondary agricultural education facility with a specialized vet science program.

community college or at a career and technical center. Postsecondary agricultural education programs are more specialized and provide students with advanced technical competencies to enter and progress in specific agricultural careers. Many states allow for dual credit agriculture courses. High school students earn credit for both high school and postsecondary courses. Instruction may be taught by the high school agriculture teacher or by the postsecondary instructor through virtual or face-to-face instruction. Some states also provide agricultural education programs for adults in the community.

At the school site, the local agriculture department may be recognized as a separate department of the school, be part of a career and technical education department, be part of a science department, or be included in some other school organizational structure established under the leadership of the school principal.

Principals operate their individual schools under guidelines provided by the school board and administered by the superintendent and other central office administrators. In a small district, principals may report directly to the superintendent, while in a larger district, there may be assistant superintendents or directors in the chain of command between the principals and the superintendent.

The head of the agricultural education department usually reports to the school principal and/or to a career and technical education director. Principals are charged with the responsibility of evaluating, managing, and leading employees in their individual schools. Teachers should be evaluated periodically by a principal or an assistant principal, and the results of these evaluations should be placed in the teachers' personnel files. Principals also make recommendations regarding the hiring and firing of their school's employees, including teachers.

Individual agriculture teachers may report to the head of the agricultural education department or, in a small school, directly to the school principal. The final decision to employ or dismiss an agriculture teacher or other school employee rests with the school board, which must follow its own policies and procedures in personnel matters. The board's decisions are usually guided by the recommendation of the principal and the superintendent. It is important for agriculture teachers to maintain positive relationships with all local school administrators and school board members.

State Organization

State policy for education is established by state legislatures and state boards of education. State education policies are administered through state departments of education, which generally have career and technical education divisions or groups.

State support and services for agricultural education programs are usually provided through the career and technical education division of the state department of education, with one individual designated as the lead or head supervisor, consultant, or coordinator of statewide agricultural education programs and activities. However, state structures for administering agricultural education vary. Some states delegate the authority for state administration of agricultural education to a state university, state department of agriculture, community college, or some other agency or institution.

Local programs receive support and direction from states in varying degrees. Some states have greater capacity to support local programs than do others. There are also differences from state to state in the amount of direction local schools receive regarding what is to be taught. Some states have required agricultural education curriculums that must be taught, while other states leave the decision about what to teach almost entirely up to the local schools.

As noted above, each state has an individual designated as the lead person for agricultural education programs. This individual may devote full time or part time to the administration of agricultural education programs and usually serves as advisor to the state FFA association.

Usually one individual has overall responsibility for leading the entire agricultural education program in a state. Often another individual coordinates the statewide FFA component, working with the state's FFA officers and coordinating the FFA activities conducted in the state. Less often, a state has several agricultural education staff members serving as consultants or supervisors for local agriculture programs and/or managing specific aspects of the state's agricultural education program. There is considerable variation among states in the location and structure of state leadership and in the amount of local support provided by state leaders. It is important for local teachers to know their state leaders personally and to understand how their state's leadership system works.

Federal Organization

Federal policies for education are established by the U.S. Congress and administered through the U.S. Department of Education. The federal government also provides some funding for agricultural education. States offer agricultural education under federal guidelines they receive through the U.S. Department of Education. The federal guidelines provide general direction to states and require submission of state plans for career and technical education. The plans describe each state's priorities for career and technical education, including how federal funds will be expended.

Federal administration of agricultural education has varied over time since the passage of the Smith-Hughes Act. Initially administration was provided through the U.S. Department of Education. However, over the decades positions were eliminated until 2019 when the National Council for Agricultural Education became responsible for appointing the director of agricultural education.

Since the 1970s, federal direction for agricultural education has decreased, as the guidelines have focused primarily on the broad field of career and technical education. With limited federal direction, state programs of agricultural education have become less consistent in their focus and delivery. To maintain some national consistency in program direction, national agricultural education leaders have collaborated with various stakeholders in establishing and refining a national vision for agricultural education, a mission statement with goals, and a strategic plan for implementing the vision and mission.

The National Council for Agricultural Education is the umbrella national organization for school-based agricultural education. The council's mission is to provide leadership and coordination to shape the future of school-based agricultural education. Representatives of 12 national organizations/entities serving agricultural education make up the council. They serve as catalysts for the development and implementation of the vision, mission, goals, national policies, and programs for agricultural education. The council recommends that states utilize the national vision, mission, goals, and strategic plans to guide the development of state and local visions, missions, goals, and strategic plans for keeping agricultural education abreast of agricultural and educational trends and issues.

In 2009 the National Council for Agricultural Education released the National Quality Program Standards for Secondary (Grades 9–12) Agricultural Education. These standards were developed to provide some consistency in delivering high-quality agricultural education programs nationwide. An online tool is available to assist agriculture teachers in analyzing their program using the National Quality Program Standards.

SCHOOL-BASED AGRICULTURAL EDUCATION IN CHARTER, ONLINE, AND PRIVATE SCHOOLS

Approximately 4% of U.S. secondary students are homeschooled, 6% attend public charter schools, and 10% of elementary and secondary students are enrolled in private schools (NCES, 2019). In the 2016–2017 school year, the most recent with data, less than 1% of students were enrolled in a virtual school. Although these are small percentages, there are more than 3 million students at these schools who could benefit from a school-based agricultural education program.

How instruction is organized and delivered in homeschool, charter schools, and private schools varies greatly. Depending on the state, instruction in agricultural education for homeschooled students can range from highly regulated to not regulated (Kararo & Knobloch, 2018). Public charter schools typically are regulated by the state, but have greater flexibility in curriculum, staffing, and expenditures (Henry et al., 2014; Trexler & Hikawa, 2001). Private school regulation varies by state with most states having few regulations except for those related to health and safety (United States Department of Education, 2009).

Alaska and North Carolina were some of the first states to expand agricultural education instruction, FFA, and SAE to homeschooled students (Kararo & Knobloch, 2018). In these and other states, the homeschooled students may be part of a local public school agricultural education program or receive instruction at home and have a homeschool FFA chapter. Several states specifically allow private school FFA chapters, while many others have these types of programs without specifically stating they are allowed. Public charter schools are organized using numerous models. However, as public schools they typically are included as being able to have a school-based agricultural education program and FFA chapter. Finally, agricultural instruction in virtual schools ranges from individual course offerings to hybrid instructional models allowing for implementation of the complete school-based agricultural education model of classroom/laboratory instruction, SAE, and FFA.

LOCAL PROGRAMS AND THE NATIONAL FFA ORGANIZATION

A high-quality agricultural education program includes all three of the agricultural education program components. The integral nature of classroom instruction, supervised agricultural experience, and the FFA and the interrelationships between these three components ensure that resources directed at any one part of the agricultural education program affect the entire program.

The National FFA constitution provides a common focus for agricultural education across the nation. With funding from members' dues, product sales, and the National FFA Foundation, the National FFA Organization has the financial and human resources to provide considerable leadership and direction for FFA chapters.

Through various partnerships and alliances, the National FFA also provides leadership for agricultural education. For example, the National FFA partnered with the National Council for Agricultural Education in developing the Local Program Success and the Reinventing Agricultural Education for the Year 2020 initiatives that provided resources to improve and enhance local agricultural education programs.

Local FFA chapters are chartered by the state association, which in turn is chartered by the National FFA Organization. To remain in good standing with the state and national levels, each local chapter must have a constitution that complies with the state and national constitutions. Local chapters must also submit state and national membership dues, submit reports as requested by the state and national levels of the organization, and conduct activities that follow the ideals and purposes of the state and national levels. Chapters in good standing are eligible to participate in all state and national events and activities according to established guidelines.

The National FFA Organization maintains a website that provides valuable information for teachers and students. Local teachers order many of their supplies for degrees and awards, chapter banquets, and FFA instruction; official dress uniforms; and various other items from the National FFA Organization. The web address is www.ffa.org. The site (see Figure 4.6) includes other resources including award applications, degree and proficiency award handbooks, and many other tools and materials needed by local agriculture teachers.

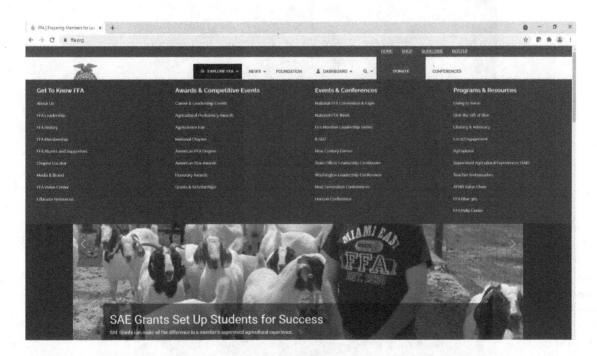

FIGURE 4.6 The FFA website contains many local program resources for teachers and students.

LOCAL AUTONOMY

Autonomy is the quality or state of being self-governing or self-directed. Local autonomy has been a part of the agricultural education program philosophy since the inception of the program. Teachers have always been encouraged to structure agricultural education programs around the needs and resources of their local communities.

As mentioned earlier in this chapter, considerable variation exists among states regarding the amount of local decision making authorized for agricultural education program management. Some states have numerous standards that programs are required to meet and consistently monitor local schools to ensure these standards are met. Other states provide general guidelines with little if any monitoring of how local schools provide agricultural education.

Strong teachers strive to involve the local community in establishing a vision and a mission for the local agricultural education program and in creating implementation strategies for achieving the vision and the mission. Representatives of all agricultural education stakeholders in a community should be invited to assist agriculture teachers in developing visions, missions, and strategic plans to guide their agriculture departments. Local agricultural education advisory committees are discussed in greater detail in Chapter 6.

Even in states that mandate statewide curriculums, local teachers can and should incorporate lessons, experiences, strategies, and other learning opportunities that address local

FIGURE 4.7 A sign indicates the facilities of an agriculture department and represents a level of autonomy for the program. (COURTESY OF LEDYARD AGRI-SCIENCE & TECHNOLOGY PROGRAM.)

community needs. This is essential for maintaining strong community support for agricultural education.

Good agriculture teachers thrive in schools where they are allowed to be innovative in determining the direction of local agriculture programs. They build on school and community strengths to ensure that high-quality instructional programs are designed for and delivered to the students. Local autonomy does not work well for weak teachers who often flounder without strong leadership, supervision, and guidance from state leaders. Strong local agriculture teachers are the most critical ingredient for maintaining high-quality agricultural education programs.

REVIEWING SUMMARY

Agricultural education is provided in the nation's schools through state-approved local programs. All agricultural education programs are expected to include the three major program components—classroom and laboratory instruction, supervised agricultural experience, and student organization (FFA). The content of the instruction and the quality of the programs vary from school to school. The most important factor in providing a high-quality program is the agriculture teacher. An agriculture program should be designed to meet the needs of the local community, and teachers should actively seek community direction and support for the program.

State leadership for agricultural education is provided through a wide variety of sources and structures. It is important that local teachers know the leaders and the leadership structure for their state. National direction and focus for agricultural education is provided through the National Council for Agricultural Education, which is a federation of national organizations/entities that represent the agricultural education community. The National FFA Organization also provides leadership through numerous programs, events, and activities.

Local autonomy works well for good teachers, because they seek and utilize school and community input in planning and delivering the agricultural education program. This strategy generates interest, enthusiasm, and support throughout the community.

QUESTIONS FOR REVIEW AND DISCUSSION

1. Why are local school boards important to agricultural education?
2. Describe the three levels of agricultural education administration and explain why each level is important.
3. In addition to public schools, where else is agricultural education instruction provided?
4. What does the National FFA Organization provide for local programs?
5. Is local autonomy an asset or a hindrance to maintaining high-quality local programs? Explain your answer.

ACTIVITIES

1. Use the National FFA Organization's website to investigate resources it provides to local agriculture programs.
2. Prepare a report that describes the state administration structure in your state. Include how the state administration interacts with local agriculture programs, how leadership

is provided for the total agricultural education program, and how leadership is provided for the FFA.

3. Prepare a report that describes the autonomy of agricultural education programs in your state.

4. Arrange to attend a local board of education meeting. (School board meetings are open to the public.) Obtain a copy of the order of business and carefully observe the proceedings of the meeting. Prepare a report that includes who chaired the meeting, how items were presented, how decision-making processes were carried out, and what financial expenditures were approved.

REFERENCES

Henry, K. A., Talbert, B. A., & Morris, P. V. (2014). Agricultural education in an urban charter school: Perspectives and challenges. *Journal of Agricultural Education, 55*(3), 89–102. https://doi .org/10.5032/jae.2014.03089

Jones, S., Doss, W., & Rayfield, J. (2020). Examining the status of middle school agricultural education programs in the United States. *Journal of Agricultural Education, 61*(2), 41–56. https://doi .org/10.5032/jae.2020.02041

Kararo, M. J., & Knobloch, N. A. (2018). An analysis of education-related policies regarding the participation potential of homeschool students in agricultural education and FFA. *Journal of Agricultural Education, 59*(3), 36–57. https://doi.org/10.5032/jae.2018.03036

National Council for Agricultural Education. (2016). *National quality program standards for agriculture, food, and natural resources education: A tool for secondary (grades 9–12) programs.* https:// thecouncil.ffa.org/

NCES. (2019). *Digest of educational statistics.* https://nces.ed.gov/programs/digest/d19/tables/dt19 _211.20.asp

Trexler, C. J., & Hikawa, H. (2001). Elementary and middle school agriculture curriculum development: An account of teacher struggle at Countryside Charter School. *Journal of Agricultural Education, 42*(3), 53–63. https://doi.org/10.5032/jae.2001.03053

United States Department of Education (ED). (2009). *State regulation of private schools.* http://www2 .ed.gov/admins/comm/choice/regprivschl/regprivschl.pdf

Part 2

*Program Development
and Management*

5

Program Planning

Brian is a beginning agriculture teacher at a newly opened program. He knows a strong program includes classroom/laboratory instruction, FFA, and SAE. He is excited to work with the local advisory committee to develop the program's vision and mission and with the administration on immediate and long-range program goals. Brian is excited that next school year another agriculture teacher will be hired to teach exploratory agriculture to seventh and eighth graders.

Agriculture teachers often have many decisions to make. The nature of a program varies with the state and the local school district. A proactive teacher actively develops and manages the agricultural education program to meet student and community needs.

TERMS

community survey
connecting activity
contextual learning
intracurricular
Local Program Success
National Quality Program Standards

needs assessment
program components
program evaluation
program planning
work-based learning

OBJECTIVES

This chapter addresses the National Quality Program Standards for Agriculture, Food, and Natural Resources Education (National Council for Agricultural Education, 2016), specifically Standard 7: Program Planning and Evaluation. It has the following objectives:

1. Explain the meaning and importance of program planning.
2. List three program components and describe the role of each.
3. Discuss the National Quality Program Standards.
4. Explain procedures in program development, including the use of community needs assessments.
5. Describe evaluation procedures used with local programs.

FIGURE 5.1 Program planning is an important responsibility of agricultural educators. This agriculture teacher is inventorying materials for the new school year.

WHAT IS PROGRAM PLANNING?

Every agriculture teacher plans, even one who seems to make decisions on a day-to-day or crisis-to-crisis basis. The old saying "If you fail to plan, then you are planning to fail" is most likely true. Successful agriculture teachers chart the course of their agricultural education programs, continuously monitor progress, and make course corrections as needed.

Program planning is, first and foremost, a process. The result is a written plan containing the agricultural education program vision and mission, short- and long-term goals and plans, and strategies for achieving these. Program planning is often said to be "all of the activities needed to design and implement local-school agricultural education" (Lee, 2000, p. 13).

Developing the Program Plan

A good program plan is developed with considerable input from stakeholders. A stakeholder is an individual who has an interest or share in an enterprise or initiative. From one perspective, all taxpayers are stakeholders in a public school agricultural education program. In program development, individuals who represent a cross section of the community are desirable on the planning group. A good program plan also conforms to state standards and meets local expectations. This section covers important concepts in program planning.

Participants

Program planning includes participation of several groups and input from numerous sources. The agricultural education program operates within the local school and community; therefore, both must be represented in the program planning process. The local agricultural education advisory committee and citizen groups, as discussed in Chapter 6, are logical groups to participate. The local school board's policies, the state agricultural education guidelines, and the national agricultural education vision, mission, and goal statements should all be sources of input. Approval, input, and guidance from school administrators are essential in the program planning process.

FIGURE 5.2 A teacher is meeting with the local advisory committee to discuss the total agriculture program.

The program planning process should involve people representing all groups affected by the local agricultural education program. Selecting these representatives will require the agriculture teacher to think broadly to ensure a wide diversity of agricultural education clients is included. Current students, parents of current students, and graduates of the local program should all be involved in the process. Representatives from all aspects of agricultural production and business in the community should be present. For example, large-scale farms, small-scale organic farms, specialty farms, and agribusiness will probably have different needs and viewpoints. The agriculture teacher must include all areas of agriculture, such as horticulture, landscaping, natural resources, small animal care, and others applicable to the local community.

Representation of the diversity within groups is important. Gender and ethnicity diversity is essential, but there are also other areas of diversity. A wide range of ages should be included. If possible, people who have been in the community a short time should be included as well as those whose families have been in the community for generations. If the community includes both rural and developed areas, representatives should be selected from both. If the community is entirely rural or entirely suburban/urban, the agriculture teacher should consider bringing in someone from outside the community to represent the missing viewpoint.

Components

There are 14 major components to a program plan. These are described in Table 5.1. Not all 14 components will be applicable to every local agricultural education program. The components are to serve as a guide, and others may be added as needed. However, some components are vital for every agricultural education program.

The first component of a written program plan is program, school, and community description. Narrative should be used for the description and observations. Demographics can be provided in tabular form. The purpose of this section is to provide the setting in which the agricultural education program operates.

TABLE 5.1

Components of a Local Agricultural Education Program Plan

Program Planning Component	Description
1. Program, school, and community description	Narrative or tables providing an overview of the setting
2. Personnel	Number of teachers, contract length(s), qualifications, professional growth plans
3. Program philosophy, vision, mission, objectives	Narrative and bullets on the direction of the program
4. Program management and planning	Administrative structure, advisory committee structure and tasks, community surveys and results, other community inputs, partnerships
5. Instructional planning and organization	Course of study descriptions, course frameworks, course outlines, teaching calendars
6. Student accounting and reports	Grading system, follow-up studies of graduates, other local and state reports
7. SAE coordination activities	Visitation/supervision plan, student record keeping, partnerships
8. FFA	Demographics, program of activities
9. Adult education program	Adult classes, Young Farmers chapter, partnerships, volunteer training
10. Recruitment and retention	Enrollment statistics, student-to-teacher ratio, marketing strategies
11. Safety training and practices	Safety instruction, inspection of facilities, safety practices
12. Summer schedule	SAE visitation/supervision, FFA activities, instructional activities
13. Physical needs and departmental budgets	List, documentation, and timeline of physical needs in instructional materials, supplies and equipment, facilities, other; instructional, capital, FFA chapter budget documentation
14. Long-range plan	5-year or longer goals

Program planning components 2, 3, and 4 develop and support the program vision. Philosophies are practiced every day but take on greater meaning and awareness when articulated and written down. A vision that is debated and agreed upon brings groups together, provides focus and clarity, and stretches people to achieve a desired future. Advisory committees, partnerships, and volunteers are vital in connecting the program to the community.

Program planning components 5, 6, 7, and 8 provide planning to the three components of the total agricultural education program: classroom and laboratory instruction, supervised agricultural experience (SAE), and FFA. In 2009, approximately 7% of agricultural education teachers had some adult or young farmer responsibility (Kantrovich, 2010). Newer supply/demand studies no longer ask about this type of agriculture teaching (Lawver et al., 2018). Even an agricultural education program that does not currently have an adult program may want to include program planning component 9 in its written plan.

Program planning components 10, 11, 12, and 13 provide support for the other components. School administrators are more accepting of changes and more willing to fund expenditures when the need for these has been documented through a program planning process. Recruitment and retention efforts are important, as agricultural education is an elective in most schools. A marketing plan should be followed to increase communications within the school and with the community.

Component 14, the development of a long-range plan consisting of goals that will take five or more years to accomplish, keeps the program focused on the future. These goals can also provide a starting point for the next program planning process.

Logistics

A written program plan is designed to guide all decisions regarding the agricultural education program, so it should be utilized on a continual basis. Annually the agriculture teacher(s), in conjunction with the local advisory committee, should revisit the plan to check for progress, make needed adjustments, and set short-term goals and objectives for the next school year. Every three to five years the entire plan should be reviewed and revised. Completing the program planning process is helpful before going through major program changes, such as adding new agriculture teachers, facilities, or curriculum areas.

The Process

The first step in developing a program plan is to assess the current situation in each of the 14 components described in the previous section. What is the program currently doing in each of the components? How well is the program currently meeting standards and needs in each of the components? If a previous program plan was in use, what are the accomplishments, what goals and objectives have been met, and what components need improvement? The National Quality Program Standards Online Assessment (National Council for Agricultural Education, 2016) is a useful tool for assessing an agricultural education program. These standards and tools are discussed in the next section of this chapter.

Next, the agriculture teacher convenes meetings with the groups described in the "Participants" section of this chapter. Each meeting must be organized, have a specific purpose, and be conducted in a timely fashion. Most people are willing to give of their time to worthy causes but do not tolerate what they perceive as wasting their time. It is recommended that each meeting cover three or fewer topics and last one to two hours. This ensures appropriate time is devoted to each topic. If possible, the meeting order of business and background materials should be sent to each participant before the meeting. Providing refreshments at all meetings helps with attendance and is a way of showing thanks to the participants.

The first meeting should be devoted to brainstorming and visioning. What changes have occurred and are occurring in agriculture and education? What knowledge, skills, and dispositions will agricultural education graduates need for agricultural occupations in the future? What should this agricultural education program look like and do in the future? What resources does this agricultural education program need to do the best job of preparing students? Subsequent meetings should be held to develop the vision and mission statement, goals and objectives, and implementation strategies.

Once the plan is developed and written, it should be implemented. The written plan now provides a guide for making programmatic decisions. The final step in the program planning process is to conduct periodic program evaluations. These evaluations feed the next cycle of revising the program plan (see Figure 5.3).

COMPONENTS OF A TOTAL AGRICULTURAL EDUCATION PROGRAM

Agricultural education involves much more than content learned without context. Agricultural education has three main *program components*: classroom and laboratory instruction, SAE, and FFA (see Figure 5.4). Each of these components is critical for students to receive full

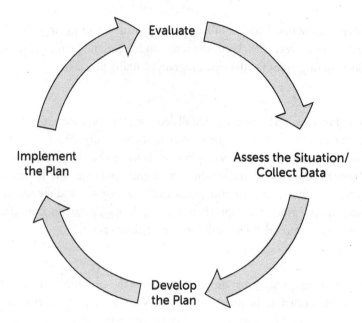

FIGURE 5.3 The program planning process is a cycle.

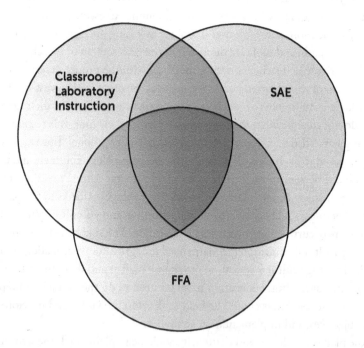

FIGURE 5.4 A model of the agricultural education program can be pictured as three interlocking circles.

educational benefit. In addition, agricultural education programs may conduct agricultural literacy efforts in elementary grades, exploratory courses in middle school, and adult education.

Classroom and Laboratory Instruction

Classroom and laboratory instruction is the foundation for everything else that occurs in the agricultural education program. Instruction can also be called contextual learning. In traditional instruction, content is taught as bits of information, such as 2 + 2 = 4.

FIGURE 5.5 This teacher is demonstrating a scientific principle using the context of agriculture.

Contextual learning is putting the instruction within a perspective to which it is easy for students to relate. Effective agricultural education instruction teaches content using the context of the plants, animals, and natural resources systems. Chapter 7 describes the process used for determining what content should be taught. Chapter 12 describes the process used for determining how the content should be taught.

Public schools are increasingly held accountable for their students' meeting state and national standards typically in academic subjects, such as English, math, and science. Agricultural education instruction should include English, math, and science concepts within the context of agriculture. For example, students learn math concepts of fractions, decimals, and calibrations within the context of agricultural power and technology. Also, within agricultural power and technology they learn physics concepts, such as the characteristics of fluids under pressure.

Supervised Agricultural Experience

Supervised agricultural experience (SAE), or work-based learning, is that part of agricultural education that allows students to practice in a workplace what they have learned in the classroom or laboratory. Ideally all agricultural education students would have SAE that directly relates to what is being learned in the classroom and that prepares the students for occupations in which they are interested. Chapter 22 describes the importance and types of SAE.

Work-based learning is a component of agricultural education that sets it apart from most other subjects. Students are able to explore areas of interest and then develop skills to a much greater depth than is possible within the regular school classroom. Tasks performed and problems encountered in SAE can be used in the classroom to provide real-world examples of concepts being learned. Foundational SAE can also be used to develop agricultural literacy.

FIGURE 5.6 Supervised agricultural experience provides students with work-based learning.

FFA

The career and technical student organization in agricultural education is FFA. FFA provides connecting activities. A *connecting activity* is one that establishes relationships between school and life. FFA is the student development component of agricultural education. It connects classroom learning to life in the areas of leadership development, personal growth, and career success. Chapter 23 describes the structure and activities of FFA.

FFA is an *intracurricular* component of agricultural education. This means that it is an integral part of the program, not an extracurricular club. Class period instructional time may be used for FFA instruction. At least on the local level, all agricultural education students must be involved in the connecting activities that FFA provides.

FIGURE 5.7 For more than 90 years, FFA has been an integral part of agricultural education. These members are preparing for an official FFA event.

Other Components of a Total Agricultural Education Program

Agricultural literacy is knowledge and understanding of the plants, animals, and natural resources systems, from production through consumption. It encompasses the idea that an agriculturally literate person can evaluate and communicate basic information about agriculture (Frick et al., 1991). Most agricultural education literacy efforts are focused at the elementary school level. These activities include Partners in Active Learning Support (PALS), a peer mentoring program; Food for America, an agricultural literacy educational program; and other limited-term efforts. Chapter 15 describes school and general-population agricultural literacy in greater detail.

Middle school agricultural education is a part of most U.S. agricultural education programs. Typically, this instruction is on an exploratory or introductory level. Depending on the state, agricultural education students in grades 7 and 8 can be a part of FFA. Chapter 16 further describes middle school agricultural education.

Adult agricultural education is a part of 10% or less of U.S. agricultural education programs. This instruction may be as adult classes or Young Farmers organization programs. A few states still have full-time adult agricultural education teachers. Chapter 18 further describes adult agricultural education.

NATIONAL QUALITY PROGRAM STANDARDS

Hughes and Barrick (1993) published a revised model for agricultural education in secondary schools (see Figure 5.8). It placed the agricultural education program within the context of the local school and community. Additionally, it restructured the three-circle model into classroom and laboratory instruction, application, employment and/or additional education, and career. This clarified that all aspects of FFA and SAE are related to classroom/laboratory instruction and that a career is the desired student outcome.

Local Program Success (LPS) was a national initiative designed to enhance the quality and success of local agricultural education programs (National FFA Organization, 2012). A national task force verified that quality and success on the local level depends on seven keys: classroom/laboratory instruction, SAE, FFA, partnerships, marketing, professional growth, and program planning. Program planning is the focus of this chapter. The LPS materials have not been updated in several years. However, they are still useful as most of the concepts remain consistent over time. The LPS materials are available in the Educator Resources section of the National FFA website.

In 2009, and revised in 2016, the National Council for Agricultural Education developed and disseminated *National Quality Program Standards*, which is a tool for local programs to use in analyzing and developing goals and objectives for program growth. The tool's 10 standards align closely with the 14 program planning components discussed in this chapter. The tool organizes the analysis into 59 quality indicators for which the reviewers place the agricultural education program at one of five levels of performance. Once completed, the quality indicator ratings are reviewed for those the program is performing at the level of "Not at Expectations" or "Approaching Expectations" to focus on as areas of growth. For those indicators the program is performing at "Meets Expectations" or above, these are areas to build upon. Next, the group selects indicators that are urgent and important to work on over the next year.

The Council developed an online version of the tool to facilitate completion and ease of analysis. Depending on the state, the National Quality Program Standards may have additional uses. For example, some states require completion of the tool to qualify for state-level

Updated Model of Agricultural Education

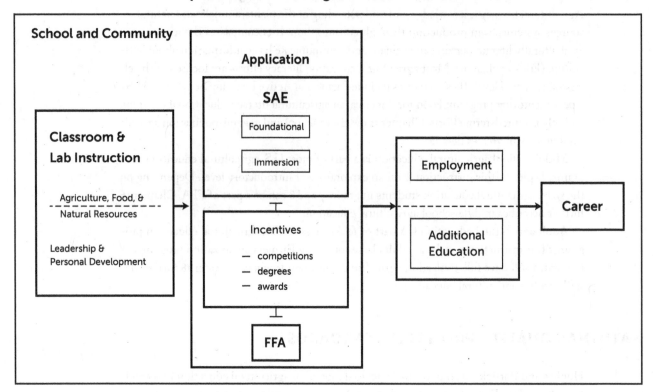

FIGURE 5.8 An updated model for school-based agricultural education positions the traditional three circles (see Figure 5.4) within the school and community with graduates in a career as the outcome. (ADAPTED BY PERMISSION FROM HUGHES, M., & BARRICK, R. K. [1993]. A MODEL FOR AGRICULTURAL EDUCATION IN PUBLIC SCHOOLS. *JOURNAL OF AGRICULTURAL EDUCATION, 34*[3], 60. HTTPS://DOI.ORG/10.5032/JAE.1993.03059.)

programmatic grant funding. Some agriculture teacher education programs use the tool to select student teaching host sites.

PROCEDURES IN PROGRAM DEVELOPMENT

Input from a number of sources is used in local program development. This includes program success materials as well as information collected about the local community.

State Standards and Curriculum Guides

Many states have established standards and provided curriculum guides for use in developing local programs. The standards and guides are known by different names such as frameworks, blueprints, and performance standards. Some states have established uniform achievement testing programs based on the content and expectations in the state-established materials.

In states where such materials have been developed, teachers in local programs are expected (and in some cases required) to use them in planning and to meet specified minimum standards as may be assessed on statewide achievement tests. Failure of a teacher to address these standards often results in student test scores below the desired level.

The state-provided guides are often on a course-by-course basis. For example, such guides have been established for introductory agriscience, horticulture, landscaping, forestry, biotechnology, and aquaculture, among other subjects. Some states may provide incentive funding to local schools that use the guides and have students who achieve certain levels of performance.

In 2009, the National Council for Agricultural Education published content standards for the Agriculture, Food, and Natural Resources (AFNR) career cluster. The standards were developed through a multistage process with input and validation from agriculture industry representatives, secondary and postsecondary agricultural education faculty, and agricultural education state staff (Hall, 2009). The content standards are organized into eight pathways: agribusiness systems, animal systems, biotechnology systems, environmental service systems, food products and processing systems, natural resources systems, plant systems, and power structure and technical systems. In 2015, the standards were revised to ensure they focused on current knowledge, skills, and career-ready practices (National Council for Agricultural Education, 2015). Many states have adopted or modified these pathways. In addition, curricular materials are mapped to these standards.

Community Needs Assessments

A hallmark of agricultural education is that the content and activities of the program reflect the agricultural industry of the local community. This provides accountability to taxpayers and curriculum relevance to careers available. Periodically, the agriculture teacher should re-evaluate the agricultural education program in light of community needs and characteristics.

A *needs assessment* is a program planning process used to collect and analyze community data. Sources of data include reports available from state and national statistical services. The U.S. Census of Agriculture provides detailed information on a county-by-county basis for all agricultural products. State and national labor departments (the official names of which vary) provide statistics on employment and employers. The local chamber of commerce, the Extension Service office, the visitors' bureau, and even the telephone yellow pages are all sources of information.

Most likely, however, complete information will not be available from these sources. The agriculture teacher will have to conduct a community survey to obtain a total local picture that is accurate. A *community survey* is a process to collect comprehensive and specific local information from various areas of the agricultural industry. The survey may include written or web-based questionnaires; telephone, virtual, or face-to-face interviews; and review of documents. Information collected includes types and numbers of agricultural industries and employers, types of occupations, descriptions of competencies performed, and current number of employees and expected future demand. (See Figure 5.9 for a sample survey form.)

The agriculture teacher, along with the advisory committee, analyzes the collected data and then evaluates the local agricultural education program. What knowledge, skills, and dispositions do students need to be employable in the local agricultural industry? What modifications need to be made to the agricultural education curriculum? What opportunities exist for student supervised experience? What FFA personal development activities, Career Development Events, and leadership events match local community needs and the agricultural education curriculum?

PROGRAMMATIC EVALUATION PROCEDURES

A *program evaluation* is an assessment of progress in achieving stated goals and outcomes of the program plan. It is a process that should be done annually to facilitate program planning for the next school year. The results of the program evaluation should be shared with the local advisory committee, school administrators and/or local school board, and state

---

Community Needs Survey: Agricultural Employment

Business name _____

Type of business _____

Business contact name _____

Contact's title _____

Business mailing address _____

Business telephone number _____

Business contact's email _____

Business website URL _____

Agricultural products or business description _____

Employees and employee needs:

Job Title	Competencies	Education Required	No. Current Employees	No. Needed Next 5 Years

---

FIGURE 5.9 Sample survey form for collecting community agricultural employment information.

agricultural education leadership as appropriate. News releases and other publicity should be used to inform the general public of program growth and successes.

Program evaluation follows many of the same steps as development of the written program plan. Data need to be collected and input sought from key groups. Results need to be compared against the goals, objectives, and outcomes set forth in the program plan. Recommendations for changes and improvement must be developed and should include action steps and implementation strategies.

Program evaluation is a vital part of the program planning process, not an afterthought. Although formal complete program evaluation can realistically be done only once a year, components of the agricultural education program should be evaluated on a continual, less formal basis. An activity should be assessed during and immediately after, with notes taken for commendations, suggestions, and things to do differently the next time. The agriculture teacher should be consistently asking if the activity is effective in its outcomes and if it is important to the agricultural education program vision, mission, and goals.

What Should Be Evaluated?

Each of the agricultural education program components (instruction, SAE, and FFA) should be evaluated. Data relating to the 14 program plan components should be included. Figures 5.10, 5.11, and 5.12 are sample survey forms indicating what data should be collected. The instructional evaluation form should be completed by students and reviewed by the agriculture

Instruction Evaluation Survey

School year _____

Course name _____

What are the best characteristics of this course? _____
_____
_____

What are suggestions for improving this course? _____
_____
_____

How has this course prepared you for a career or further education in agriculture? _____
_____
_____

For the next questions, circle the number that best matches your opinion on the statement.
1 = Strongly disagree, 2 = Disagree, 3 = Agree, 4 = Strongly agree

1 2 3 4   This course has challenged me to think.
1 2 3 4   This course is interesting.
1 2 3 4   I would take another agriculture course.

FIGURE 5.10 Sample survey form for student feedback on agricultural instruction.

Supervised Agricultural Experience (SAE) Evaluation Survey

School year _____
Number of students enrolled in agricultural education _____

Number of students with a Foundational SAE _____
Number of students with an Immersion SAE _____

Number of students with a Placement/Internship SAE _____
Number of students with an Ownership/Entrepreneurship SAE _____

Number of students with a Research SAE _____
Number of students with a School-Based SAE _____

Number of students with a Service-Learning SAE _____
Total gross dollars earned by students in SAE _____

Number of Proficiency Awards given on local level _____
Number of Proficiency Awards submitted above local level _____

Number receiving State FFA Degree last year _____
Number receiving American FFA Degree last year _____

Percentage of graduates entering workforce in agriculture _____
Percentage of graduates pursuing further education in agriculture _____

Ask students these two questions. Summarize results and report.
What is the best part of the SAE program at _____ Agriculture Department?
What needs to be improved about the SAE program at _____ Agriculture Department?

Ask supervisors/employers at SAE sites these two questions. Summarize results and report.
What is the best part of the SAE program at _____ Agriculture Department?
What needs to be improved about the SAE program at _____ Agriculture Department?

FIGURE 5.11 Sample survey form for supervised agricultural experience evaluation survey.

FFA Evaluation Survey

School year _____
Number of students enrolled in agricultural education _____

Number of FFA members (not including graduates) _____
Number of out-of-school (graduated high school) FFA members _____

Number Greenhand Degrees awarded last year _____
Number Chapter FFA Degrees awarded last year _____

Number receiving State FFA Degree last year _____
Number receiving American FFA Degree last year _____

Number Proficiency Awards given on local level _____
Number Proficiency Awards submitted above local level _____

Percent members participating in at least one local FFA activity _____
Percent members participating in at least one activity above the local level _____

Percent members participating in a CDE _____
Percent members participating in an LDE _____

Ask students these two questions. Summarize results and report.
What is the best part of the FFA program at _____ Agriculture Department?
What needs to be improved about the FFA program at _____ Agriculture Department?

Ask Advisory Committee members these two questions. Summarize results and report.
What is the best part of the FFA program at _____ Agriculture Department?
What needs to be improved about the FFA program at _____ Agriculture Department?

FIGURE 5.12 Sample survey form for FFA evaluation survey.

teacher. The data for the supervised agricultural experience form is easily obtainable if a database-type record keeping system is used. If a paper-based system is used, the agriculture teacher will need to summarize the data. The FFA evaluation form is completed by the agriculture teacher.

Making Interpretations

Once information has been collected, the data must be tabulated and analyzed. A summary of the findings is then prepared. Some findings will be numerical and can be summarized with frequencies, means, ranges, and other descriptive statistics. Other findings will be qualitative and can be summarized through patterns, themes, and quotes. Afterward, synthesis of the findings is needed to draw conclusions and prepare recommendations for improving the local agricultural education program. As the data are synthesized, these questions may be helpful in making interpretations. What trends emerged? What priorities were identified? What were lesser priorities that could be eliminated or scaled back to allow resources to be better utilized? What is a realistic timeline for implementing changes? What resources will be needed? Who is responsible for implementing the recommendations and what accountability measures will be put in place? These are important responsibilities. Teachers may get the assistance of school district evaluation specialists as well as advisory committee members in this process. Administrators and school board members may wish to be informed. Local media can be used to inform citizens of the worthwhile findings.

REVIEWING SUMMARY

Program planning is essential to the success of the local agricultural education program. Although program planning takes time up-front, it will greatly save the agriculture teacher time as it is implemented. Program planning is not done in a vacuum by the agriculture teacher alone but instead involves input and participation from several school and community groups.

Every local agricultural education program consists of three components. Classroom and laboratory instruction is the foundation of the program. Without strong instruction, the program cannot succeed. Supervised agricultural experience is the work-based learning part of an agricultural education program. Students take what is learned in the classroom and apply it in the workplace. They also bring their workplace experiences back into the classroom. FFA is the intracurricular student organization that provides connecting activities between instruction and the workplace. FFA is also the primary student development component of agricultural education.

National Quality Program Standards are designed to assist agriculture teachers in having successful total agricultural education programs that meet local community needs. NQPS can be used to identify indicators that are urgent and important to work on for the next year.

Program evaluation is vital to the success of a local agricultural education program. A systematic evaluation should be conducted annually. The results should be shared with the school, community, and general public. To be effective, program evaluation must have an impact on the program plan.

QUESTIONS FOR REVIEW AND DISCUSSION

1. Why is program planning important to the local agriculture teacher?
2. Who should be involved in developing the local program plan?
3. What are the 14 major components of a program plan?
4. What are the steps in program planning?
5. Why is it important that all agricultural education students be involved on the local level in the three main program components?
6. What are other components that may be included in the local agricultural education program?
7. What are the seven standards identified in NQPS that are critical to achieving local program quality and success?
8. What is included in a community survey?
9. How does program evaluation lead to local program success?
10. What should be included in a local program evaluation?

ACTIVITIES

1. Investigate the program planning process used when starting a new agricultural education program at a high school. Prepare a report on your findings.
2. Explore the National Quality Program Standards resources on the National Council for Agricultural Education website. Report about the impact of NQPS on agricultural education in your state.

3. Using the *Journal of Agricultural Education, The Agricultural Education Magazine*, and the proceedings of the National Conference of the American Association for Agricultural Education, report on research about program planning within agricultural education. Current and past issues of these journals can be found in a university library. Selected past issues can be found online at www.aaaeonline.org or www.naae.org.

REFERENCES

Frick, M. J., Kahler, A. A., & Miller, W. W. (1991). A definition and the concepts of agricultural literacy. *Journal of Agricultural Education, 32*(2), 49–57. https://doi.org/10.5032/jae.1991.02049

Hall, D. (2009). National agricultural education content standards. *The Agricultural Education Magazine, 81*(4), 13–14.

Hughes, M., & Barrick, R. K. (1993). A model for agricultural education in public schools. *Journal of Agricultural Education, 34*(3), 59–67. https://doi.org/10.5032/jae.1993.03059

Kantrovich, A. J. (2010, October). *The 36th volume of a national study of the supply and demand for teachers of agricultural education 2006–2009.* http://aaaeonline.org/Teacher-Supply-and-Demand

Lawver, R. G., Foster, D. D., & Smith, A. R. (2018). *Status of the U.S. supply and demand for teachers of agricultural education, 2014–2016.* http://aaaeonline.org/Teacher-Supply-and-Demand

Lee, J. S. (2000). *Program planning guide for agriscience and technology education* (2nd ed.). Pearson Prentice Hall Interstate.

National Council for Agricultural Education. (2015). *National agriculture, food and natural resources (AFNR) career cluster content standards.* https://thecouncil.ffa.org/

National Council for Agricultural Education. (2016). *National quality program standards for agriculture, food, and natural resources education: A tool for secondary (grades 9–12) programs.* https://thecouncil.ffa.org/

National FFA Organization. (2012). *Local program success.* https://www.ffa.org

6

Advisory and Citizen Groups

At the spring meeting of the Spruce Pine High School Agricultural Education Advisory Committee, members of the committee toured the agricultural mechanics lab and horticultural facilities. Sally Evans, Rex Watson, Joe Smith, and John Lee are all well established in the school community. Sally owns Evans Farm Supply, Rex operates Watson's Greenhouses and Nursery. Joe works as a crop consultant, and John farms cotton and soybeans.

Bob Wells, the new teacher at Spruce Pine, just completed his first year following the retirement after 35 years of the previous agriculture teacher. For much of the last 10 years, the curriculum centered on production agricultural and agricultural mechanics. Last year, the school system provided funding for construction of a greenhouse and headhouse facility.

The committee gathered in the new greenhouse where one-third of the table space had bedding plants, but the rest of the greenhouse was empty. Chairperson Lee said, "Bob, I can't help but notice that these bedding plants aren't doing too well." Bob replied, "Yes, I agree. I am having a hard time getting the spring plant sale going. Horticulture was not my strong suit in college, and I have been trying to figure things out as I go." Sally Evans spoke up: "It's tough running a greenhouse for the first time. There are bound to be problems along the way." Joe Smith said, "But your poinsettia crop this past fall was excellent. I have never seen more beautiful plants grown in a school greenhouse."

Bob's face turned red. He replied, "Well, I didn't exactly grow those in the greenhouse. The poinsettias I transplanted all developed a white fly problem, and I lost the whole crop. I bought replacement plants and just had to keep them healthy and alive until the sale."

After a good laugh, Rex Watson spoke up: "If I had a nickel for every plant I killed in my greenhouse and nursery, I'd be a rich man. I would be glad to coach you through the remainder of the spring growing season and help you get started on a fall plant sale. I'd also recommend that you start with something easier than poinsettias."

TERMS

advisory committee
craft committee
FFA Alumni and Supporters

layperson
mission statement
stakeholders

OBJECTIVES

This chapter addresses the National Quality Program Standards for Agriculture, Food, and Natural Resources Education (National Council for Agricultural Education, 2016), specifically Standard 4: School and Community Partnerships. It has the following objectives:

1. Explain the meaning and importance of citizen participation.
2. Describe the role of advisory groups.
3. Discuss how to organize and use an advisory group.
4. Describe how to promote and use an FFA Alumni group.

After more discussion, the advisory committee developed a plan to coach Bob through the management and operation of the horticulture facility. At the end of the meeting, Bob told the committee, "Thank you for helping me figure out a plan for the horticulture operation. I have lost a lot of sleep worrying about it. Now I feel I have the plan for this to be a successful learning experience for my students." Sally Evans spoke up: "That is what we are here for, to help you and the school develop a high-quality agricultural education program. You should always feel welcome to call on us. That is why we signed on to this committee in the first place."

As the committee members enjoyed cookies and soda provided by the FFA Alumni and Supporters chapter, the discussion continued about the learning opportunities for students in the horticulture program. As the discussion progressed, Bob Wells felt a genuine sense of relief that his advisory committee was willing to help him be a successful agriculture teacher.

CITIZEN PARTICIPATION IN SCHOOLS

In 1942, Harl Douglass, director of the College of Education at the University of Colorado, wrote in the *Phi Delta Kappan*,

> While the redirection of the program of education will take and is taking many new directions, none is more important or more badly needed than the great emphasis upon the training of young people for the intelligent consideration of the problems which confront American democracy. (Douglass, 1942, p. 6)

Douglass thought, as countless scholars have done before and after him, that schools were created by society as a means for solving societal problems. Citizen participation is an essential component in any school because, as Douglas reminds us, citizens have the most to gain when the education system works properly and the most to lose when schools fail in their education enterprise. Schools are a product of their community. Community members, through their government, arrange for a system of schooling that provides for the development of productive residents who contribute to society. Community members expect schools to help youth develop social and cognitive skills that lead them to responsible citizenship and a successful vocation. One of the inherent characteristics of a highly evolved society is that its members are able to communicate and reason together in such a way as to increase understanding and learn from one another (Darling-Hammond & Bransford, 2005). In this sense, the school is an extension of the community. As such, the community has the responsibility to be involved in the work of schools.

Parent Involvement in Schools

Parents of students enrolled in agricultural education courses serve as the foundation of citizen involvement in schools. Parental involvement can be categorized in six areas:

> **Basic legal and civic obligations**—American society views the education of children as a basic parental obligation. Parents, and the public in general, see schools as essential for preparing youth for life after high school.
> **Communication**—Schools typically communicate with parents on matters of importance ranging from academic performance to invitations to school events.
> **Volunteerism**—Parents, at the encouragement of teachers and school administrators, volunteer to assist with instructional activities. This includes attending field trips and providing support for in-class learning activities.

FIGURE 6.1
Parents can provide advice as members of an advisory committee.

Teaching at home—Parental involvement may include assisting students with homework and providing the encouragement to complete assignments in study sessions.

Parents as decision makers—Parents may assume elected leadership roles in school boards and may serve on advisory boards. This provides an outlet for parents to be involved in decision making and key policy matters.

Parents and the community—Parental activism on important school issues may involve participating in parent–teacher organizations or civic-based education organizations.

(Epstein, 2018; Richardson et al., 2017)

Citizen Participation in Agricultural Education

Agriculture teachers depend upon the agriculture and business segments of the community to provide support for instruction, supervised experiences for students, FFA program support, technical expertise on a variety of subjects, and external funding. For its efforts, the community will have an improved supply of workers entering the workforce upon graduation from high school and postsecondary school. Citizens support agricultural education because they want to provide high-quality educational experiences for their children. This relationship requires that agricultural education teachers be attentive to the need for citizens to be involved in educational programs and encourage local citizens to support their agricultural education programs proactively.

Among the strongest proponents of using citizen advisory groups in agricultural education was Herbert M. Hamlin (1894–1968). While a professor of agricultural education at the University of Illinois, and later at North Carolina State University, Hamlin wrote extensively about community input into educational policy formulation. He felt decisions about local agricultural education programs should only be made with citizen input. One of Hamlin's best-known treatises on the subject is *Public School Education in Agriculture: A Guide to Policy and Policy-Making* (Hamlin, 1962). He reminds us, "In characteristic American fashion, public agricultural education originated in communities. State initiative followed. National action came later" (Hamlin, 1962, p. 5). What the local agricultural education program was, and is destined to become, is derived from the wishes of the community in which it resides. The agricultural education program is a national endeavor, but it resides in and is most responsible to local communities. It is important to note the key function of an advisory group is to advise. Advisory groups should not assume leadership roles or decision-making roles within a school or agricultural education program (Castillo, 2003). To do so would usurp the authority of school officials or lead to confusion in the development of goals for the agricultural education program.

The Parent–Teacher Relationship

The parent is the child's first teacher. Yet, parents perceive that teachers provide parents with a superficial connection to the school, engaging them only at a few selected times during the year and only when necessary to address serious concerns with student learning. Parents are welcome at open house events and special programs, but are rarely invited into a partnership with teachers in the teaching and learning process. For parents who wish to be involved in the work of the school there is a strong need to be recognized and accepted as partners (Miretzky, 2004). Teachers for the most part have indicated one of the major reasons why they leave the teaching profession is because of a perceived lack of support by parents.

Herein lies the problem. Parents who want to be involved in their child's education want to be involved at a deeper level. Teachers need parents to be involved in a meaningful way in the teaching and learning process, but have few tools with which to make that happen. Parents and teachers both feel as if they are underappreciated by each other.

Parents choose to become involved in the work of schools because they perceive it is their responsibility to be involved at some appropriate and meaningful level. Parents receive personal satisfaction and a sense of efficacy from participating in their child's education. Because of the complex and multifaceted task of providing instruction, schools have a variety of opportunities for parents to be involved. Parents have the duty, motive, and opportunity to participate in a child's education (Hoover-Dempsey & Sandler, 1995). Teachers have the duty as well to involve the students' parents in a meaningful way in the instructional program.

Hoover-Dempsey and Sandler Model for Citizen Participation in Schools

Hoover-Dempsey and Sandler (1995) examined reasons why parents choose to participate in the school activities of their children and defined the social mechanisms at play. Parents become involved in their child's education and on school advisory committees because of their perceptions of the parenting role. If they see the education of their child as an elemental responsibility of parenting, they will demonstrate interest in it. However, the amount of parental involvement in school functions is a factor almost wholly decided upon by the parent. Some parents may take their parenting obligation toward education to mean making sure their child gets to school on time and is prepared for schoolwork each day. Still other parents may take their responsibility to include volunteering in the school at every available opportunity (Epstein, 2018).

Because parental involvement is the prerogative of the parent, schools rely on communication strategies designed to engage parents in meaningful ways (Epstein, 2018). Schools use newsletters, direct contact, electronic messaging, and news/social media to inform parents for the purpose of engaging with them in a positive manner.

FIGURE 6.2 A craft advisory committee working on leadership programming for an FFA chapter.

Parents also may derive satisfaction from the academic achievement of their children. They may devote time and attention to helping their children continue in those activities that promise to lead to more achievement in the future. Finally, parents may participate in schooling simply because they were asked to do so by teachers and school administrators. This leads parents to serve on advisory committees and volunteer in schools.

Volunteerism

Parents, at the encouragement of teachers and school administrators, volunteer to assist with instructional activities. These include chaperoning field trips, providing support for in-class learning activities, and serving as substitute teachers for a fee.

Parental involvement also includes assisting students with homework and providing the encouragement to complete homework assignments and study sessions at home (Epstein, 2018). In agricultural education, this would also include asking for parental support and supervision of supervised agricultural experience projects.

Citizen and Community Scientists, and Technical Expertise in Agriculture

Community scientists may be a relatively new concept to agricultural educators, but the principle of community-based or citizen science goes back to before the creation of schools, universities, and research labs. Citizen science is a form of informal science education conducted outside of the formal research environment. Citizen scientists, or the more modern term "community scientists," assist the formal science community in many types of research endeavors (Cooper, 2016).

For example, the Cornell Lab of Ornithology relies on community-based amateur birdwatchers to identify and track bird distribution and abundance through the eBird website (Cornell Lab of Ornithology, 2021). As of 2021, the eBird website has collected data from more than 36 million bird-watchers around the world (Cornell Lab of Ornithology, 2021).

FIGURE 6.3 Members of this advisory committee tour the agricultural education facility.

Community scientists have taken their interest in a hobby and developed enough knowledge and expertise in the subject to become "subject matter experts" (Cooper, 2016). These scientists can conduct lectures in agriculture classes, help conduct field trips, and provide resources the agricultural education teacher can use to create reusable learning objects for the classroom.

Community scientists also connect agricultural education to the network of people who are studying various agriculture subjects because of a deep personal interest. These scientists are motivated more by their love of the science than by earning a paycheck. Community scientists connect teachers to the people who have deep knowledge of agricultural subjects.

Technical expertise in the agricultural sciences takes many forms when applied to agricultural education. Experienced agriculture teachers possess a good degree of technical expertise, but it is virtually impossible to know all there is to know about agriculture subjects. Subject matter experts in the school community can provide one-on-one in-service training on new developments in the industry. Local agricultural producers provide information on the latest innovations in production agriculture. Agribusinesses provide instructors with information on new products and services. These specialists may also be utilized directly in agricultural education classrooms to teach the students.

AGRICULTURAL EDUCATION ADVISORY ORGANIZATIONS

An advisory organization, or committee, is a group of laypersons who provide advice on educational programs. A *layperson* is an individual who is not a professional agricultural educator or other professional or certificated person employed by the school. An *advisory committee* may also be identified as an advisory group, a citizen advisory group, an advisory board, a lay advisory committee, or a citizens committee. Regardless of the exact wording of the

FIGURE 6.4 An advisory committee can address highly technical and complex issues, as long as time and resources are provided for the committee to accomplish its work.

committee name, the intent is the same—to provide advice to decision-makers on the quality and direction of the local agricultural education program. This is also an important outlet for parental involvement in important decisions that influence education policy (Epstein, 2018).

An advisory group does not have the power to make decisions, but its input can be used by a legally empowered board in making decisions or in a less authoritative way as a source of advice for an agriculture teacher or administrator. Advisory groups should be representative of the community served by the school. It is best if an advisory group is formally organized and has a lay chair, who directs the affairs of the group.

An agricultural education program may have several advisory groups. A program advisory committee may serve to provide advice for the overall program. It may have several subcommittees that serve various needs or areas. These subcommittees are in the specialty areas within agricultural education. They are sometimes known as craft committees. A *craft committee* is an advisory group that is specific to a particular area, such as horticulture or forestry. The focus of a craft committee is on a technical field being served by the instruction. A craft committee may be known by the name of the area in which it provides advice—for example, the biotechnology advisory committee.

The Work of an Advisory Committee

What exactly does an advisory group do? The two basic functions of an advisory group are to assist professional agricultural educators in program development and to help in evaluating the local agricultural education program.

Program Development

The advisory committee for an agricultural education program has a major role in helping develop the program so it meets most effectively the needs and interests of the community served. Of course, all the input from the committee must be adapted to curriculum standards established by the school system or the state. Further, the local school board has authority over the agricultural education program through the appropriate administrative channels of the school district. Examples of areas in program development where an advisory committee (including subcommittees) can be useful are found in Box 6.1.

BOX 6.1 *Program Development Areas for Advisory Committee Input*

Career Development

Determining needs for preemployment instruction.

Guiding the career planning process.

Providing resources for career orientation.

Providing professional development opportunities for students and faculty.

Providing placement services for graduates and students with SAE programs.

Conducting job analyses in specialized areas to determine what skills are needed by graduates in these areas.

Providing instructors with information on industry innovation and trends.

Keeping instructors informed of current events in agricultural education.

Providing job shadowing and other career observation experiences.

Curriculum Development

Determining the courses to be offered and the course sequence.

Finding resource persons in the community.

Assisting with projects or activities of the agricultural education program.

Providing sources of teaching materials and equipment.

Making recommendations on physical facilities, equipment, and supplies needed for the curriculum.

Overall Program Development

Identifying and addressing the needs of disadvantaged students.

Conducting cost/benefit studies.

Conducting future planning of the agricultural education program.

Recruiting students.

Providing learning resources for instructional purposes in classrooms.

Developing partnerships with other agencies working with agricultural education.

Communicating with educational and political leaders in the community.

Identifying the equipment, tools, and supplies needed.

Identifying facility modifications needed to achieve curriculum goals.

Program Evaluation

Evaluating the effectiveness of CTE program(s).

Utilizing assessments used to measure whether or not instructional competencies are being met.

Reviewing courses of study for content relevance and accuracy.

Reviewing textbooks and other instructional materials as to their suitability for classroom use.

———————

Adapted from *Advisory Council Manual: Bureau of Career and Technical Education*. (2017). Iowa Department of Education, Bureau of Career and Technical Education Services.

Establishing a Network of Community Support

The best agricultural education programs have useful and viable advisory groups to help them carry out effective education. The push to increase educational accountability in schools has increased the need for community involvement instead of diminishing it. The curriculum may be the result of a state mandate, but the teacher has flexibility in how that curriculum is taught. An advisory committee can help the teacher and school administration fit the curriculum to the unique needs of the community. As more and more students seek post-secondary education after high school graduation, the agricultural education program must consider broadening the scope of its mission. An advisory committee can provide valuable support and assistance to the school so the best decisions are made and agricultural education continues to serve the changing needs of youth.

Program Evaluation

The advisory committee plays an important role in program evaluation. Program evaluation is the process of assessing the extent to which a program is meeting the objectives established for it. It answers the question: Is the program effectively and efficiently doing what it should be doing? An agriculture teacher is usually evaluated by a school official solely on the basis of teaching skills. The school principal or assistant principal visits a class in progress and completes an assessment of the instructor's ability to teach. Often, there is no additional review of the program except in cases where an accrediting agency visits the school and performs a periodic evaluation. The advisory committee can evaluate the agricultural education program in a timely manner and provide specific recommendations for the school administration to consider. Specific items that a well-established advisory committee can evaluate include the following:

- The validity of the curriculum to economic need within the community and adjacent communities
- The quality and availability of instructional materials
- Long-range plans for curriculum and instruction
- The quality and sophistication of facilities and equipment as measured against the current state of agricultural technology
- The overall perception of the agricultural education program in the community
- The effectiveness of supervised experience in meeting the vocational needs of youth

Once the advisory committee has completed an evaluation of the agricultural education program, it can recommend changes in curriculum, instruction, and the physical plant that address deficiencies or magnify strengths in the program (Doerfort, 2003).

Other Functions

The advisory committee has a unique view of the present economic situation in a local community. Members of the committee know where local needs are, based upon their interaction with others in the community, and can recommend suitable locations for students to have supervised experience in agriculture. In addition to finding potential sites for supervised experience, advisory committee members can make recommendations to the agricultural education faculty on policy and procedures. Committee members can also identify problem areas in the supervised experience program and head off trouble before it escalates. The advisory group can also serve as an independent voice for agricultural education in promoting and defending supervised experience to school officials.

Becoming Proactively Involved in the Politics of Education

Public education in the United States was created in the political arena and is subject to political pressures and decisions. The advisory committee can help school leaders make informed decisions about the direction of the agricultural education program. Advisory committees have no administrative policymaking or legislative authority, and therefore do not serve a political function in schools. By providing accurate information about the mission and goals of the local agricultural education program to school administrators, the school board, and civic leaders, the advisory committee encourages continued and increased support of the program. The advisory committee can help "tell the story" of agricultural education in the local community. The local community is more likely to support agricultural education when it understands its mission and its purpose. A high-quality advisory committee ensures that agricultural education is effective and serving the school and the community.

Organizing and Using an Advisory Group

Advisory committees can do good things for agricultural education, but rarely can they establish themselves and get to work without leadership and guidance from the agriculture teacher. Developing an advisory committee is a relatively easy process; however, it requires more than just calling a meeting and asking for volunteers. Setting up an effective advisory group requires careful thought and attention to detail.

Before beginning the process of establishing an advisory committee, the agriculture teacher should carefully analyze the community. Every community has its unique characteristics, and the advisory committee should be representative of these characteristics. Otherwise, the agriculture teacher runs the risk of not getting accurate advice and input on the future direction of the agricultural education program. A community survey provides valuable information to assist the teacher in selecting the right members for the advisory group as well as essential data for making good decisions. Box 6.2 lists the components of a good community survey.

BOX 6.2 *Information Needed for a Community Survey*

Demographic Information
Ethnicity
Educational level
Home ownership
Poverty levels
Population trends

Economic Development
Number and types of businesses
New and emerging industries
Unemployment rates
Labor force
Manufacturing layoffs
Percent working age
Working population

Agricultural Industry Information
Types of agribusiness
Farm ownership, type, and size
Natural resources
Total agricultural receipts

Local Education Trends
Graduation and dropout rates
Enrollment trends
School budgets
Per student expenditures
Educational attainment rates

Six Steps to a Productive Advisory Group
Step 1: Develop a Mission and a Purpose

Advisory committees serve a multitude of roles and purposes. The agricultural education teacher should develop an initial mission and operating guidelines for the committee. One of the first questions often asked by school administrators and potential advisory committee members is, "What will the advisory committee do?" This question can be answered easily if the appropriate groundwork has been laid. Although the advisory committee can amend its mission to meet the changing needs of the agricultural education program, the agriculture teacher should provide a basis for establishing the committee (Mosley, 2017).

In Lewis Carroll's (2015) *Alice's Adventures in Wonderland*, Alice asks the Cheshire Cat,

"Would you tell me please, which way I ought to go from here?"
"That depends a good deal on where you want to get to," said the Cat.
"I don't much care where . . . ," said Alice.
"Then it doesn't matter which way you go," said the Cat.

An agricultural education advisory committee may be a far cry from *Alice's Adventures in Wonderland*, but only if care and attention has been placed on developing the mission and purpose of the committee. Without this guiding mission, a committee may drift from one problem to another without meaningful discussion or effective results.

The *mission statement* should define the work the committee is being asked to do. It should provide clear direction for the measurable results to come later after the committee has begun its work, and it should delineate the parameters by which the committee will operate. For example:

The mission of the Midway High School Agricultural Education Advisory Committee is to advise the agricultural education teachers and school administrators on the following:

- Technical content to include in the curriculum
- Planning and execution of the FFA program
- Development of the supervised agricultural experience program
- Improvement and growth of the total agricultural education program
- Current trends and issues in the agriculture industry

Note that this mission statement delineates the work of the committee and states the committee's work is advisory in nature. The mission statement also addresses all aspects of a total agricultural education program and states that the committee is accountable to the school administration.

Step 2: Gain Support of School Administrators and the Board of Education

An agricultural education program exists as part of a whole school. It is not a separate entity and no more or less important than science, mathematics, English, history, or the fine arts. It is an essential part of the whole, and it functions under the leadership of school administration. The school principal has been delegated authority and responsibility by an elected school board to effectively operate the school. While advisory committees may provide direct and practical advice for the operation of an agricultural education program, they serve for the ultimate purpose of advising school leaders. Therefore, the natural second step in creating an advisory committee is to meet with the appropriate school administrators and inform them of the purpose of the advisory committee and the need for the group. School administrators can provide advice on the selection of committee members, assist in getting the advisory group officially recognized by the school system, and provide administrative and clerical assistance where needed. It is a good idea to get the approval of the board of education in establishing such a committee.

Step: 3: Choose the Members

After securing administrative support for an advisory group, the next step is to choose the members. One of the most common mistakes an agriculture teacher can make in creating this committee is choosing favorite parents or supporters to serve on it. That ensures that the committee will not address the difficult issues that arise in the agricultural education program. The committee might be reluctant to disagree on the issues. On the other hand, a committee staffed with people who do not trust each other will be paralyzed by continual disagreement. A good advisory committee should include people who

- Are genuinely interested in the success of the agricultural education program
- Are from diverse backgrounds and have varied interests
- Have significant experience in agriculture
- Understand their purpose on the committee
- Are committed to participating actively on the committee
- Have the time and resources necessary to serve
- Understand their role on the advisory committee

(Mosley, 2017)

It is very important that the advisory committee represent the diverse aspects of the community. An advisory committee that has only one or two segments of the community represented on it will be hindered by its narrow focus.

Once the prospective advisory committee members have been identified, the agriculture teacher then submits the list to the school principal to seek official approval. The principal invites the prospective members to serve on the advisory committee. If the teacher and school principal have a good working relationship built on trust, this process should be easy. The agriculture teacher drafts the letter for signature by the school principal, and then the prospective members are contacted and asked to serve.

Step 4: Choose or Elect a Chair and Other Officers

The chair plays a very important role in the success of the advisory committee. The chair should be someone with excellent leadership skills and the ability to build a consensus on issues. The chair serves as the liaison between the committee and the agriculture teacher and school administrators. Some responsibilities of the advisory group chair include developing agendas for meetings, following up on committee decisions to ensure that they are acted upon, and attending school meetings to represent the advisory committee. In some cases, the chair may be asked to serve on school-wide advisory committees.

Once the advisory committee is established, it may choose to appoint or elect a member to serve as the secretary of the committee. It is essential that the committee maintain files and records of meetings and documents so institutional memory is preserved. While agriculture teachers attend all meetings of the advisory committee, it is not recommended that agriculture teachers serve on the advisory committee. Stakeholders are the most appropriate members for an advisory committee (see Figure 6.5).

- School administrator (ex officio)
- Agricultural education teacher (ex officio)
- FFA Alumni chapter representative
- FFA chapter president
- Agriculture industry representatives
- Parents

FIGURE 6.5 The agricultural education advisory committee membership.

Step 5: Organize the First Meeting

The first meeting of the advisory committee will set the tone and pace of the committee's work in the agricultural education program. The first meeting should firmly set the mission and purpose of the committee and establish members as owners in the committee's work. Figure 6.6 shows an example of the first meeting agenda of an advisory committee. The first meeting should involve school administrators, who help establish the importance and validity of the committee. A good way to involve the committee quickly in its duties would be to make plans for a program evaluation.

AGENDA
Advisory Committee Meeting
Spruce Pine High School Agricultural Education Program
January 28
2800 Hall Building
6:30 p.m.

Reception	*Hosted by the FFA chapter officers*
Welcome and committee mission	*Ms. Sharon Dean, assistant principal*
Recognition of the Advisory Committee chair	*Mr. Bob Wells, agriculture teacher*
New business	*Ms. Erin Jones*
Program evaluation plan	
Next meeting date	
Tour of agriculture facility	*Mr. Bob Wells*
Adjournment	

FIGURE 6.6 Sample agenda for an advisory committee meeting.

Step 6: Establish Operating Procedures for the Committee

For the committee to function successfully, some procedures need to be established governing its activities. Sometimes a committee develops a constitution to serve as a guide, but usually a committee does not need anything this elaborate. A simple list of guidelines may be all that is needed for the committee to carry out its mission. Following are some items that need to be included in the guidelines:

- Time, location, and general dates of meetings. A committee should probably not meet more than four times nor less than two times annually.
- Rotation schedule for members of the committee.
- Process for selecting new members to serve on the committee.
- Process for selecting the chair and other officers.
- General rules for handling business. Most committees adopt parliamentary procedure as the method of choice.

Members of the advisory committee should be selected to serve three-year terms. A system can be established so only a few members rotate off the committee every year. In this manner, the institutional memory of the committee is preserved.

Advisory Committees and Program Stakeholders

Advisory committee members bring a wealth of technical knowledge to the table, but members also bring a wealth of information about the community, its citizens, and key stakeholders. If an agriculture teacher needs technical assistance, and the advisory committee does not have the expertise to help, committee members might be able to help by putting the agriculture teacher in contact with individuals in the community who have the needed technical expertise. If the advisory committee cannot answer an agriculture teacher's questions, they often know someone in the community who does. This knowledge of the community leverages the power of the advising ability of the advisory committee.

The term "stakeholder" is used frequently in this chapter, but what is a stakeholder? The answer is complicated. *Stakeholders* consist of those people within the organization who stand to win or lose when new policies or programs go into effect. These "agents" are the agricultural education teachers and school administrators. Stakeholders can also be the people who benefit directly or indirectly from the actions of the organization. These beneficiaries are students, their parents, and citizens in the community (Guba & Lincoln, 1989). In an agricultural education program beneficiaries also include local agricultural businesses.

Additionally, stakeholders could be victims who suffer in one way or another because of the actions or policies of an organization. An example from agricultural education history would be the reduced opportunities of Black Americans under segregated schooling in the early 1900s. Black agricultural education students were not provided with the same educational opportunities in agricultural education or through their participation in the New Farmers of America (NFA) (Wakefield & Talbert, 2003). In today's school-based agricultural education, it would be rare to have someone suffer negatively from participation in agricultural education programs. Perhaps one example is students who have career interests in animal science, but the agricultural education program only offers courses in plant science and horticulture. These students may suffer because they do not have access to the educational experiences they need to enter the career of their choice.

There are stakeholders who benefit from participation in the program or in some way benefit indirectly from the program. An example would be participation in the local FFA chapter. FFA members benefit from career development and leadership development as a result of participating in the activities of the local FFA chapter. Another example would be local businesses that benefit from students who are employed by these local businesses through their supervised agricultural experience.

Program Evaluation by the Advisory Committee

Advisory committees can play a distinct and valuable role in agricultural education program evaluation. School systems, through their membership and involvement in accreditation organizations, must periodically evaluate the education program. Some accreditation agencies examine the credentials of the faculty, to assure the teachers are prepared to execute the curriculum and instructional program. Accreditation agencies are most interested in the ability of the instructors to plan and conduct an effective instructional program. This includes an examination of the curriculum, the instructional facilities, and administrative program support (Southern Association of Colleges and Schools Commission on Colleges, 2016). An effective advisory committee can be a valuable resource in the accreditation process because of its ability to validate instructional goals and provide insight in the agricultural education program's resources—facilities, funding, teacher quality, and administrative program support (Allen & Allen, 2021).

Conducting the Evaluation

It is essential that the advisory committee begin its work with an in-depth examination of the agricultural education program. The advisory committee needs data to make good decisions. The worst thing that can happen is for the advisory committee to meet and make arbitrary recommendations about curriculum, planning, and goals. Sometimes such arbitrary recommendations are inconsistent with reliable data. To guard against arbitrary decision-making, the advisory committee should perform a program evaluation that answers the following questions (Martin, 1994):

- Does the curriculum adequately reflect the career needs of students?
- Are agricultural education courses sufficiently challenging and relevant to the best practices in the industry?
- Do all students have the opportunity to take agricultural education courses, provided the courses meet the students' educational and vocational goals?
- Does the agricultural education program administer a supervised experience program, and do all agricultural education students have valid supervised experience?
- Do graduates have viable job opportunities upon completion of the agricultural education program?
- Are facilities and instructional equipment current with technology in the agricultural industry?
- Do adequate resources exist for carrying out the instructional program?
- Does the agricultural education program have a well-prepared and highly skilled and motivated agriculture teacher?
- Does the agricultural education program have an active FFA program, and do all students have the opportunity to participate in the FFA chapter?
- Is there an active FFA Alumni group affiliated with the agricultural education program?

In some cases, this work will have been done largely through a school accreditation program. Most schools are required to be periodically evaluated by an accrediting agency for quality assurance purposes. The questions above may have been answered by the reports of the accrediting agency. However, if the accreditation process has not answered some or all of the questions, then the advisory committee should take on this task.

Curriculum Development

Most advisory committee members do not know how to develop a curriculum for an agricultural education program, nor should they be expected to know how. Curriculum development is the responsibility of the agriculture teacher and other education professionals in the school. It is not essential that the advisory committee understand how to write course and lesson objectives and plan the scope and sequence of a curriculum. However, advisory committee members can inform the curriculum development process through their knowledge of agriculture and natural resources. Committee members can validate the technical content choices proposed in the curriculum.

General Suggestions on Advisory Committees

When organizing an advisory committee for the agricultural education program, there are some ways to make it run more smoothly. Perhaps the most important element to make advisory committee service meaningful and productive is to establish an environment where members are free to address the important issues. The committee should operate in a hospitable environment where all opinions are welcomed and discussed.

For agricultural education programs to grow, difficult problems must be solved with new and innovative ideas. New ideas are often met with resistance because "That's not the way we've always done it." The agriculture teacher must be sensitive to the need to maintain a hospitable environment and must help the committee move through the difficult issues with everyone's emotions in check.

Rewarding Members

As members of the advisory group complete their terms of service, make certain they are adequately rewarded. It is recommended that each retiring member receive a thank-you letter from the agricultural education program and the school administration and the FFA chapter consider awarding the member the Honorary FFA Degree or other similar honor (Doerfort, 2003).

The people who serve on advisory groups do so for a number of reasons. Rewards that address the interests of advisory group members may be more meaningful (Castillo, 2003).

Making Use of Resource Persons and Data

It may be necessary from time to time for the advisory committee to bring in resource people to provide information about issues important to the agricultural education program. For instance, in determining the effectiveness of supervised experience programs, the county economic development director may be brought in to talk about trends in the industry (Doerfort, 2003). Without viable data, the committee is essentially guessing about the direction the agricultural education program should take. The agriculture teacher should provide reports and resources of interest to the advisory committee to validate its decisions.

Developing a Rotational Plan for Terms of Service

The work in an agricultural education program is ongoing. It is never complete. There is something to do all the time to adapt and improve the program. Steps must be taken to prevent those individuals who donate their time, effort, and expertise to the advisory committee from becoming overburdened with duties. If the advisory committee is well organized and has a rotational plan for terms of service, then it is unlikely any one individual will be overworked. A three-year term of service is recommended for every committee member. At the end of three years, it is time for the person to retire from the committee and allow someone else to bring some new ideas to the table. The individual may serve again in the future, but terms should not be consecutive.

Keeping Committee Members Productive

It is often said the busiest people make the best workers because of their ability to get things done. That is not always the case. Sometimes the busy people are so very busy because they are disorganized, are unable or unwilling to delegate, and procrastinate in their work.

Select individuals who are adept at carrying out plans and projects to completion. They do not have time to waste, nor do they like having their time wasted in meaningless meetings and poorly planned activities. If task-oriented committee members are given substantial work to do, they will have ownership in the agricultural education program and be its champions for many years to come.

FFA Alumni and Supporters

One of the most productive methods for developing community support for an agricultural education program is involving the FFA Alumni and Supporters. Formed in 1971, the *FFA Alumni and Supporters* is an adult group within the National FFA Organization that supports agricultural education by raising funds, promoting community awareness, and providing resources needed by the programs. In 2011, there were approximately 52,555 alumni members in local affiliates of FFA chapters in the United States.

Contrary to the way most alumni organizations work, membership in the FFA Alumni and Supporters is open to anyone who has a sincere interest in supporting FFA. Because the National FFA Alumni Association is a subordinate entity of the National FFA Organization, its mission is very similar to that of FFA.

While it may seem that advisory groups and FFA Alumni Affiliates could serve the same function, they do not. The FFA Alumni and Supporters is FFA for adults—providing for social interaction, FFA participation, and opportunities for involvement in community service. The advisory group is more mission-oriented and focused on agricultural education program improvement. Members of the FFA Alumni Affiliate can serve on the advisory group, but the functions of the two organizations are different (McGrew, 2017).

Mission Statement

A mission statement is a written statement of the purpose of an organization. It reflects the best efforts of an organization to go about achieving its purpose.

The mission of the National FFA Alumni Association is to secure the promise of FFA and agricultural education by creating an environment where people and communities can develop their potential for premier leadership, personal growth, and career success.

Organizational Structure of the FFA Alumni and Supporters

The most active component of the FFA Alumni and Supporters organization is the local alumni affiliate. The affiliate works closely with the local FFA chapter to conduct joint activities and plan independent activities that support the local FFA chapter.

The affiliates within a state form the state alumni association, which provides overall leadership and direction. The state association elects officers to serve according to the alumni constitution in that state. The state associations are chartered by the National FFA Alumni Association, which is headquartered in the National FFA Center in Indianapolis, Indiana.

Membership in the FFA Alumni and Supporters

Membership is open to anyone with an interest in supporting FFA. The local FFA Alumni and Supporters affiliate can create community awareness of FFA activities. As a civic organization, it is a natural advertising agency for the activities of the FFA chapter. Alumni members also provide valuable support of supervised experience because of their close relationship to the agricultural education program. Some alumni members are excellent teachers and trainers and provide instructional resources and ideas for improving instruction and for improving FFA chapter programs. Alumni members can provide opportunities for teachers to learn more about the subject matter they teach. Alumni groups are even a natural audience for adult education in agriculture.

Alumni members can accomplish many things in support of the FFA chapter and still be able to hold separate activities that appeal to the older generations found in the FFA Alumni and Supporters. Examples of alumni activities include the following:

- Fundraisers for the local FFA chapter
- State and national alumni conventions
- Big Brother/Big Sister programs
- Coaching and judging of Career Development Events/Leadership Development Events
- Substitute teaching in the agricultural education program
- Celebrations, parades, and holiday activities

Initiating an FFA Alumni and Supporters Chapter

The first step is to select a nucleus of highly motivated and interested individuals. Choose one person to lead the effort to create an alumni chapter, and set up an organizational committee to share in the work. Members of this committee should be knowledgeable about the agricultural education program and FFA and should have effective leadership and organizational skills. Call an organizational meeting and have the FFA members advertise it to parents and the local community. An FFA chapter can create an interest in alumni membership through the following activities:

- An "Each one, reach one" campaign. Each member should bring one prospective alumni member to a meeting.
- Personal contacts. The members and the FFA advisor can contact individuals and ask them to join.
- Mailings and telephone calls. The agricultural education teacher can send letters to parents and other supporters of the program.
- Newspaper announcements. The local newspaper can run an advertisement about the upcoming FFA Alumni and Supporters meeting.

At the meeting, provide an agenda, a draft constitution, and information about the benefits of FFA Alumni and Supporters membership. Once a constitution has been agreed upon by the membership, the FFA Alumni and Supporters chapter can then elect officers, set a budget, and begin developing a plan of activities. After the officers are elected, the organizational committee will cease to exist, and the officers will begin their service.

Recruiting and Retaining Members

The FFA Alumni and Supporters is a social organization by nature. Including things in the program of activities that are relevant, fun, and exciting for members is imperative.

FFA Alumni and Supporters members need a cause in which they can become involved and to which they can make a useful contribution. Respecting the needs and wishes of individuals is important. For instance, the FFA Alumni and Supporters should meet regularly but not too often. If the group meets too frequently, the members might become "burned out" and inactive. Politics, like some religious issues, can divide the members as well as the community. Avoid involvement in political activities through the FFA Alumni and Supporters chapter.

The FFA Alumni and Supporters can help the agricultural education teacher and the advisory committee carry out their mission (Agnew & Jumper, 2003). By engaging the support of the community through the alumni, an agricultural education program has the human resources necessary to accomplish great things for young people.

REVIEWING SUMMARY

The purpose of advisory committees is to assist the local school administration and instructor in their efforts to plan, develop, evaluate, and keep contemporary agricultural education program (Iowa Department of Education, 2017). Such groups should reflect the nature of the community served by the school and be formally organized to carry out their work more effectively. The groups must be aware of their roles and that they do not make policy, as that is a function of the local board of education. Citizens groups can assist in program development and evaluation, supervised experience, and the gaining of overall support for agricultural education.

Thorough knowledge of a community is needed before initiating an advisory committee. A survey can be made if the needed information is not otherwise available. Several important steps should be followed in establishing and operating an advisory committee:

1. Develop a mission and a purpose.
2. Gain support of school administrators and the board of education.
3. Choose the members.
4. Choose or elect a chair and other officers.
5. Organize the first meeting.
6. Establish operating procedures for the committee.

Teachers are responsible for getting an advisory group organized and helping it progress. This requires keeping members focused on their roles and providing recognition for their service. A teacher is an ex officio member of an advisory group.

FFA Alumni groups can also provide support for the local agricultural education program. The National FFA Alumni Association is affiliated with the National FFA Organization. Members are not required to be former FFA members but should be committed to supporting the local FFA chapter.

QUESTIONS FOR REVIEW AND DISCUSSION

1. What are the steps to establishing an advisory committee in an agricultural education program?
2. What are some activities of the FFA Alumni that can support the mission of the local agricultural education program?

3. Why are the FFA and FFA Alumni missions similar?

4. How does the FFA Alumni and Supporters group and the advisory committee serve parents who wish to support the agricultural education program? Explain your answer.

ACTIVITIES

1. Contact a local agricultural education program and ask to attend the next advisory committee meeting or FFA Alumni and Supporters meeting. Talk to the alumni president or the advisory committee chair about their leadership role. Interview the agricultural education teacher for suggestions on how to establish a successful advisory committee or FFA Alumni and Supporters chapter.

2. Investigate the mission and programs of the National FFA Alumni Association. Begin your study with the National FFA website: www.ffa.org. You may also gather information from the state supervisor of agricultural education in your state. Prepare a report on your findings.

3. Investigate the beliefs of Herbert M. Hamlin about citizen input into agricultural education programming. Refer to one or more of his publications, which include *Public School Education in Agriculture: A Guide to Policy and Policy-Making* (1962) and *The Public and Its Education* (1955). These books should be available from the library at your college or university.

REFERENCES

Agnew, D. M., & Jumper, P. (2003, July–August). Managing successful alumni relations. *The Agricultural Education Magazine, 76*(1), 8–9.

Allen, J. M., & Allen, D. C. (2021). *Maximizing program advisory committees* (ACCSC Monograph Series). Accrediting Commission of Career Schools and Colleges.

Carroll, L. (2015). *Alice's adventures in wonderland.* https://doi.org/10.1515/9781400874262

Castillo, J. (2003). Managing advisory committees. *The Agricultural Education Magazine, 76*(1), 20–21.

Cooper, C. (2016). *Citizen science: How ordinary people are changing the face of discovery.* Abrams Press; eBook Collection (EBSCOhost).

Cornell Lab of Ornithology. (2021, April 29). *EBird.* EBird. https://ebird.org/home

Darling-Hammond, L., & Bransford, J. (2005). *Preparing teachers for a changing world.* Jossey-Bass.

Doerfort, D. L. (2003). Discussing the future with advisory committees. *The Agricultural Education Magazine, 76*(3), 20–21.

Douglass, H. R. (1942). The basic responsibility of public education. *Phi Delta Kappan, 25*(1), 4–6.

Epstein, J. L. (2018). *School, family, and community partnerships: Preparing educators and improving schools.* http://search.ebscohost.com/login.aspx?direct=true&scope=site&db=nlebk&AN=1835821

Guba, E., & Lincoln, Y. S. (1989). *Fourth generation evaluation.* Sage.

Hamlin, H. M. (1955). *The public and its education: A citizens' guide to study and action in public education.* Interstate.

Hamlin, H. M. (1962). *Public school education in agriculture: A guide to policy and policy-making.* Interstate Printers & Publishers.

Hoover-Dempsey, K. V., & Sandler, H. M. (1995). Parental involvement in children's education: Why does it make a difference? *Teachers College Record, 97*(2), 310–331.

Iowa Department of Education. (2017). *Advisory council manual: Bureau of career and technical education.* Iowa Department of Education.

Martin, R. A. (1994). *Advisory committees: Questions in search of answers.* Unpublished manuscript.

McGrew, W. (2017). Advisory committee provides a firm foundation for an agricultural education program. *The Agricultural Education Magazine, 89*(6), 5–6.

Miretzky, D. (2004). The communication requirements of democratic schools: Parent-teacher perspectives on their relationships. *Teachers College Record, 106*(4), 814–851.

Mosley, C. (2017). A beginner's guide to creating an effective advisory committee. *The Agricultural Education Magazine, 89*(6), 7–9.

National Council for Agricultural Education. (2016). *National quality program standards for agriculture, food, and natural resources education.* https://thecouncil.ffa.org/

Richardson, J., Heller, R., Bucheri, C., Jacques, M., Langer, G., Holyk, G., Filer, C., Sinozich, S., & De Jong, A. (2017). *The 49th annual PDK poll of the public's attitudes toward the public schools.* https://journals.sagepub.com/doi/pdf/10.1177/0031721717728274

Southern Association of Colleges and Schools Commission on Colleges. (2016). In S. L. Danver (Ed.), *The SAGE encyclopedia of online education.* Sage. https://doi.org/10.4135/9781483318332.n330

Wakefield, D. B., & Talbert, B. A. (2003). A historical narrative on the impact of the new farmers of America (NFA) on selected past members. *Journal of Agricultural Education, 44*(1), 95–104. https://doi.org/10.5032/jae.2003.01095

7

Curriculum Development

OBJECTIVES

This chapter addresses the National Quality Program Standards for Agriculture, Food, and Natural Resources Education (National Council for Agricultural Education, 2016), specifically Standard 1A: Program Design and Instruction—Curriculum and Program Design. It has the following objectives:

1. Explain the meaning of curriculum and the importance of curriculum development.
2. Describe the characteristics of a high-quality curriculum.
3. Differentiate among UbD, differentiated instruction, and UDL curriculum models.
4. Discuss the role of national, state, and local standards in curriculum development.
5. Identify factors influencing curriculum development.
6. Explain how Perkins V influences the development of programs of study and the courses embedded within them.
7. Prepare curriculum maps.
8. Relate class scheduling to curriculum development.
9. Describe the importance of accountability.
10. Explain the meaning and role of academic integration.

Jay Jorgenson, then a new college graduate, began as teacher of agriculture at George Washington High School in 1982. He set about designing a curriculum that was in line with community needs and his interests. His students were almost all boys, and most lived on farms or had families who worked in agriculture. Except for a few adjustments from time to time, including new course names, he teaches much the same content as he did when he began teaching.

Meanwhile, the nearby city has sprawled increasingly toward George Washington High School. Farmland has been bought and converted to residential areas, malls, parks, and athletic fields. New ordinances have been enacted that make it nearly impossible to keep livestock and poultry. Enrollment in the agriculture program has gone down to only a couple dozen students. All this has caused Mr. Jorgenson concern. He has checked with the state retirement system and may end his teaching career.

Assume you were following Mr. Jorgenson's retirement as the new agriculture teacher at George Washington High School. What would you do to modernize the curriculum, recruit students, and, in general, breathe life into the agriculture program?

TERMS

academic subjects
accountability
articulation agreement
backwards design
block scheduling
career pathways system
Carnegie unit
coherent
CTE concentrator
CTE participant
CTE programs of study
curriculum
curriculum alignment
curriculum development
curriculum guide
curriculum integration

curriculum map
differentiated instruction (DI)
dual credit
educational standard
hidden curriculum
learning goals
prepared curriculum
program of study (POS)
scope
sequence
stackable credentials
teaching calendar
teaching schedule
traditional scheduling
transfer goals
trimester scheduling

FIGURE 7.1
Teachers are working cooperatively on a curriculum project guided by the National AFNR Standards.

WHAT IS CURRICULUM?

Who determines what courses are offered in a local agricultural education program? How does the agriculture teacher decide what content to teach in each course? How are the scope and sequence of content within a course determined? Curriculum development addresses these questions.

Meaning of Curriculum

The term *curriculum* is used to describe several concepts. To some, curriculum is what is taught in a single course. To others, it is the sequence of courses taught within a particular area, such as agricultural education, while others describe curriculum as a tangible map of how to accomplish designated learning outcomes, along with suggested assessments and learning activities to achieve those aims (Wiggins & McTighe, 2005). Those with a broad view see curriculum as encompassing everything related to "learning activities and experiences that a student has under the direction of the school" (Finch & Crunkilton, 1999, p. 11).

For the purposes of this chapter, *curriculum* is a broad overview of what students should learn, including learning goals and transfer goals. The curriculum of an entire agricultural education program is comprised of learning activities and experiences that students have within classes and across their experiences in the entire program, including course offerings, content within courses, supervised agricultural experience (SAE), and FFA. These learning experiences should complement each other and help students advance toward designated program *learning goals* (what students should know and be able to do), and program *transfer goals* (what students should be able to do beyond the scope of the program). Curriculum for an individual course should be developed with these program learning goals and transfer goals in mind, so learning activities within the course specifically help sequence students toward developing those culminating goals.

Curriculum focuses on planning, while instruction focuses on implementing those plans through individual lessons. Curriculum is a broad overview of what students should learn throughout a course or program, whereas lesson planning looks at specifics and how content is taught on a day-to-day basis. Lesson planning is discussed in Chapter 12.

Sources of Curriculum

Curriculum should be planned using educational standards. Since teachers have a variety of responsibilities, sometimes they are not individually tasked with writing or designing curriculum. At times, curriculum is developed by district or statewide curriculum consultants and provided to teachers. At other times, not-for-profit organizations, government agencies, or industry organizations develop curriculum materials to share with teachers free of charge. In the age of digital learning, some companies develop digital curriculum materials and provide them to teachers for a one-time fee, or a yearly subscription fee. Further discussion of these types of resources is found in Chapter 10. More detailed description of prepared curriculum materials specific to agricultural education can be found later in this chapter.

Quality Curriculum

Given that curriculum can vary so much, it is important that teachers ensure the curriculum they use to guide their daily instruction is of high quality. According to Sousa and Tomlinson (2011), a quality curriculum has five important characteristics (listed below and shown in Figure 7.2):

1. Curriculum is organized around essential content goals that represent the essence of the discipline studied.
2. All content goals, assessments, and learning experiences are aligned with each other.
3. Curricular activities are focused on cultivating and extending student understanding of the big ideas or concepts.
4. Learning experiences are relevant and engaging for students, building upon the innate desire for discovery. Learning activities are purposeful, interesting, and useful to students.
5. Authentic learning activities and assessments engage students in real-world settings in which students are not just consumers of knowledge, but also creators of knowledge.

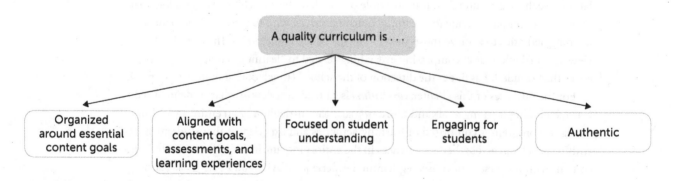

FIGURE 7.2 Characteristics of a quality curriculum. (ADAPTED FROM SOUSA & TOMLINSON, 2011.)

The impact of a quality curriculum cannot be overstated. Sousa and Tomlinson (2011) sum it up best: "What students learn will shape who they become, how they view learning itself, and how they interact with the world around them." What a weighty responsibility! Planning learning experiences that align with these five characteristics helps ensure that students are prepared to meet the demands of life head-on.

The National Quality Program Standards (NQPS) also identify five key indicators of a quality curriculum, but specifically addressing the curriculum and design of an agricultural education program. These indicators are used to assess the quality of the program as part of a regular program evaluation review process. The five quality indicators are as follows:

1. A *program of study* (POS) reflecting the needs of the community has been developed in accordance with the state requirements.
2. The courses in the program of study are organized logically and sequentially from introductory to advanced levels.
3. The technical content is aligned with core academic content standards.
4. The program of study allows students to gain postsecondary education credits through dual or concurrent enrollment programs or other means.
5. Each program of study includes knowledge and skill development through a balance of the three components of agriculture, food, and natural resources education.

WHAT IS CURRICULUM DEVELOPMENT?

Curriculum development primarily looks at content, depth of coverage, sequencing of content, standards, and related issues (Finch & Crunkilton, 1999). It is the process of deciding what should be taught, how to teach it, how to assess learning, and an estimated time frame and sequence in which to teach it. A *curriculum map* or a *curriculum guide* outlines this information in alignment with appropriate educational standards. The format and layout of these documents may vary significantly depending on the preference of the school district or software used to write the curriculum.

Teachers may be provided with previously developed curriculum when hired to teach, or they may be tasked with developing their own curriculum. Curriculum maps or guides are documents that are intended to be used by teachers to plan instruction; they are not documents that are typically shared with students like a class syllabus. While they are not typically shared, some schools are now posting complete curriculum maps on their school web page in an effort to be transparent with stakeholders.

There are a variety of approaches to the process of developing curriculum. Your school district may prefer a different process than a neighboring district, or your state might specify a preferred approach to curriculum development. This chapter is intended to be an overview primer on the essentials of curriculum development, and not a specific manual on the details of the process.

Curriculum Development Models

Many of the details and steps that go into developing, delivering, and revising curriculum are specific to individual teachers, departments, and schools. Sometimes it can feel daunting not knowing where to begin. In this chapter, we will focus on three different models or systems for developing a curriculum that fosters an inclusive learning environment to support all students in meeting established learning goals. These three models include understanding by design

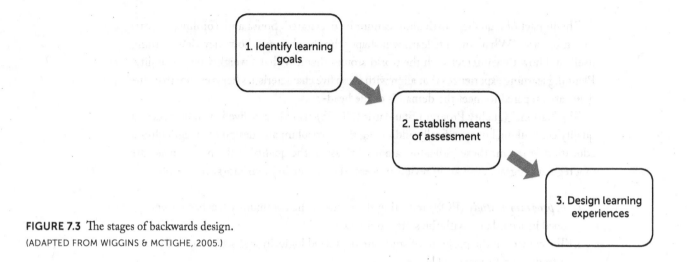

FIGURE 7.3 The stages of backwards design.
(ADAPTED FROM WIGGINS & MCTIGHE, 2005.)

(UbD), differentiated instruction (DI), and Universal Design for Learning (UDL). Working in tandem with one another, these three models help ensure all students have opportunities to engage in authentic learning that is suitable for their specific needs and learning context.

Understanding by Design (UbD)

UbD is a model that provides a road map for how to move through the process of designing learning experiences. The process of developing curriculum using an understanding by design approach outlines what we want students to learn, how we know that they have learned it, and how we intend to engage students in working toward the learning goals.

When following an UbD approach to designing instructional experiences, a teacher, program, or school begins by identifying their long-term goals for their students first, and then works in reverse to design learning assessments and activities that help students accomplish these goals. This is referred to as a ***backwards design*** approach (Wiggins & McTighe, 2005). In this approach, there are three major stages (listed below and shown in Figure 7.3):

1. Identify desired results (including what students should know, understand, and be able to do).
2. Establish assessment evidence of successful attainment of these results.
3. Design learning experiences that will prepare students to demonstrate their mastery of the desired results.

Working backwards to plan learning activities and assessments that are aligned with learning goals ensures students are engaged in meaningful work that leads to the desired outcomes. Ensuring all learning and assessment activities are aligned with the established learning goals ensures clarity and helps promote student success.

Differentiated Instruction (DI)

Coined by Carol Ann Tomlinson (Tomlinson & Strickland, 2005), the term ***differentiated instruction*** refers to a systematic approach to how teachers design instruction to meet individual learning needs of academically diverse students. Differentiated instruction places student needs at the center of instructional planning, building on student readiness to learn, interests, and individual learning profiles. Teachers differentiate instruction through intentional design choices, as well as in the moment actions to respond to student needs. There are four major ways teachers can differentiate their instruction:

Content—The teacher adjusts what the students need to learn, or how they access the content.

Process—The teacher adapts which learning activities the students engage in to learn the content or enrich their understanding of the content.

Products—The teacher adapts the format of assessment that demonstrates a student's mastery of unit knowledge and skills.

Affect/learning environment—The teacher adjusts the way the classroom looks and feels to better engage students in the learning process.

Universal Design for Learning (UDL)

Universal Design for Learning is an approach to curriculum in which barriers to learning are removed or minimized so all learners can engage in learning. UDL ensures learning experiences are accessible to an increasingly diverse array of learners, building on neuroscience principles of learning. UDL curriculum design seeks to help students become expert learners who are purposeful and motivated, resourceful and knowledgeable, and strategic and goal-directed. Those designing curriculum from a UDL perspective believe the following:

- Variability and student choice are the rule, not the exception.
- All students can learn and meet learning goals when provided appropriate supports.
- All students can be expert learners if barriers to learning are removed.

The concept of "universal design" is rooted in architecture. For example, an airport may install escalators instead of stairs to assist people who have difficulty using stairs. This design feature benefits those folks who need it, as well as those who do not rely on escalators. This same concept is applied to learning through UDL. For example, a teacher might provide students with a graphic organizer to help them organize their notes while watching a video. Some students may not need the graphic organizer, but others may rely on this support to help them make sense of their thoughts on the video content. While the graphic organizer is not necessary for some students, it can provide beneficial supports for them.

When designing instruction or curriculum using a UDL framework, designers should begin by identifying the learning goals they want to establish for their students. Immediately after identifying the learning goal, they should consider what barriers may exist that could prevent their diverse group of learners from being able to reach the learning goals. Designers can then use three main strategies that tap into various neural networks in the brain to remove existing barriers to learning:

- Provide multiple means of engagement.
- Provide multiple means of representation.
- Provide multiple means of action and expression.

The UDL Guidelines (CAST, 2018) are an outstanding tool for implementing a UDL framework and can be accessed at https://udlguidelines.cast.org/.

Curriculum Coherence and Alignment

Embedded within each of these curriculum design approaches is a focus on ensuring that all stages of the development process align to create a coherent learning experience. A *coherent* curriculum is intentionally organized to ensure student learning builds upon prior learning and leads to attainment of established goals. *Curriculum alignment* refers to the process of ensuring the skills and learning outcomes within and across courses build upon each other, creating program coherence. The agricultural education curriculum will have some degree of alignment between agriculture courses. The introductory course should prepare students for the more specialized courses at the next level. Concepts learned in a horticulture class, for example, should prepare students for applications in a landscaping class. Management practices learned in an animal sciences course can be used by students in their entrepreneurship supervised agricultural experience.

As a part of the overall school curriculum, agricultural education should have alignment with other subjects. For example, content in soil science should be articulated with content learned in chemistry and geology. The scientific concepts learned in chemistry can help students better understand the processes discussed in agriculture. If students have studied soils in agriculture first, they benefit in chemistry by being able to connect the abstract concepts and principles to real-world situations within agriculture. Students benefit by learning similar content in different settings, thereby taking advantage of the learning principles discussed in Chapter 12.

Levels of Curriculum Development

To coordinate coherent learning experiences, it is important for teachers to collaborate within and across various groups of stakeholders. Deciding which level to begin from largely depends upon how long your program has been established. Chapter 5 outlined program planning processes. If your program is newly established, or has not engaged in any program planning for some time, it is wise to begin by following the suggestions in Chapter 5. Curriculum development can occur at many levels, which are outlined below.

State   Some states develop curriculum specifically for courses approved by the state education agency. Usually a combination of teachers, administrators, and community stakeholders are involved in this process. Teachers may be provided this curriculum through professional development workshops, through online portals, or through teacher preparation programs.

School district   Depending on the size of your school district and how it is administratively organized, there may be curriculum development specialists that write curriculum or assist teachers in developing curriculum. This curriculum may be accessible to teachers across schools in the districts. The school district is usually concerned with development of common learning outcomes or learning expectations of students, particularly with the end goal focusing on competencies that students should demonstrate upon high school graduation. Much of the curriculum work at the district level is focused on coordination and alignment between elementary and middle school, and middle school to high school.

School   Your school may establish school-wide learning expectations that they expect all students to attain. To ensure students have multiple opportunities to advance toward these end goals, curriculum may be revised or written.

Department or program   Depending on the size of your school, you may be an agriculture teacher within a science department, CTE department, or electives department. If you are in a multiteacher program, it is possible you may have your own agriculture department. It is important that to create a complete three-component agriculture program experience, coordination and development of agriculture curriculum occurs at this level. Programs should establish a clear mission and vision statement to guide development of their learning goals and transfer goals for all students enrolled in the program.

Course   Teachers may be able to design curriculum for their own classes. Since agriculture teachers are frequently the only teachers teaching agriculture courses, they may do much of the curriculum development themselves. In multiteacher programs, teachers may be able to collaborate to develop curriculum for commonly taught classes, such as an introductory foundations course in agriculture.

Curriculum Decisions: Who Should Be Involved?

Decisions regarding the curriculum should not be made by the agriculture teacher alone. Just as program planning, discussed in Chapter 5, involves the school and the community, so should curriculum development. The agricultural education advisory committee can provide community feedback regarding the agricultural education curriculum. School administrators and other teachers can provide valuable input, especially in the articulation of agricultural education with other subjects within the school. Involving students in the development of the curriculum can ensure that your curriculum content is not only aligned with the needs of the community and agricultural industry, but builds upon the interest and motivation of your learners.

HOW DO WE DECIDE WHAT SHOULD BE IN THE CURRICULUM?

A variety of factors influence and shape the direction of curriculum content of public schools. These include educational standards, federal and state legislation, local community needs, and foundational philosophical beliefs about teaching and learning.

Educational Standards and Curriculum

Agriculture teachers should not teach only what they want to teach. National, state, and local educational entities require that certain educational standards be met. An *educational standard* is the expectation of what students should know and be able to do after completing a given area of instruction. The teacher is responsible for using their professional knowledge and experience to determine how to meet the standard.

Standards are typically enforced by student standardized testing, funding, or accreditation. Typically, these standards have been developed with input from educators, business and industry, and the general public. They also may be benchmarked against content covered in standardized tests. Finally, they are officially adopted by the applicable governing body.

The Every Student Succeeds Act (ESSA) was signed into federal law in 2015. This legislation required that every student in the United States experience education that is aligned with academic standards that would guide them toward college and career success. The act also ended federal influence over state academic standards, allowing states complete control over what academic standards it would adopt. Career and technical education was also identified as part of a complete education, indicating high levels of support for CTE experiences for all students.

Academic Standards

Academic subjects are core subjects required of all students and include English/language arts, mathematics, science, and social studies. The curriculum in these subjects is mostly prescribed and fairly consistent among all schools within a state. In 2009, the Common Core State Standards Initiative developed standards in English/language arts and mathematics that were adopted by 41 states, the Department of Defense Education Activity, Washington,

D.C., Guam, the Northern Mariana Islands, American Samoan Islands, and the U.S. Virgin Islands (Common Core State Standards Initiative, 2021). Recently, several states have moved away from Common Core State Standards (CCSS) and have developed their own state academic standards, some of which are very similar to the CCSS, but with a different name (Lee, 2021). Regardless of the type of academic standards, subjects outside of core academic subject areas are increasingly being expected to assist students in learning and meeting academic standards. Agricultural education is uniquely situated to contribute to knowledge and skill development across core academic subject areas.

Through the study of the food, fiber, plants, animals, and natural resources systems, agricultural education addresses standards in all areas of science. Examples and problems used in the classroom can come from the real world and the workplace of agriculture. Computational math is used almost every day in agriculture. However, the content of algebra, geometry, and trigonometry is also found throughout agricultural areas of study. English/language arts are a part of every classroom, but the agricultural education components of SAE and FFA provide additional opportunities to learn and practice these skills. Finally, agriculture has been and is a vital part of our nation's history, geography, culture, politics, and economics.

If agricultural education is to assist students in achieving academic standards, alignment and integration must occur on a conscious and consistent basis. Alignment works to have students learn similar concepts in different courses at an appropriate time and in an appropriate sequence. *Curriculum integration* is the process of combining academic curriculum with career and technical education curriculum so learning is more relevant and meaningful to students (Finch & Crunkilton, 1999; National FFA Organization, 2012). Whereas alignment allows students to build on prior or concurrent learning, integration is designed to eliminate the rigid distinction between academic and career and technical education. Integration is discussed in more detail in a later section of this chapter.

Agricultural Education Standards

Educational standards specific to agricultural education guide the implementation of agriculture curriculums in local programs. Agricultural education standards exist at the national and state levels. The National Council for Agricultural Education developed national content standards for each of the eight career pathways in the Agriculture, Food, and Natural Resources career cluster (career clusters are discussed in Chapter 8). Each pathway has one broad standard, multiple performance elements that state what content should be provided, and performance indicators that are measurable at three levels. These are designed to guide the development or revision of state and local content standards (Pentony & Hall, 2008), and serve to mold the design of complete agricultural education programs. Not only do the standards shape classroom and laboratory instruction, but they also provide structure for student experiences through career and technical student organizations (CTSOs) such as FFA, and work-based learning experiences such as SAE. See Figure 7.4 for an illustration of an overview of the National AFNR Standards organizational model.

The National AFNR Content Standards were revised in 2015 to align with the current needs of the AFNR industry, and to ensure that they were an ideal foundation to design career and technical education (CTE) coursework and assessment. To illustrate how agricultural education coursework helps strengthen skills and knowledge in other subject areas, the National AFNR Content Standards include alignments to other national academic standards including Common Core English and Math, Next Generation Science Standards

FIGURE 7.4 National AFNR Content Standards Organizational Model. The National AFNR Content Standards are comprised of two foundational pieces and eight career pathways aligned with career clusters (https://thecouncil.ffa.org /afnr/). (REPRODUCED BY PERMISSION FROM THE COUNCIL, HTTPS://THECOUNCIL.FFA.ORG/AFNR/.)

(NGSS), Green/Sustainability Knowledge and Skill Statements, and the National Standards for Financial Literacy. To download a copy of the National AFNR Content Standards, visit https://thecouncil.ffa.org/afnr/. Standards for individual states are usually available online through the website of the state education agency.

States may tie funding to documentation that standards are being used to design local agricultural education courses. Agricultural education programs may be required to show a minimum percentage, maybe 70% to 80%, of state standards are being covered. This allows a local program some flexibility in meeting local community agricultural needs. Other states administer end-of-course testing based on the state agricultural education standards. Depending on the state requirements, only students who pass the test receive credit for the agriculture course, and their pass rate may even be linked to program funding.

The Influence of Perkins V on Curriculum

Federal and state legislation also influences the development of agricultural education curriculum. Federal funding to support career and technical education (CTE) was reauthorized through passage of the Strengthening Career and Technical Education for the 21st Century Act in 2019. Also known as Perkins V, this legislation provides $1.3 billion annually for CTE across the United States. The legislation requires that academic, technical, and employability skills should be integrated within both secondary and postsecondary CTE programs, leading to the development of work attitudes and employability skills, as well as higher-order reasoning and problem-solving skills. The updates present in Perkins V will have ripple effects on CTE programming at both the secondary and postsecondary level.

Organization of Career and Technical Education

Perkins V utilizes a *career pathways system* to organize CTE. Career pathways help provide structure to students enrolled in secondary and postsecondary CTE to help them develop skills and industry-recognized credentials (referred to as "*stackable credentials*") that lead to postsecondary education attainment and consequently financially viable jobs. This system requires that industry, postsecondary education, and secondary education all coordinate their work so that students can more easily enter the workforce as prepared, skilled workers. Career pathways are intended to streamline training and education so that students can chart out a path to their intended career. See Figure 7.5 for an illustration of the *integrated career pathways model* on how learners can move through the career pathways system to accrue credentials that lead to employment and/or postsecondary education degree credit.

Perkins V provided an updated definition for *CTE programs of study*: a "structured sequence of academic and career and technical courses leading to a postsecondary-level credential" (U.S. Department of Education, 2021c). States are provided flexibility to determine what programs of study will result in in-demand or emerging jobs. Consequently, this means each state may have a different interpretation of how they intend to structure approved programs of study, including the number and types of courses that comprise each program of study. Since agricultural production varies so widely across the country, not all states will have the same programs of study. Figure 7.6 illustrates the relationship among the various levels of curriculum coordination under Perkins V provisions.

Students who have successfully completed at least one course in any CTE program of study are called *CTE participants*. Students who have successfully completed at least three courses within a single CTE program of study are called *CTE concentrators*. If a student takes three courses, but all three are from different programs of study, then the student is still considered a CTE participant, but not a CTE concentrator. This distinction is important, as the accountability requirements of Perkins V only apply to CTE concentrators. Some states may tie program funding to the number of CTE participants or concentrators.

Effects of Perkins V Legislation on Curriculum Planning

So, how does all of this influence your curriculum? Assessing the needs of your local community and state will inform what programs of study you offer at your agriculture program. Since the programs of study and pathways are aligned with industry credentials and postsecondary credentials, planning the curriculum offered in your local program will be tied to the performance outcomes or competencies affiliated with these credentials. For example, a landscaping or horticulture course may be aligned with the requirements of the National Association of Landscape Professionals' Landscape Industry Certified credential (National Association of Landscape Professionals, 2020). The skills and knowledge in your course should prepare students to successfully pass the credentialing requirements.

In addition to industry credentialing, the course might be eligible to be taught for *dual credit* with an area community college, junior college, or university. Dual credit allows students to earn college academic credit while enrolled in high school. For example, a student enrolled in a high school veterinary science class may be eligible to receive dual credit for an introductory veterinary assisting course at a local community college. They earn credits needed at both the high school level and the community college. Each academic institution may have its own requirements for how to establish dual credit, which may include following the syllabi and final exams their instructors use.

Establishment of dual credit between institutions is usually documented through an *articulation agreement*. An articulation agreement serves as a formal agreement between public

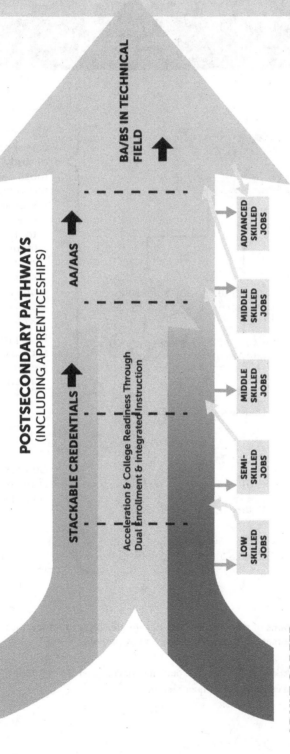

SECONDARY TO POSTSECONDARY PROGRAMS OF STUDY

SYSTEM OUTCOMES

Financially sustainable, aligned career pathways systems for youth and adults

Increased number of skilled workers with credentials of value to the labor market

Greater cost efficiencies by reducing duplication of services

POSTSECONDARY PATHWAYS
(INCLUDING APPRENTICESHIPS)

STACKABLE CREDENTIALS

Acceleration & College Readiness Through Dual Enrollment & Integrated Instruction

AA/AAS

BA/BS IN TECHNICAL FIELD

LOW SKILLED JOBS

SEMI-SKILLED JOBS

MIDDLE SKILLED JOBS

MIDDLE SKILLED JOBS

ADVANCED SKILLED JOBS

ADULT CAREER PATHWAYS

FIGURE 7.5 Integrated career pathways model. Middle and high school students enrolled in school-based agricultural education enter career pathways beginning at the top left corner of this model. (REPRODUCED BY PERMISSION FROM U.S. DEPARTMENT OF EDUCATION, OFFICE OF CAREER, TECHNICAL, AND ADULT EDUCATION. [2015, DECEMBER]. *ADVANCING CAREER AND TECHNICAL EDUCATION [CTE] IN STATE AND LOCAL CAREER PATHWAYS PROJECT: FINAL REPORT.* WASHINGTON, D.C.: CLAGETT, MARY GARDNER, HTTPS://CTE.ED.GOV/INITIATIVES/CAREER-PATHWAYS-SYSTEMS#ELEMENT2MODALRESOURCE.)

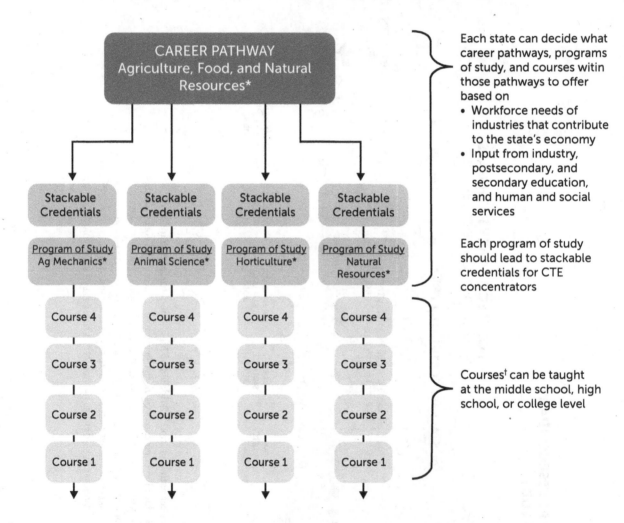

CAREER CLUSTERS FRAMEWORK

- Agriculture, Food, and Natural Resources
- Architecture and Construction
- Arts, A/V Technology, and Communications
- Business Management and Administration
- Education and Training
- Finance
- Government and Public Administration
- Health Science
- Hospitality and Tourism
- Human Services
- Information Technology
- Law, Public Safety, Corrections, and Security
- Manufacturing
- Marketing
- Science, Technology, Engineering, and Mathematics
- Transportation, Distribution, and Logistics

CAREER PATHWAY
Agriculture, Food, and Natural Resources*

| Stackable Credentials | Stackable Credentials | Stackable Credentials | Stackable Credentials |

| Program of Study Ag Mechanics* | Program of Study Animal Science* | Program of Study Horticulture* | Program of Study Natural Resources* |

Course 4	Course 4	Course 4	Course 4
Course 3	Course 3	Course 3	Course 3
Course 2	Course 2	Course 2	Course 2
Course 1	Course 1	Course 1	Course 1

Each state can decide what career pathways, programs of study, and courses witin those pathways to offer based on
- Workforce needs of industries that contribute to the state's economy
- Input from industry, postsecondary, and secondary education, and human and social services

Each program of study should lead to stackable credentials for CTE concentrators

Courses[†] can be taught at the middle school, high school, or college level

* Sample titles of pathways and programs of study. Each state has the flexibility to design these to fit their needs.
† Not all programs of study have 4 courses. This varies by state and by pathway.

FIGURE 7.6 Levels of curriculum coordination in Perkins V. Performance outcomes or competencies established at higher levels guide the development of curriculum content at subsequent levels.

high schools and postsecondary institutions to document the alignment of the high school curriculum with the postsecondary institution, resulting in equivalent postsecondary academic credit earned. An articulation agreement is considered a defining trait of career and technical education. Articulation agreements may specify the content and assessment of agricultural education courses at the local level.

Local Community Needs

As described in Chapters 5 and 6, the agriculture teacher should be sure to design and deliver a program that meets the needs of the local community. The variability and diversity of agricultural crops produced across the United States is impressive, but teaching students about the production of all crops may not be relevant to your location. For example, many agriculture programs across New England incorporate instruction on maple syrup production into their curriculum. Maple syrup can only be produced in regions where maple trees grow and the climate produces appropriate combinations of freezing and thawing temperatures that promote sap flow in the early spring. Teaching students in Arizona about the details of maple syrup production would perhaps be interesting to them, but not particularly relevant to their location.

One way agriculture teachers can ensure they are meeting community needs is to conduct a needs assessment. Details on conducting a needs assessment can be found in Chapter 5. In addition to a needs assessment, the agricultural education program's advisory committee can be invaluable in providing insight and guidance on the needs of the local community. Agricultural employers can provide feedback on what skills and knowledge new employees at their businesses need to have as well as information on the trajectory of their specific industry. The workforce needs of the area and the state can have a tremendous influence on what skills are emphasized within your program. For more information on how to best utilize an advisory committee, see Chapter 6.

Philosophical Foundations Guiding Curriculum

Finally, philosophical beliefs about education influence the content and learning methods included in curriculum. Determining what subjects are of value, how to design meaningful learning experiences, and how to designate learning outcomes are all practices directly related to the philosophical beliefs of those writing and delivering curriculum. Chapter 2 describes the philosophical foundations of career and technical education in agriculture. In addition to the philosophy of the profession, the teacher's personal beliefs about the purpose of education and the best ways to carry out education influence the curriculum. If you have not already done so, it is a useful and important exercise to deeply consider your own philosophical beliefs regarding education and to compose a teaching philosophy statement. Knowing what your beliefs are and where they come from is an important first step in recognizing your own biases, as well as in building a foundation upon which to make decisions as an educator.

While curriculum does outline content to be learned and suggested learning activities, it often implicitly includes historical, social, and cultural values. These values might not be explicitly articulated within curriculum, but are communicated through what is considered acceptable or unacceptable behaviors. This is what is referred to as *hidden curriculum* (Jackson, 1968). For example, if a teacher only gives tests that elicit rote responses memorized from lecture notes, then students quickly learn that the only way to experience "success" in that class is to memorize. This in turn, may communicate that critical thinking, application of knowledge, or creativity are not skills to be developed or not deemed important.

Hidden curriculum contributes to the oppression of historically marginalized groups, further exacerbating societal and educational inequities (Brownell, 2017). It is essential that teachers reflect on how their teaching may be communicating hidden curriculum and actively take steps to change hidden curriculum that causes harm.

PREPARING CURRICULUM MAPS

Curriculum development is an art requiring application of principles to unique programs and local situations. A part of curriculum development is the preparation of programs of study. A curriculum map is a written guide that aids in teaching a particular subject or course. It typically contains objectives, an outline of content, a list of learning activities, means of assessment, and suggested instructional materials. Content includes known facts and subject matter; however, it is continually updated as research creates new knowledge and discoveries. Since agriculture is a constantly changing field, it is necessary to regularly revisit the content of curriculum in agricultural education to ensure it is up-to-date and relevant to the needs of today's industry and students.

Some guiding principles of curriculum planning as is done in preparing course curriculum maps are presented in Box 7.1. These principles will help in determining the scope and sequence of the agricultural education curriculum. Scope and sequence are two important concepts in preparing a curriculum map.

BOX 7.1 *Curriculum Planning Principles in Agricultural Education*

1. The agriculture teacher is responsible for developing the curriculum for the local program.
2. The curriculum should reflect the present and future importance of agricultural situations and resources in the community and larger geographic area. Input from the local advisory committee and others should be used.
3. The curriculum should be planned in advance but be flexible to allow adaptability to student situations and conditions.
4. Teaching and learning activities should be planned as sequenced and systematic units.
5. Appropriate state and national standards must be included. The objectives and content that are most significant to the local situation and/or mandated by a state curriculum guide should be included.
6. Content, along with teaching and learning activities, should be planned based on student developmental levels, student learning modes, and principles of learning.
7. A balanced program of instruction should include agricultural literacy courses, agricultural career preparation courses, specialized courses, supervised experience, and FFA. Depending on the program and local situation, middle school exploratory courses and adult agricultural education are included.
8. Each year (or semester if on alternative scheduling or using semester courses) is articulated with other courses guided by state and/or national career pathways.

Adapted from Lee, 2000.

Determining Scope and Sequence of Course Content

Scope is the extent, depth, or range of the curriculum. Will courses be taught for grades 6 through 12 or grades 10 through 12? Will five class periods be devoted to a unit of instruction, or 15? Will emphasis be placed on a wide breadth of topics, or will fewer topics be taught in greater depth?

Sequence is the order or progression of the material covered. Sequence in agriculture may be determined in any of several ways, such as from simple to complex or by production cycle (crop planting to harvest). Will an introductory course be required before students can take an advanced course? Are general courses required before specialized courses? How will units be arranged within a course? Will content proceed from known to unknown, from simple to complex, from whole to parts to whole again? Will units be arranged according to seasons of the year?

The scope and sequence of course content can be affected by a variety of factors, especially in agriculture. Since an agriculture teacher may be teaching content that is contingent upon the weather or season, the order of topics may be more dependent upon the whims of Mother Nature than perhaps other subject areas. Possible factors affecting the scope and sequence of course content in an agricultural education course include

- Type of assessments
- Capturing student interest
- Student background knowledge and experience
- Teacher's personal experience
- Foundational to more advanced skill development
- Season/weather
- Breeding cycles
- Planting/harvest dates
- Supplies available
- CDE/LDE dates
- FFA and school calendar
- Conferences and workshops
- Field trip and guest speaker opportunities
- Community-specific events or traditions

More specific discussion on sequencing daily instructional content is discussed in Chapter 12. When mapping your course scope and sequence, it is helpful to make a chart of these factors and the time of year they may occur. For example, a teacher who wanted to take students on a field trip during lambing season would need to make sure to align course content to coincide during that time of year. The chart in Figure 7.7 can be a useful tool for mapping significant events that might impact the design and delivery of the course content.

Selecting and Sequencing Content

To select course content, the teacher should use the various sources of input, along with state curriculum guides. If a state curriculum guide mandates content, then that content must be included in the course. Any local-option content should be based on community needs and endorsed by the local advisory committee. Content articulated and integrated with academic subject areas should also be considered. Finally, articulation between agricultural education courses and their content must also be taken into account.

Developing a Scope and Sequence

Using the chart below, map out significant events that might impact the delivery of your curriculum. If you teach subjects that involve lots of outdoor labs, weather events and production seasons will likely play a bigger role in your teaching.

Month	Major School Activities/Testing	Weather Events	Production Session	Field Trips/Guest Speaker Availability	FFA CDEs & LDEs
August					
September					
October					
November					
December					
January					
February					
March					
April					
May					
June					

FIGURE 7.7 This chart can be a useful tool for mapping potential factors that will impact the scope and sequence of your course delivery.

The agriculture teacher is responsible for the scope and sequence of the content taught in agricultural education courses. Determining the scope of the content for a course should involve answering these questions:

1. What is the purpose of the course? An introductory or exploratory course should introduce students to a variety of topics and content but may not go into much depth on any one topic. An advanced-level or career preparation course, on the other hand, may cover a limited number of topics but go into a greater depth of content so that students completing the course are extremely proficient in the area.

2. What is the level of standards in the curriculum guide? Content to be learned on the analysis, synthesis, and evaluation levels (see Chapter 12 for an explanation of these terms) may require more class periods than content to be learned on the knowledge level.

3. What is the background of the students? What prerequisites have the students already taken? Learning totally new content requires more class periods than learning content that expands previous learning or relates to previous learning. If the students are unfamiliar with basic terms, concepts, and practices, then class time will need to be devoted to learning these before moving on to more complex content.

4. What content that relates to this course is covered in other classes? Content covered in other classes, whether other agricultural education classes or not, deserves special consideration. Is it critical content that should be repeated and reinforced in this course? Is it content that needs to be covered only once and therefore should be left out of this course? Is it content that provides a foundation for further learning in this course?

Determining the sequence of the content for a course should involve answering these questions:

1. What content that relates to this course is covered in other agricultural education and academic courses? If it is complementary, then the agricultural education instruction in this course should occur close in time to the instruction in the other courses.
2. What prerequisite knowledge do students need for learning this content? In general, content should be sequenced from the known to the unknown. Prerequisite knowledge, by definition, should be sequenced first.
3. What is the order of complexity of the content? In general, content should be sequenced from simple to complex.
4. Is there a season or other time of year that is most logical for content to be learned? In agriculture, content should be sequenced so it occurs just before or concurrently with the agricultural practice. For example, in a horticultural science course, drawing landscaping designs might be sequenced first. This would be followed by selecting plants and landscaping materials, which would come before site preparation, which would come before landscaping the site.
5. What are the interests and what is the readiness level of the students? Younger students typically are more interested in small animals than in livestock. If small-animal care is sequenced first, this may build interest and motivation for learning about livestock next.
6. What conditions or special activities occur throughout the school year that may influence content sequencing? Weather may be a deciding factor. For example, in a horticultural science course, drawing landscaping designs might be done in the winter or rainy season when no outside work can be accomplished. School activities, such as science fairs, homecoming, or prom, may influence what time of the year certain content is taught. The nature and timing of student supervised experience should be taken into consideration when sequencing content. Finally, FFA Career Development Events and other activities may provide input into the sequencing of course content.

Developing a Teaching Schedule and a Teaching Calendar

A *teaching schedule* (sometimes called a pacing guide) shows the scope and sequence of content within an individual agricultural education course. Broad units of instruction are listed in the order in which they would be taught during the semester or school year. A suggested number of class periods to spend on each unit is also given. Other information that can be on a teaching schedule includes grade level, length of course, and citations for textbooks used in the course. Some school corporations require that standards and assessments also be included.

A *teaching calendar* is a specialized type of teaching schedule that provides additional details on the time when units of instruction will be taught. The teaching calendar may be on a yearly, semester, monthly, or weekly basis, depending on the specificity needed. If known, testing days, holidays, and other noninstructional activities should be placed on the teaching calendar. The teaching calendar provides advantages over the basic teaching schedule. With a teaching calendar, an agriculture teacher can order supplies and materials, schedule field trips, and involve resource persons in a more timely and effective manner. A teaching calendar also serves to keep the agriculture teacher on track for wiser use of instructional time. Finally, the teaching calendar serves as a public relations tool with administrators, the local advisory committee, and others.

INFLUENCE OF CLASS SCHEDULING ON CURRICULUM DEVELOPMENT

The amount of time teachers see their students as well as the frequency with which they meet for instructional time largely depends on the schedule used by their school. The class schedule has a large impact on what learning activities can be realistically accomplished within a set period of time, and it should be considered when designing curriculum.

Carnegie Unit System

Credit for high school courses is typically based on the Carnegie unit system. A *Carnegie unit* has traditionally been the amount of credit a student is awarded for a course that meets one instructional hour per day for 180 school days (36 weeks). An instructional hour may range from 45 to 60 minutes of actual clock time.

Carnegie units are important for determining high school graduation and college entrance requirements. So, a course that meets for 50 minutes per day for one semester (18 weeks) would be worth one-half Carnegie unit. On the other hand, a course that meets for two 50-minute periods per day for the entire 180-day school year would be worth two Carnegie units.

In *traditional scheduling*, high schools are organized with six or seven one-instructional-hour class periods for five days per week for the entire school year. Under this system, students earn six or seven Carnegie units per school year. If classes are on a semester basis, students can take up to 14 semester classes. Within the last years, other organizational systems have emerged, with block scheduling and trimester scheduling being the two most prevalent.

Block Scheduling

Block scheduling is a class scheduling system in which students take fewer classes each semester than in a traditional system but in which each class period occupies a larger block of time. Each class period may be 85 to 100 minutes. Students take four classes each day and earn eight Carnegie units per school year. If classes are on a semester basis, students can take up to 16 semester classes.

This system has several advantages over a traditional schedule. Students are in fewer classes, so they can be more focused in their thinking and studying. Block scheduling allows more time for laboratories, in-class field trips, and projects. There is the potential for greater efficiency; half as much time needs to be spent on housekeeping duties at the beginning and ending of classes since there are half as many class periods. Pedagogically, one class period can be broken into three to five instructional variations. Potentially, this allows the teacher to demonstrate a concept, break it into its component parts, have students do a laboratory experiment, and debrief all in the same class period. Teachers have to prepare for fewer classes, as they will teach three or four class periods per day rather than five to seven. With four classes per day instead of seven, the chances for integration and articulation occurring are greater.

Block scheduling also has some disadvantages. The total amount of time for instructional purposes may be less than in a traditional scheduling system if the class meeting time is not doubled. Although teachers have fewer classes to prepare for, they must be better prepared for each class since it meets for a longer time. Students who miss a class period because of illness, a field trip, or some other situation miss the equivalent of two instructional periods and can fall behind very quickly. Finally, teachers who do not modify their instructional techniques to fit the longer time span lose student attention due to boredom, must deal with greater misbehavior problems, and experience reduced student learning.

There are two common forms of block scheduling: 4 × 4 and 4 × 2. A 4 × 4 block schedule has four class periods each day, with all classes meeting every day. An instructional year is completed in one semester. A 4 × 2 block schedule, often called an A/B schedule, has four class periods each day, but a different set of classes is offered on alternate days. Classes meet the entire instructional year. A modified block scheduling system combines block scheduling with traditional scheduling. For example, Monday and Wednesday may be A schedule, Tuesday and Thursday may be B schedule, and Friday all eight classes meet.

Trimester Scheduling

Trimester scheduling is another alternative class scheduling system that divides the school year into three 12-week trimesters. Each class period may be 70 to 80 minutes in length. Students take five classes each day during the trimester. Classes taken for two trimesters equal one Carnegie unit each, so students can earn seven to seven- and one-half Carnegie units per school year. Classes taken for one trimester are the equivalent of one semester, so students can take up to 15 semester classes. A trimester scheduling system is designed to include the advantages of a block system without some of the disadvantages.

Alternative Scheduling Systems and Agricultural Education

With its rich history of hands-on learning, agricultural education has adapted well to alternative scheduling systems. Concepts can be taught at the beginning of a class period and then put into action in the same class period through experiments or laboratory projects. Agriculture teachers who once had to break up laboratories and projects into multiple segments and days can now do them in one class period. Agricultural land laboratories and nearby facilities can be utilized more easily, more readily, and more often. In addition, some agricultural education programs have experienced increased enrollments, because students have 15 or 16 semester slots to fill, as opposed to 12 to 14 under a traditional scheduling system.

Alternative scheduling systems have presented more challenges for SAE and FFA than for classroom instruction. It is possible under a block 4 × 4 system for an FFA member to have an agricultural education class the first semester of the sophomore year and not have one again until the second semester of the junior year. In a trimester system, it is likely that students will have an agricultural education class only two of the three trimesters. This causes problems of continuity and contact with students. Students may be less likely to participate in FFA and SAE if they are not in an agricultural education class at the time. Agriculture teachers may teach twice as many students under an alternative scheduling system. This puts strains on a teacher's ability to make SAE visits and to involve all FFA members in FFA activities.

Research has shown many of these difficulties can be overcome with planning and ingenuity. Moore, Kirby, and Becton (1997) recommended better and more frequent communications regarding SAE and FFA be used to help keep students engaged and participating. Other suggestions include having two sets of officers (one in the fall and another in the spring), continually collecting FFA dues, and repeating Career Development Events in both the fall and the spring.

ACCOUNTABILITY

Increasing pressure is felt from parents, the private sector, and government agencies for schools, programs, and teachers to be accountable for student learning. *Accountability* means the school, program, or teacher is held answerable or responsible for student learning and achievement. Most people agree the primary role of schools is education, so holding schools accountable for this makes sense. However, what are not agreed upon are how to measure student learning and achievement and what the standard for accountability should be. In many states, standardized tests in the areas of English/language arts, mathematics, science, and social studies (core academic areas) have become the instruments for measuring accountability.

The No Child Left Behind Act of 2001 had accountability as the centerpiece of the legislation. A significant accountability measure in the law was mandatory academic testing of all students. Schools were accountable for closing the achievement gaps between students of different racial, ethnic, and economic groups. Schools had to demonstrate they were staffed by teachers qualified to teach in their subject areas. In 2015, the Every Student Succeeds Act (ESSA) undid some of the requirements of No Child Left Behind. ESSA still requires statewide assessments in science, language arts, and math, but instead of accountability being controlled at the federal level, schools are accountable to statewide assessments.

Under Perkins V, states are required to annually report to the federal government on a variety of performance indicators at the secondary level, including four-year graduation rate; academic proficiency in reading/language arts, math, and science; postprogram placement; attainment of postsecondary credentials or credits; participation in work-based learning, and many more. Of particular note to agricultural education programs are tracking postsecondary placement and earned postsecondary credentials. It will be important for agricultural educators to develop a system for tracking this information to share with their CTE director as part of accountability measures.

Accountability is the job of all teachers, not just English, math, science, and social studies teachers. Agricultural education teachers are also accountable for student learning in the four core academic areas. This means agriculture teachers may have to give more writing assignments and more closely grade those assignments for grammar, spelling, punctuation, and sentence structure. They may have to provide a certain amount of time per week for silent reading, either in literature or agricultural publications. The use of math concepts in agriculture may become a more explicit part of lessons. Scientific concepts and principles will need to be explained more thoroughly, and experiments debriefed rather than just conducted for the "gee whiz" effect. When teaching about international trade, the agriculture teacher may need to bring in the social studies concepts of government and politics. These changes should not only result in improved learning in the core academic areas but should enhance learning in agricultural education as well.

PREPARED CURRICULUM

Teachers typically use a combination of prepared curriculum (see Figure 7.8) and teacher-created curriculum. *Prepared curriculum* materials are instructional materials created by sources other than the teacher. They can include curriculum maps, lesson plans, and instructional materials needed to implement the lessons. They might be developed to be delivered as a complete curriculum, or there may be curriculum materials that are designed to be used à la carte, where teachers can pick and choose which lesson materials to use as they see fit.

FIGURE 7.8 A variety of prepared curriculums are available to agriculture teachers.

Prepared curriculum materials can be obtained in a variety of ways. Teachers traveling to the National FFA Convention and Expo or to the National ACTE Conference often pick up sample curriculum materials distributed by vendors. For example, the U.S. Poultry & Egg Association has provided curriculum kits to agricultural educators at their booth at the National FFA Convention. Sometimes state FFA associations or agricultural education agencies maintain websites with curricular materials. Other times, teachers can purchase prepared curriculum. Some companies, such as iCEV, AgEdNet, One Less Thing, and MyCAERT specialize in the preparation and sale of prepared agricultural education curriculum. Curriculum materials can also be accessed through postsecondary institutions, Extension, or the USDA. More details about curricular resources can be found in Chapter 10.

Curricular materials may be distributed to teachers through CDs, flash drives, or more commonly cloud-based or web-based distribution. To gain access to these materials, teachers may have to create an account with the agency distributing the materials, while other times they are downloadable with no registration requirement. Typically, curriculum materials that are more comprehensive, such as curriculum for an entire course, are only available through the combination of purchasing the curriculum and participating in professional development sessions to learn how to best use the curriculum. Perhaps the most prevalent example of this type of curriculum resource in agricultural education is CASE curriculum.

In 2007, the National Council for Agricultural Education launched the Curriculum for Agricultural Science Education (CASE) initiative as a means to increase the rigor and relevance of agricultural education curriculum through inquiry-based learning (Bird & Rice, 2021). This national curriculum is comprised of a sequence of courses designed to build upon each other through spiraling and scaffolding content knowledge and technical skills.

While CASE is not a required curriculum, it is a resource that thousands of agriculture teachers across the United States utilize. CASE includes courses in animal science, plant science, agricultural engineering, and natural resources, which all build off a foundational Introduction to AFNR course. Each course is designed with units and lessons that contain activities, projects, and problems, along with support materials. Each lesson is connected to FFA and SAE. To obtain and use CASE curriculum, teachers must participate in an intensive training on the course curriculum in which teachers experienced in using CASE curriculum serve as lead teachers. More information on CASE can be found on its website, www.case4learning.org.

INTEGRATING ACADEMICS AND AGRICULTURAL EDUCATION

Earlier sections of this chapter have discussed the role standards and accountability play in education today. Also discussed was how articulation and integration can help students learn more effectively. This section discusses levels of integrating academics and career and technical education (Finch & Crunkilton, 1999).

The first level of integration involves putting more academic content into career and technical education courses. Academic courses may include agricultural topics in examples and problems. With this level of integration, academic and agricultural education teachers may or may not be discussing curricular issues with each other.

The next level of integration involves formally bringing academic and career and technical education teachers together to devise ways to enhance the curriculum. Strategies are devised to infuse academics into agricultural education and agriculture into academic subjects.

The third level of integration involves aligning the two curriculums. Teachers meet regularly and plan content and instructional activities to supplement what is happening in the other area. Both curriculums may be modified to better facilitate integration.

The highest level of integration eliminates the hard distinction between academic and career and technical education. Content may be organized around career clusters. In this model, the content of each subject is taught using the career cluster as a theme and motivating factor. Academies and magnet schools demonstrate this level of integration.

Since so many of the topics in agricultural education can be easily connected to other subjects and applications beyond the classroom, integration of other subject area knowledge and skills can be accomplished. One area that has been of increasing interest to address within agricultural education includes the STEM fields (science, technology, engineering, and mathematics). The combination of the demand for STEM-related jobs, the Next Generation Science Standards' emphasis on integrated learning, and increased funding available to support STEM education has led to increased interest in integrating STEM through agricultural education.

While many teachers see the value of STEM integration into their classes, it can also be challenging. Intentional cross-disciplinary collaborations between agriculture teachers and science teachers can be hindered by scheduling and bureaucratic barriers in schools. Clarity on the definition of integration is also needed. To address this lack of clarity, Wang and

Knobloch (2018) developed a rubric to assess the levels of STEM integration in AFNR lessons and units. This tool can be helpful for agriculture teachers looking to evaluate the presence of STEM concepts and methods in their instructional plans to promote student skill and knowledge development. Figure 7.9 (p. 146) shows this rubric. Research continues to evaluate and identify how STEM integration in agricultural education can be more effective for student learning, and how to better prepare agriculture teachers to integrate STEM knowledge and skills into their lessons.

REVIEWING SUMMARY

The agricultural education curriculum includes courses, content within courses, SAE, and FFA activities. The curriculum may be specified by a state agency, be totally under local control, or be at some point on a continuum between the two. Regardless, curriculum development should occur with input from the school and the local community. The local agricultural education advisory committee should be involved in curriculum decisions.

Various curriculum development models exist to guide the process of designing curriculum that meets the needs of students and the community. Backwards design emphasizes beginning with establishing learning goals as a first step, followed by designing assessments, and then instructional activities. Differentiated instruction and UDL are useful tools to serve as a guiding framework when designing instructional activities to center the diverse needs of students and eliminate barriers to learning.

As with other content areas, agricultural education is bound by standards. National AFNR standards exist to guide implementation of school-based agricultural education at the state and local level. These national standards are cross walked to other national standards including Common Core English and Math, Next Generation Science Standards, Green/Sustainability Knowledge and Skill Statements, and the National Standards for Financial Literacy. Since much of public education is under the purview of state government, each state or territory might have their own set of state educational standards for career and technical education in agriculture. Teachers and agricultural education programs are accountable for students achieving to required levels on standards.

Curriculum development is influenced by several factors. Educational standards, federal legislation, community needs, and philosophical foundations all influence the curriculum content and the ways in which students learn and are assessed. As a career and technical education subject (CTE), agricultural education is guided by the structures outlined in the most recent Perkins V legislation. As such, secondary instruction in agriculture needs to be aligned with postsecondary credit earning opportunities as well as industry credentialing experiences.

Agricultural education teachers are responsible for determining the scope and sequence of their courses. Teachers should follow curriculum planning principles when developing curriculum models and teaching schedules. Although teaching schedules should be detailed, they should leave room for flexibility and student interests.

Schools have a variety of class scheduling systems. The three most popular scheduling systems are traditional, block, and trimester. Each system has advantages and disadvantages for students and teachers. Changing from one class scheduling system to another typically requires the agriculture teacher to revise curriculum plans and evaluate teaching strategies.

Teachers utilize a variety of resources when preparing curriculum. At times they may be provided prepared curriculum and cannot deviate far from the prescribed instructional

	Levels of Integration		
	Exploring STEM Integration	Developing STEM Integration	Advancing STEM Integration
Role of Integration in Learning Objectives	Learning objectives create awareness of STEM connections.	Learning objectives develop STEM learning content/skills.	Learning objectives apply STEM knowledge to solve problems.
Role of the STEM Concepts, Content, Knowledge, and Skills Presence Usage	Core disciplinary STEM concepts and skills are mentioned to point out the connections in different disciplines or one of the STEM disciplines is predominantly present. No strong evidence of using STEM content knowledge to solve problems. It is activity-driven. For example, the activity focuses on practicing engineering design process or problem solving, but no explicitly stated STEM content knowledge is needed to solve the problem.	Core disciplinary STEM concepts and skills are taught and/or practiced to bridge different disciplines or multiple STEM disciplines are distinctly present. Use of STEM content knowledge are explicitly taught to solve the problem. Content knowledge is fixed, students do not go beyond the knowledge as it exists in its disciplines.	Core disciplinary STEM concepts and skills are considered as prior knowledge, and are naturally and meaningfully used/applied to solve problems or multiple STEM disciplines are difficult to distinguish as separate disciplines because they are closely interdependent. Use of STEM content knowledge is used to analyze and interpret the problem. Content knowledge is integrated, synthesized, or transformed into some kind of tools or solutions that can be transferred beyond the knowledge used to solve the problem.
Role of Learning Outcomes	Learning outcomes merely focus on one discipline (one concept, and/or one skill).	Learning outcomes mainly focus on one discipline (one concept, and/or skill), but other disciplines are used to support the understanding of the core learning outcomes.	Learning outcomes focus on interdisciplinary concepts and skills that are woven throughout when solving problems.
Role of the Instructor and Type of Instruction	The instructor merely gives directions or guidelines. Students follow "cookbook" type of instruction to complete the task.	The instructor mainly gives directions or guidelines, but students have some freedom to determine the direction to complete the task in a controlled environment.	The instructor is a facilitator and provides enough directions/guidelines to engage students to solve a problem. Students determine the direction of the task that needs to be completed.
Role of AFNR Content Knowledge	AFNR content is the primary focus of the lesson.	AFNR provides a context for STEM learning or experiential learning process.	AFNR serves as an integrator of STEM learning by focusing on a real-world problem that blends disciplines.
Role of Students' Thinking	Thinking merely stays inside of the box (the discipline, or the concepts/skills that need to be learned), but may see outside of the box upon completion of the problem-solving process.	Thinking is mainly inside of the box, but occasionally steps outside the box to draw connections from other disciplines to solve the problem.	Thinking is predominantly outside of the box with few to no boundaries that limit thinking. Students demonstrate systems thinking, critical thinking, creative thinking, and/or complex problem-solving.

FIGURE 7.9 Rubric of levels of STEM integration and features.

(ADAPTED BY PERMISSION FROM WANG & KNOBLOCH, 2018.)

content and activities. Often, teachers pull ideas from available prepared curriculums to integrate into their own instruction that meets the needs of their students. One prepared curriculum popular with thousands of agricultural educators across the United States is the Curriculum for Agricultural Science Education (CASE), which requires an intensive period of training for teachers to use their copyrighted curriculum.

The integration of academics and career and technical education is important in meeting standards and being accountable for students' educational experiences. Schools may desire to integrate at various levels, ranging from the infusion of some content to the development of highly integrated curriculums involving career clusters, academies, or magnet schools. Agricultural education increasingly has focused on how to effectively integrate STEM knowledge and skills into AFNR instruction.

QUESTIONS FOR REVIEW AND DISCUSSION

1. What is the curriculum for a local agricultural education program?
2. What are the sources of input for curriculum development in agricultural education?
3. How does curriculum influence students?
4. How does the philosophical lens of an educator designing curriculum influence the curriculum content, assessments, and learning activities?
5. How does the structure of Perkins V influence the programs of study and courses offered in a local agricultural education program?
6. Why should agricultural education be articulated with other subjects?
7. How does your state organize standards for agricultural education?
8. What purpose do standards serve in agricultural education?
9. What curriculum resources are available in your state?
10. What are the advantages and disadvantages of each scheduling system for agricultural education?
11. Why are scope and sequence important in student learning?
12. What are the levels of integration between academic and career and technical education?

ACTIVITIES

1. Investigate the degree of integration of academics and career and technical education at a selected school. You may want to interview the principal, the agriculture teacher, and one or more academic teachers.
2. Develop a scope and sequence for an agricultural education course. Remember to include time for testing, special days, and holidays. Provide a strong rationale for your scope and sequence.
3. Using the *Journal of Agricultural Education*, *The Agricultural Education Magazine*, and the proceedings of the National Conference of the American Association for Agricultural Education, write a report on block scheduling research in agricultural education. Current and past issues of these journals can be found in the university library. Selected past issues can be found online at www.aaaeonline.org or www.naae.org.
4. Research the foundational literature regarding curriculum and instruction. Make sure to include a review of Ralph W. Tyler's work in comparison to Grant Wiggins and Jay McTighe's work on backwards design.

5. Collaborate with a science teacher to develop a lesson that integrates STEM skills and knowledge within an agriculture class. Utilize the rubric developed by Wang and Knobloch (2018) to evaluate the degree to which you have been able to integrate STEM into your lesson.

REFERENCES

Bird, T. D., & Rice, A. H. (2021). The influence of CASE on agriculture teachers' use of inquiry-based methods. *Journal of Agricultural Education, 62*(1), 260–275. https://doi.org/10.5032/jae.2021.01260

Brownell, C. J. (2017). Starting where you are, revisiting what you know: A letter to a first-year teacher addressing the hidden curriculum. *Journal of Curriculum and Pedagogy, 14*(3), 205–217. https://doi.org/10.1080/15505170.2017.1398697

CASE. (2021). *Pathways and courses. Curriculum for Agricultural Science Education.* https://www.case4learning.org/curriculum/pathways-and-courses/

CAST. (2018). *Universal design for learning guidelines version 2.2.* http://udlguidelines.cast.org

Common Core State Standards Initiative. (2021). *Standards in your state.* http://www.corestandards.org/standards-in-your-state/

Finch, C. R., & Crunkilton, J. R. (1999). *Curriculum development in vocational and technical education: Planning, content, and implementation* (5th ed.). Allyn and Bacon.

Jackson, P. W. (1968). *Life in classrooms.* Teachers College Press.

Lee, J. S. (2000). *Program planning guide for agriscience and technology education* (2nd ed.). Pearson Prentice Hall Interstate.

Lee, N. (2021, August 5). *States are implementing new educational standards, signaling the end of Common Core.* CNBC. https://www.cnbc.com/2021/08/05/states-are-implementing-new-educational-standards-signaling-the-end-of-common-core.html?__source=sharebar|email&par=sharebar

Moore, G., Kirby, B., & Becton, L. K. (1997). Block scheduling's impact on instruction, FFA, and SAE in agricultural education. *Journal of Agricultural Education, 38*(4), 1–10.

National Association of Landscape Professionals. (2020). *Landscape industry certified.* National Association of Landscape Professionals. https://www.landscapeprofessionals.org/LP/Certification/Certification-NALP.aspx

National Council for Agricultural Education. (2016). *National quality program standards for agriculture, food, and natural resources education.* https://thecouncil.ffa.org/

National FFA Organization. (2012). *Local program success* (online). Retrieved November 5, 2012, from http://www.ffa.org

Pentony, D., & Hall, D. (2008). *ANFR national content standards: Project Roundtable Sessions* [PowerPoint presentation]. Retrieved June 3, 2008, from http://www.ffa.org

Sousa, D. A., & Tomlinson, C. A. (2011). *Differentiation and the brain: How neuroscience supports the learner-friendly classroom.* Solution Tree Press.

Tomlinson, C. A., & Strickland, C. A. (2005). *Differentiation in practice: A resource guide for differentiating curriculum, grades 9–12.* Association for Supervision and Curriculum Development.

U.S. Department of Education Office of Career, Technical, and Adult Education. (2021c, June 30). *CTE programs of study.* Perkins Collaborative Resource Network. https://cte.ed.gov/initiatives/octaes-programs-of-study-design-framework

Wang, H. H., & Knobloch, N. A. (2018). Levels of STEM integration through agriculture, food, and natural resources. *Journal of Agricultural Education, 59*(3), 258–277. https://doi.org/10.5032/jae.2018.03258

Wiggins, G., & McTighe, J. (2005). *Understanding by design* (Expanded 2nd ed.). Pearson Education.

ADDITIONAL SOURCES

Advance CTE : State Leaders Connecting Learning to Work. (2021). *Career clusters*. https://career tech.org/career-clusters

Canales, J., Frey, J., Walker, C., Walker, S. F., Weiss, S., & West, A. (Eds.). (2002). *No state left behind: The challenges and opportunities of ESEA 2001*. Education Commission of the States. (ERIC Document Reproduction No. ED468096).

Cushing, E., English, D., Therriault, S., & Lavinson, R. (2019, March). *Developing a college- and career-ready workforce: An analysis of ESSA, Perkins V, IDEA, and WIOA*. College & Career Readiness & Success Center at American Institutes for Research. https://ccrscenter.org/implementa tion-tools/developing-college-and-career-ready

Lynn, R. L., Baker, E. L., & Betabenner, D. W. (2002). *Accountability systems: Implications of requirements of the No Child Left Behind Act of 2001* (CSE Technical Report 567). Los Angeles: Center for the Study of Evaluation, UCLA. (ERIC Document Reproduction No. ED467440).

Novak, K., & Tucker, C. R. (2021). *UDL and blended learning: Thriving in flexible learning landscapes*. Impress.

Ornstein, A. C. (1990). Philosophy as a basis for curriculum decisions. *High School Journal, 74*(2), 102–109. https://www.jstor.org/stable/40364829

U.S. Department of Education Office of Career, Technical, and Adult Education. (2021a, June 30). *College- and career-ready standards*. U.S. Department of Education. https://www.ed.gov /k-12reforms/standards

U.S. Department of Education Office of Career, Technical, and Adult Education. (2021b, June 30). *Non-regulatory guidance for accountability*. U.S. Department of Education. https://cte.ed.gov /accountability/nonregulatory-guidance-for-accountability

8

Student Enrollment and Advisement

OBJECTIVES

This chapter addresses the National Quality Program Standards for Agriculture, Food, and Natural Resources Education (National Council for Agricultural Education, 2016), specifically Standard 5: Marketing. It has the following objectives:

1. Explain the meaning and importance of student enrollment.
2. Discuss student recruitment and identify useful strategies.
3. Discuss student retention and identify useful strategies.
4. Describe the role of school counselors and school administrators in student enrollment.
5. Discuss barriers to recruitment and retention.
6. Describe the meaning and importance of student advisement.
7. Describe the role of career guidance.
8. Describe the role of personal guidance.

Isabella Francisco is the agriculture teacher at Etowa High School. She is a popular teacher and always has good enrollment in her classes. Once students take one of her classes, they return for others. She carefully plans what she is to teach and has an abundance of instructional resources. Most years her classes rapidly reach peak enrollment, and students are turned away. Last year Ms. Francisco was named teacher of the year in the school district.

Elizabeth Lee is the agriculture teacher at a high school near Etowa. Her classes are small, and she has to struggle each year to have enough students to offer the classes. Those who take one class usually fail to return for a second class. The agriculture curriculum and FFA program of activities have not been updated for many years. At a recent board of education meeting, the superintendent mentioned money for teacher salaries could be saved by eliminating the agriculture classes.

What makes the difference between these two programs? If you were advising Ms. Lee, what would you say?

TERMS

career cluster	promotional material
career guidance	recruitment
enrollment barrier	retention
feeder school	student advisement
free elective	student enrollment
guidance	

FIGURE 8.1 A school counselor discusses class scheduling with students. (COURTESY OF EDUCATION IMAGES.)

STUDENT ENROLLMENT

Secondary agricultural education courses are elective courses. An elective course is a course that is not specifically required for high school graduation. In some states, agricultural education courses count as free electives. A *free elective* is a course that does not fulfill any requirements. In these instances, agricultural education competes with all other electives for students. In other states, agricultural education courses may fulfill an open requirement or restricted elective. For example, three science courses may be required for graduation, with biology and chemistry specified to be taken. The third course can be selected from a list of choices that include agriscience courses. Another example is a sequence of agricultural education courses used to fulfill a career pathway requirement.

Student enrollment is the total number of students in the agricultural education program. This is important, as it justifies the number of agriculture teachers employed as well as gives an indication of the impact the agricultural education program has within the school. How do students enroll in agricultural education? What factors influence students' decisions about whether to enroll? Why do students reenroll during subsequent years? These questions will be addressed in the following sections.

Because of its elective status, students must be recruited into agricultural education. *Recruitment* is the process of seeking and soliciting students to enroll in agricultural education courses. There are six key variables in the successful recruitment of students into an agricultural education program. As shown in Figure 8.2, the agriculture program, the recruitment program, student characteristics, parents, school support, and community support all influence whether a student enrolls in an agricultural education course (Myers et al., 2003).

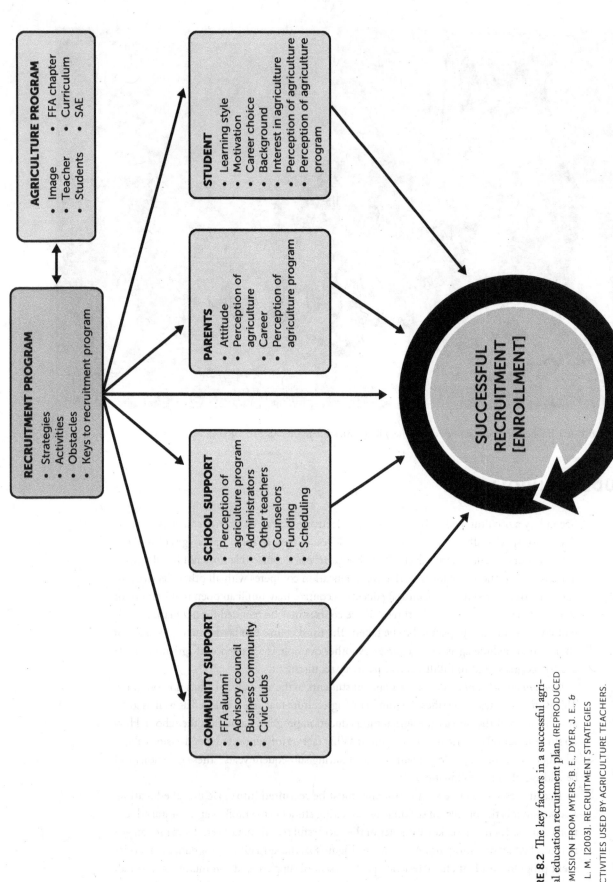

FIGURE 8.2 The key factors in a successful agricultural education recruitment plan. (REPRODUCED BY PERMISSION FROM MYERS, B. E., DYER, J. E., & BREJA, L. M. [2003]. RECRUITMENT STRATEGIES AND ACTIVITIES USED BY AGRICULTURE TEACHERS. *JOURNAL OF AGRICULTURAL EDUCATION, 44*[4], 97. HTTPS://DOI.ORG/10.5032/JAE.2003.04094.)

STRATEGIES FOR RECRUITING AND ENROLLING STUDENTS

Student enrollment in agricultural education comes from three sources: (1) students who enroll as a result of the recruitment program, (2) students who as a result of retention reenroll in agricultural education, and (3) students who enroll for reasons outside of their control, also known as disavowance reasons. The first two sources will be discussed separately. Disavowance reasons such as "there were no other courses to take," "the counselor put me in the course," or "I thought this course would be easy" are typically not considered when developing recruitment and retention strategies.

Hoover and Scanlon (1991), using literature (Quey, 1971) on group dynamics, stated students take a class such as agricultural education for three reasons:

1. Students take an agricultural education class if they like the curriculum or activities in the class.
2. Students enroll in agricultural education if they like the people, such as the teacher and their peers, in the class.
3. Students enroll in agricultural education if they are able to satisfy a need, such as competition or career preparation, through the class.

Students who have knowledge, observations, and information relating to agricultural education courses may have a greater tendency to enroll in those courses (Fishbein & Ajzen, 1975). Several strategies consider this and are effective for student recruitment. These include feeder school contact, agriculture teacher–student contact, FFA chapter events, publications and other promotional materials, curriculum, parents and other support groups, and recruitment events (Myers et al., 2003). For successful student recruitment, it is important to reach out to students and provide support to them, parents, and the community (Anderson, 2013). These strategies will be discussed next.

Feeder Schools

A *feeder school* is a middle or junior high school whose students will attend the high school where the agricultural education program is located. A strong recruitment tool is to have an agricultural education program in the middle school. (Middle school agricultural education is further discussed in Chapter 16.)

Regardless of whether agricultural education is taught in the middle school, there are other ways the agriculture teacher should connect with students in grades 5 through 8. Whenever possible, agriculture students/FFA members should conduct activities in the middle school. These may include demonstrations in science classes, presentations to student organizations, or the hosting of a middle school science fair. Middle school FFA programs, with developmentally appropriate activities, can get students excited about agricultural education and prompt them to enroll in high school agricultural education.

Another way to reach students in this age group is through involvement with the 4-H program. This involvement may be as a club leader, in joint activities, or through sponsorship of 4-H activities.

Contact

Another recruitment strategy is personal contact with prospective students and visibility to them. Some agriculture teachers make home visits to eighth graders who are prospective agriculture students. They may also use telephone calls, cards, or face-to-face meetings as recruitment tools.

FIGURE 8.3 FFA activities and events can be effective recruitment tools.

The agriculture teacher should be visible throughout the school. Important for establishing contact with prospective students is attendance at sporting events, class activities (such as dances), and academic competitions (such as science fairs), along with daily visibility in the hallways and cafeteria.

Although not a substitute for personal contact, well-constructed agricultural education program websites and social media sites are beneficial. Social media sites are useful for information sharing, community building, and asking for user action (Myer et al., 2017). Students who are exploring classes to take may use these sites during this process. Parents and other influential adults may use social media sites to better understand a program and its potential benefits for their student. While final responsibility rests with the agriculture teacher, it may be helpful for a high school student to develop and maintain the sites, possibly as their SAE. The school's use policy for websites and social media sites must be followed.

FFA

FFA can be a tremendous recruitment tool. The activities associated with FFA are highly visible and can fulfill student needs for achievement, recognition, and self-improvement.

Some of the most visible are Career Development Events, Leadership Development Events and activities, and out-of-school opportunities, such as trips and scholarships. A caution is that whatever the image students have of FFA, it is typically transferred to agricultural education classes. For example, if students perceive FFA to be for rural students only, then the agriculture teacher will have a difficult time recruiting into agricultural education students who do not fit this image. To alleviate this problem, the agriculture teacher should strive to involve all students in FFA and have an FFA program that appeals to a diversity of student backgrounds. Additionally, the agriculture teacher needs to develop relationships with other teachers, administrators, parents, and the overall

FIGURE 8.4 Student enrollment should reflect the diversity of the school. (COURTESY OF LEDYARD AGRI-SCIENCE & TECHNOLOGY PROGRAM.)

community as these groups have influence on nontraditional student enrollment and retention (Anderson, 2013).

Publications and Other Promotional Materials

Publications and other promotional materials are what most people think of when they hear about recruitment. *Promotional material* is a product or approach used to gain desired responses from people. It is somewhat like advertising.

Publications and promotional materials include brochures, bulletin boards, newspaper articles and ads, radio and television spots, posters, web pages, social media sites, videos, school announcements, roadside signs, and school display cases. School-appropriate social media filters are a fun way to promote the agricultural education program. Promotional materials also include giveaways, such as calendars, USB flash drives, and other items with the agricultural education program name on them. The success of these is difficult to gauge. They definitely raise awareness of the program name and provide valuable information regarding the agricultural education program. In fact, other than by word of mouth, promotional materials are how most people become aware of something. However, these are rarely cited as the number one reason for a student taking an agricultural education class.

Curriculum

Ultimately students and parents must be satisfied with the curriculum for enrollment to take place. What in the agricultural education curriculum adds value for the student?

The agricultural education curriculum must be current and address local community priorities. Ideally it should be career-focused, even if it leads to additional education at the postsecondary level. It is a strong selling point if agricultural education courses count as science, communication, or economics credit. A student-centered instructional approach with student choice of assignments/projects is important in students deciding to enroll in agricultural education courses (Anderson, 2013).

For recruitment of the broadest range of students, at least part of the curriculum must have an agricultural literacy and career awareness focus that appeals to students without an agricultural production background. This includes broadening FFA activities and opportunities in SAE (Myers et al., 2004). As discussed in Chapter 22, the five components of foundational SAE are for all students enrolled in agricultural education.

Parents and Others

Parents and other support groups can serve as voices in the community for agricultural education recruitment. For this reason, they should be involved in the agricultural education program as advisory committee members, Young Farmers or FFA Alumni and Supporters members, chaperones for trips, and possible classroom resource persons. It is critical for the agricultural education program to be inclusive and involve adults from diverse backgrounds in the program.

Events

Recruitment events include open houses, agricultural awareness days, National FFA Week activities, orientation nights, and eighth-grade assemblies. These are usually large events and must be planned well in advance of the dates of occurrence. They serve an awareness function similar to publications but hold the potential for personal contact that publications cannot give. Several questions must be answered when planning a recruitment event. Who is the audience? What needs does the audience have? What is the message the audience should get from the event? Who will be the speakers? What activities will be conducted? How will success be gauged? Typically, the agriculture teacher does not plan a recruitment event alone. Depending on the activity, FFA chapter committee(s), the advisory committee, and/or volunteers/supporters should assist in planning and implementing the event. School counselors and administrators need to approve the event and provide their support for it.

RETENTION

Retention is gaining repeated enrollment of a student in agricultural education classes. Several practices support students reenrolling in the agricultural education program. Many of the practices used in recruitment aid in student retention. Once students are in agricultural education courses, they need to see themselves fitting into the image of agricultural education and FFA at their school. Because of this, agriculture teachers need to be constantly aware of the image portrayed and compare that against what they want to be reality. What is the focus of the curriculum? Is it current, and does it meet community/student needs? Is the agricultural education program/FFA welcoming, inclusive, and open to a diversity of student backgrounds? Are students of various backgrounds accepted and encouraged to maintain their uniqueness? Does the teacher work to reduce cliques and exclusionary practices?

Just as with recruitment, personal contact and word of mouth are very effective for retention. Students want to know that they belong and that the agriculture teacher cares about them. With FFA and SAE, the agriculture teacher has two vehicles to get to know students better and to affect them on a personal level.

Retention is important for programmatic and student reasons. Obviously, from a programmatic perspective, student retention means increased enrollments. It also means reenrolled students can serve as ambassadors for the program and recruit new students. Finally, retention makes program planning easier because the agriculture teacher knows there will be a cohort of students who need specific agriculture courses.

Retention is also good from a student perspective. Supervised agricultural experience is designed to build from one year to the next. Students who continue in agricultural education courses expand their SAE and gain valuable career skills. Continuers also have the benefit of the personal growth, leadership development, and career success activities available to them through involvement in FFA. In addition to this, they have the opportunity for scholarships, awards, recognition, travel, and other benefits of FFA. Finally, students who continue in agricultural education courses increase their agricultural literacy and build their skills and knowledge for agricultural careers.

ROLE OF SCHOOL COUNSELORS AND ADMINISTRATION

School counselors and school administrators are critical for student enrollment in agricultural education. These are the people who meet with students to register them for classes, so they have a great influence on what classes students take. Counselors and administrators who are knowledgeable about the value of the agricultural education program are more inclined to advise students to enroll in its courses. They are also cognizant of the benefits of agricultural education for all students, including those who are college-bound, those with career interests in agriculture-related careers outside of production, and even those without specific career interests in agriculture.

Counselors and administrators are typically involved in designing the master course schedule. This schedule charts where each course in the school will be taught during the school day and how many sections of a course will be offered. Where an agricultural education course is placed within the daily schedule may make the difference between students taking the course and not being able to take it. This is especially true in a small-enrollment school where section offerings of required courses may be limited. An agricultural education course placed at the same time as the only section of a required course for a particular grade level eliminates a group of students from taking agriculture.

FIGURE 8.5 School counselors can be valuable allies in recruiting students into the agricultural education program.

BARRIERS TO RECRUITMENT AND RETENTION

Agricultural education researchers have identified several barriers to successful recruitment and retention of students. An ***enrollment barrier*** is a perception or other element in the school or community environment that promotes nonenrollment in agricultural education. Some barriers may be beyond the immediate control of the agriculture teacher. Phelps, Henry, and Bird (2012) identified negative perception, apathy, and scheduling conflicts for why students would not want to become involved with FFA and by association not take an agriculture course.

Breja, Ball, and Dyer (2000) also identified schedule-related problems as some of the greatest barriers to student enrollment. College-bound students who have schedule conflicts between courses required for college admission and agricultural education courses will not be able to take the agriculture courses. If increased graduation requirements reduce the number of available slots for elective courses, then enrollment in agricultural education suffers. Finally, if school counselors and school administrators have negative perceptions of the agricultural education program, they will be less likely to encourage students to enroll.

Student Perceptions of Image

Image can also be a barrier to enrollment. This includes both the image of agriculture and of agricultural education. Some students, particularly students from ethnic minority groups, may view agricultural careers as low-paying menial labor with no opportunities for them. Others may view agriculture as just production agriculture. It takes effort on the agriculture teacher's part to overcome both of these stereotypes.

Some view agricultural education as being only for rural, White students. This image of agricultural education is inaccurate. According to the National FFA Organization (2021), FFA membership is about 40% nonfarm/nonrural, 45% female, and 35% ethnic minority. Much work still needs to be done for agricultural education to be more inclusive and more closely reflect national student demographics.

Program Quality

Another barrier may be the quality of the agricultural education program and teacher. Eliminating this barrier is within the control of the agriculture teacher. Why is the program of poor quality? Does the curriculum need an update? Are the facilities inadequate? Does the program focus fit community and student needs? Why is the agriculture teacher not performing up to standards? Does the teacher need technical retraining? Is there a mismatch between teacher expertise and program emphasis? Is the teacher receiving appropriate parental and administrative support? Are there personal reasons for the teacher performing below standard? Does the teacher need intercultural competence awareness and training? These questions should be addressed thoughtfully.

Student Expectation Barriers

Other identified barriers include a mismatch between student interests and the focus of the agricultural education course and a mismatch between student career goals and agricultural education. These barriers may be difficult to overcome and may be legitimate reasons for students not to enroll in agricultural education courses.

Student Advisement

Agriculture teachers are in a unique position to offer their students meaningful advice. ***Student advisement*** is the process of offering assistance to students regarding their career and educational decisions or actions. As stated earlier in this chapter, students may take

FIGURE 8.6 An agriculture teacher should be trusted by students and provide guidance on career and certain personal issues as a part of instruction.

agricultural education courses for multiple years. This allows the agriculture teacher to know students better academically and personally. In the role of FFA advisor, the agriculture teacher gets to know students even more on academic, personal, and social levels. Finally, through supervised experience, the agriculture teacher meets students' parents or employers and develops a deeper understanding of the students' career goals.

Advisement is not telling students what they should do or making decisions for them. Rather, it is providing information, asking thought-provoking questions, and recommending possible solutions. However, before any of these, the foremost requirement is listening. Students want to be heard before they can hear. Sometimes students only want a sounding board and are not expecting the agriculture teacher to give advice.

Agriculture teachers are often asked to give guidance and counseling. *Guidance* is the process of providing assistance to help people make wise decisions regarding choices or changes (Phipps & Osborne, 1988). One technique within guidance is counseling. Counseling typically involves an individual, face-to-face meeting with the purpose of addressing a specific situation. A counselor is professionally trained and licensed to provide proper support to students. Depending on school size and location, a school may have one or more counselors, nurses, psychologists, psychiatrists, social workers, and law enforcement officials who are trained to provide various types of counseling.

Because of the relationships developed with students, an agriculture teacher provides guidance and acts as a counselor in many instances. This is appropriate as long as the teacher is acting as a listening and caring individual who knows when to refer a student to others. However, most agriculture teachers are not trained as counselors, so they should limit the scope of the counseling they provide. For example, students who are depressed or suicidal require the careful intervention that only a trained counselor can provide. Other issues that are best referred to trained counselors include drug or alcohol use, school violence, child abuse, sexual abuse/harassment, and bereavement.

Agriculture teachers provide guidance in two general areas. One is assisting students to choose a career path and plan the preparation necessary to achieve that goal. The other is assisting students in numerous personal issues common to almost all teenagers. Both of these are discussed in the following sections.

CAREER GUIDANCE

Career guidance is the process of helping people proceed through career development stages. It involves providing information regarding careers and helping students understand their interests, abilities, preferences, and values so they can make wise decisions regarding career planning. Career guidance traces its modern roots to Frank Parsons, who in 1909 wrote of the three factors influencing a person's wise choice of a vocation (Brown, Brooks, & Associates, 1990). In choosing vocations, people should (1) understand themselves and their abilities, interests, and circumstances, (2) have knowledge of requirements, advantages and disadvantages, and opportunities of various career paths, and (3) have sound reasoning regarding the relationship between these factors (Parsons, 1909).

Vocational Life Stages

Super et al. (1957) theorized that people go through vocational life stages. These stages are further divided into substages. Although Super assigned an age range to each stage, it is understood there are transitional phases between stages and people proceed through the stages at differing rates. The five stages are growth (birth–age 14), exploration (15–24), establishment (25–44), maintenance (45–64), and decline (age 65 and on). Middle school and high school students are in the first two stages.

The growth stage has three substages. During the fantasy substage (4–10), children are involved in role-playing. In the interest substage (11–12), children's likes and dislikes influence what they believe they will do as careers. At the capacity substage (13–14), children focus more on their abilities and the requirements of jobs/careers. It is important throughout the growth stage to provide opportunities for students to role-play, explore their interests, and discover job requirements.

FIGURE 8.7 Adolescents in the exploration life stage try out different possible careers. (COURTESY OF LEDYARD AGRI-SCIENCE & TECHNOLOGY PROGRAM.)

In the exploration stage, students examine themselves and try different roles. They explore jobs/careers through school activities, leisure activities, and part-time employment. The exploration stage also has three substages. The tentative substage (15–17) includes most high school students. Students examine their interests, abilities, values, and needs in light of their circumstances. They make tentative career decisions based on many factors and experiences. They try out these decisions through fantasy; discussions with peers, significant persons, and others; courses in school; clubs and activities; work; and other experiences. The transition substage (18–21) involves more consideration of reality. Students either enter the job market or pursue postsecondary education and training. In the trial substage (22–24), people typically decide on a field, obtain a beginning job, and try it out for their life's work.

Throughout the exploration stage, students continue to need chances to learn about and try out careers. They also need guidance in matching their interests and abilities with the requirements of the careers. When young adults choose careers that are either too demanding or not demanding enough for their abilities, this causes mismatches that result in frustrations. The agriculture teacher, through classroom and laboratory instruction, FFA, and SAE, can help students explore various careers and find ones that fit their interests and abilities.

Although the original research (Super et al., 1957) was conducted with only boys as subjects, the concept of vocational life stages has been applied to a diversity of groups (Vondracek et al., 1986). Continuing work in career development has expanded the focus to include individual, social, and societal aspects (Lent et al., 2002; Porfeli & Savickas, 2012; Savickas et al., 2009). Social cognitive career theory is widely used to understand how students explore and choose careers. Lent et al. proposed three aspects of career development: basic academic and career development, making educational and career choices, and obtaining academic and career success. Through agricultural education, students can develop a belief that they are capable of agriculture-related career behaviors, the career will be advantageous to them, and they can be successful in the chosen career. Additionally, agricultural education can contribute to students career readiness (Mouser et al., 2019).

Career Clusters

The number of careers available for students to enter is staggering. It is also important to note that careers are changing constantly. This means continual retraining and lifelong learning will be essential for students as they progress through their working lives. An effort to assist students in this task is the use of career clusters.

A *career cluster* is a broad grouping of occupations to help students decide on and prepare for further education and careers (NASDCTEC, 2012). Career clusters help organize curriculum so students learn academic and technical skills in an area of their interests to lead to success in a career (Advance CTE, 2021). All possible occupations are included in one of 16 career clusters (see Figure 8.8). For example, one of the clusters is Agriculture, Food, and Natural Resources. The career cluster concept is an organizational tool for student learning. Career clusters link classroom instruction with career-based learning. They also outline pathways for students to obtain additional education and training needed for the occupations within a cluster. Finally, the career cluster concept encourages collaboration and partnerships between schools, business and industry, and the community.

Career Guidance Tools

A school counselor can be an ally to the agriculture teacher in advising students regarding their career development. School counselors are trained professionals who can work with students in group or individual settings. They can also advise the agriculture teacher on specific resources and tools available to assist students. The agriculture teacher must keep

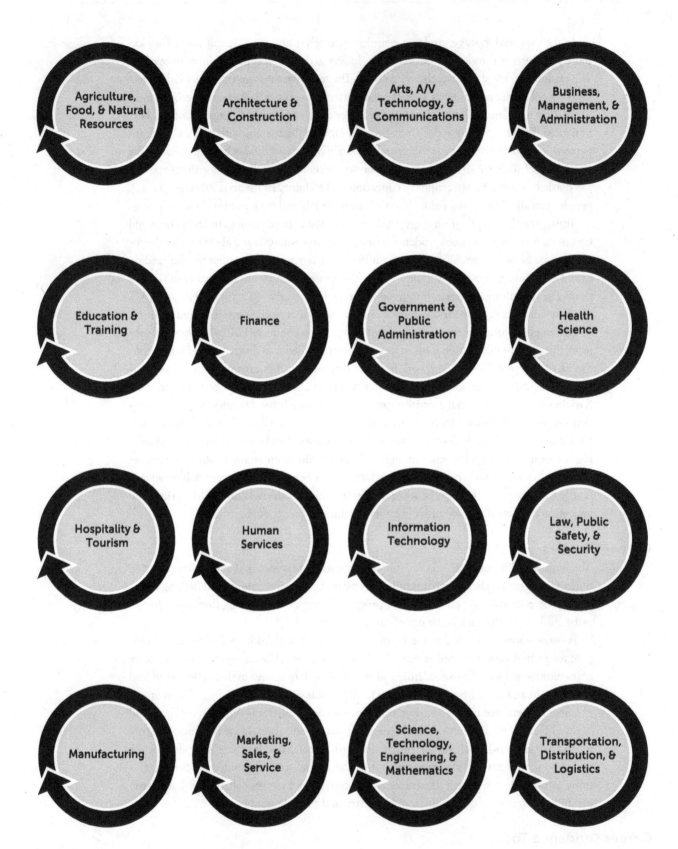

FIGURE 8.8 Career clusters organize all possible occupations into similar groupings.

the school counselors updated on the agricultural education curriculum, FFA and SAE activities, and career opportunities within the Agriculture, Food, and Natural Resources career cluster.

There are numerous career guidance tools available to school counselors and agriculture teachers. These include web-based and software programs, the *Occupational Outlook Handbook* maintained by the U.S. Bureau of Labor Statistics, aptitude tests, and interest surveys. The school counseling department will have many of these tools and can arrange for administering others. The main advantage of these tools is they allow students to narrow their focus down to those occupations that meet needs, interests, and abilities. Many of these tools also describe working conditions with pay ranges, training/education required, and employment outlook. Several tools specific to agricultural careers are available through universities, curriculum providers, and the National FFA Organization, such as AgExplorer (https://agexplorer.ffa.org/).

Career development interventions in secondary schools fall into four categories (Dykeman et al., 2001). Work-based interventions include apprenticeships, service-learning, job shadowing, and job placement. Advising interventions include academic planning and counseling, conferences, tutoring, career library, and various assessments and testing. Introductory interventions include career days and fairs, career field trips, and short-term lessons on development. Curriculum-based interventions include longer-term interventions, such as career information infused into the school curriculum, and various career-based models, such as magnet schools and career academies. Career and technical education courses, such as agricultural education, fall into this fourth category.

The Agriculture Teacher in Career Guidance

The agriculture teacher is in a unique position to assist students in career guidance. The agriculture teacher is one of the few school personnel who follows students from possibly the fifth grade through post–high school graduation. This gives the agriculture teacher a broad and deep perspective of students' experiences, academic abilities, interests, and skills. The agriculture teacher can assist students in career guidance as SAE supervisor, FFA advisor, provider of information, and career counselor. Many agriculture teachers arrange career fairs specifically in the career cluster of Agriculture, Food, and Natural Resources. This allows students to focus on one career cluster when obtaining information and other resources. Agriculture teachers also host guest speakers as resource persons from a number of occupational areas. In this way students are exposed to people from the community representing various occupational areas.

Personal Guidance

As mentioned earlier, students may experience certain personal issues the agriculture teacher must refer to trained counselors. However, the agriculture teacher is qualified to counsel in numerous other situations. A student must feel trust and comfort in sharing with the agriculture teacher. Listening on the part of the agriculture teacher is often more important than talking and offering advice. Some of the best personal guidance is given when the agriculture teacher listens, asks questions, and affirms the student has been heard.

Students will share information with their agriculture teacher about relationships, self-concept, family, financial concerns, and other personal issues. This information must be treated as confidential. The teacher must not gossip or get in the middle of disputes between students. Doing so would lead to mistrust, accusations of favoritism, and a lessening of teacher authority.

For privacy reasons, personal guidance should be given in an office or other setting where a student's peers cannot hear. The student may become emotional and could become embarrassed if heard by peers. In all situations the agriculture teacher should strive to maintain the dignity of the student.

The agriculture teacher should exercise caution when meeting individually with students. The conference needs to be arranged in such a way as to remove any hint of impropriety. For example, the agriculture teacher should never meet alone with a student before or after school. During the day, it is best to meet in an office with an inside window or open door.

REVIEWING SUMMARY

Recruitment and retention are important components of an agricultural education program plan. Successful recruitment and retention strategies depend on a number of variables. It is imperative that the agriculture teacher maintain a quality program.

Students enroll in a class because of (1) curricular factors, (2) people factors, and (3) need factors. In addition, providing information and observation opportunities will increase the likelihood that students will enroll. Successful recruitment strategies are built around these basic reasons for enrolling.

Student retention is an important issue for an agricultural education program. It is desirable that students who take one agricultural education course subsequently enroll in additional courses. This has benefits for both the agricultural education program and the students.

School counselors and school administrators affect enrollment in agricultural education on two fronts. First, they are the point of contact for students when decisions are made regarding what courses to take. A positive recommendation from a school counselor could encourage a student to take an agricultural education course. Conversely, a negative comment or no information at all is a detriment to enrollment. Second, counselors and administrators put together the master schedule. It is important that they understand how agricultural education courses are sequenced and the target class grade for each course.

Among the barriers to successful recruitment and retention of students are some that are out of the agriculture teacher's control. These are ones that are schedule-related or related to high school graduation or college admission requirements. Other barriers can be overcome with planning and dedication.

Agriculture teachers often provide advisement to students. Students share with their agriculture teachers because they trust the teachers. Agriculture teachers provide guidance to students in two areas: career and personal. Career guidance involves assisting a student in choosing a career and gaining the preparation necessary to enter the career. Personal guidance involves all the issues that are a part of the student's life. The agriculture teacher may not be qualified to deal with some of these issues. In that case, the student should be referred to a professional counselor.

QUESTIONS FOR REVIEW AND DISCUSSION

1. Should every student be required to take at least one agricultural education course sometime in either middle school or high school? Why or why not?
2. What are the variables in the successful recruitment of students into agricultural education programs?

3. Why do students enroll in an agricultural education course?
4. What are several strategies for effective student recruitment?
5. Why is a middle school agricultural education program an effective recruitment strategy?
6. Why do certain students respond to particular recruitment strategies while others do not?
7. Why is retention so important for agricultural education programs?
8. What are the benefits of retention for the students?
9. What are the major barriers for student enrollment in agricultural education?
10. Why do students frequently come to their agriculture teacher for guidance?

ACTIVITIES

1. Investigate the confidentiality laws of your state and how these are practiced in the schools. Include sharing of student information, such as grades. Also, investigate requirements regarding teachers reporting child abuse and other harmful situations to authorities.
2. Develop a recruitment and retention strategy for an agricultural education program in a local high school. Include an electronic presentation that could be used in a middle school to promote enrollment. Give a report in class on what you develop.
3. Using the *Journal of Agricultural Education*, *The Agricultural Education Magazine*, and the proceedings of the National Conference of the American Association for Agricultural Education, report on career guidance research in agricultural education. Current and past issues of these journals can be found online at www.aaaeonline.org or www.naae.org.
4. Design a brochure or infographic that would be useful in student recruitment for a high school agricultural education program. Write the content, develop line art or photographs, and lay out the brochure or infographic.

REFERENCES

Advance CTE. (2021). *Career clusters.* https://careertech.org/career-clusters

Anderson, J. C., II. (2013). An exploration of the motivational profile of secondary urban agriculture students. *Journal of Agricultural Education, 54*(2), 205–216. https://doi.org/10.5032/jae.2013.02205

Breja, L. M., Ball, A. L., & Dyer, J. E. (2000). Problems in student retention: A Delphi study of agriculture teacher perceptions. *Proceedings of the 27th Annual National Agricultural Education Research Conference*, San Diego, CA, 27, 502–513.

Brown, Brooks, D. L., & Associates. (1990). *Career choice and development: Applying contemporary theories into practice* (2nd ed.). Jossey-Bass.

Dykeman, C., Ingram, M., Wood, C., Charles, S., Chen, M., & Herr, E. (2001). *The taxonomy of career development interventions that occur in America's secondary schools.* (ERIC Digest No. CG-01-04).

Fishbein, M., & Ajzen, I. (1975). *Beliefs, attitudes, intentions, and behaviors.* Addison-Wesley.

Hoover, T. S., & Scanlon, D. C. (1991). Recruitment practices—A national survey of agricultural educators. *Journal of Agricultural Education, 32*(3), 29–34. https://doi.org/10.5032/jae.1991.03029

Lent, R., Brown, S. D., & Hackett, G. (2002). *Career choice and development.* Jossey-Bass.

Mouser, D. M., Sheng, Z., & Thoron, A. C. (2019). Are agriculture students more career ready? A comparative analysis of Illinois juniors. *Journal of Agricultural Education, 60*(2), 15–27. https://doi.org/10.5032/jae.2019/02015

Myer, D. C., Holt-Day, J., Steede, G. M., & Meyers, C. (2017). A content analysis of the 2016 national teach ag day's Facebook posts. *Journal of Agricultural Education, 58*(3), 120–133. https://doi.org/10.5032/jae.2017.03120

Myers, B. E., Breja, L. M., & Dyer, J. E. (2004). Solutions to recruitment issues of high school agricultural education programs. *Journal of Agricultural Education, 45*(4), 12–21. https://doi.org/10.5032/jae.2004.04012

Myers, B. E., Dyer, J. E., & Breja, L. M. (2003). Recruitment strategies and activities used by agriculture teachers. *Journal of Agricultural Education, 44*(4), 94–105. https://doi.org/10.5032/jae.2003.04094

National Association of State Directors of Career Technical Education Consortium (NASDCTEC). (2012). *Career clusters at a glance*. Retrieved November 7, 2012, from http://www.careertech.org

National Council for Agricultural Education. (2016). *National quality program standards for agriculture, food, and natural resources education: A tool for secondary (grades 9–12) programs*. https://thecouncil.ffa.org/

National FFA Organization. (2021). *Growing leaders: 2021–22 fact sheet*. FFA. Retrieved February 7, 2022, from https://ffa.app.box.com/s/lgrbxeltnzsnmaw4agsz08mnuq1vqto1

Parsons, F. (1909). *Choosing a vocation*. Houghton Mifflin.

Phelps, K., Henry, A. L., & Bird, W. A. (2012). Factors influencing or discouraging secondary school students' FFA participation. *Journal of Agricultural Education, 53*(2), 70–86. https://doi.org/10.5032/jae.2012.02070

Phipps, L. J., & Osborne, E. W. (1988). *Handbook on agricultural education in public schools* (5th ed.). Interstate Printers & Publishers.

Porfeli, E. J., & Savickas, M. L. (2012). Career adapt-abilities scale-USA form: Psychometric properties and relation to vocational identity. *Journal of Vocational Behavior, 80*(3), 748–753. https://doi.org/10.1016/j.jvb.2012.01.009

Quey, R. (1971). Functions and dynamics of work groups. *American Psychologist, 26*(10), 1081.

Savickas, M. L., Nota, L., Rossier, J., Dauwalder, J. P., Duarte, M. E., Guichard, J., Soresi, S., Van Esbroeck, R., & van Vianen, A. E. M. (2009). Life designing: A paradigm for career construction in the 21st century. *Journal of Vocational Behavior, 75*(3), 239–250. https://doi.org/10.1016/j.jvb.2009.04.004

Super, D., Crites, J., Hummel, R., Moser, H., Overstreet, P., & Warnath, C. (1957). *Vocational development: A framework for research*. Teachers College Press.

Vondracek, F. W., Lerner, R. M., & Schulenberg, J. E. (1986). *Career development: A life-span developmental approach*. Lawrence Erlbaum Associates.

9

Classroom and Laboratory Facilities

Lucy Stabler is teaching in a high school where student numbers are rapidly increasing. The original school building, including the agriculture facility, was constructed 40 years ago. Temporary classrooms are used for some classes. The board of education has obtained land for building a new, larger, modern school facility. The voters in the school district have approved the sale of bonds to finance a new building.

Planning for the building is underway by the superintendent, principal, and others. Ms. Stabler is delighted that a new agriculture facility will be included. She has been invited to a meeting at the superintendent's office with the architect so that she can share her ideas. Now, she is pondering how to prepare, what to take, and what to say. She is even wondering about the nature of the classroom and laboratory facilities that are needed to ensure all students have opportunities to develop and apply agricultural knowledge and skills.

For a teacher to have input into facility design is unusual. Most teachers are at established schools, and they work with what exists. To offer ideas on a facility is a wonderful opportunity, but the teacher must be prepared.

OBJECTIVES

This chapter addresses the National Quality Program Standards for Agriculture, Food, and Natural Resources Education (National Council for Agricultural Education, 2016), specifically Standard 1C: Program Design and Instruction—Facilities and Equipment. It has the following objectives:

1. Describe the role of facilities in teaching.
2. Identify the kinds of facilities needed for agricultural education.
3. Explain the organization and maintenance of facilities.
4. Develop a plan for the continual updating of facilities and equipment.
5. Maintain a complete inventory of equipment and supplies.
6. Practice safety in instructional environments.

TERMS

agriscience laboratory
aquaculture laboratory
classroom
facility
inventory

laboratory
learning stop
safety
Safety Data Sheets (SDS)
storage facility

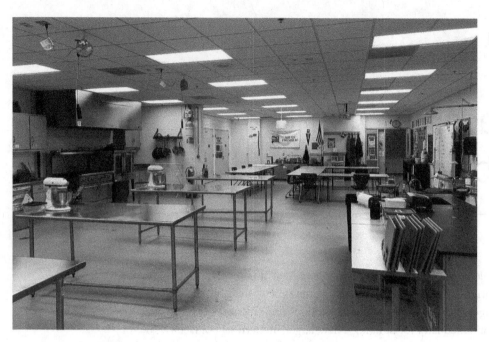

FIGURE 9.1 This food science lab has workstations for students to work in groups and a demonstration bench for the instructor. Drop-down power cords are available above each workstation to allow for flexibility. (COURTESY OF RACHEL HOLDEN.)

ROLE OF FACILITIES IN TEACHING AND LEARNING

Agricultural education is a facilities-intensive program. A *facility* is a building, a specialized area within a building, an outside area, or a large item that is not movable or that is attached. Although students are capable of learning in any environment, facilities make a difference in the quality, breadth, and depth of instruction. It is very difficult to demonstrate, experiment with, and practice concepts without the proper facilities.

As you learned in Chapter 3, agricultural education has its roots in vocational education, but today utilizes a blend of training in specific agricultural skills and development of critical thinking and problem-solving skills to help students become agriculturally literate citizens prepared for life beyond high school. To accomplish these aims, agricultural education uses a combination of classroom and laboratory-based instruction in which students have opportunities to learn and apply new knowledge and skills. Agriculture teachers use a diversity of laboratory facilities and spaces to facilitate this approach to learning. Nationwide, access to agricultural laboratory facilities varies, depending on the needs and resources of the community. In a 2012 study, Shoulders and Myers identified the top five most frequently reported laboratory facilities available to study participants: mechanics/carpentry/welding facilities (76.8%), greenhouses (72.2%), landscaping areas, (51.0%), gardens (38.7%), and aquaculture tanks/ponds (33.0%).

Bear in mind that not all laboratory-based instruction occurs in school laboratory facilities. Teachers also utilize field trips to farms, businesses, and other locations in the community to provide students a learning laboratory experience. Considering a laboratory experience provides students the chance to apply their knowledge and skills in a practical way, involvement in various FFA activities also can be considered a laboratory for leadership skill development. In this chapter, we will focus on the laboratory facilities available through the school's campus.

Facilities and Program

Facilities must be matched to the type and size of program desired. The program plan, as discussed in Chapter 5, should guide the facilities planning process. Ideally, a program is planned first and the facilities are established afterward so that the program can be efficiently and effectively implemented.

Here are a few questions to consider: What is the program emphasis? What are the program goals? What are the needs of the community? What agricultural education courses will be offered? What is the expected student enrollment? How many agriculture teachers will the facilities need to support?

When planning your agriculture program, it is helpful to use the National Quality Program Standards (NQPS; National Council for Agricultural Education, 2016) to ensure the program facilities and equipment support the implementation of the curriculum so that all enrolled students have opportunities to develop agriculture skills and knowledge. The following statements are NQPS (2016) indicators of quality program design and instruction as it relates to facilities and equipment available in the agricultural education program:

- Facility and layout provides for effective delivery of all programs of study (POS) offered.
- Facility is in compliance with existing local, state, and federal safety and health standards.
- Training and evaluation are in place so individuals using the facility create a safe working environment.
- Facility is clean, organized, and maintained to provide an environment conducive to learning.
- Facility is designed to be accessible and accommodating to all students.
- Storage space is sufficiently sized and organized for both student and teacher materials, supplies, and equipment.
- An inventory of equipment, tools, consumable items, and instructional technology is completed and includes a plan for new purchases and replacements.
- Equipment, tools, and instructional technology are safe, adequately maintained, and current to industry standards.
- The quantity of tools, equipment, and consumable supplies is adequate for equipping all students enrolled at all times.
- Equipment, tools, and instructional technology are current, available, and used effectively for delivering instruction.

The Teacher's Role

When a new school building or new agricultural education facilities are being planned, the agriculture teacher must inform planners, decision makers, and architects of the needs of the agricultural education program. Along with the local advisory committee and the state agricultural education specialist (state supervisor), the agriculture teacher should give input on items of which the others may not be aware. Information to be provided includes type and size of facilities needed; layout of rooms and equipment; special electrical, ventilation, or plumbing needs; and agricultural considerations, such as optimal location for a greenhouse. Input given before construction begins will save time and money and will enhance teaching effectiveness for the life of the facilities.

Building codes, state and local standards, and safety and fire codes must be met when any facility is being constructed. In addition, facilities must be accessible to persons with disabilities. In some states, facilities must meet state education standards if state funds are to be available for construction and equipment.

KINDS OF FACILITIES NEEDED FOR AGRICULTURAL EDUCATION

Agricultural education facilities are typically in two major areas: classroom and laboratory. Classrooms often have many similar qualities. Laboratories vary widely based on the instruction to be provided and the needs of the local community served by the school. Storage and work areas are parts of each.

Facility Location

Agricultural education facilities are typically found in one of four locations.

In a separate building   A school may have the agricultural education facilities as a separate building or complex close to the main school building. This arrangement has the advantages of providing space for the program to expand, removing noise and odors from the main school building, and providing program visibility to the community.

Within the main building   A school may locate the facilities within the main school building, usually within a wing housing other career and technical education programs. This arrangement has the advantages of efficiently utilizing the building infrastructure, involving agricultural education students and teachers in the total school, and providing for sharing of space between programs.

Within other departments   A school may sometimes place the agricultural education facilities within other school areas, such as the science wing. This has the advantage of providing for sharing of space between programs, viewing the agricultural program as an integral part of the school, and recruiting students who might never see the program otherwise.

Off school grounds   Some programs have laboratory facilities or school farms off the school campus, within a short drive from the school. Locating laboratory facilities off-campus has the advantage of being embedded within the community, and may allow for expansion of facilities if the school grounds have no additional space for expansion. Using laboratory facilities that are off-campus does mean that agriculture teachers need to plan for how students will be transported from the main campus to the off-campus site, which may prove to be costly in both time and money.

• • •

A teacher who has input into design should promote a location that enhances instruction. Access is very important for some instructional areas, such as farm machinery, livestock, and horticulture. Location can also enhance student participation by convenience of access as related to locations of other classes at a school facility.

Categories of Facilities

There are four broad categories of agricultural education facilities: classroom, laboratory, storage, and office/teacher workroom. Within each of these categories can be specialized areas for different agricultural emphases.

Classroom

A *classroom* is a facility designed for group instruction. It typically has individual student desks or tables and chairs. It is important for the agricultural education program to have its own classroom(s). Experiments may need to be left in place for several days. The room may contain specialized equipment, such as GPS units, microscopes, aquariums, or food processing machines. FFA activities, student projects, and supervised experience records all require special room arrangements or storage places within the classroom.

An agricultural education classroom should have a minimum of 33 square feet of floor space per student. Ideally, the largest agricultural education classroom should have 45 square feet of floor space per student to allow for special activities. This translates into 825 to 1,125 square feet of floor space for a classroom designed for 25 students. Tables and chairs are preferable to tablet arm desks because they provide space and flexibility for experiments, student presentations, and group projects.

At least one wall of the classroom should be equipped with a writing surface (chalkboard or whiteboard). In some classrooms, the writing surface has been replaced or supplemented with a digital interactive whiteboard, such as a SMART Board or Clevertouch. This should be paired with a digital video projector and internet-connected computer capable of playing Blu-ray and DVDs. While the interactive nature of these digital whiteboards allows teachers to more easily integrate tech-savvy lessons into their instruction, they may not be ideal for classrooms that may have dust and debris (such as in a shop facility) or which have excessive sun glare. Having a physical chalkboard or whiteboard allows for instruction to continue even when experiencing temporary network or power outages. If a classroom does not have an interactive whiteboard, then a projection screen should be located so all students can easily see it from wherever they are seated. Natural light through windows is desirable in the classroom, but blinds or other means must be available to control light level for audiovisual use. If applicable, stations with water, electricity, and gas are important additions to the classroom.

Other walls of the classroom can be equipped with bulletin boards and storage cabinets. Inclusion of useable counter space above these storage cabinets can allow the teacher to set up manipulatives as part of a demonstration or stations for students to rotate through. Sometimes an agricultural education classroom is equipped with a few desktop computers or classroom sets of laptops. These computers should be networked to a printer and other peripherals such as scanners. With the rise in one-to-one device use, desktop computers or classroom sets of laptops may not be as common anymore, but are still in use in some districts.

Laboratory

Some classrooms may have built-in *laboratory* facilities. Possible examples are those for food processing, small-animal care, computer technology, hydroponics, or aquaculture. However, most laboratory facilities are separate rooms or buildings. With either arrangement, an agricultural education facility should have restrooms, sinks for cleaning up, and a water fountain. For all facilities applicable National Electrical Code (NEC) requirements (National Electrical Installation Standards, 2021), National Fire Protection Association (NFPA) codes and standards (2021), and Occupational Safety and Health Administration (OSHA) laws and regulations must be met. Agricultural education programs require several types of laboratory facilities. A laboratory is an area for individual or group student experiments, projects, or practice in agriscience, food science, aquaculture, agricultural mechanics, horticulture, plant and soil science, animal science, and natural resources. Each of these will be discussed in further detail.

Agriscience   An *agriscience laboratory* is a facility used in teaching the science and math principles and concepts associated with agriculture. This type of facility should have 160 square feet of floor space per student. For a 25-student class, this would mean 4,000 square feet. This amount of space allows for bench-type experiments, demonstrations, and projects. The agriscience laboratory may need water, electricity, and gas at each workspace. Ventilation and floor drainage must be provided.

Equipment storage is a necessity. Microscopes, measurement devices, and other equipment need storage cabinets, ideally with the ability for the teacher to lock them. Depending on the community, the laboratory will need cages for small animals, aquaculture tanks, hydroponics units, grow lights, soil-erosion experiment tables, and other agriscience-related equipment.

Food science   Food science laboratories are becoming increasingly popular. One type of food science laboratory includes the equipment and facilities for processing meat and other foods. The other type of facility provides for experiments, cooking, and preparation of food products. State guidelines should be followed when designing and equipping food science laboratories. Food safety, sanitation, and processing regulations vary by state and locality and must be followed.

Aquaculture   An *aquaculture laboratory* is a facility used for providing learning experiences in fish farming and related areas. The laboratory may be inside, using tanks or vats, or outdoors, using ponds or raceways. Some programs pair the production of edible fish with the production of vegetables using a variety of aquaponics systems. Still other programs have aquaculture facilities to raise fish for the pet or research industry.

Agricultural mechanics   An agricultural mechanics laboratory should have 150 to 200 square feet of floor space per student. For a 25-student class, this would mean 3,750 to 5,000

FIGURE 9.2 This indoor aquaculture facility allows for the production of a variety of species of fish. (COURTESY OF LEDYARD AGRI-SCIENCE & TECHNOLOGY PROGRAM.)

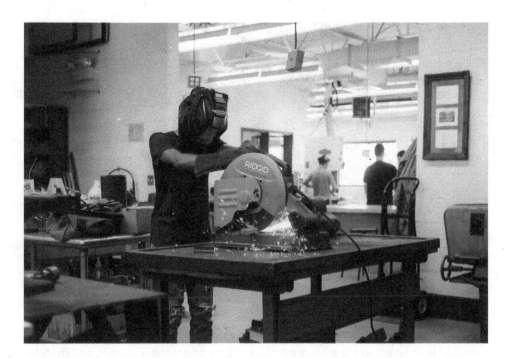

FIGURE 9.3 Agricultural mechanics laboratories should be designed to allow for students to work safely in various stations, using appropriate personal protective equipment. (COURTESY OF LEDYARD AGRI-SCIENCE & TECHNOLOGY PROGRAM.)

square feet. The type of agricultural mechanics program needed by the community will determine the design of the facility. Possible types include woodworking and construction, metals and welding, large engines and hydraulics, small engines, electricity, plumbing, concrete, and paint and bodywork. Each type requires a different layout and different equipment and tools. Adequate storage for projects, tools, and supplies is a necessity. An agricultural mechanics laboratory must be designed with student safety in mind. Building and fire codes must be adhered to strictly. Student health considerations are important when designing agricultural mechanics facilities. Dust, welding fumes, volatile organic compounds, exhaust fumes, and other contaminants must be properly handled.

Phipps and Osborne (1988) provided useful guidelines for designing an agricultural mechanics laboratory. The central portion of the laboratory should be clear to allow space for large projects. An overhead door at least 24 feet wide and 18 feet high will permit tractors and such to be brought into the facility. A regular-sized entrance door should be provided so people can enter and exit without having to open the overhead door. The ceiling should be 20 to 24 feet high. Concrete floors provide for numerous uses of the space. The electrical wiring should provide for both 110- and 220-volt current. Depending on the equipment needs, three-phase current will also need to be provided. An outside, fenced-in patio can provide both work and storage space. Inside and outside water service, including appropriate floor drains, is needed for many projects.

Horticulture   Horticultural laboratories include greenhouses, headhouses for preparation and sales, indoor or outdoor landscaping areas, floriculture areas, and aquaculture/hydroponics areas. (*Note:* Depending on purchasing regulations, greenhouses are sometimes designated as equipment rather than facilities.)

A primary greenhouse should be well constructed of glass or hard plastic and have 60 square feet per student, but no less than 1,600 square feet total. The headhouse should have

FIGURE 9.4 A headhouse with plenty of workspace allows students to easily arrange flowers. (COURTESY OF LEDYARD AGRI-SCIENCE & TECHNOLOGY PROGRAM.)

30 square feet per student, but no less than 1,150 square feet total. The agriculture teacher must decide on bench arrangement, drainage, walkways, the watering system, and the temperature regulation system. Cost is almost always a limiting factor, but in all decisions, functionality as a teaching facility must take priority.

Floriculture can be taught using a classroom or the headhouse but requires some special equipment. Various-sized coolers and display cases are necessary to maintain the flowers and plants. Storage cabinets are important for the care and maintenance of tools and equipment.

Landscaping facilities can be both indoors and outdoors. Indoor facilities can be used for water garden displays, displays of landscaping materials, and construction of items such as gazebos. Ideally, an outdoor landscaping laboratory will be at least one acre in size. The area can be sectioned into smaller plots as needed. For many agricultural education programs, the school grounds or a town park becomes the outdoor landscaping laboratory. With this arrangement, care must be taken that tasks and projects conducted are for educational purposes.

Plant and soil science Land laboratories may be located on the school grounds or a distance from the school. These may range in size from as little as one acre to 100 acres or more.

FIGURE 9.5 An outdoor landscaping laboratory can include both landscapes and hardscapes to allow students opportunities to develop skills related to planning, installation, and maintenance. (COURTESY OF LEDYARD AGRI-SCIENCE & TECHNOLOGY PROGRAM.)

Land laboratories can be used for a variety of purposes including production of hay or corn, serving as a test plot for seed trials, a school garden, or agriscience research projects. Larger land laboratories may also serve as moneymakers for the agricultural education program, as supervised experience for students, or as demonstration plots for agricultural producers in the community. Some programs own the planting, tillage, and harvesting equipment themselves, while others contract with community farmers for these services. Land laboratories are most useful when used for year-round agricultural education instruction. Best utilization requires summer employment for the agriculture teacher and extended class periods, such as in block scheduling, during the school year.

Animal science Animal science laboratories range in size from what is needed for housing small animals indoors, to barns for animals, to several acres for livestock grazing. Some programs even include small animal grooming and boarding facilities, offering students the opportunity to run a "doggie daycare" business from the school. Animal facilities must follow health and safety guidelines in addition to humane animal-care guidelines. Provisions must be made for food and water, medical care, waste disposal, odor control, and dispersal of the animals at the end of instruction. Note that when housing animals in the classroom or indoor laboratory, any state or local indoor air quality regulations must be followed. Working closely with area veterinary professionals can help teachers develop facilities and management plans to account for the health and well-being of the animals housed in program facilities. Consideration should be given to external access to the animals for feeding on weekends, holidays, or inclement weather days, so the students or staff responsible for feeding can do so without needing access to the entire school building.

For more detailed guidance on appropriate housing, husbandry, health care, and biosecurity for agricultural animals, please see the *Guide for the Care and Use of Agricultural Animals in Research and Teaching*, published in 2020 through a collaborative effort of the American Dairy Science Association, the American Society of Animal Science, and the Poultry Science Association (available at https://www.asas.org/services/ag-guide). If you are teaching in a facility that uses small animals commonly used in laboratory animal research, please consult the National Research Council's *Guide for the Care and Use of Laboratory Animals*, 8th ed. (2011; available at https://www.aaalac.org/the-guide/).

FIGURE 9.6 This chicken coop not only served to help students develop skills in an animal science class, but also was utilized as part of the Ledyard Regional FFA's Living to Serve project in which they raised laying hens to produce eggs for a local food bank. (COURTESY OF LEDYARD AGRI-SCIENCE & TECHNOLOGY PROGRAM.)

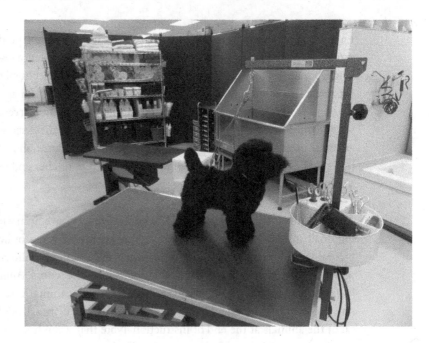

FIGURE 9.7 Some animal laboratory facilities focus on the care and grooming of companion animals as part of a veterinary science course. This dog grooming facility, pictured at a Connecticut high school, allows students to connect with animal science concepts that are relevant to the urban and suburban communities served by the program.

Natural resources Natural resources laboratories include Christmas tree farms, maple sugar bushes, wildlife areas, ponds, rivers and streams, nature areas with trails, and timber production areas. These laboratories are excellent for teaming with other high school teachers and with middle school and elementary school teachers. These laboratories, more so than others, require signage and learning stops. A *learning stop* is a location in a lab where students stop and read information, do a task, or observe some phenomenon. Students can be included in the creation of learning stops as one way of sharing responsibility for curation of laboratory facilities in the agriculture program. Land laboratory facilities should be designed using a master plan to build in as many learning experiences as possible.

Storage

Agricultural education programs need an assortment of tools, equipment, textbooks, audio-visuals, resource materials, and consumables. When not in use, these items must be properly stored. A *storage facility* is a secure location where materials and equipment can be inventoried, organized, and protected from loss and damage. Storage facilities in agricultural education include separate rooms, as well as cabinets, bookcases and racks, lockers, and various types of shelving.

The agricultural education classroom should have storage for textbooks and other resources, FFA materials, student supervised experience record books, and student notebooks. If the classroom is dedicated to a specific function, such as agriscience, then locking storage for frequently used tools, equipment, and consumables should also be provided in the classroom.

A separate storage room is best for tools and equipment that are valuable, used infrequently, or bulky. A guideline is to have 10 to 16 square feet of storage space per student. Wall and cabinet storage should have tool silhouettes so the agriculture teacher can easily see if any tools are missing.

If possible, planning for an area for students to store a change of clothing and shoes is helpful in the agriculture program. Given that students may frequently be outdoors or getting dirty through the planned learning activities, providing students locker space to store a change of clothing and shoes can ensure students are prepared for a variety of hands-on learning activities at a moment's notice.

Office/Teacher Workroom

Dedicated office space allows the agriculture teacher to be organized, professional, and responsive to student and community needs. Each agriculture teacher should have a minimum of 120 square feet of office space. Teachers can share office areas when a school has more than one agriculture teacher. The office can be combined with a workroom or conference room.

In a teacher office, each agriculture teacher should have, at a minimum, a desk, a desk chair, a worktable, computer with internet access, and a filing cabinet. The office should have at least one telephone with an outside line. A telephone on each desk is preferred. An agriculture teacher is required to communicate with employers, adult agricultural education students, suppliers, and others. This justifies an agriculture teacher having an outside line when other teachers may not.

ORGANIZING AND MAINTAINING FACILITIES

Facilities should be properly organized and maintained to assure their efficient use and long life. A neat facility represents the agriculture teacher(s) and program well with the public who sees it.

Organization

Agricultural education facilities should be organized for effective and efficient instruction. In addition, they should be attractive and professional in appearance. Agricultural education facilities may be used by the community more often than other school instructional facilities and can be good public relations tools.

The state supervisor of agricultural education in each state will likely have sample facility plans for various curriculum offerings. Safety is a major concern in establishing facilities. Plan doors, windows, ventilation systems, and other features to promote student and teacher safety.

Maintenance

Equipment and tools must receive both daily and periodic maintenance. Daily tasks, such as cleaning, checking for wear and tear, and storing properly, help maintain appearance and functionality. Periodic maintenance schedules should be followed for replacing parts, lubricating, making adjustments, and cleaning thoroughly. Well-maintained equipment and tools will provide years of quality service. Often the maintenance of tools and equipment is included as a component of curriculum, but sometimes agriculture teachers will need to conduct maintenance activities outside of classroom time for efficiency and student safety.

UPDATING FACILITIES AND EQUIPMENT

A replacement and updating plan should be developed for the facility and equipment. This will extend the life of the facility and equipment while preventing them from becoming completely outdated. Local, state, and federal funding sources will have various requirements for documenting the need for new or replacement facilities and equipment. The agriculture teacher should develop one-, three-, and five-year plans.

The one-year plan should include items needed immediately and included in the budget. It is good to have additional items in the one-year plan in case more funds become available during or at the end of the school year. Safety equipment, such as eye, hearing, and body protection, should be replaced annually. The three-year plan should include replacement items as well as items needed for new curricular thrusts. The five-year plan should be a capital improvements plan that will give school administrators and the school board time to budget for larger-expense items.

ACCOMMODATING ACCESS TO FACILITIES

It is important to ensure laboratory facilities can be made accessible to all students. This may require updating facilities to include ramps, widening door openings, or more spacing between tables or equipment to allow for increased maneuverability. According to federal law, learning activities must be made accessible for students. Work with your special education department and building and facilities manager to determine appropriate access to your existing laboratory facilities. If you are in a position to design new facilities, be sure the architect with whom you are working accounts for accessibility in the design. Resources related to making agricultural equipment more accessible for people with physical disabilities can be found from the National AgrAbility Project (2021) at www.agrability.org.

INVENTORYING EQUIPMENT AND SUPPLIES

An *inventory* is a complete list (by kind and number) of equipment, tools, and supplies. An inventory should be taken as soon as possible upon initial employment and yearly thereafter. The results should be communicated to administrators. A sample inventory form is shown in Figure 9.8. The teacher can use this form when walking around viewing the items. Then, the information can be entered into a computer database created in a software program such as Microsoft Excel or Google Sheets. Note that in addition to counting the number of items, documenting their condition and usefulness is also necessary. Items that are in less than desirable condition should be repaired or replaced. Consideration should be given to getting rid of items that were not used the previous year and are not anticipated to be used the next year. Not only do these take up space that could be used for other purposes, but they contribute to an environment of clutter and disorganization.

SAFETY IN INSTRUCTIONAL ENVIRONMENTS

All agricultural facilities require safety be practiced. *Safety* is the reduction of risk and of the likelihood of personal injury. Many items used in the agricultural education curriculum can cause injury if misused or if safety is not practiced. Chemicals, power tools, heavy objects, pinch points, electricity, motorized vehicles, hot objects, and animals all have the potential to cause injury or harm to students.

Safety is a vital concern for the agriculture teacher. Instructional time must be devoted to safety instruction so students both know and practice safe habits. Safety equipment must be provided in working condition and its use strictly enforced. Finally, the agriculture teacher must be a role model in the use of safe practices.

Agriculture teachers may be held liable for actions they may or may not take within the classroom and laboratory (Lee, 2000). An agriculture teacher is expected to do what a reasonable, prudent person would do. The teacher should provide instruction and practice in the

INVENTORY FORM

School year _____ – _____

Agriculture program _____

Person completing form _____

Item Name and Brand	Number of Items	Model No./ Serial No.	Condition	Used Last Year or Will Be Used Next Year?

FIGURE 9.8 An inventory form for equipment, tools, and supplies.

safe use of equipment, tools, and supplies. It is advisable that prior to any students entering a laboratory facility, they have received instruction in safety and have passed a safety knowledge test. Documentation must be kept that shows students have passed a safety knowledge test and have demonstrated safe operation and practices. Safety guards and equipment must be installed, and students must understand their use. Tools and equipment must be properly maintained or taken out of use until repaired. Finally, tools and equipment should be used for designed purposes. Other uses may place students at risk of harm or injury.

Teachers are responsible for the safety of their students. Many teacher actions and practices contribute to a safe instructional environment. Lee (2000) summarized these into 10 essential safety practices for teachers (see Table 9.1).

Power Equipment Safety

Power equipment presents safety considerations for the eyes, hearing, and the extremities. Whether using power hand tools or larger stationary power equipment, students must follow safe practices. Eye protection is not only important for the person operating a tool but also for those in the immediate area. A good practice is to require the use of eye protection at all times within the laboratory space. This eliminates students' forgetting to put their eye protection on and the need to define the dividing line for when eye protection must be worn.

Hearing protection is important for the student operating a power tool or power equipment. It may also be required for those within the immediate area. Teenagers may not realize the extent to which exposure to loud noises can cause damage, so it is critical that the agriculture teacher be a role model and enforce use of hearing protection.

Often overlooked is the need to keep long hair from contacting moving surfaces. Long hair should be pulled back or put under a hat so that it is not loose.

TABLE 9.1

Safety Practices in Agricultural Education

Practice	Explanation
Plan and practice.	Teachers should plan learning activities and try out what students will be doing.
Instruct students.	Teachers should instruct students in potential dangers to avoid and safe practices to follow. Documentation is critical.
Supervise the learning environment.	Teachers should always be present with their students and be attentive to what is happening in the learning environment.
Keep up the facilities.	Teachers should keep laboratory facilities clean, in good repair, and free of potential hazards. Broken equipment, dangerous chemicals, and safety devices should always be dealt with properly.
Be a good example.	Teachers should set a good example by dressing and grooming properly and wearing proper protection. They should never violate safety procedures.
Use safety cleanup materials.	Teachers should have the appropriate cleanup materials readily available in case of an accident.
Know and follow school policies.	Teachers should know and follow school policies. The hazardous waste officer should be asked to review the laboratory and teaching content.
Obtain professional educator liability insurance.	All educators need to have insurance coverage that protects them while they are conducting their professional duties. However, insurance does not provide protection if the teacher is negligent.
Be a good housekeeper.	Keeping the classroom and the laboratory in good condition helps avoid some hazards.
Post emergency numbers.	Teachers should post emergency telephone numbers near all telephones.

Other safety considerations around power equipment include proper clothing and footwear and removal of jewelry. Sweatshirts with long hood strings, long necklaces, and various rings or piercings can cause safety concerns when working with tools or animals. When welding, it is important that in addition to wearing appropriate welding coats or coveralls, students wear cotton or wool-based fabrics, as man-made fabric blends tend to be highly flammable. Students should also be aware that slick-bottomed shoes and open-toed shoes are not appropriate for the laboratory. There may be instances when athletic footwear is also not appropriate. The agriculture teacher should conduct a safety inventory to identify potential dangers and hazards before using laboratory instruction.

Chemical Storage

Chemicals are used throughout the curriculum in agricultural education. The agriculture teacher should be trained in the proper storage, handling, and use of all chemicals used in the program. A system should be established for filing *Safety Data Sheets (SDS)* and for ensuring that out-of-date chemicals are removed and properly disposed of. Previously known as Material Safety Data Sheets (MSDS), the SDS include information about hazardous materials in a consistent, easily understood format. For more on what information can be found on an SDS, please visit https://www.osha.gov/sites/default/files/publications/OSHA3514.pdf (Occupational Safety and Health Administration, 2012).

Often, facilities and maintenance departments in your school district will request a copy of the SDS be sent to them, and a binder of SDS be kept in the location where the chemicals are stored so that in event of a spill or exposure, the SDS are readily accessible. Lee (2000, p. 77) gave the following suggestions for properly storing chemicals:

FIGURE 9.9 Chemicals such as pesticides should have their own locking storage cabinet that is designed to store them properly.

- An appropriate storage area should be provided. Often known as the stockroom, this area should be locked and kept off-limits to students and other school personnel.
- Only compatible chemicals should be stored together. Some chemicals are incompatible and will react violently if they come together when accidentally spilled.
- Chemical containers should always be labeled appropriately. Chemicals that are unlabeled invite trouble.
- Chemicals that are corrosive and flammable should be stored in approved fireproof cabinets.
- All containers should be properly covered. Tops should be appropriate for the materials being stored.

Safety Equipment

It is important that students have access to and use proper safety equipment. Lee (2000, pp. 79–80) gave the following recommended safety equipment list:

Emergency eyewash equipment—A handheld wash bottle that can be disposed of after one use is likely best; several are needed in each laboratory area.

A fountain, permanently mounted, may work well (minimum of one per laboratory, with additional fountains in larger laboratories). An alternative is a unit that requires no plumbing and uses a special eyewash solution. Either should be capable of flushing both eyes at the same time.

Caution: Keep eyewash facilities clean. Microorganisms that cause infection can build up in facilities that are not properly maintained.

Drench shower—freestanding or mounted (minimum of one per laboratory).

Fire blanket—fiberglass; 43 × 39 inches (one per laboratory).

First-aid kit—metal cabinet complete with standard first-aid supplies (one per laboratory).

Tethered buoy ring (needed for aquaculture involving ponds, large tanks, and streams)—rope 30 feet or longer attached to sturdy post (one per worksite or tank).

Sanitizing goggle cabinet with goggles—wall-mounted, with ultraviolet lamp; 20-goggle minimum capacity (one or more per laboratory).

Goggles—clear; scratch-resistant (one pair for each student).

Fire extinguisher—Halon or dry chemical charge (one or more per laboratory).

Particle masks—bulk quantity.

Safety storage cabinet—approved double-wall construction; 18-gauge welded steel; must conform to OSHA regulations (one per laboratory).

Sand bucket—metal; five-gallon size, with sand (one per laboratory).

Smoke or heat detector—standard detector (one per laboratory).

Aprons—vinyl, rubberized, or leather, depending upon the requirements of the laboratory (one per student or 20 per laboratory, depending on the number of students).

Safety gloves—disposable latex gloves (quantity supply for laboratory); leather gloves as appropriate (one pair per student or 20 pairs per laboratory, depending on the number of students).

Trash containers—metal- or plastic-lined; 10-gallon size (one per workstation).

Broom with dustpan—for cleaning nonhazardous spills (one set per workstation).

Mitts—nonslip; used for handling hot materials (one pair for each workstation).

Safety charts—wall-mounted signs or charts appropriate for laboratory area (one set per laboratory).

Cleanup kits—approved kits for caustic, solvent, and acid spills (one of each for each laboratory).

Other safety equipment—as needed for laboratory activities.

Vandalism and Violence

Classroom management techniques are discussed in Chapter 14. There are additional considerations regarding student management in the laboratory. In all cases, the agriculture teacher must follow local procedures regarding violence, vandalism, and student behavioral problems. Some students may be prone to threaten or cause harm to other students or the teacher. These students may be denied access to the laboratory. Vandalism can render some equipment inoperable and, depending on budgets, not repairable. Vandalism also disrupts the aesthetics of the laboratory and if left untreated can lead to further vandalism. Theft, besides being illegal, creates a need for expensive replacement of items stolen and disrupts learning, as students may not be able to work until items are replaced.

The agriculture teacher can take some actions to alleviate potential vandalism and violence problems. First, tools and other materials should be kept in locked storage rooms or cabinets. See Chapter 21 for suggestions on end-of-class cleanup procedures. Second, the laboratory should be arranged so the teacher can see all students as they work. And finally, the teacher should never leave the laboratory or classroom unsupervised. If the teacher must be out for a time, another licensed agriculture teacher should be in the room to supervise.

REVIEWING SUMMARY

Facilities are important for instruction in agricultural education. The facilities must match the type and size of program the school and the community need. The agricultural education program plan should guide the facility planning, maintenance, and updating processes, using the National Quality Program Standards as an evaluation and planning tool. Building agricultural education facilities involves input from the community, school administrators, the agriculture teacher(s), architects, and others.

Four types of facilities are typically found in an agricultural education program. One or more dedicated classrooms assist in the instructional program. Laboratory facilities are a must and need to match the instructional program. The agricultural education program requires storage facilities for the various equipment, tools, and supplies needed for the instructional program. Finally, an office and workspace facilitate the agriculture teacher doing their job.

Organization of agricultural education facilities aids the instructional process and gives them an attractive and professional appearance. Careful thought must be put into whether the facilities will be designed for a one-teacher or multiple-teacher agricultural education program. If possible, the design should allow for future additions or modifications as the program focus expands or changes.

The agriculture teacher is responsible for selecting and ordering equipment, tools, and supplies. Using local suppliers is good for positive public relations and community support but may not always be possible or allowed. The teacher is also responsible for the maintenance and inventorying of these items. Regular maintenance greatly extends the life and usefulness of equipment and tools.

Safety is critical when using agricultural education facilities. Teachers may be held liable for student injuries because of actions they did or did not take. Teachers must be familiar with safety rules and practices themselves and must instill these in their students. Most important, agriculture teachers must model safe practices and must not violate safety procedures.

QUESTIONS FOR REVIEW AND DISCUSSION

1. What are the questions to be asked when designing agricultural education facilities?
2. What are the four categories of agricultural education facilities? Why are these needed?
3. Why are tables and movable chairs preferable for an agricultural education classroom?
4. What are the areas of agricultural education for which laboratories are needed?
5. What are special storage needs in agricultural education?
6. Why does an agriculture teacher need an outside telephone line and an Internet-connected computer?
7. What questions should be answered when purchasing equipment, tools, and supplies?
8. Why is maintenance of equipment and tools important in agricultural education?
9. Why is a beginning inventory upon initial employment important? Why is an annual inventory important?
10. Why is safety so important in agricultural education?

ACTIVITIES

1. Investigate the guidelines and requirements for agricultural education facilities in your state. Begin with the website of the state department supervising CTE education in agriculture. Contact the individual in charge of school facilities at the state level and request information, including copies of relevant state laws.
2. Debate the necessity for agriculture teachers to have equipment and facilities that may not be available to other teachers. Examples include a private office, an outside telephone line, additional storage space, a workroom or conference room, vehicles, and specialized equipment or rooms.
3. Using the *Journal of Agricultural Education*, *The Agricultural Education Magazine*, and the proceedings of the National Conference of the American Association for Agricultural

Education, write a report on research about safety within agricultural education. Current and past issues of these journals can be found in the university library. Past and current issues can be found online at www.aaaeonline.org or www.naae.org.

4. Consider an ideal facility for the area of agricultural education you want to teach. Sketch the layout of the classroom and laboratory areas. Be sure to include storage and office areas. Be sure to include doors, windows, bench arrangements, and other details. Make your sketch on poster paper for display and discussion.

5. Using the National Quality Program Standards quality indicators, assess the program facilities and equipment at a program of your choice. Identify examples of evidence and develop an action plan for how to improve the quality of facilities and equipment available at the program. The NQPS can be accessed at https://thecouncil.ffa.org/program-standards-tool/.

6. Interview a variety of agriculture teachers to develop a list of equipment and consumables that would be needed to stock a laboratory facility of your choice. While developing this list, identify sources where you could purchase this equipment, and the estimated cost. As part of your interview with your selected teachers, describe the source of funds used to support the cost of maintaining their facilities.

REFERENCES

American Dairy Science Association, American Society of Animal Science, & Poultry Science Association. (2020). *Guide for the care and use of agricultural animals in research and teaching* (4th ed.). https://www.asas.org/services/ag-guide

Lee, J. S. (2000). *Program planning guide for agriscience and technology education* (2nd ed.). Pearson Prentice Hall Interstate.

National AgrAbility Project (2021, June 28). *AgrAbility: Cultivating accessible agriculture*. http://www.agrability.org/

National Council for Agricultural Education. (2016). *National quality program standards for agriculture, food, and natural resources education: A tool for secondary (grades 9–12) programs*. https://thecouncil.ffa.org/

National Electrical Installation Standards. (2021, June 28). *National electrical installation standards*. http://www.neca-neis.org

National Fire Protection Association. (2021, June 28). NFPA website. http://www.nfpa.org

National Research Council. (2011). *Guide for the care and use of laboratory animals* (8th ed.). National Academies Press. https://doi.org/10.17226/12910

Occupational Safety and Health Administration. (2012, February). *OSHA brief—Hazard communication standard: Safety data sheets*. https://www.osha.gov/sites/default/files/publications/OSHA3514.pdf

Phipps, L. J., & Osborne, E. W. (1988). *Handbook on agricultural education in public schools* (5th ed.). Interstate Printers & Publishers.

Shoulders, C. W., & Myers, B. E. (2012). Teachers' use of agricultural laboratories in secondary agricultural education. *Journal of Agricultural Education, 53*(2), 124–138. https://doi.org/10.5032/jae.2012.02124

ADDITIONAL SOURCE

U.S. Environmental Protection Agency. (2021, June 28). *Indoor air quality (IAQ)*. United States Environmental Protection Agency. http://www.epa.gov/iaq

10

Instructional Resources

Have you ever attempted to do a job without adequate tools or materials? Hard to do, right? Think of the situation you were in as related to that of a teacher without needed instructional resources.

Achieving goals in education requires resources just as much as making achievements in other areas of our lives and society. A student needs "tools" in order to learn; a teacher needs "tools" in order to promote learning. Teachers need resources to efficiently develop skilled and knowledgeable students just as veterinarians need instruments and equipment to promote animal health and floral designers need flowers and supplies to construct an arrangement. Without the needed "learning tools," the teaching–learning process will be inefficient and fail to reach its potential in promoting goal achievement.

Some schools provide an abundance of resources for teaching and learning; other schools may not. Teachers are in the important position of using resources to promote student learning in an attempt to gain the desired outcomes of education. Further, teachers have the responsibility of obtaining the needed resources to the fullest extent possible. This will often require extra initiative to make needs known to school administrators and promote the allocation of school resources so that students have the "tools" they need in order to learn.

TERMS

academic software
activity manual
ancillary instructional resource
basal instructional resource
Chromebook
cloud storage
computer-based module
consumable
desktop computer
display panel
document camera
e-book reader
e-learning
electronic-based material
equipment
facility
instructional materials
instructional materials adoption
instructional resource guide

instructional resources
instrument
interactive whiteboard
laptop
lesson plan library
netbook
paper-based material
reference
school management software
smartphone
student-use material
supply
tablet computer
teacher's manual
teacher-use material
textbook
tool
webcam

OBJECTIVES

This chapter addresses the National Quality Program Standards for Agriculture, Food, and Natural Resources Education (National Council for Agricultural Education, 2016), specifically Standard 1C: Program Design and Instruction—Facilities and Equipment. It has the following objectives:

1. Describe the roles of instructional resources in accountability.
2. Identify kinds and sources of instructional resources.
3. Discuss e-learning media in agricultural education.
4. Apply selection criteria in obtaining instructional resources.
5. Explain the management of instructional resources.

FIGURE 10.1 Agriculture teachers use a range of instructional resources to help students learn agricultural knowledge and skills.

ROLES OF INSTRUCTIONAL RESOURCES IN ACCOUNTABILITY

A resource-rich learning environment in the classroom and laboratory promotes teaching and learning. An agricultural education teacher has the responsibility of requesting and obtaining appropriate instructional resources. A wise teacher always maintains a list of needed materials just in case funds become available. A teacher also has the responsibility of preparing precise requests for instructional materials (similar to bid specifications) and justifying their importance and usefulness.

Instructional resources are the materials used in the teaching process. They are among the "tools" teachers use in providing instruction and students use in learning. In a broad sense, instructional resources include many kinds of materials, ranging from those that are published on paper or as computer-based materials to plants, animals, lab instruments, supplies, and safety equipment.

Some materials and teaching approaches are much more effective than others. Good assessment of materials is needed before obtaining and using them in a teaching–learning environment. Just because certain materials are available does not mean they will make a strong contribution to student achievement. Teachers should be prudent in making resource choices. Careful investigation of materials is needed before obtaining and using them with students. Determine the experiences of other teachers, the approaches used in the development of the material to assure authenticity, correlation to content standards, and contribution to high assessment scores of students.

Defining Instructional Materials

Some instructional resources are known as *instructional materials*. Instructional materials are materials that have the content or skill information being taught. They include textbooks, activity/lab manuals, transparencies, electronic presentations, brochures, magazines, reference manuals in either print or electronic format, and many others, but not laboratory facilities, equipment, tools, and supplies. Students primarily learn through their interactions with teachers and instructional materials. Selection of these instructional materials varies by subject and school district.

A teaching and learning environment rich in relevant, up-to-date instructional materials promotes increased achievement. As tools in the teaching and learning process, instructional materials are sometimes compared to the tools carpenters use in construction, such as hammers, levels, and squares. Without these tools, construction cannot progress; carpenters cannot achieve their goals. Likewise, without instructional materials, learning cannot progress; teachers and students cannot achieve their educational goals.

Students in an environment that is deficient in instructional materials will not likely perform well on accountability testing. Low scores by students may result in the teacher not being viewed as accountable in promoting student achievement. The school administrators may also not appear accountable as instructional leaders. The bottom line: A resource-rich learning environment promotes student educational achievement and provides motivation for future success.

Indicators of High-Quality Instructional Materials

Instructional materials have a direct impact on student learning and achievement. In some cases, the quality of instructional materials has been shown to have at least as much if not more impact on student learning as the quality of the teacher using those materials. It is imperative the teacher access and utilize high-quality instructional materials to support students' learning and skill development.

Identifying what makes instructional materials high quality varies from state to state, with the state department of education often providing a definition. Typically, instructional materials are deemed to be of high quality when they are

- Aligned with educational standards
- Implemented through rigorous instructional experiences that provide students opportunities to learn through meaningful and authentic activities
- Developed based on research on the science of teaching and learning
- Implemented by teachers who are engaged in ongoing professional learning on how to best implement the materials

(Bucolo, 2020)

Recent research during the COVID-19 pandemic suggested additional indicators of high-quality instructional materials:

- Technology enabled throughout to allow learning to occur across modalities
- Culturally relevant and responsive to the needs of students and families
- Designed to help families assist in student learning

(Chu et al., 2021)

When deciding what instructional materials will be used to implement your agriculture curriculum, it is essential to evaluate your potential materials for evidence of the above characteristics.

Agricultural Education Instructional Materials

Many kinds of instruments, tools, equipment, and supplies are instructional resources that may be used in agricultural education. These vary with the nature of the instructional program. The curriculum outline and courses are established first. Appropriate instructional resources are then obtained to implement the curriculum. A particular course should not be offered just because a school has the facilities for doing so. Courses are offered to fill educational needs.

Facilities are essential in carrying out the instruction of a course of study. A *facility* is the school building (including classrooms and laboratories), grounds around the building, and any structures on a permanent foundation. Greenhouses and other useful structures are sometimes defined as facilities and other times as equipment. As a facility, a greenhouse, livestock barn, or other structure would need to be situated on a permanent foundation. Facilities should provide a modern and safe educational environment for students and teachers. Some states have facility plans for agricultural education facilities. In constructing facilities, local school boards engage the services of an architect. Agriculture teachers and advisory groups should provide input to the architect in the design of facilities for agricultural education.

More information on agricultural education laboratory and classroom facilities can be found in Chapter 9.

Equipment includes instructional resources that are larger, often stationary, and frequently operated with motors or engines. Equipment with motors or engines may be known as power equipment. In an agricultural mechanics lab, equipment may include welding machines, radial arm saws, and air compressors. Equipment in horticulture may include trimmers, mowers, soil mixers, and components of irrigation systems. With care, equipment is long lasting and can be used repeatedly. Equipment often poses hazards to students and teachers. Proper instruction in safety and safe usage is essential. In addition, instructional equipment includes items used in the classroom or laboratory to provide information and reinforce learning.

FIGURE 10.2
Instructional resources should support the intended outcomes of teaching and learning, such as those used in extracting DNA from strawberries. (COURTESY OF EDUCATION IMAGES.)

Instruments, tools, and equipment are used repeatedly in teaching and learning. An ***instrument*** is a device used for a particular purpose. An example is a caliper used to make precise measurements. Instruments are often small and require careful storage to maintain them and protect them from damage. Some instruments are fairly expensive; others are inexpensive.

A ***tool*** is a handheld device that aids in doing a task. An example is a claw hammer. A claw hammer is designed to drive and pull nails. Without a hammer, nail driving is difficult. Proper hammers, for example, are needed so students can be taught selection, use, and care of them. Tools used in agricultural mechanics classes vary from those used in horticulture, veterinary science, or forestry, though a few may be the same.

Schools that fail to allocate funds for instructional resources are planning to fail in student achievement. Quality instructional resources allow students to learn on their own and help develop self-motivation toward learning. Relevant instructional resources designed for student use empower students as learners to continue learning on their own without the continuous intervention of a teacher.

The instruments, tools, and equipment needed vary with the course and curriculum outline. A general agriscience class will have instructional resources that support achieving the educational objectives of the course. These may include microscopes, slides, soil test kits, and numerous other items.

FIGURE 10.3 Equipment and supplies should be appropriate to the area of instruction, such as animal science.

Supplies are needed to use most instruments, tools, and equipment. A *supply* is a consumable material used to carry out instructional activities and/or work. *Consumables* are teaching supplies that are used up each year during instruction. Supplies vary with the nature of the instruction. Horticultural supplies may include potting soil, seeds, fertilizer, herbicide, insecticide, and other materials used to grow plants in the greenhouse or land lab. In agricultural mechanics, supply needs vary with the areas taught. For example, welding requires rods or wire, metal, and/or other supplies; construction requires lumber, plywood, nails, and other supplies; and machinery maintenance requires oil, filters, and other supplies.

A list of instruments, tools, equipment, and supplies for a particular course is often included with the curriculum map for a course. Otherwise, a teacher would get input from local advisory council members, other experienced teachers, and state supervisory personnel. Teachers gaining employment with existing programs will often go to schools that have some resources remaining from previous years. A new teacher should inventory what is on hand and in workable condition and then request resources needed to implement the curriculum. In all cases, teachers should follow school procedures in making requests. More information about ordering consumables for laboratory-based instruction can be found in Chapters 9 and 21.

KINDS OF INSTRUCTIONAL MATERIALS

Instructional materials can be classified in a number of ways, such as by intended user and by medium of presentation. Teachers choose materials that provide the best fit in their schools and with their students. (*Note:* This section of the chapter refers to materials used in communicating subject matter and achieving subject-matter objectives. It does not include instruments, tools, and equipment used in laboratory skill development. See Chapters 9 and 21 for more information about these items.)

Commonly used materials include textbooks and the ancillaries that accompany them. An *ancillary instructional resource* is a secondary or supplementary material that accompanies another material, most often a textbook. The material an ancillary accompanies is known as basal. A *basal instructional resource* is a material that provides the base or foundation for the organization of a subject. All else revolves around the basal resource. Textbooks are often considered basal instructional resources.

In selecting an ancillary, ask the following questions:

- How well does the material correlate with the basal textbook/material?
- Does the material promote systematic learning?
- Does the material reinforce important concepts in the basal textbook?
- Are the activities appealing to learners?
- Has the material been field tested?
- Is the material current?
- Is a companion instructor's guide available? (This may include online support for the ancillary.)

User Classification

Instructional resources can be classified by who uses them. Some materials are designed for use by students as important resources in learning. Other materials are designed for use by teachers in efficiently directing the learning of students often focused on using student materials. Either may be published on paper or available electronically.

Student-Use Material

Student-use material is an instructional resource prepared for use by students. Usually, a great deal of care has been taken to make the material appealing to students. The writing level is within the range of readability of the students. Numerous photographs and line drawings are used to depict important concepts. Textbooks are the most common student-use materials, though activity manuals, lab exercise manuals, and self-paced CD materials are also used.

Textbooks A *textbook* is a book designed for use by students that deals with a specific subject. Textbooks have usually been prepared following a systematic format, carefully reviewed to assure technical accuracy, and correlated to appropriate course objectives or standards. Textbooks are usually organized into parts or units. Each part or unit is composed of several chapters. The chapters are further divided with headings and subheadings. Motivational approaches to gain student attention and interest are used. End-of-chapter evaluations and hands-on activities are presented.

A textbook is usually prepared for students at a particular grade level or range of grades, such as grade 10 or grades 9 through 11. Hard copy textbooks are typically printed in color and bound with hard covers. Textbooks are increasingly available in digital formats. At times these digital textbooks, sometimes called eTextbooks, are available in PDF form, while at others they are more interactive in nature, allowing students to highlight, flag, and take notes on the text. eTextbooks are often accessed through an app or website login.

Typical subjects of textbooks used in agricultural education classes include agriscience, horticulture, biotechnology, landscaping, veterinary science, and plant and soil science. Teachers who choose to use hard copy textbooks might purchase enough copies for an entire class to check out for the year, or they may purchase classroom quantities of books that remain in the agriculture classroom for use during class time only.

With care, paperback and hardbound textbooks are not destroyed after one use and can be used repeatedly with several classes. Students should be encouraged to take good care of books so the books have long, useful lives. Paper books can be used in locations where internet service and computer equipment are not available, but eTextbooks are increasingly becoming available to use in off-line settings.

Activity manuals An *activity manual* is an ancillary resource that has activities organized around a basal resource, such as a textbook. The activities often focus on several goals and approaches in learning. Overall, the activities reinforce the achievement of objectives in the textbook, review terms, and provide hands-on experiments or other investigations. Students carry out the investigations and record their observations or experiences in space provided in the manual. Some of the investigations may involve using precise laboratory instruments. Other activities may involve collecting data from crops, animals, streams, forests, and other phenomena. Activity manuals often have instructor's guides to aid the teacher in directing student learning. Each student needs their own activity manual. Most activity manuals are consumable; they can be used only once because students answer questions and record information in them. Some activity manuals may be available as online materials. Some teachers have students create interactive notebooks using cloud-based storage systems like Google Classroom, which replace hard copies of activity manuals.

Computer-based modules A *computer-based module* is instructional material that involves the use of units or modules guided by a computer program. The computer program directs students from one activity to another and through the activities. Textbooks, video clips, and hands-on activities may be part of the instruction. Some observations show students with this type of instruction lose interest after a few weeks.

Record books   Supervised experience and FFA participation accomplishments should be recorded. This may be with paper-printed record books or computer-based record systems. Records are essential for students to apply for advancements in FFA. States and/or local school districts may have record books or systems developed specifically for their use. National FFA award applications for proficiency awards, American Degrees, and SAE grants all require use of the Agricultural Experience Tracker (AET) software to upload records of SAE and FFA involvement.

Online reinforcement and assessment   Online approaches for reinforcing rote learning, providing student tutorials, and assessing student learning are available. Most widely used is the online assessment of student learning. Some are offered through state education agencies; others are from commercial assessment sources.

Teacher-Use Material

Teacher-use material is an instructional resource designed for use by a teacher. Its purpose is to aid the teacher in planning, delivering, and evaluating instruction. Some teacher-use materials stand alone, such as program planning materials. Other teacher-use materials are companions to student-use materials.

Following are a few examples of teacher-use materials.

Teachers' manuals   A *teacher's manual* is a teacher-use material prepared to accompany a student-use material, such as a textbook. A teacher's manual or teacher's edition provides useful information for the teacher to use in delivering instruction using a student edition textbook. A teacher's manual will typically summarize content, offer teaching suggestions, include additional recommended resources, and provide answers to questions in the student edition. Teachers' manuals may be provided free of charge by publishers to teachers who adopt and purchase classroom sets of student edition textbooks. Teachers' manuals may be printed on paper or available in electronic form as a CD/DVD or for downloading from the provider's internet site. (In some cases, editions of textbooks are published that incorporate the teacher's manual in the same binding as the student edition. These may be known as teachers' editions.)

Instructional resource guides   An *instructional resource guide* is a collection of material that helps a teacher plan and deliver instruction. Instructional resource guides often contain detailed lesson plans, sample tests, transparencies or electronic-presentation images, lab sheets, and other material for use in delivering instruction. The guides may be prepared as loose-leaf notebooks, available to access through an online portal, on CD-ROMs, or a combination of formats. Increasingly, such materials may be available on the internet by using access codes. Such materials are usually sold, though they are occasionally offered free of charge if teachers adopt and use specific student materials.

Lesson plan libraries   A *lesson plan library* is a collection of teaching plans focused on a fairly broad area, such as landscaping or wildlife management. The lesson plans are complete with a summary of content, suggested instructional strategies, sample tests, lab sheets, and transparencies or electronic-presentation images. Lesson plan libraries are often correlated to state standards or course blueprints. They may also be closely correlated to end-of-course or other forms of competency tests. These lesson plans may be available on a CD or by subscription to an online site. Some state teacher organizations and education agencies provide lesson plan libraries; others are available through commercial sources. Most commercial sources have rigorous review processes to assure that materials have been field-tested and are technically accurate and educationally sound.

Reference materials   *Reference* materials are needed for use by the teacher in preparing lesson plans and answering questions. A reference is material that is more advanced than students would normally use. References may be in a variety of formats, including books, manuals, brochures, notebooks, websites, CDs, and videos. They are sometimes publications from the Extension Service. References on science, agriculture, horticulture, veterinary medicine, and other areas are often needed, depending on the technical areas included in the instructional program and the nature of the agricultural industry in the local community. References on FFA are needed in all programs. Websites can often be used as sources of information. An example is the publication of the National Agriculture, Food, and Natural Resources Career Cluster Content Standards, which is available at https://thecouncil.ffa.org/afnr/.

Medium Classification

Medium classification refers to the form in which instructional resources are used by students and teachers. Some instructional resources are printed on paper; others may be in electronic form on CD or available through the internet.

Paper-based material is instructional and learning material that is printed on paper. This medium of material has long been used and has yielded good educational results. The material may be printed in color or in black and white. Almost all are bound in some way—for example, as loose-leaf notebooks, hardback textbooks, or softback activity manuals.

Electronic-based material is instructional and learning material that is used in conjunction with a computer or computer network system. While once primarily depending on material available through CDs or DVDs, most electronic materials are now accessible through the internet. Each student needs access to a computer that has a strong connection to the internet. Students may work alone or work together in pairs. Some schools establish instructional technology labs, where groups of 20 or so students work at the same time.

Electronic materials available through the internet typically involve the teacher having an access code to a website. The availability of an access code may require payment of an annual fee to the website provider. Computer hardware is needed for students to use these instructional resources. (The next section of this chapter, "Materials for E-Learning," provides more information about electronic-based materials.)

MATERIALS FOR E-LEARNING

E-learning is electronically based teaching and learning that usually involves some application of computer technology. It may be used by a teacher to supplement classroom instruction or as a stand-alone source of instruction. In most cases, e-learning approaches are integrated as supplements or enhancements in the traditional agricultural education classroom and laboratory. In agricultural education, they typically do not replace the instruction of the teacher. As Roblyer and Dowering (2010) indicated, "simply having students use technology does not raise achievement." They further indicated the impact of technology depends on the ways it is used.

Essentially, e-learning involves an internet-enabled computer or smart device (such as an iPad) to transfer knowledge and skills. E-learning approaches are also used in the evaluation or assessment of student performance. Interaction may involve a mainframe computer thousands of miles away from the school, home, or wherever it is being used. The fundamental principles of how people learn apply to e-learning just as they do to other forms of instruction.

E-learning, and all that it embraces, has received considerable attention in recent years. Schools have made major financial investments to obtain and install e-learning materials and train teachers in the use of e-learning approaches. During the COVID-19 pandemic, e-learning became the primary mode of instruction for millions of schools around the world.

Agriculture teachers often find computer-based and online materials and approaches excellent supplements to face-to-face methods. Visual materials are increasingly computer-based and have mostly replaced older materials such as transparencies, 35 mm slides, and videotapes. Online information can be accessed that is up-to-date and relevant to instructional needs.

With e-learning, the quality of the content and the engagement of learners in actively mastering the information are very important. The teacher must be quite involved in delivering instruction and using e-learning approaches to enrich, enhance, review, and assess student learning, not to be the major source of instruction. Technology is not to be used in teaching just for the sake of having technology. This chapter focuses on basic categories of instructional resources and not on the approaches in their use. Chapter 13 highlights approaches to implementing e-learning or digital learning in agricultural education.

Presentation Media

Classroom approaches may involve computer-based presentation materials. Two major ways are available for presentation enrichment of classroom learning using computer-based materials: interactive whiteboards and electronic display panels or monitors.

An *interactive whiteboard* is a presentation medium that uses a computer connected to a projector to display materials on a white surface that acts as a touch screen. (*Note:* An interactive whiteboard is sometimes referred to as a Smart Board, which is a registered trademark of Smart Technologies.) The components are connected wirelessly or with USB or HDMI cables. A laptop or PC at the teacher's workstation often serves as the source of information. The information may be from a CD/DVD, online sources, or entered by the teacher. A projector displays a computer's video output on the interactive whiteboard. It becomes interactive when special pens or other devices are used to write or draw on the presentation. (See Figure 10.4.)

Projectors for interactive boards are typically mounted to the ceiling. They are positioned to assure the best possible projection without distortion. They are typically connected to the computer with a cable, though some operations may involve wireless technology. It is important to follow the manufacturer's recommended maintenance and care of the projector, particularly in regularly cleaning an air filter, as neglecting this can lead to the untimely demise of the very expensive projector light bulb.

A *display panel* is an electronic presentation device connected to a computer, camera, or other equipment for the presentation of visual information. Large tube-type devices have been replaced with liquid crystal display (LCD) flat panel displays. The images may be prepared by the teacher or obtained from a commercial source on a CD/DVD, from an online source, by way of satellite or broadcast television, or from other sources. In some cases, display panels may have audio capabilities. For classroom use, the area of display should be large enough for the most distant students to readily view what is being shown. In larger

FIGURE 10.4 This interactive whiteboard allows the teacher to draw on the screen as if it were a whiteboard, but it also operates as a large touchscreen that controls the computer to which it is connected.

rooms, display panels may be positioned around the room from the ceiling or wall. Most are rather permanently attached to the wall or ceiling, though some are on carts that make them transportable. With individual or small group use, smaller versions such as personal computer or laptop screens may be adequate.

Arrangements may be established so that teachers have access to smaller screens at their desk or station. Portable devices with wireless connectivity are sometimes used. Information may be recorded or saved for future use.

A teacher station may also include a document camera to present visual information to the group. A **document camera** can be positioned to capture live video of the teacher writing on a piece of paper, demonstrating how to perform a skill, or showing where to find information in a textbook. Document cameras can be used for face-to-face instruction to more easily show a large room of students how to perform a skill such as making a corsage and in e-learning environments when teaching online during live or on-demand instruction.

Many laptop computers come equipped with a built-in webcam. A **webcam** is a camera that can capture live video and transmit it to the internet in live time. Webcams can also be added externally to perch atop a desktop screen, or they can be stand-alone cameras that sit on a desk. Webcams allow teachers to record videos that show their faces or to participate in online virtual meetings. Imperative to the success of using a webcam are microphones, which capture the sounds of the person speaking on the webcam. These microphones might

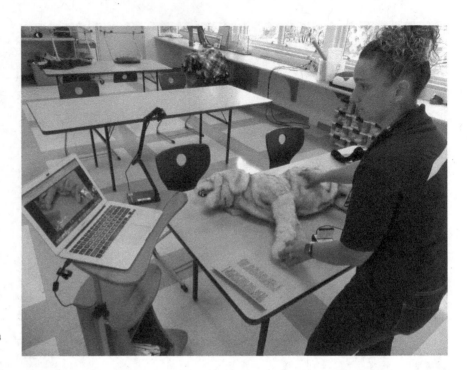

FIGURE 10.5 This teacher is demonstrating how to perform CPR on a canine CPR dummy to students learning at a distance through online learning. A document camera positioned to the right of the laptop makes this type of demonstration possible.

be internal to the computer, or can be plugged in externally. Microphones are also sometimes found built into earbuds or headphone devices. It is important to ensure that appropriate software is installed on your computer to ensure the webcam and microphone are properly working.

Various remote devices are available to operate and prepare interactive materials. Some of these are used by the teacher; others are used by students. Wireless pens may be used on whiteboard surfaces or other devices. A common wireless device used in educational settings is an Apple iPad tablet, which can be used to control the image on the projection screen from a distance.

The "heart" of an interactive system is the computer software. The software may be designed for teachers and/or students to prepare and present information. It may be used to prepare visuals and save presentations for future use. This software is typically stored in the computer located in the classroom near the equipment. The software selected must be compatible with the interactive surfaces and other devices that are used. One presentation software program that is often used in education is PowerPoint, a product of Microsoft. This software allows teachers and students to prepare presentations using bulleted summaries of the content and includes various illustrations such as line art and photographic images. Increasingly, many schools are utilizing educational resources from Google, which are freely accessible through cloud-based programs such as the Google Suite. The Google Suite includes Google Docs, Sheets, and Slides, which are analogous to the Microsoft Office Suite programs of Microsoft Word, Excel, and PowerPoint.

Software programs may also make virtualization possible. These often use virtual materials prepared by commercial or government sources. The processes or items depicted are not real but are realistic. Mostly, it is modeling through the use of a computer, such as the virtual tail docking of a pig or construction of a flower arrangement.

Networking of all school rooms may be used to share useful information with all students and teachers. In a school district, all school sites may be networked so multiple schools are involved. School districts often employ IT specialists to establish and operate networks and

FIGURE 10.6 A teacher's workstation in a classroom can often have a variety of electronic presentation media. How many different digital devices can you identify at this teacher's desk?

assist teachers with individual classroom needs. In some cases, a school program in radio or television may allow students to prepare and present information by way of the network.

Learner-Focused Media

Several electronic devices are available for individual e-learning and/or ready access to information. Individuals may purchase these devices or schools may purchase them in bulk and check them out to individual students. Changes frequently occur with these devices as new applications are developed. Several examples are included here.

Desktop computer   A *desktop computer* is a personal computer with limited portability that is essentially for use in a single location. Schools typically place networked desktops in agricultural education classrooms, though some are stand-alone. They may be connected to a modem for internet access and to a printer for producing copies on paper. Programs may be loaded into each individual computer. In some cases, a central server might be used for several keyboards and screens. Students would be provided access codes for networked desktop computers. Use by students would occur at school and during times made available for computer use. The uses might be for internet research, instructional program use, preparation of presentations, and entering record information. Schools often restrict access on a computer to certain websites and other information.

Laptop   A *laptop* is a personal computer that is easily transported for mobile use and has the capabilities of a typical desktop computer. A rechargeable battery system is located inside a laptop. Such computers can also be operated by plugging into a 110-volt receptacle, provided an adapter is used. Laptops may be issued to students for use at school or to take home for use. Schools would typically obtain a quantity of laptops so each student has one for their use. An IT specialist would typically be in charge of school-issued laptops. Students can generally view the same information and perform the same operations on a laptop as on a desktop. Internet access may be wireless or by cable depending on how the laptop is configured. School websites and FFA chapter websites may have educational information or present links for accessing educational sites. Laptops can be used in vehicles using battery power or 110-volt plug-in as found in some vehicles. Vehicles with "hot spots" may also allow mobile access to the internet, and most smartphones can be turned into a wireless hot spot as well.

The *netbook* is a category of small and inexpensive laptop computer that has found some favor for educational purposes. Netbooks are often purchased by schools and issued to students. A popular device increasingly used by schools is a *Chromebook*. These devices operate much like a netbook but do not have the ability to install Microsoft Office products, as the Chromebook utilizes Google's Chrome operating system. Since Chromebooks rely on the Google Suite of programs for browsing the internet, word processing tasks, and preparing presentations, they depend on a strong connection to the internet for maximum functionality.

Smartphone   A *smartphone* is a mobile telephone with a computing platform that allows features such as internet and email capability far beyond telephone conversations. Smartphones allow the downloading of apps (software for specific purposes) that make it possible to access agricultural or other information that is not otherwise available on a mobile basis. Agriculture apps allow individuals on farms—in pastures and fields—to access information for a specific use. For example, the USDA has the app known as SoilWeb, which is a GPS-based program that provides soil survey information. Another app is the PictureThis plant identifier, which aids in the identification of plants as well as diagnosis and treatment of plant diseases. A QR (quick response) code reader app may be installed on a device for scanning QR codes to gain near-instant connectivity to a site or source of information. Many smartphones now have QR reader capabilities built into their camera and do not need to download an additional app. Smartphones are often the only source of internet for students in rural and urban areas who do not have access to high-speed internet.

Tablet computer   Best known as the iPad by Apple, Inc., tablet computers have made major advances in use and popularity. A *tablet computer* is a device that has features of a small computer and of a smartphone. The operation platform is similar to that of a smartphone. Apps can be added to achieve specific goals. Some schools are issuing iPads to students for access to online information, including textbooks and other e-learning materials. The devices are small, lightweight, resilient, and yet have an adequate-sized screen. Email and web work can be done with an iPad in a location with wireless internet access. Mini iPads are also used with connectivity to Wi-Fi or cellular sources for sharing information.

E-book reader   Also referred to as an e-reader, an *e-book reader* is a mobile electronic device designed for reading digital e-books and periodicals. E-book readers are similar to tablet computers. The Amazon Kindle is the most popular e-book reader. Others include the Nook by Barnes and Noble and the Kobo Clara HD.

School Software

The two kinds of software, according to Lever-Duffy and McDonald (2011), most widely found in schools are administrative software and academic software. Administrative software may be referred to as school management software.

School management software is a computer-based program used in schools to achieve a wide range of functions related to delivering education, recording data, and reporting information. Such software is typically networked throughout a school system, with limited access by authorized individuals. Access is granted to various sites in the system to teachers, students, parents, and others using access codes and passwords restricting what can be seen and done. The system may include administrative functions of the school, student enrollment and class scheduling, posting announcements and homework, and recording and reporting the performance of students. For example, teachers may report student attendance, grades, and other information via online software management. Some systems have built-in parental notifications if students are absent, fail to achieve a minimum performance level on a test, or do not turn in homework.

An example is PowerSchool, which has web-based school management software programs. In addition to school and education management, PowerSchool also has reporting features for parents and students. Alerts can be quickly communicated to all school personnel, parents, and students through posting and calling systems. Access codes restrict use to authorized individuals. A feature gaining attention is that of an app for mobile devices such as iPhones, iPads, and similar devices. In addition to PowerSchool, examples of other systems are RenWeb, Rediker Software, and QuickSchools.

School management software is used in one way or another in almost all school districts. Links within school systems may direct students to e-learning sites, and in some cases, the electronic gradebook is embedded within a learning management system. Some school systems are linked to regional and state education agencies. FFA record systems are typically separate from school management.

Academic software is software that is useful to teachers and learners in the teaching and learning processes. This software is used to enrich the education environment. Examples of uses of academic software, according to Lever-Duffy and McDonald (2011), include desktop publishing, graphics, reference, tutorials and drill-and-practice, educational games, and simulations. Agricultural educators have some software applications with unique roles for agricultural education, such as supervised experience record keeping and curriculum planning.

In the use of online e-learning materials, all educators should be aware of the provisions of the Children's Online Privacy Protection Act of 1998 (COPPA). In short, this act is administered through the Federal Trade Commission (FTC) and prohibits the online gathering of information of children under the age of 13 unless parental consent has been provided. This would particularly apply to some middle school agricultural education programs. Many school systems have local policies and procedures that also apply.

SELECTING AND OBTAINING INSTRUCTIONAL RESOURCES

Choosing the instructional materials to use in teaching is an important decision. Recent agricultural education research revealed teachers do not rely on a single source of instructional materials and instead tend to select resources that meet their specific program needs (Easterly & Simpson, 2020). The content of materials should be closely correlated to the state or local curriculum requirements. These should carefully follow end-of-course or other testing programs used in local schools and may be mandated by the state education agency. It would be unwise to use materials that do not cover content in the tests that are administered. The materials must be appropriate for the levels of the learners and up-to-date in content. Teaching out-of-date information is a serious compromise of student time and other resources.

Instructional materials adoption is the process used by school districts and state education agencies in selecting materials and allocating resources. Local school boards or school administrators often establish procedures to be followed. Some states (particularly those that provide funds for purchasing textbooks) have statewide adoptions carried out under the direction of their state boards of education. A key part of the process at all levels is to obtain the input of laypeople. The laypeople are stakeholders in the schools and may be parents, business leaders, and others who are not educators. A selection committee or advisory group is often involved. Public hearings may be held for community members to provide input.

Selection Criteria

Teachers make thousands of decisions each day during in-the-moment instruction, and during their instructional planning time. There are a myriad of factors that influence these instructional decisions. Bugler et al. (2017) interviewed teachers to identify how they obtained, evaluated, and selected instructional materials to be used in their classrooms. Figure 10.7 illustrates their findings.

Selecting instructional materials is an important decision. School funds are limited, and teachers want to get the best possible materials. The same criteria are typically applied in the

TEACHERS' CRITERIA FOR DETERMINING THE QUALITY OF INSTRUCTIONAL MATERIALS

Accuracy, visual appeal
- No errors; correct information
- Well written
- Strong visual appeal

Alignment to standards, depth of knowledge
- Aligned to standards
- Efficiently addresses standards
- Appropriate depth of knowledge, questions, and activities

Ease of use, support
- Easy for teachers, students, and parents to use
- Complete set of instructions, materials, activities, assessments, and answers
- Appropriate support for new teachers

Engagement, ability to meet student needs
- Engagement: Sparks student interest; relevant
- Differentiation: Appropriate material by skill level, language ability, cognitive capability, and learning style
- Cultural and background knowledge: Culturally relevant; aligns with prior background knowledge
- Diverse activities: Group and individual, hands-on, requires movement, longer investigations

Trusted Sources for Instructional Material
- Made by and for teachers
- Include teacher comments, opinions, and reviews
- Ratings based on use by teachers (with information about student characteristics)

FIGURE 10.7 Teachers' criteria for determining the quality of instructional materials. (FROM *HOW TEACHERS JUDGE THE QUALITY OF INSTRUCTIONAL MATERIALS* BY DAN BUGLER, STACY MARPLE, ELIZABETH BURR, MIN CHEN-GADDINI, AND NEAL FINKELSTEIN. REPRODUCED WITH PERMISSION FROM WESTED. RETRIEVED FROM HTTPS://WWW.WESTED .ORG/RESOURCES/SELECTING-INSTRUCTIONAL-MATERIALS-BRIEF-1-QUALITY.)

selection of computer-based and e-learning materials as to those that are printed on paper. Here are some questions to ask in the selection process:

- Is the material correlated with the objectives and end-of-course test for a course? (In other words, does the material contribute to the achievement of state-specified objectives and good performance by students on end-of-course tests?)
- Is the content coverage in sufficient detail?
- Is the content technically accurate?
- Is the content current (up-to-date)?
- Is the content sequenced properly?
- Is the content written on an appropriate level?
- Is the material appealing to students?
- Is the material appropriately illustrated with photographic images and line art?
- Is the material durably constructed/prepared?
- Has the material been field tested and reviewed as part of development?
- Are ancillaries available to support use of the material?
- Is the material priced within range for the funds available?
- Is the material appropriate to the values of community citizens?

Specific criteria may be used in evaluating textbooks, activity manuals, and other materials. (See Figure 10.8 for a sample textbook evaluation form.) It is important to be aware of local school district and state procedures on materials acquisition.

Types of Instructional Resources

Several different types of instructional resources are available to teachers including paid and free access. No matter what type of material teachers decide to use when planning instruction, they should be sure it helps students accomplish the learning objectives. Be sure to reflect upon the indicators of high-quality instructional materials outlined earlier in this chapter. Ideally, teachers should select materials that help them conduct a student-centered activity-based lesson that engages students' curiosity and motivation.

Information-Based Resources

Information-based resources are materials that communicate content knowledge such as a textbook, Extension document, or peer-reviewed journal articles. Teachers can curate intentional selection of information-based resources to provide students as part of their exploration on a subject instead of providing lecture notes, or these resources could be used to create accurate class notes. Teachers should be sure to curate information-based materials from reliable sources so that content is up-to-date and accurate.

File-Sharing Resources

Teachers frequently share materials that they created through various platforms. Platforms that provide a space for teachers to share educational materials are file-sharing resources. Examples of file-sharing resources include NAAE Communities of Practice and a popular Facebook group called Ag Education Discussion Lab. While file-sharing resources can be a great source of mentoring and quick information, not all content shared on these platforms is automatically high quality. Teachers should use their judgment when determining if the materials obtained from file-sharing sites are accurate and appropriate for use in their classrooms to meet the needs of their diverse students.

Agricultural Education
TEXTBOOK EVALUATION FORM

Title of book: _____ Edition: _____

Date published: _____ Class quantity cost (per copy): _____

Name and address of publisher/source: _____

_____

Instructions: Rate the following items as acceptable (YES) or unacceptable (NO) as they relate to the textbook. Place an x in the column that represents your evaluation.

The textbook	YES	NO
1. Is written at the appropriate level	_____	_____
2. Has a durable binding	_____	_____
3. Is technically accurate	_____	_____
4. Is technically up to date	_____	_____
5. Has modern photographs	_____	_____
6. Has helpful and good-quality line drawings	_____	_____
7. Has a companion teacher's manual	_____	_____
8. Has a student-friendly layout	_____	_____
9. Facilitates systematic instruction	_____	_____
10. Has useful questions at the end of each chapter	_____	_____
11. Suggests activities to apply the content	_____	_____
12. Equitably treats all individuals	_____	_____
13. Emphasizes safety and ethics	_____	_____
Overall, the book would be useful in my program	_____	_____

Comments: _____

_____

_____

Signature of reviewer: _____ Date: _____

FIGURE 10.8 Sample form to use in evaluating textbooks.

Lesson Plans Provided by Organizations

Many nonprofit organizations feature education and outreach as a component of their mission. Consequently, some organizations may provide free educational materials for teachers as long as they register to have an account with their page. At times, some materials will be made available for free, while other components are available for purchase. Examples of organizations that provide lesson plans include the National FFA Organization, National

4-H, Agriculture in the Classroom, and the National Good Agricultural Practices Program at Cornell University. Depending on the source, teachers may have to evaluate whether the materials are appropriate for their students and their needs.

Fee-Based Instructional Materials

Some companies or organizations seek to generate income from the development and sale of instructional materials. Fee-based instructional materials may be available for a one-time fee or a subscription-based program in which the teacher is either charged by the number of students utilizing a program, or a time frame of use. Common fee-based instructional material services that provide agricultural education curriculum materials include CASE, MyCAERT, One Less Thing, iCEV, and AgEdNet. While these resources are usually of high quality and often integrate easily into online learning platforms, they can be expensive.

Web-Based Inquiry Resources

Finally, teachers might choose to use web-based inquiry resources that serve as tools to virtually simulate various natural phenomena. Many web-based inquiry resources are primarily focused on science concepts and skills instead of production agriculture. Examples of web-based inquiry resources include Labster, the National Science Digital Library, and ExploreLearning Gizmos.

Obtaining Materials

Always follow the procedures of a local school in obtaining materials. The school principal, director, or department chair can provide information on how the process works. Some schools may have textbook coordinators that stay current with the situation on adoptions and bidding processes.

School systems vary, but many solicit instructional materials in the spring for use the following year. The requests are obtained from all teachers and compiled into a master list. Administrators and others review the list and set priorities on what is to be obtained. Once purchasing decisions have been made, the recommendations may be taken to the local school board for approval before a purchase order is issued. Depending on the budgetary year, purchases are often made in early July with funds budgeted for the next school year.

In some states, state funding is available for instructional materials in local schools only if materials on state-adopted lists are selected. Such lists may include textbooks, CD-based materials, web-based materials, and other kinds of instructional materials. A state list may have two or more alternatives for a particular course or subject. State adoptions often occur on a cycle of five years or so. Local school districts appoint local adoption committees. Local districts that buy materials not on the state-adopted list must usually use local school funds for these purchases.

In any school, developing a strong justification for instructional materials is often needed. It is a teacher's responsibility to provide a rationale for obtaining new materials. Developing support from advisory committee members is usually helpful. As mentioned earlier, it is always wise to have a list of needed instructional materials available just in case you are informed at the last minute that some funds are available. It is the teacher's responsibility to be aggressive in obtaining quality, up-to-date instructional materials for agricultural education.

While it is the teacher's responsibility to procure quality instructional materials for their students, they should avoid making a habit of purchasing these supplies using their personal funds. The school budget should support the acquisition of supplies and not rely on individual teachers to regularly purchase expensive supplies.

MANAGING INSTRUCTIONAL RESOURCES

Instructional materials need to be managed so that they are readily available and are protected from unnecessary wear and tear. The appearance of an agricultural education office, classroom, or laboratory can be greatly influenced by how materials are filed. Piles of worn books, stacks of out-of-date brochures, broken computer equipment, damaged CDs and DVDs, and dilapidated racks of magazines detract from appearance and lower the overall perception of agricultural education facilities by students and others. Keep facilities neat, clean, and well organized!

Books

If a textbook is used as the basis of the planned curriculum, greatest learning efficiency occurs when there is a textbook for every student and every student has a textbook checked out. The textbook may be taken home or to a study area as well as used in the classroom. Having students use textbooks as an integral part of learning has a major upward impact on student achievement scores. In the case of online textbooks, students may be provided access codes for use at home.

When not in use or checked out to students, textbooks and other books used by students can be organized on bookshelves in the classroom or storage room. Stored books should be arranged by title, and all turned the same way. The place where books are stored should be dry and away from direct sunlight.

Reference books used by students may be kept on bookshelves in the classroom area. Students should be instructed to return each book to its proper location after use. Reference books used by the teacher may be kept in the agricultural education office or on the teacher's desk at the front of the classroom.

Brochures and Magazines

Brochures and magazines related to the instruction should be displayed in the classroom. Extension Service bulletins, as well as handbooks on agricultural chemicals, plant and/or animal diseases, and related topics, may also be included. These are often in racks or on shelves that allow students to easily see titles or subjects and return the items after use. Brochures and magazines for teacher reference can be kept in the agricultural education office.

Brochures and magazines should be periodically checked for damage and age. Tattered materials should be discarded, and new ones obtained. Out-of-date materials should also be discarded. Older materials may promote the use of practices that are no longer permissible, such as the common use of methyl bromide as a soil fumigant. Magazines older than one year should be discarded unless there is a very special reason to retain them. Out-of-date magazines might be saved for future learning activities such as making a collage, or for use as reference material.

School libraries or media centers can often support the agricultural education program by obtaining brochures and magazines. Libraries and media centers have facilities for organizing such materials and making them available to students. The agriculture teacher can work with the librarian or media center director to see that appropriate materials are obtained and made available.

Electronic Media

While much of media storage is being moved to the cloud and other digital repositories, some programs still use CDs, DVDs, and in some cases, videotapes. *Cloud storage* is a network of online storage where data is stored in virtualized pools by third parties. It can be helpful to have copies of CDs, DVDs, and videotapes to use when the school internet may not be working.

All electronic materials used in an agricultural education program should be carefully and securely stored to provide protection, prevent loss, and facilitate locating them when needed.

Organizer cabinets or shelves can be obtained for CDs and DVDs. Videotapes, though not widely used any more, can be stored on regular shelving. Turn all materials so the titles are easily seen. Materials may be organized by subject in alphabetical sequence or by the class in which they are used. To assure security, such materials are often kept in the agricultural education office or stored in an instructional materials room. Conditions of storage should protect the materials from damage. Such materials may be in locked storage cabinets or rooms to assure protection from theft and vandalism.

Records

An agricultural educator has several roles related to records. In some cases, records are kept in the central office of the school, and the teacher must provide the information that is to be recorded. In other cases, the teacher enters the information in a networked computer information system. Some information is recorded on a daily basis, such as class attendance. Other information is recorded on a weekly, semester, or annual basis.

Records of FFA participation and supervised experience activities are often kept in the agricultural education office or classroom. Students may have access to their records and regularly record information to keep them up-to-date. Good records are essential for the teacher in preparing state reports on the program. They are important for individual students in working toward FFA advancements. Students may save SAE records on a USB flash drive (thumb drive) for transportability and reduce information on the hard drive of a computer. (It is wise to keep backup files in case a USB flash drive is misplaced.)

Increasingly, schools are using online forms of electronic record keeping for SAE and FFA engagement. The Agricultural Experience Tracker (AET) is an example of an online record keeping system that allows students to access their records anywhere that they have an internet connection, including through an app on their smartphone. Some states have state-specific record keeping software for SAEs. Be sure to check with your state affiliate of the National Association of Agricultural Educators to find out what type of record keeping system is the expectation in your state.

Some records should be retained indefinitely in an agricultural education department. Records of FFA membership may need to be retained for at least 10 years but may be kept longer depending on school procedures. In all cases, records must be accurate and protected from tampering. Inaccurate records result in later problems and in decision making based on false information.

Agriculture teachers need to be aware of FERPA guidelines in dealing with student records. FERPA is an abbreviation of the Family Educational Rights and Privacy Act, or Buckley Amendment. The federal act protects the privacy of student education records. FERPA gives parents certain rights until the student reaches the age of 18 or attends a school beyond the high school level. The law provides that parents may inspect records, ask that errors be corrected, and sign a written release for any information to be disclosed. In some instances, uses of student records are exempt from FERPA regulations, such as when a student transfers to another school or the school prepares a directory. FERPA applies to all schools that receive funds through the U.S. Department of Education.

Laboratory Materials

Laboratory materials must be managed properly to assure safety and security and to prevent unauthorized use and damage. How materials are stored and organized varies with the kinds of tools, instruments, equipment, or supplies. (*Note:* Chapters 9 and 21 provide additional information related to tools and equipment in laboratory settings.)

Tools and instruments are typically stored in a locking cabinet or tool room that is opened and made available to students as needed. Each item should have a specific place in a cabinet or on a wall panel or shelf. The name of the item or a drawing may be used to identify the location.

Equipment is managed to assure a long, useful life. Covers may be put on some types of equipment, such as microscopes, when they are not in use. Other equipment may be locked in cabinets, with voltage meters and DO (dissolved oxygen) meters being examples. Larger equipment, such as welding machines and table saws, may be stationary on a counter or lab facility floor. Management also includes being sure that the equipment is cleaned after use, protected from damage, and kept in a safe operating condition. All safety guards, goggles, and other personal protective equipment (PPE) must be in good condition and accessible before some laboratory learning activities.

Supplies are stored based on the nature of the material. Gasoline, for example, must be stored in approved containers and in locations away from open flames or sources of ignition. Chemicals may be kept locked in cabinets or closets to prevent unauthorized access. Other materials may need only to be protected from the weather, such as some potting media that should be protected from rain. Living forms, such as seeds, should be stored to assure viability.

REVIEWING SUMMARY

Many kinds of instructional resources (materials used in teaching) are used in agricultural education. Student achievement is promoted by having a resource-rich learning environment. It is the teacher's responsibility to have appropriate materials available for efficient teaching and learning. Further, these materials should support the achievement of learning standards established for the agricultural education program and promote student performance on assessments.

Instructional resources include both published and nonpublished materials. Each type is obtained to promote the achievement of the goals of the instructional program. Some materials are designed for student use; other materials are for teacher use.

Instructional materials are published kinds of instructional resources. These include textbooks, activity manuals, teachers' manuals, references, and similar materials in paper or electronic formats. Materials that provide the base or foundation are known as basal materials. A textbook is an example. Materials that accompany and enrich basal materials are known as ancillary materials. An activity manual is an example.

Nonpublished instructional resources are used to add hands-on learning opportunities for students. A wide range of instruments, tools, equipment, and supplies are needed. Instruments, tools, and equipment are lasting and can be used over and over. Supplies tend to be consumable and include such things as seeds, flowers, animal feed, paint, lumber, nails, and welding rods. These are obtained to implement instruction as specified in the curriculum guide provided by the state or local school district.

E-learning is the use of computer technology to promote teaching and learning. In most cases, it should supplement and enhance instruction provided by the teacher—not replace the teacher. Presentation media include the interactive whiteboard with associated computer and projector or a display panel or set of panels, depending on room size. Individual learner-focused media may be used with desktops, laptops, smartphones, and tablet computers.

Materials are selected after the courses and the curriculum have been identified. The materials should help implement the curriculum and help achieve the instructional objectives. All materials should have appropriate content and be up-to-date, technically accurate, sequenced properly, written at the appropriate level, and appealing to students. Illustrations, durable construction, evidence of field testing, and availability of ancillaries are also factors in selection.

Once materials are obtained, the teacher is responsible for managing them. This includes organizing and storing them, protecting them from damage or loss, and issuing them to students. It also includes obtaining, organizing, maintaining, and promoting use by students of the appropriate personal protective equipment (PPE).

QUESTIONS FOR REVIEW AND DISCUSSION

1. What are instructional resources?
2. Distinguish between consumable and nonconsumable instructional materials.
3. Discuss the responsibility of a teacher in obtaining instructional resources.
4. What is instructional material? Give three examples.
5. Distinguish between instruments, tools, and equipment used in agricultural education.
6. What is a supply? How do supplies vary with the instructional program?
7. What is student-use material? Name and briefly explain four examples.
8. What is teacher-use material? Name and briefly explain four examples.
9. Distinguish between paper-based and electronic-based instructional material.
10. What is e-learning? How is it used in agricultural education classes?
11. What presentation media may be used in e-learning?
12. What questions should be answered in selecting instructional materials?
13. What practices may be used in obtaining instructional materials?
14. Discuss the educational benefits of issuing textbooks to students.
15. What practices should be followed in managing instructional materials?

ACTIVITIES

1. Investigate the instructional materials selection and purchasing procedures in a local school. Interview an administrator, the chair of the materials adoption committee, or a teacher. Determine the practices in adopting materials, the use of citizen input, the schedule followed, and other details relating to agricultural education. Prepare a written report on your findings.
2. Assess the sources of instructional materials for agricultural education. Visit the websites of at least three sources and determine the kinds of materials available, the costs, and other details. Following are a few sample websites:
   AgEdNet—https://www.agednet.com/
   Ag Educational Solutions—https://ageducationalsolutions.com/
   Curriculum, Content and Assessments for CTE—www.mycaert.com
   CEV Educational Multimedia—https://www.icevonline.com/
   ITCS Instructional Materials—www.aces.uiuc.edu/IM
   Lab Aids Lab Kits—https://lab-aids.com/
   NASCO Education Supplies—https://www.enasco.com/c/Education-Supplies
   NASCO Farm and Ranch—https://www.enasco.com/c/Farm-Ranch
   National FFA Organization—www.ffa.org
   Pearson (career and technical–agriscience)—www.pearson.com
   Cengage Learning—https://www.cengage.com/
   Curriculum for Agricultural Science Education—www.case4learning.org
   One Less Thing—https://www.onelessthing.net/

3. Select a secondary agriculture textbook (basal material) of your choosing. Any title or subject will be fine. Assess the book. Apply criteria stated in this chapter. Write a one- or two-paragraph assessment of the merits of the book for use in a high school class on a subject in line with that of the book.

4. Assume you have just taken an agriculture teaching position at a school that has not previously had an agriculture program. Your administrator has indicated funds are available to purchase materials. What would you request? Choose a typical class (such as introduction to agriscience, livestock, or horticulture) and develop a specific list for a class of 20 students. Prepare details such as a list of bid specifications. Include estimated total cost, not to exceed $3,000.

5. Investigate the possible impact of the Children's Online Privacy Protection Act of 1998 on students in agricultural education. Prepare a brief report on your findings. The act can be accessed at www.ftc.gov/ogc/coppa1.htm

6. Develop a list of instructional materials needed to teach a unit of your choice. Identify the vendor, price per unit, estimated quantity needed, estimated shipping, and estimated total cost.

REFERENCES

Bucolo, D. (2020, October 6). 4 signs of high-quality instructional materials. *Shaped*. https://www. hmhco.com/blog/4-signs-of-high-quality-instructional-materials

Bugler, D., Marple, S., Burr, E., Chen-Gaddini, M., & Finkelstein, N. (2017). *How teachers judge the quality of instructional materials*. WestEd. https://www.wested.org/wp-content/uploads/2017/03/resource-selecting-instructional-materials-brief-1-quality.pdf

Chu, E., Clay, A., & McCarty, G. (2021). *Fundamental 4: Pandemic learning reveals the value of high-quality instructional materials to educator-family-student partnerships*. Center for Public Research and Leadership at Columbia University. https://cprl.law.columbia.edu/sites/default/files/content/Publications/CPRL_2021_Fundamental%204_Final.pdf

Easterly III, R. G. (Tre), & Simpson, K. A. (2020). An examination of the curricular resource use and self efficacy of Utah School-Based Agricultural Education teachers: An exploratory study. *Journal of Agricultural Education*, *61*(4), 30–45. http://doi.org/10.5032/jae.2020.04035

Lever-Duffy, J., & McDonald, J. B. (2011). *Teaching and learning with technology* (4th ed.). Pearson Education.

National Council for Agricultural Education. (2016). *National quality program standards for agriculture, food, and natural resources education: A tool for secondary (grades 9–12) programs*. https://thecouncil.ffa.org/

Roblyer, M. D., & Dowering, A. H. (2010). Integrating educational technology into teaching (5th ed.). Allyn & Bacon.

ADDITIONAL SOURCES

Chingos, M., & Whitehurst, R. (2012). *Choosing blindly: Instructional materials, teacher effectiveness, and the Common Core* (Brown Center on Education Policy). Brookings Institution. https://www.brookings.edu/wp-content/uploads/2016/06/0410_curriculum_chingos_whitehurst.pdf

D'Angelo, T., Barry, D., Bunch, J. C., & Thoron, A. (2018). Selecting educational resources. *EDIS*, *2018*(4). https://doi.org/10.32473/edis-wc303-2018

Steiner, D. (2017, March). Curriculum research: What we know and where we need to go. *StandardsWork*. https://standardswork.org/wp-content/uploads/2017/03/sw-curriculum-research-report-fnl.pdf

Part 3

Instruction in Agricultural Education

11

The Psychology of Learning

Mr. Franklin Kowalski teaches Agriculture to students in 7th through 12th grades. He has noticed students in the same grade are at different maturity and cognitive levels and wonders why. He wonders why he can teach the same the lesson he did last year, the same way, but where last year's group fully understood the material and wanted to go deeper into the subject, this year's class struggles with the basics.

Mr. Kowalski remembers the learning theories he studied in college. Fortunately, he can recall the important areas that will help him improve on student learning. In recalling the learning theories, he asked himself three questions:

1. How do students learn?
2. Do students have different ways of learning?
3. Why are some students "naturally" motivated to learn?

He is looking back in his college notes and books to brush up his answers. He is beginning with his *Foundations of Agricultural Education* book.

TERMS

affective
behavioral learning theory
brain-based learning theory
cognitive
cognitive learning theory
constructivism
hierarchy of human needs
learning
learning style

mindset
motivation
operant conditioning
Premack principle
psychomotor
schema
self-efficacy
theory of multiple intelligences

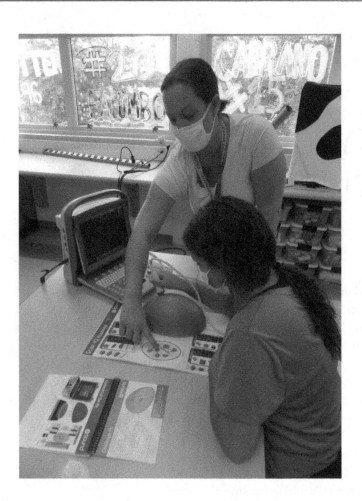

FIGURE 11.1 Understanding how people learn helps teachers be more effective.

WHAT IS LEARNING?

Learning is a permanent change in behavior as the result of an experience. The change can be in thinking, actions, and problem-solving. Learning results in people viewing situations differently in the future. They adapt their behavior based on what they have learned and the demands of their new environment (Martin & Torok-Gerard, 2019).

Behavior is more than a physical, outward activity. It is broadly defined and includes the *cognitive* (knowledge or "stored" information), *affective* (attitude), and *psychomotor* (manipulative skill) domains. Learning must also be the result of experience. So, although students change as they age, such as growing in height and weight, this is not learning.

Learning is not limited to school-based experiences. Students learn from their environment, parents, peers, and numerous other sources. Agricultural education, through supervised agricultural experience and FFA, is designed to expand school-based learning beyond the classroom and laboratory. Through this structure, agricultural education places learning within real-world social settings.

MAJOR LEARNING THEORIES

In the 20th century, two dominant types of theories were held of how people learn: behavioral learning theories and cognitive learning theories. These theories continue to impact education in the 21st century.

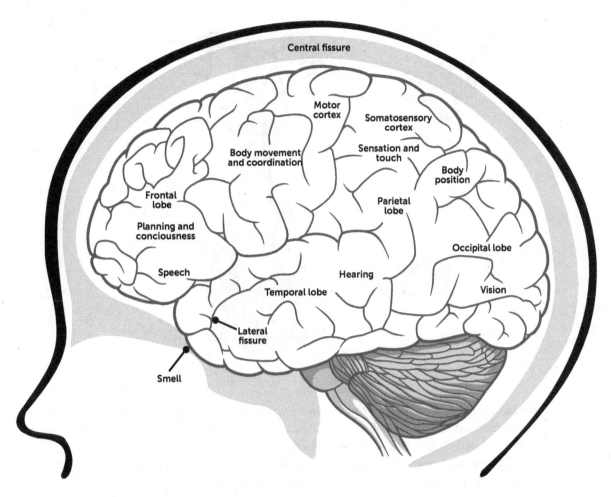

FIGURE 11.2 The brain is the most important learning tool.

A *behavioral learning theory* is a theory of learning that focuses on observable changes in outward behavior and on the impact of external stimuli to effect change. From this type of theory, we get student learning objectives, as discussed in Chapter 12.

A *cognitive learning theory* is a theory of learning that focuses on the internal mental processes, how they change, and how they affect external behavior changes.

Behavioral Learning Theories

An influential behavioral theory explaining how students learn is operant conditioning, based on the work of Edward Thorndike and B. F. Skinner (Parsons et al., 2019). This theory expands on two earlier behavioral principles: contiguity and classical conditioning.

Contiguity states that if two things, a stimulus and a response, are paired together often enough, the learner will make an association between them. This occurs in the classroom with drill-and-practice activities. A student who pairs a picture of a steer "red with white face and no horns" with the beef breed "polled Hereford" is following the principle of contiguity.

Classical conditioning is best known by the work of Pavlov and his experiments with dogs salivating. Dogs naturally salivate when food is placed in front of them. However, Pavlov found dogs could associate other things, such as a sound, with getting food. Eventually, the dogs would salivate upon hearing the sound although no food was present. Classical conditioning occurs regularly in the classroom. A student who becomes nervous or anxious when a test is given is an example. Teachers who have open, inviting classrooms are using classical

FIGURE 11.3 Classical conditioning explains how external stimuli can trigger involuntary responses.

conditioning by making students feel safe and accepted and associating that with learning. In classical conditioning, the behavior is involuntary and typically emotional or physiological. The stimulus is first, followed by the behavior.

In *operant conditioning*, the student has control of the behavior. The stimulus is a consequence of the behavior. In the classroom, a teacher who praises correct student answers to questions is using operant conditioning. The consequence (praise) follows the behavior (correct response), leading to greater classroom participation. Teachers use reinforcement to increase or strengthen behaviors and punishment to decrease or weaken behaviors.

An important principle regarding reinforcement is called the *Premack principle*. This states that a preferred activity can be used as a reinforcer for a less preferred activity (Santrock, 2008). The Premack principle works by placing the less preferred activity before the preferred activity. Agricultural education examples abound. Mr. Jones's animal science class dislikes calculating feed rations and mixtures for the aquaculture tanks but enjoys testing and analyzing the water and recording the results. Mr. Jones is using the Premack principle when he says to the class, "Let's do a good job of accurately calculating feed mixtures during the first part of the class period, then we'll conduct water-quality tests in the aquaculture laboratory."

Cognitive Learning Theories

Cognitive learning theories look at how people process information, organize information, and construct knowledge. In the cognitive view, students are active learners who seek information and are constantly reorganizing new and old information. Associated with cognitive learning theories is the information processing model, which focuses on how people gather, code, and store information in their memory and subsequently retrieve it (Martin & Torok-Gerard, 2019). The information processing model compares human information processing to the computer and how it works. Teachers use cognitive learning theories when they direct student attention, enhance student perception, and provide students with rehearsal tools. Teachers who use frequent examples, clear objectives and directions, and varied interest approaches are assisting students in cognitive learning. Organizational tools such as notes and learning webs are other examples of this theory in practice.

Bandura's social cognitive theory (1986) added social factors to learning theories. Bandura's theory recognizes behavior, person/cognitive, and environment factors all play a reciprocal part in learning (Santrock, 2008). Teaching through Career Development Events (CDE) can use social cognitive theory. The agriculture teacher establishes an environment of modeling, guided practice, and peer teaching. Students develop cognitive strategies allowing them to problem solve the CDE, which improves performance. Mock events develop behavior that is culminated at the CDE competition. The person/cognitive factor most recently emphasized is self-efficacy. *Self-efficacy* is the belief that one can accomplish something and have a positive outcome (Bandura, 1997). Self-efficacy influences behavior, so increasing student self-efficacy should lead to positive behavior and more effective learning.

Constructivism is a set of learning theories that emphasize how students actively make sense of the information they receive (Parsons et al., 2019; Woolfolk, 1998). One view of constructivism is concerned with how students use internal schemas to represent accurately the outside world. A *schema* is an abstract guide used to organize an experience or concept. A second view of constructivism focuses less on the accurate representation of the outside world and more on how old knowledge and new knowledge are transformed to be useful to the individual. A third view of constructivism states knowledge reflects the outside world but is filtered and influenced by culture, language, teachers, and other factors.

Agriculture teachers use constructivism when they follow a "learning by doing" approach. They are encouraging their students to make sense of the outside world through actively engaging in their environment. An example is the agriculture teacher who teaches parliamentary procedure by organizing the class as a mini-FFA chapter with activities and with decisions to be made. Some students may construct the parliamentary procedure knowledge around a schema of "This is how my youth group could operate." Other students may construct a schema of "This is what it means to be an FFA member or officer."

Brain-Based Learning

The *brain-based learning theory*, developed by Caine and Caine (n.d., 1991, 1994, 1997), resulted from a synthesis of theories and literature regarding the brain from multiple disciplines, including biology and psychology. From this synthesis, Caine and Caine developed 12 principles of brain/mind learning. These are listed in Box 11.1, and a brief description of each follows.

Principle 1 states the brain and body both are involved in learning. Teachers using this principle will engage students in activities that require them to use their senses and bodies. Principle 2 states that learning has social aspects and is influenced by social interactions. FFA, with its student, chapter, and community activities, provides multiple opportunities to combine learning with social interactions and relationships. Principles 3, 4, and 5 are concerned with how learners acquire and store information. Learners need to make sense of their environment and experiences (Principle 3). They do this by organizing information into unique mental schemas (Principle 4). Emotions and mindsets are a part of learning and are critical in establishing mental patterns or linkages (Principle 5).

Principle 6 states our brains perceive information both in parts and in its wholeness at the same time. In addition to facts and information, agriculture teachers should organize instruction so real-world examples and stories are also included. Principle 7 states although the brain perceives what it is focused upon, it also perceives information that is outside of

---

BOX 11.1 *The 12 Mind/Brain Learning Principles*

1. Learning engages the physiology.
2. The brain/mind is social.
3. The search for meaning is innate.
4. The search for meaning occurs through patterning.
5. Emotions are critical to patterning.
6. The mind/brain processes parts and wholes simultaneously.
7. Learning involves both focused attention and peripheral perception.
8. Learning always involves conscious and unconscious processes.
9. We have at least two ways or organizing memory: a spatial memory system and a set of systems for rote learning.
10. Learning is developmental.
11. Complex learning is enhanced by challenge and inhibited by threat.
12. Each brain is uniquely organized.

Adapted from Caine & Caine (n.d.).

---

this focus. Teachers use this principle when they place posters and pictures on the classroom walls. Principle 8 states although learning involves conscious activity, much learning occurs afterward in unconscious processes. Therefore, learning is facilitated when time is provided for reflection. Kolb's experiential learning theory (1984), used in 4-H and agricultural education, emphasizes the need for reflection within the learning process. Principle 9 states we organize in two different categories of memories. One, rote memorization, is most useful for isolated information. The other, dynamic, is engaged best through experiences and infotainment.

Principle 10 states the brain develops in a predetermined way during childhood yet never loses its capacity for learning. Educators who espouse a lifelong learning approach are following this principle. Principle 11 states the best learning environment is one that appropriately challenges learners. Caine and Caine used the term "downshifting" to describe what happens to students who are threatened in the learning environment. Students who downshift feel helpless and fatigued and then give up. Principle 12 states every student is unique. In structuring the learning environment, the teacher must recognize different learning styles, multiple intelligences, and student diversity in its broadest definition. Chapter 12 explains the teaching processes involved in meeting the unique needs of all students.

ROLE OF LEARNING THEORIES IN AGRICULTURAL EDUCATION

Agricultural educators use learning theories to organize and structure subject matter, to motivate students, and to determine what teaching methods to use. The ultimate goal is to provide the optimal learning environment for students of agricultural education. Learning theories help explain why supervised experience and FFA are critical components of the total agricultural education program.

As discussed in Chapters 2 and 3, agricultural education over the years has fluctuated between a vocational focus and a science/business focus. Another way of characterizing this is between instructing students in agriculture for future careers and instructing students about agriculture for agricultural literacy. It is important to consider the focus of the agricultural instruction in our view of student learning.

To have the optimal impact on educational practice and student learning, learning theories must be translated into language that is relevant and applicable to teachers. Newcomb et al. (2004) described 16 principles of teaching and learning to guide teachers in the instructional process. These are summarized in Box 11.2.

WAYS OF VIEWING STUDENT INTELLIGENCE

Howard Gardner (1983, 2008), in his *theory of multiple intelligences*, proposed there are eight dimensions of intelligence. These are bodily-kinesthetic (skillful use of the body and handling of objects), interpersonal (ability to sense emotions, needs, and motivations in others), intrapersonal (ability to use self-awareness to guide decisions and behavior), linguistic (skillful use of words and language), logical-mathematical (skillful use of numbers, logic,

---

BOX 11.2 *Principles of Teaching and Learning*

1. When the subject matter to be learned possesses meaning, organization, and structure that are clear to students, learning proceeds more rapidly and is retained longer.
2. Readiness is a prerequisite for learning. Subject matter and learning experiences must be provided that begin where the learner is.
3. Students must be motivated to learn. Learning activities should be provided that take into account the wants, needs, interests, and aspirations of students.
4. Students are motivated through their involvement in setting goals and planning learning activities.
5. Success is a strong motivating force.
6. Students are motivated when they attempt tasks that fall in a range of challenge such that success is perceived to be possible but not certain.
7. When students have knowledge of their learning progress, performance will be superior to what it would have been without such knowledge.
8. Behaviors that are reinforced (rewarded) are more likely to be learned.
9. To be most effective, reward (reinforcement) must follow as immediately as possible the desired behavior and be clearly connected with that behavior by the student.
10. Directed learning is more effective than undirected learning.
11. To maximize learning, students should "inquire into" rather than "be instructed in" the subject matter. Inquiry-oriented approaches to teaching improve learning.
12. Students learn what they practice.
13. Supervised practice that is most effective occurs in a functional educational experience.
14. Learning is most likely to be used (transferred) if it occurs in a situation as much like that in which it is to be used and immediately preceding the time when it is needed.
15. Learning is most likely to take place when what is to be transferred is a generalization, a general rule, or a formula.
16. Students can learn to transfer what they have learned; teachers must teach students how to transfer learning to laboratory and real-life situations.

Adapted from Newcomb et al. (2004).

reasoning, and patterning), musical (ability to sense sound, tone, pitch, and rhythm), spatial (ability to perceive the visual world and recreate based on one's own perceptions), and naturalistic (ability to identify and distinguish elements of the natural world). Gardner believed everyone possesses intelligence in all eight dimensions; however, most people excel in only a few. Students are more successful when engaged in learning using the intelligence they excel at than when there is a mismatch. Because of this, teachers must provide learning opportunities that engage students' multiple intelligences.

Carol Dweck (2006) proposed two mindsets that can be applied to learning. A *mindset* is what a person believes about the underlying nature of their abilities. A person with a fixed mindset equates success in learning with their innate characteristics. Therefore, their abilities are fixed and learning is validation of their abilities. A fixed mindset can lead to someone not reaching their full potential. Individuals with a growth mindset challenge themselves and see failure as an opportunity to enhance learning and their abilities. Having a growth mindset can lead to someone seeking to learn and achieve more. Teachers who want to build students' growth mindset should praise students for effort rather than innate ability, challenge students to try more difficult tasks, and encourage students to persevere.

Learning styles are often included in how we view teaching. A *learning style* is a description of how a person prefers to learn and under what environment he or she prefers to learn. Some students are intrinsically motivated and approach education as a "learn for the sake of learning" situation. These are the students who easily take notes, ask questions on their own initiative, and delve deeper into subjects than required by assignments. These students appreciate teachers who provide time for class discussions and who give assignments with flexible requirements. Other students are more extrinsically motivated and approach learning from a "what is required" perspective. These students are motivated by grades, rewards, and structure. Extrinsically motivated students appreciate teachers who use a businesslike approach and who give assignments with specific requirements.

Another way to characterize learning styles is as auditory, visual, or kinesthetic. Auditory learners prefer to be in an educational setting where there is talking, speaking, and auditory clues regarding the content. Visual learners like to read, see notes, and observe. Kinesthetic learners like to touch, manipulate, and do in order to learn. Regardless of their primary learning style, students learn best when instruction is varied and includes all three learning styles.

STUDENT MOTIVATION

Motivation is the energy and direction given to behavior. Some students appear to be highly motivated; others appear to lack motivation. Of course, this interpretation of motivation is in terms of learning a particular subject or skill.

There are two sources of motivation for students in the classroom: those things internal to them (intrinsic motivation) and those things external to them (extrinsic motivation). Although a teacher has the most control over extrinsic motivational factors, understanding intrinsic motivational factors is also important.

Intrinsic Motivation

Maslow (1970) developed a *hierarchy of human needs*, which is depicted in Figure 11.4. The most basic needs (physiological) are those of shelter, food, sleep, water, and survival. Transferring this to the school setting, the classroom should be in good repair, have adequate furniture, and provide light, heating, and cooling. Students who come to school hungry or sleep-deprived are less motivated to learn until these needs are met.

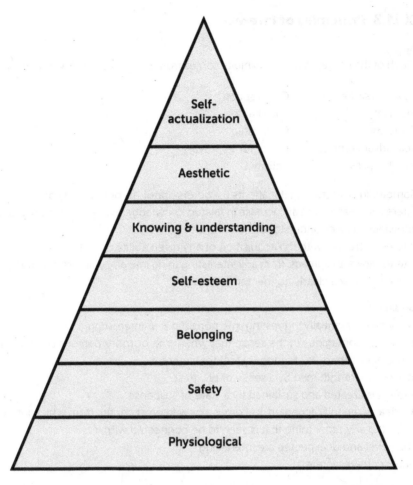

FIGURE 11.4 Maslow's hierarchy of needs. (ADAPTED FROM MASLOW, 1970.)

Next in the hierarchy are safety needs. Personal safety concerns related to gangs, bullies, or crime may detract from student learning. It is important that all students feel safe in the classroom.

Belonging is the third step in the hierarchy. Students need to feel they belong in the classroom, are accepted by their peers, and have something of worth to contribute. The teacher's attitude, words, and actions contribute to or detract from students' sense of belonging.

The next step is self-esteem. Students need opportunities to build their self-confidence, obtain approval, and be recognized. Teachers must proactively provide opportunities for all students to develop their self-esteem.

The top three steps in the hierarchy are called growth needs. When these needs are met, the person is motivated to seek even more. For example, when knowing and understanding needs are met, the student has an even greater motivation to know and understand more in that area.

Extrinsic Motivation

Crunkilton and Krebs (1982) described practices teachers can use to affect student motivation. These are outlined in Box 11.3 as primary, or universal, principles and as secondary principles. These principles can be used to gain student attention and motivation at the beginning of the class period. However, teachers should use these principles throughout their instruction.

Agricultural education is well suited for teachers to use all the principles of interest but especially the natural impulses. Anything living, plants and animals, or other real objects from nature touch on the love of nature within students. Hands-on learning taps into students'

BOX 11.3 *Principles of Interest*

Primary

1. Much of the interest a person exhibits comes from natural impulses within the person:

Love of nature	Competition
Altruism	Sociability
Curiosity	Ownership
Self-advancement	Desire for approval
Creativeness	Activity

2. Something is interesting if it affects us, others about us, or humanity at large.
3. Interest increases with an increase in related knowledge of any subject, provided such knowledge is well understood.
4. Interest increases with the acquisition of any given ability or skill.
5. Interest flows, or spreads, from any interesting thing into any uninteresting thing whenever the two are clearly connected in thought.

Secondary

1. Thinking is essentially interesting; memorization, uninteresting.
2. Interest is contagious in the sense that when one or more persons show interest in something, others will tend to "catch" that interest.
3. Interest is strengthened by a sense of progress.
4. Interest is created and sustained by a state of suspense.
5. An idea, when fully accepted, becomes a new interest center, from which interest will spread to any other thing that is seen to be connected with it.
6. The novel and unexpected are interesting.
7. Humor creates interest.

_____

Adapted from Crunkilton & Krebs (1982).

FIGURE 11.5 A variety of realia (real things or likenesses of real things) stimulates interest and helps motivate student learning. (COURTESY OF EDUCATION IMAGES.)

creativeness, activity, and self-advancement. When structured appropriately, hands-on learning can also involve ownership, competition, and approval. Community service activities are one example of altruism. Group projects allow students to be sociable but also involve activity, creativity, and ownership.

REVIEWING SUMMARY

To be effective, agriculture teachers need to know how students learn. The goal of instruction is to engage all students actively.

Educational practice in agricultural education has been influenced by both behavioral and cognitive learning theories. Brain-based learning theory gives an explanation of why some agricultural education tenets work and an impetus to change those teaching practices that do not work.

Students are intelligent in multiple ways. Agriculture teachers need to structure instruction so students of differing intelligences can all succeed. The best teachers are cognizant of student mindsets and learning styles and tailor instruction to meet student needs.

Motivation is a key factor in the learning environment. Without motivation, there will be no learning. The goal is to achieve intrinsic motivation.

QUESTIONS FOR REVIEW AND DISCUSSION

1. What is learning?
2. Where does learning occur?
3. Describe how an agricultural education class is conducted by a teacher who is a behavioral learning theory adherent.
4. Describe how an agricultural education class is conducted by a teacher who is a cognitive learning theory adherent.
5. What is the Premack principle?
6. What is downshifting?
7. Explain Gardner's theory of multiple intelligences.
8. Explain Dweck's growth mindset theory.
9. Explain the impact of Maslow's hierarchy of needs on agricultural education classrooms.
10. How are the primary and secondary principles of interest used in agricultural education classrooms?
11. What is motivation? Differentiate between the two types.
12. How does classical conditioning occur in the classroom? Explain.

ACTIVITIES

1. Investigate mindset theory. Chose an agricultural topic and develop a lesson incorporating activities and challenges that encourage students to enhance their growth mindset.
2. Investigate Gardner's theory of multiple intelligences. Report on the eight dimensions using a different FFA Career Development Event to highlight each dimension.
3. Using the *Journal of Agricultural Education, The Agricultural Education Magazine,* and the proceedings of the National Conference of the American Association for Agricultural

Education, report on learning theory and learning styles research in agricultural education. Current and past issues of these journals can be found in the university library. Selected past issues can be found online at www.aaaeonline.org or www.naae.org.

REFERENCES

Bandura, A. (1986). *Social foundations of thought and action*. Prentice Hall.

Bandura, A. (1997). *Self-efficacy: The exercise of control*. W. H. Freeman.

Caine, R. N., & Caine, G. (n.d.). *12 brain/mind natural learning principles*. Retrieved in 2013 from https://www.nlri.org/

Caine, R. N., & Caine, G. (1991). *Making connections: Teaching and the human brain*. Association for Supervision and Curriculum Development.

Caine, R. N., & Caine, G. (1994). *Making connections: Teaching and the human brain* (rev. ed.). Addison-Wesley.

Caine, R. N., & Caine, G. (1997). *Education on the edge of possibility*. Association for Supervision and Curriculum Development.

Crunkilton, J. R., & Krebs, A. H. (1982). *Teaching agriculture through problem solving* (3rd ed.). Interstate Printers & Publishers.

Dweck, C. S. (2006). *Mindset: The new psychology of success*. Random House.

Gardner, H. (1983). *Frames of mind: The theory of multiple intelligences*. Basic Books.

Gardner, H. E. (2008). *Multiple intelligences*. Basic Books.

Kolb, D. A. (1984). *Experiential learning: Experience as the source of learning and development*. Prentice Hall.

Martin, J. L., & Torok-Gerard, S. (Eds.). (2019). *Educational psychology: History, practice, research, and the future*. ProQuest Ebook Central. https://ebookcentral.proquest.com

Maslow, A. (1970). *Motivation and personality* (2nd ed.). Harper & Row.

National Council for Agricultural Education. (2016). *National quality program standards for agriculture, food, and natural resources education: A tool for secondary (grades 9–12) programs*. https://thecouncil.ffa.org/

Newcomb, L. H., McCracken, J. D., Warmbrod, J. R., & Whittington, M. S. (2004). *Methods of teaching agriculture* (3rd ed.). Prentice Hall.

Parsons, T., Lin, L., & Cockerham, D. (2019). *Mind, brain and technology: Learning in the age of emerging technologies*. Springer.

Santrock, J. W. (2008). *Educational psychology* (3rd ed.). McGraw-Hill.

Woolfolk, A. E. (1998). *Educational psychology* (7th ed.). Allyn & Bacon.

12

The Teaching Process

Juanita Rivera is an excited beginning agriculture teacher. It's the week before school begins, and her facilities are clean and organized. Her classroom has tables and chairs for 30 students and is equipped with an interactive whiteboard and internet-connected computer. Countertops with base storage cabinets and strategically placed sinks are along two of the walls. Against another wall are five internet-connected computer stations.

A walk through her office and into the laboratory reveals two aquaculture tanks with a hydroponics station and grow lights. Spaced throughout the remainder of the laboratory are workbenches and tool/equipment storage cabinets for various agricultural mechanization experiments and projects.

Juanita is continually thinking about one overriding question: "How will I teach my students to best help them learn?" Now, that is a good question. It is time she started planning!

OBJECTIVES

This chapter addresses the National Quality Program Standards for Agriculture, Food, and Natural Resources Education (National Council for Agricultural Education, 2016), specifically
Standard 1B: Program Design and Instruction—Instruction
Standard 1D: Program Design and Instruction—Assessment
It has the following objectives:

1. Discuss the principles that guide learning in the 21st century.
2. Explain teaching methods.
3. Identify considerations in selecting a method.
4. Describe Bloom's taxonomy and relate its use to teaching.
5. Compare and contrast student-centered and teacher-centered instructional methods.
6. Discuss examples of teacher-centered instructional methods.
7. Discuss examples of social interaction instructional methods.
8. Discuss examples of student-centered instructional methods.
9. Describe the process of developing and/or adapting appropriate lesson plans in teaching.
10. Discuss the use of instructional technologies in teaching.

FIGURE 12.1 The teacher facilitates classroom learning activities and empowers students to take responsibility for their learning.

TEACHING AND LEARNING IN THE 21ST CENTURY

Before selecting a teaching method, it is important to consider foundational principles of designing learning experiences to help students be prepared for the demands of the modern workforce. Darling-Hammond et al. (2008) identified the following ideas that should guide teaching and learning in the 21st century:

Students possess diverse experiences and knowledge they bring with them to classroom learning. Teachers should be sure to activate and build upon this prior knowledge to maximize learning.

To cultivate deep understanding and application of learning, students must have a foundation of factual and conceptual knowledge they know how to draw upon in practical applications in the real world.

Learning is more effective for students when they are aware of their own learning processes and learn how to identify and reflect on their own metacognition.

Remember that no matter what content you are teaching or what method you use to engage students in learning about the content, you are first teaching *students*. Building positive relationships with them; learning about their communities, values, and beliefs; and grounding your teaching within these contexts is absolutely critical for student success. The diverse array of learners in your classroom need a teacher who can create a positive and welcoming learning environment that embraces their identities. Utilizing a ***culturally responsive*** approach to teaching, teachers can build upon the cultural knowledge, experiences, and frames of reference of culturally diverse students to create more relevant and meaningful learning experiences that are validating, empowering, and transformative.

CHOOSING THE TEACHING METHOD

Several teaching methods or strategies are available to agriculture teachers. A ***teaching method*** is the overall means a teacher uses to best facilitate student learning. Most methods comprise a number of techniques or details for delivering the instruction.

Taxonomy of Learning Experiences

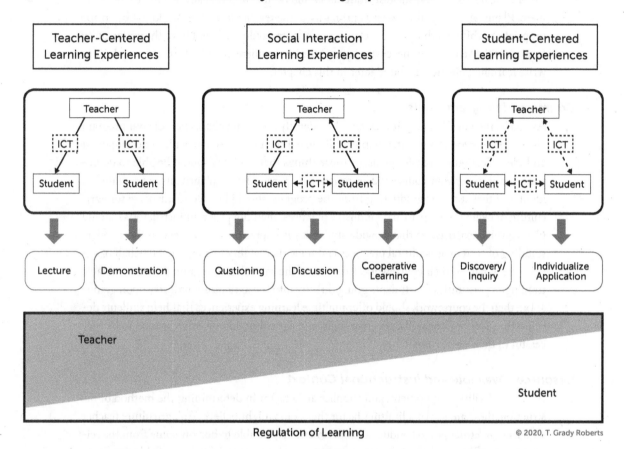

FIGURE 12.2 Teaching methods can be broadly categorized into teacher-centered, social interaction, and student-centered learning experiences. In the diagram, "ICT" refers to "information communication technologies." (COURTESY OF T. GRADY ROBERTS.)

Major Methods

Teaching methods can be broadly categorized based on who is the focus of the learning activities: teacher-centered, social interaction, or student-centered. Roberts et al. (2010) outlined major learning methods, as shown in Figure 12.2.

Factors Affecting Selection of Teaching Method

Selection of appropriate teaching methods should be guided by the established learning objectives for the lesson, but are also influenced by the overall goals of the course and program, resources available and educational context, and the experience of the instructor as well as the instructor's philosophical beliefs about teaching and learning.

Learning Objectives

Learning objectives outline what students should know, do, or be able to understand by the end of a lesson. Learning objectives should be aligned with your state or national educational standards frameworks to ensure daily instruction addresses the major concepts and skills deemed important by your state educational agency. To provide students with learning experiences that help them move toward the established learning objectives, agriculture teachers need to select an appropriate teaching method. For example, a learning objective that states, "Students will be able to independently create a boutonniere using the supplied materials" would suggest students should be able to physically assemble a boutonniere. Planning a lesson that consists primarily of the teacher lecturing would not be an appropriate teaching method in this setting, as students would likely not have the opportunity to work with the materials necessary to craft the boutonniere. More information on how to write learning objectives is found later in this chapter.

Course and Program Goals

As you learned in Chapter 5, it is essential to identify program goals as part of your program vision and mission statement. Chapter 7 discussed the process of curriculum development and identification of transfer goals, or those things that students should be able to do beyond the scope of your course and/or program. For example, if an agriculture program has identified that students graduating from the program should be able to advocate for agriculture drawing upon up-to-date industry practices, then the program instructor(s) should plan learning experiences that provide students with opportunities to deepen their understanding of current agricultural practices and hone their ability to advocate on behalf of agriculture as part of their coursework. Likewise, if the agriculture program provides valuable training opportunities for students going directly into the agriculture industry upon graduation, then the coursework should offer multiple learning experiences that help students develop the psychomotor skills, affective skills, and agricultural knowledge needed to be successful in their employment.

Resources Available and Instructional Context

Although facilities, equipment, and supplies are a factor in determining the method of instruction, these are less of a limiting factor than some might believe. An agriculture teacher who lacks microscopes to conduct an experiment may be able to borrow some from the science teacher. The problem of outdated equipment might be solved through field trips, computer simulations, or supervised experience at businesses with up-to-date equipment.

The abilities to improvise, borrow, and create are extremely valuable ones for an agriculture teacher to possess. Chapter 6 discussed utilizing advisory and citizen groups. These are helpful in getting needed resources for effective teaching. Chapter 9 discussed in further detail the role facilities play in teaching, and Chapter 21 discusses laboratories. Chapter 24 explains the importance of community resources for an effective agricultural education program.

Finally, considerations such as class size, time available to devote to a topic, and class structure influence the decision on what method of instruction to choose. As detailed below, some methods are best suited for individual or small-group instruction. Other methods are designed to cover a large amount of information as efficiently as possible. Class structural factors include time of day, time of year, proximity to holidays or vacations, and type of schedule system. Class periods under a block system, which can range from 85 to 100 minutes in length, require great variety in both instructional methods and learning tasks.

Experience and Philosophical Beliefs of the Instructor

Finally, the teaching philosophy and experience of the teacher influence what instructional approaches might be used so as to best reach the diverse array of students whom they teach. Teachers should be always striving to build their repertoires of teaching methods. Beginning teachers tend to utilize those methods they are most comfortable with and feel are the strongest. However, these teachers are cautioned not to overuse a particular method.

Chapter 11 outlined the theories that undergird the science of teaching and learning. Two broad categories of educational theories fall into behaviorist approaches and constructivist approaches. Teaching methods and instructional design that rely primarily on acquisition of discrete and compartmentalized skills and knowledge, and position the teacher as in control of the learning, draw from behaviorist principles. In this approach, the focus of teaching is on transmission of content. A teacher drawing from constructivist principles would focus more on helping students make connections to the content across different settings, and tends to craft lessons that utilize instructional methods that encourage students to have varying degrees of control over the learning experience.

Teaching methods that draw from behaviorist principles are typically teacher-centered approaches, while those that draw from constructivist principles are student-centered or social interaction approaches. While there are numerous teaching methods, this chapter will focus on the major approaches utilized by school-based agriculture teachers within each of these categories.

Choosing the appropriate method is not merely using a formula but involves numerous considerations. What principles of learning should be applied to the lesson? At what developmental level are the students? What are their learning styles and dimensions of multiple intelligences? Are there physical, mental, or emotional considerations? What resources are available to use in implementing a teaching method? What should students know and be able to do at the end of the lesson? At what level of learning are students expected to demonstrate knowledge, skills, or dispositions?

BLOOM'S TAXONOMY

The question of level of learning expected of students is important. In 1956, Benjamin S. Bloom classified educational objectives into what is commonly referred to as ***Bloom's taxonomy***. He first stated there were three domains of learning. The cognitive domain is those things involving mental activity. The affective domain is those things involving attitudes, emotions, and values. The psychomotor domain is those things involving body movements, mechanical skills, and manual skills. Figure 12.3 illustrates the three domains of Bloom's taxonomy.

Categories of Cognitive Skills

Within Bloom's taxonomy for the cognitive domain are six categories. Although Bloom did not present these as a hierarchy, the term "higher-order thinking skills" is sometimes applied to the last three of the six. The categories of the original taxonomy (from lowest to highest) are Knowledge, Comprehension, Application, Analysis, Synthesis, and Evaluation. Krathwohl (2002) presented a revised taxonomy. The three lower levels were renamed as Remember,

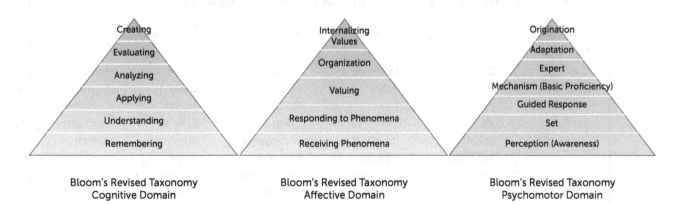

FIGURE 12.3 The cognitive, affective, and psychomotor domains of Bloom's revised taxonomy.

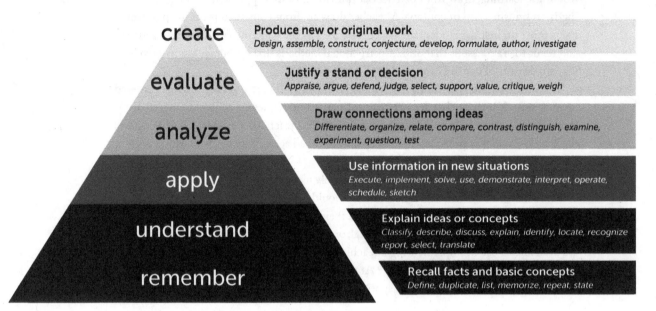

FIGURE 12.4 Bloom's revised taxonomy for teaching, learning, and assessment.
(COURTESY OF VANDERBILT UNIVERSITY CENTER FOR TEACHING.)

Understand, and Apply. The three higher-level categories were renamed as Analyze, Evaluate, and Create, with the top two category places switched. Figure 12.4 illustrates Bloom's revised taxonomy for teaching, learning, and assessment.

Boxes 12.1, 12.2, and 12.3 give examples of verbs associated with each of the domains.

Writing Objectives

Teachers use Bloom's taxonomy to write learning objectives (cognitive, affective, and psychomotor) to guide instruction, inform students, ensure coverage of subject matter, and guide evaluation. A *learning objective* is a carefully written statement of the intended outcomes of a learning activity. Several variations are used, including behavioral objectives. Teaching proceeds best if the learning objectives have been carefully specified.

BOX 12.1 *Verbs Associated With Bloom's Cognitive Domain*

Remember	Understand	Apply	Analyze	Evaluate	Create
Define	Classify	Execute	Differentiate	Appraise	Design
Duplicate	Describe	Implement	Organize	Argue	Assemble
List	Discuss	Solve	Relate	Defend	Construct
Memorize	Explain	Use	Compare	Judge	Conjecture
Repeat	Identify	Demonstrate	Contrast	Select	Formulate
State	Locate	Interpret	Distinguish	Support	Author
	Recognize	Operate	Examine	Value	Investigate
	Report	Schedule	Experiment	Critique	
	Select	Sketch	Question	Weigh	
	Translate		Test		

BOX 12.2 *Verbs Associated With Bloom's Affective Domain*

Receiving Phenomena	Responding to Phenomena	Valuing	Organization	Internalizing Values (Characterization)
Asks	Answers	Completes	Adheres	Acts
Chooses	Assists	Demonstrates	Alters	Discriminates
Describes	Aids	Differentiates	Arranges	Displays
Erects	Complies	Explains	Combines	Influences
Follows	Conforms	Follows	Compares	Listens
Gives	Discusses	Forms	Completes	Modifies
Holds	Greets	Initiates	Defends	Performs
Identifies	Helps	Invites	Explains	Practices
Locates	Labels	Joins	Formulates	Proposes
Names	Performs	Justifies	Generalizes	Qualifies
Points to	Practices	Proposes	Identifies	Questions
Replies	Presents	Reads	Integrates	Revises
Selects	Reads	Reports	Modifies	Serves
Sits	Recites	Selects	Orders	Solves
Uses	Reports	Shares	Organizes	Verifies
	Selects	Studies	Prepares	
	Tells	Works	Relates	
	Writes		Synthesizes	

BOX 12.3 *Verbs Associated With Bloom's Psychomotor Domain*

Perception (Awareness)	Set (Mindsets)	Guided Response	Mechanism (Basic Proficiency)	Complex Overt Response (Expert)*	Adaptation	Origination
Chooses	Begins	Copies	Assembles	Assembles	Adapts	Arranges
Describes	Displays	Traces	Calibrates	Calibrates	Alters	Builds
Detects	Explains	Follows	Constructs	Constructs	Changes	Combines
Differentiates	Moves	Reacts	Dismantles	Dismantles	Rearranges	Composes
Distinguishes	Proceeds	Reproduces	Displays	Displays	Reorganizes	Constructs
Identifies	Reacts	Responds	Fastens	Fastens	Revises	Creates
Isolates	Shows		Fixes	Fixes	Varies	Designs
Relates	States		Grinds	Grinds		Initiate
Selects	Volunteers		Heats	Heats		Makes
			Manipulates	Manipulates		Originate
			Measures	Measures		
			Mends	Mends		
			Mixes	Mixes		
			Organizes	Organizes		
			Sketches	Sketches		

*Key verbs are same as Mechanism, but are performed at a quicker or higher quality level.

Learning objectives are always written in terms of what the student will be able to do at the conclusion of the instruction. Some school districts may expect teachers to format their learning objectives as "I can ..." statements so that the objectives are written in student-friendly language. Writing objectives using Bloom's taxonomy is as simple as ABCD.

Consider this example: "Students will be able to quote the FFA Creed from memory with 80% word accuracy."

- "A" stands for audience and is almost always students.
- "B" stands for behavior, or what is expected of the students. In this example, the students are to quote the FFA Creed. (This is sometimes known as the "terminal behavior.")
- "C" stands for the condition under which the behavior is to be done. In this example, students are to quote from memory.
- "D" stands for the degree or level to which students are to perform to succeed. In this example, students must be 80% accurate, or can misspeak only 20% of the creed.

TEACHER-CENTERED METHODS

Different methods of instruction assist students in accomplishing the learning objectives. Teacher-centered instructional methods center the focus of the learner's attention primarily on the instructor. The teacher is almost exclusively responsible for structuring the learning environment and serves as the primary conveyor of information. These methods tend to be utilized for whole group direct instruction. Lecture, demonstrations, and teacher-led experiments are commonly used teacher-centered methods in school-based agricultural education.

Lecture

Very few agriculture teachers lecture in the formal definition of the methodology. An accomplished lecturer should be an expert in the subject, be skilled in oratory, and have experience and presence that earn respect from the audience.

However, the use of lecture in its many forms is a frequently used instructional method. As commonly defined, *lecture* is an oral presentation by the teacher or other individual. It may include a wide range of techniques. Lecture involves these five qualities:

1. More teacher talk than student talk.
2. Greater degree of flow of communication from teacher to students than from students to teacher or from student to student.
3. Linear flow of information, with the structure determined by the teacher.
4. Use of illustrative materials to supplement what is being spoken. These include PowerPoint presentations, chalkboard/whiteboard notes and illustrations, overhead transparencies, slides, real objects, audio and video presentations, and pictures.
5. Use in combination with discussion to create lecture-discussion. This is typified by periods of lecture with note-taking followed by class discussion, analysis, or application of the preceding content.

When does a teacher choose lecture as a preferred instructional method? Lecture is best suited to large groups as is often seen in college classrooms. Lecture is also best when the content is such that all students need to have the same information on the same level. Within

---

BOX 12.4 *Characteristics of a Good Lecturer*

1. Thoroughly prepares for the lecture
2. Sees that the room is properly arranged for the lecture
3. Has knowledge of the audience the helps in relating to their interests
4. Is enthusiastic, promoting student interest
5. Speaks clearly and fluently; varies rate and intensity of speech
6. Avoids distracting mannerisms
7. Emphasizes major points and uses appropriate realia (real things) to promote learning
8. Chunks learning into appropriately sized segments
9. Properly introduces and summarizes the subject

Bloom's taxonomy, knowledge and comprehension objectives are well suited for the lecture methodology. When time is a concern, lecture allows for disseminating a large amount of information in the least amount of time.

A good lecture requires organization. Unprepared teachers who believe they can just "throw together" a lecture at the last minute are fooling themselves and stealing instructional time from their students. Class needs to begin with a well-thought-out introduction. What are the students expected to know and be able to do after the instruction; in other words, what are the objectives? Why should students want to learn the content? What do students already know in this area, or what have they studied in previous lessons related to this area? The introduction should transition into a structured presentation of new content. If the topic is new, the lecture should first present the whole picture, then break it into its components, then present the whole again.

High school students should be provided an organizational outline to guide their note-taking from the lecture. This outline could be done on the chalkboard/whiteboard, as a PowerPoint presentation, or in the form of a note-taking guide. When lecturing to high school students, the teacher needs to check for understanding through questions, exercises, and frequent reviews.

Box 12.4 lists the characteristics of a good lecturer. Teachers should evaluate themselves on each of these characteristics and work to improve those in which they have a deficiency.

Demonstration

The agricultural education curriculum has many topics that are well suited for demonstrations. There are skills in each agricultural area—mechanics, horticulture and landscaping, plant and soil science, natural resources, agricultural business, animal science, and food science—in which demonstrations are appropriate. A *demonstration* is a teaching process that involves showing students how to do something before they do it for themselves or that involves showing them what would happen as the result of a particular action. Practice by students of the skill demonstrated is essential in order to assure their skill development.

As in other group teaching methods, planning is essential to the success of a demonstration. All materials need to be gathered ahead of time, the teacher needs to have practiced the demonstration, and care must be taken to ensure all students can see the demonstration. The teacher may want to have parts of the demonstration completed in advance, especially if time is a consideration. For example, demonstrating proper painting techniques may require letting the primer dry for 24 hours before putting on the coat of paint. Instead of taking multiple days to do the demonstration, the teacher may demonstrate applying primer, then pull out a previously primed wall section that has dried and use it to demonstrate applying the coat of paint.

When planning a demonstration, it is important for the agriculture teacher to consider the layout of the room or space in which they will be demonstrating a skill so they can ensure all students will be able to see the skill demonstrated. The demonstration may be aided by the use of a document camera that projects a live video feed to a larger screen so students can view the small details of what the teacher is demonstrating, such as when demonstrating how to make a bow or corsage. Demonstrations can also be helpful when the teacher has limited resources for all students to engage with the materials directly, such as when conducting a dissection of a ruminant digestive system. These systems, which are quite substantial in size, are not easily obtained in large quantities, nor would many agriculture classrooms be suited to large quantities of these tracts laid out around the room. Through a demonstration with an overhead document camera, an agriculture teacher can show students the fine details of the various parts of the digestive system in an efficient manner.

Teacher-Led Experiments

Because agriculture is an applied science, it is especially important experiments be a part of instruction. All areas of agricultural education lend themselves to experiments. An ***experiment*** is a trial using a definite and planned procedure. The outcome of the trial is predicted as part of the scientific method. Experiments can be carried out using very formal procedures or less formal ones. Teacher-led or teacher-directed experiments are those experiments in which the teacher determines what research question will be explored and what procedures will be followed. Often these experiments follow a step-by-step format that students follow to achieve the teacher-planned outcomes, leading to this approach sometimes being referred to as a "cookbook" method of experimentation. Often, this approach is used when a teacher wants to have students see a scientific principle in practice, or help students familiarize themselves with following laboratory procedures, using laboratory equipment, and engaging in the scientific method.

What is the value of experiments? Experiments are intertwined with the scientific method, which is the basis for the problem-solving teaching approach. Experiments teach observation skills, measurement, analysis, and reasoning. Experiments utilize a "learning by doing" approach that actively engages the students physically as well as mentally. While teacher-directed or cookbook experiments have been criticized for not actively engaging student thinking in the laboratory setting, recent research suggests teacher-directed experiments can positively influence student learning about the nature of science and positively influence student enjoyment of science. When blended with inquiry-based instruction opportunities (discussed in the student-centered methods section), teacher-directed laboratory instruction may benefit student learning in science even more than when used alone (Areepattamannil et al., 2020).

Teacher planning and preparation are keys to the success of using experiments as a teaching method. An experiment needs to be planned thoroughly. The teacher must have done the experiment beforehand in order to anticipate student questions and frustration areas. The teacher must decide whether supplies and equipment will be needed for each student or whether students will work together in groups. In addition, the students need to be prepared for the experiment. What are the procedures to follow? What are students to observe? How are students to record and analyze data? Students also need to know experiments do not always yield perfect data. Sometimes experiments that yield outlier data are the most interesting. It is not enough to simply do the experiment; the students need to understand why the experiment yielded the results.

> **BOX 12.5 *Steps in Conducting an Experiment***
>
> 1. Interest is developed. The appeal may flow from a student's SAE, an FFA CDE, or personal interests. The teacher may also create interest through techniques discussed in Chapter 11.
> 2. The problem is defined. What do the students want to discover?
> 3. Information is gathered. Students find out what is already known about this problem. This is also known as a literature review.
> 4. Hypotheses are formed. What possible explanations are there for the observed phenomena?
> 5. The experiment is planned. Supplies and equipment are gathered. Procedures are determined. A timeline for data collection is developed.
> 6. The experiment is conducted. Data are collected through observations and measurements.
> 7. Data are analyzed, and conclusions are drawn.
> 8. A lab report is written.

Box 12.5 shows the typical steps in conducting an experiment. Experiments in agriculture may last for multiple days or even weeks, so several data collection sheets may be necessary.

SOCIAL INTERACTION METHODS

Social interaction methods of learning are those that involve interaction not only between students, but between the teacher and the students, so all parties are interacting with another throughout the learning experience. The teacher may not be the focus of the learning activity, but has assuredly spent much time up front scaffolding the learning experiences to allow for more interaction among students and the instructor. Social interaction methods include questioning and discussion, cooperative learning, project-based learning, role-plays, field trips, and guest speakers.

Discussion and Questioning

Discussions are commonly used in agricultural education. A *discussion* is a teaching process that involves student sharing of information. Typically, discussion might involve the agriculture teacher posing questions to the students, which they answer directly, offer their opinions about, or debate. Discussion may be combined with lecture, as described in the previous section, or be the sole instructional method used.

A well-conducted discussion requires extensive planning by the teacher. The teacher needs to determine the objectives of the discussion ahead of time and structure the discussion to accomplish the objectives within the time frame selected. Nothing is more frustrating than an endless discussion with no conclusion or resolution of the topic discussed. A good discussion involves student-to-student interaction in addition to student-to-teacher and teacher-to-student.

Box 12.6 presents qualities of a good discussion. All participants—the students as well as the teacher—need to be ready for the discussion. The teacher should have a lesson plan for the discussion, just as he or she would for any other lesson using another instructional method.

A properly organized discussion will flow from point to point logically yet allow for student insights and questions. Students need to possess the necessary background information to make informed comments. This may require students to have read materials for homework

---

BOX 12.6 *Qualities of a Good Discussion*

1. An appropriate situation exists for discussion.
2. Students possess the background knowledge necessary to give informed answers or opinions.
3. The classroom climate is such that students feel safe in voicing opinions and answering questions; they do not fear ridicule or embarrassment.
4. The teachers asks appropriate-level questions, asks probing questions when students give one-word answers, and provides answer prompts when necessary.
5. Incorrect information is corrected while the dignity of the student who gave the wrong information is maintained.
6. All students are actively engaged in the discussion.
7. Students have ample opportunities to contribute to the discussion, with no one monopolizing the time.
8. The teacher actively listens, provides feedback, and clarifies when necessary.
9. Key points are highlighted during the discussion and are summarized at the end.
10. The discussion is brought to a conclusion.

---

or during previous class instruction. During the discussion, the teacher serves as a facilitator, prompter, and conversation referee. Depending on the class size, the teacher may need some system for making sure all students participate in the discussion. Before the allotted time expires, the teacher needs to bring the discussion to closure, highlight key points, and assist the students in formulating conclusions.

The types of questions asked in a discussion are critical to the types of answers students will give and the level of their learning. There are five basic types of questions: closed, open, probing, higher-order, and divergent. Closed questions are used to regulate answers or obtain specific answers. For example, "Lecture is best used in what situations?" Open questions are useful for starting a discussion and seek a variety of answers. Such questions many times allow students to describe their own experiences or express opinions, so there are no wrong answers. Probing questions are used when students give one-word answers, more information is needed, or there is a desire to bring other students into the discussion. Sometimes probing questions mirror what a student has said. For example, "Are you saying you think lecture is overused?" Higher-order questions require students to think rather than recite memorizations. Higher-order questions are designed to make students discover rules and principles as they express their ideas. These questions use verbs from the upper three levels of Bloom's taxonomy, such as analyze, generalize, and critique. Divergent questions have no "right" answers. These require students to use their imaginations and think beyond set borders or parameters. Divergent questions lead to creativity and new ideas and are sometimes referred to as open-ended-answer questions.

Cooperative Learning

In *cooperative learning*, students work together in a group small enough to allow all members to collectively contribute to a central, clearly defined task without necessarily needing the direct supervision of the teacher. Learning in a group setting can help students learn how to work with others, practice communication skills, and promote problem-solving. A recent survey of agricultural employers indicated being dependable, ability to think critically, engage in strategic planning, clear communication, active listening, and problem-solving skills as being the most desirable workforce competencies of agriculture and natural resources college graduates (Easterly et al., 2017). Planning opportunities for students to work toward development of these competencies at the high school level through cooperative learning experiences can help your students be much more prepared for the demands of the workforce.

When planning to have students work cooperatively in groups, it is very important to establish structured expectations of how the groups should function. The teacher may assign specific students to roles that promote on-task and collaborative behavior, or provide students a method through which to assign students to various roles. For example, one student might serve as the discussant who facilitates the conversation of the group, another might be the note taker, and another might be assigned to intentionally bring up opposing views to drive discussion and analysis. When determining the arrangement of students into such groups it is wise to consider the social dynamics that exist between students, demographic differences that may result in inequitable group behaviors, and specific learning needs of students who may have an IEP or 504 Plan.

Project-Based Learning

Though not always occurring in a group setting, *project-based learning* is an approach to learning and student engagement that can be used to guide a particular unit or course, or even as a framework for how an entire school may facilitate learning. Project-based learning or PBL provides opportunities for students to investigate an authentic and complex question, problem, or real-world challenge in a personally meaningful way.

While many teachers may use projects in their classes, "doing a project" is not the same thing as project-based learning. In PBL, the project *is* the unit; it is the main focus of the unit and requires students to engage in critical thinking, problem-solving, and various forms of communication to answer a driving question that guides the path of the project. Conversely, "doing a project" in a class differs in that the project is often completed at the conclusion of a unit, after the content has been taught by the teacher.

For example, an agriculture teacher might teach a welding fabrication unit in which students craft a welded project determined by the teacher, such as a hay rack, creep feeder, gate, or truck bed liner. Students would typically create these projects after having participated in lengthy instructional experiences led by the teacher on how to construct the selected project. If this same unit were to be redesigned to utilize a project-based approach, the teacher might present students with a problem that can be solved through welding fabrication, such as a local horse farm seeking to redesign their hay racks after experiencing challenges with the currently existing design. Students might visit the farm, or examine sample hay racks from the farm to troubleshoot the source of their problem, and then work together to design new racks that provide a better solution. Once all hay racks have been designed and fabricated, the class might install the hay racks for the barn owner and communicate how their new designs will address the problem identified.

Role-Play

Role-playing is the act of an individual assuming the real or imagined role of another individual. It is a useful instructional methodology for the agricultural education classroom when teaching concepts related to human relations, developing leadership skills, or teaching students other basic agricultural skills. Role-playing allows students to practice these skills in a safe and simulated environment so they are better prepared when encountering similar situations outside the classroom.

Newcomb et al. (2004) identified the following steps to planning and implementing role-play in the agriculture classroom:

1. Select appropriate subject matter for role-play (do *not* use culturally sensitive topics that demean, diminish, or dehumanize the experiences of others, particularly those of historically oppressed groups).
2. Determine the learning objectives for the lesson.
3. Design student roles to be played so they are aligned with the learning objectives.
4. Determine how many students will actively participate (will it be all students in the class, or a select few "on stage" in front of the class?).
5. Prepare the students who are participating in their roles for what they should be prepared to demonstrate.
6. Plan how you will help students reflect upon their experience and observations, and identify the key takeaways.

The FFA component of agricultural education provides many opportunities for student role-playing. When studying parliamentary procedure, students are better able to comprehend officer roles and FFA ceremonies by role-playing officers and members and conducting FFA business. Role-playing is also a useful tool in the classroom when discussing sensitive topics. A debate on the place animal rights have in the animal production industry could quickly become heated and unruly. If, instead, students are assigned to role-play various characters, such as animal rights activist, rancher, feedlot owner, and consumer, then the situation becomes less personal and more objective.

Field Trips

Field trips are a useful instructional method for making the "real world" part of the planned learning experiences. A *field trip* is a learning experience that involves traveling away from the agriculture classroom. An example is a field trip to an agricultural experiment station to observe research projects that are underway.

Field trips can be used for one of three purposes. At the beginning of an instructional unit, field trips are beneficial as an interest approach and to help the students to see the whole of a topic. Field trips during the middle of a unit can serve as fact-finding missions for problem solving or provide opportunities to test hypotheses. At the end of an instructional unit, field trips are useful for demonstrating theory into practice and bringing the parts back into a whole picture.

Field trips that involve travel must be planned and approved well in advance. School systems sometimes restrict when field trips can be taken and how many can be taken within the school year. School administrators may insist trip planning and approval forms be submitted at the beginning of the school year for any field trips taken that year. This requires the teacher to be organized and to prioritize which trips are crucial to student learning. Another consideration is transportation. Details such as cost, a bus driver, parental permission, and notification of other teachers must all be planned.

Field trips do not have to involve taking buses to faraway sites. Mini-trips can be taken right on the school grounds. For schools on block scheduling, the longer class periods allow mini-trips to and from sites close to the school within the agriculture class period. Virtual field trips may not involve the sensory experiences real field trips provide, but they can be an acceptable substitute. Videos, the internet, and student presentations can all provide some sense of "being there."

Guest Speakers

Agricultural education can be a difficult curriculum to teach. Very few agriculture teachers, if any, feel comfortable in their knowledge of every agricultural subject area. *Resource people*, who are experts in their area, can provide the knowledge and insights the agriculture teacher may not have as a guest speaker. Resource people can also lend credibility to information or instruction students doubt or about which they have heard conflicting opinions. In addition, by utilizing resource persons in the classroom, the agriculture teacher is introducing their students to key community contacts and potential employers.

Resource people may be used in several ways. An individual resource person may be invited to present to the class in a lecture, discussion, or demonstration format. Another common procedure is to invite three to five resource persons to conduct a panel discussion. Given the flexibility virtual video conferencing platforms offer, the teacher might also be able to include guest speakers over a virtual format, opening up the possibility to feature experts and perspectives not just from the local community, but across the country and the world! No matter the approach, the agriculture teacher should contact each resource person several weeks in advance of the presentation date. This should be followed by an email, letter, or phone call describing the class, the objectives of the class period, and suggested topics of discussion. A day or two before the presentation, the agriculture teacher should contact each resource person to confirm date, time, and location and answer any last-minute questions. If the speaker will be coming to school, it will also be important to review any procedures such as checking in at the main office, parking, and bell schedules.

STUDENT-CENTERED METHODS

Methods that engage students as the primary agent of learning are student-centered. Development of interpersonal relationships between the teacher and the student are essential to the success of student-centered methods. Students may be involved in developing the learning goals in collaboration with the teacher, with the teacher serving as a guide and resource. While teacher-centered methods are sometimes described as "sage on the stage" approaches, student-centered methods require the teacher to be a "guide on the side" to allow students to take leadership in their own learning. Common student-centered methods used in agricultural education include inquiry-based instruction, case studies, scenarios, simulations, games, supervised individual study, independent studies, learning modules, and projects.

Inquiry-Based Instruction

A common practice in science classrooms, *inquiry-based instruction* is a teaching method that meshes the natural curiosity of students along with the scientific method to cultivate critical thinking skills and scientific knowledge. Inquiry-based approaches to learning attempt to replicate the natural problem-solving processes we engage in when observing new phenomena or asking questions about the natural world.

There are varying levels of student inquiry that rely on instructional scaffolding to support student learning. The most teacher-supported type of inquiry is structured inquiry. A structured inquiry is a teacher-led or teacher-directed experiment, in which the entire class engages in exploration of the same research question. In controlled inquiry, the teacher selects the topics to be explored and identifies resources that students will use to answer the selected question; in this manner the teacher can intentionally curate the learning experience of students. In guided inquiry, the teacher might pose a question to students and then allow students to design a product or solution using their choice of resources. Finally, the most loosely structured form of inquiry is free inquiry or open inquiry. In this approach, students decide what to research and how they will conduct their research. Integrating the National FFA Agriscience Fair into your instruction is one way students can easily engage in inquiry. The degree to which students have freedom to make decisions is determined by how much scaffolded support is provided by the teacher.

When utilizing inquiry-based approaches it is important to consider how you will support student learning. Knowledge of student learning preferences, self-efficacy, and experience with more autonomy in the classroom can help the teacher determine to what degree they should structure a learning experience using inquiry. Facilitating inquiry-based instruction can be challenging for teachers accustomed to teacher-centered classrooms. Several professional development resources exist for teachers looking to expand their ability to facilitate inquiry-based learning, such as NAAE's National Agriscience Teacher Ambassador Program and Curriculum for Agricultural Sciences Education (CASE) institutes.

Case Studies

Case studies are stories with an educational message. They put students in the position of the problem solver situated in the story, but do not provide answers, generating an action-oriented teaching environment. According to Herreid (1998), a good case:

- Tells a story, focusing on an interest-arousing issue
- Is set in the past five years and is relevant to the reader
- Creates empathy with the central characters
- Includes quotations
- Has pedagogical use
- Is conflict-provoking, forcing the reader to make a decision
- Is general enough that it can apply to a myriad of situations
- Is short enough to hold the reader's attention but long enough to introduce the facts and complexity of the case

Case studies can be used to explore a topic or kick off a unit. For example, an agriculture teacher teaching a natural resources class might craft a case study about an invasive species affecting the local area or state to introduce a unit on management of invasive species. To guide student learning, they might chunk the case into several pieces and selectively distribute it at key moments across the unit to propel student decision-making and drive instruction.

Scenarios

Similar to case studies, *scenarios* are short stories focused on a particular group of individuals that describe the nature of the individuals involved, the context, and the goals they want to achieve. Scenarios are generally shorter than case studies, ranging from a few sentences to a paragraph long. They may be used to activate prior knowledge, provide opportunities for students to apply their learning, or be used to assess learning. There are five types of scenarios: skill-based, problem-based, issue-based, speculative, and gaming. For more information on these five types of scenarios, see Chapter 13.

Simulations

Simulations are learning activities that are used to replace and amplify real experiences with guided ones in a controlled setting. They are often immersive in nature and are closely aligned with role-plays, but can also be used when the teacher has students act out various abstract concepts in a concrete, tangible way. For example, in an animal reproduction unit, the agriculture teacher might have students represent different structures and hormones within the bovine reproductive tract to kinesthetically illustrate the estrous cycle.

Simulations can also occur in virtual settings. For example, students might practice welding using welding simulation software, or they might conduct dissections using a fetal pig dissection website. Other virtual simulations allow students to remove or add various variables to illustrate the outcomes through a computer-generated animation. Simulations such as these can be very helpful tools to replace actual immersive experiences or enhance students' learning.

Games

Learning games can have several practical uses within an agriculture classroom, as described in Chapter 13. They involve students in competition against themselves, a computer, or other students, and highlight problem solving and decision making. Games can be used to introduce or reinforce content and can occur in a face-to-face setting or virtual setting.

Supervised Study and Individualized Application

Supervised study is a method in which students are directly responsible for their own learning while under the direction of the teacher. When used effectively, supervised study enhances student interest and develops students' problem-solving abilities. To guide the students, most supervised study involves the use of worksheets or of questions to be answered.

Although supervised study is typically thought of as students reading textbooks, students' use of references can take many forms. Videos, pictures, extension publications, commercial brochures, journals and magazines, internet resources, and resource people are all examples of valid references.

Care should be taken when the internet is used as a resource. Students should understand that on the internet:

- Not all websites are appropriate.
- Not all information is accurate.
- Not all information is unbiased.

The agriculture teacher should work with the school media specialist in determining appropriate resources and teaching students information-searching procedures.

The supervised part of supervised study is critical. The agriculture teacher is responsible for directing the supervised study. The teacher should be in a location where they can see all the students and should not be using supervisory time for tasks such as making telephone calls or grading assignments. Occasionally the teacher should circulate around the room and monitor student progress. This does not mean interrupting students with chitchat or needlessly giving answers. Instead, the teacher can provide encouragement, assist with misunderstandings, and assess individual comprehension of the topic.

Although supervised study is categorized as individualized instruction, there may be times when students can work together—for example, on group projects or presentations and during peer-assisted instruction. Grouping students is also appropriate when reference materials are limited.

Guided practice is a special form of supervised study. Whereas supervised study is typically associated with cognitive and affective learning objectives, guided practice is used for psychomotor objectives. In guided practice, the agriculture teacher demonstrates a skill or activity and then provides opportunity for students to practice the skill while under the watchful eye of the teacher. The teacher can quickly correct bad habits or safety problems individually or bring the students together for group instruction if appropriate.

Independent Study

There are occasions when it is appropriate to teach an agriculture course or topic as an independent study. An *independent study* is a learning activity that is not part of an organized class learning experience. Maybe a student wants to take an advanced level of an agriculture course and the number enrolled is not enough to justify a class period for the course. In such a case, the student would take on more responsibility for their instruction.

To be successful, independent study must have clear objectives, guidelines, and timelines in place at the beginning. Although the student assumes greater responsibility for finding resource materials and studying those materials, the agriculture teacher is still responsible for evaluation and overall direction.

Increasingly, independent study opportunities are available through distance education. Instruction may be delivered via packets of printed material, videos, the internet, or combinations of these. Students may be eligible to receive high school or college credit or both.

Learning Modules

A *learning module* is a self-contained activity that a student can complete independently in a face-to-face or online setting. Learning modules are most often used for cognitive and psychomotor objectives. Learning modules may be purchased or may be developed by the agriculture teacher. Modules placed around the room can be combined into stations that students rotate through as they complete the activities. An advantage of learning modules is that students can proceed at their own pace and instruction can be modified to meet students' individual learning needs. Learning modules are also useful in meeting FFA membership requirements for nonenrolled students.

Projects

Agricultural laboratories are ideal for individual student projects. A *project* is a significant, practical activity of educational value with one or more definite goals. An advantage of projects is they can maximize student interest because the student has flexibility in choosing a project to complete. The project can also be geared to the ability of the student. Thus, instruction is available at the appropriate level for all students.

As with other methods discussed, projects require planning and active participation by the agriculture teacher. The teacher must instruct students on wise use of time; responsibility for safety, cleanliness, and productivity; and the importance of doing a quality job. The agriculture teacher must answer several questions before using individual projects as an instructional method. What is the learning purpose behind a project? What are the minimum requirements for an acceptable project? What is the maximum time allowed for project completion? What are the procedures for using tools/equipment, cleaning up the laboratory, and obtaining materials/supplies? What safety and behavior rules need to be emphasized?

DEVELOPING A LESSON PLAN

A *lesson plan* is the road map an agriculture teacher uses to ensure effective, efficient, and empowering instruction. It is much like the plans used in building a house or the script used in making a movie or television program. Doctors performing surgery devise plans of action, sometimes even placing dotted lines where incisions should be made. Football coaches develop plays, playbooks, and game plans. Teachers also need direction and vision! It should be made clear, however, that teachers are not movie actors and lesson plans are not scripts. A well-developed lesson plan allows the teacher to be flexible and creative and to take advantage of the "teachable moment."

There are many styles and formats of lesson plans, yet they all have some common components. Following are the essential components of a lesson plan.

Preparatory block   The first essential component is the preparatory block. This contains organizational information useful for filing and indexing, documentation, and quick referencing. The block also includes unit and lesson titles, the length of the class period for which the lesson plan is designed, as well as the intended grade level of the lesson.

Desired results block   Beginning your instructional planning by identifying your educational goals and objectives is important. Here, the educational standards, key ideas/questions explored by the lesson, and key vocabulary terms are listed. As part of this block are the assessments of learning aligned with the learning objectives. The teacher might also detail in this section how they plan to differentiate their lesson to meet the needs of specific learners in their class. For more information on differentiated instruction, see Chapter 7.

Materials and resources block   This block contains a list of what materials are needed to deliver the lesson, including worksheets, handouts, printouts of PowerPoint presentations, and note-taking guides.

Lesson initiation block   The next component is the lesson initiation, or beginning of the lesson, block. This part of a plan contains the interest approach, review of previous material, pretests, and other activities that occur at the beginning of a lesson.

Content and teaching–learning activities block   The content and teaching–learning activities block is next. The two sections of this block may be set up one on top of the other or side by side. Depending on the lesson plan style, the sections of this block may be very detailed or serve as outlines. At a minimum, the content section contains an outline of the subject matter with key information highlighted. There is no need to repeat word for word information that is already found in a textbook or on a worksheet.

The teaching–learning activities section describes the strategies used to teach the content. These include instructions to the teacher on methods, questions to be asked of students, student activities, and procedural instructions, such as when or how to do a task. This section should include checks for understanding and scaffolding techniques. When designing the learning activities for a lesson, it is important to consider the three primary elements of the Universal Design for Learning (UDL) framework: (1) How will the teacher plan to provide students multiple means of engagement with the content? (2) How will the teacher plan to provide multiple means of content representation? (3) How will the teacher provide students with multiple means of action and expression of their learning?

Lesson closure block   The lesson closure block provides details on how the lesson will be summarized and what conclusions can be drawn from the lesson. This block may also contain details on how this lesson links back to previous lessons and forward to the next lesson. It is important to include suggestions on how students can apply the lesson content. The evaluation section contains a description of how the students will be assessed, and copies of the evaluation instrument(s) and answer key(s) are attached to the plan.

In general, a lesson plan will cover one to five instructional periods. A lesson plan that takes less than one instructional period to cover means the topic is probably too narrow for a complete lesson. A lesson plan that takes more than five instructional periods to cover may mean the topic is a unit of instruction that needs to be divided into two or more lessons. Also, the teacher may develop daily plans to guide the introduction, teaching–learning, checks for understanding, and evaluation for each class session. These are broad guidelines and do not apply to all situations. A sample lesson plan template can be seen in Figure 12.5.

Preparatory block

Lesson topic: _____ Grade level: _____

Unit & course: _____ Length of class period: _____ minutes

Prepared by: _____

Desired results block

Desired Results
Educational standards addressed by this lesson: • •
What key problem(s) or concepts are students investigating during this lesson?
Key vocabulary terms used in this lesson:

Student Learning Objectives and Associated Assessments		
Objective Students will be able to:	**Formative Assessment(s)**	**Summative Assessment(s)**
Note: Attach all formative assessments and student handouts to this lesson plan. Include an answer key.		

Differentiation
How will you differentiate the <u>content</u> of your lesson?
How will you differentiate the learning <u>process</u> of your lesson?
How will you differentiate the expected learning <u>products</u> of your lesson?
How will you differentiate the <u>learning environment</u> during your lesson?

Materials & resources block

Materials and Resources Needed		
Source	**Material**	**Quantity**

FIGURE 12.5 Sample format of a lesson plan showing relationships of plan parts to the major blocks.

	Content block	Teaching–learning activities block		
Lesson segment	**Agriculture terms/ concepts/skills content outline** *(What ag knowledge/skills are students learning in this segment of the lesson?)*	**Activity to engage learners in the content** *(What are the students doing?)*	**Anticipated management strategies** *(What is the teacher doing?)*	
Lesson initiation (est. time)		Students will be... • •	The teacher will... • •	Lesson initiation block
Transition (est. time)	*What will you say to cue desired behavior and link the initiation to the next segment?*			
Activity 1 (est. time)	Students will learn and know... • • • •	Students will be... • •	The teacher will... • •	
Transition (est. time)	*What will you say to cue desired behavior and link this activity to the next segment?*			
Activity 2 (est. time)	Students will learn and know... • • • •	Students will be... • •	The teacher will... • •	
Transition (est. time)	*What will you say to cue desired behavior and link this activity to the next segment?*			
Activity 3 (est. time)	Students will learn and know... • • • •	Students will be... • •	The teacher will... • •	
Transition (est. time)	*What will you say to cue desired behavior and link the learning from today to the closing activity?*			Lesson closure block
Lesson closure (est. time)		Students will be... • •	The teacher will... • •	

FIGURE 12.5 *Continued.*

In a traditional schedule, the agriculture teacher instructs students for 180 days and has five to six class periods a day. If the average lesson plan lasts for three days, then this typical agriculture teacher will need to develop 60 lesson plans for each class, or up to 360 separate lesson plans. Of course, time frames and lesson plan needs vary somewhat with semester-long classes and alternative scheduling. Why should an agriculture teacher develop their own lesson plans if these can be purchased? How should an agriculture teacher use commercially developed lesson plans?

Many states, either through curriculum centers or private companies, provide lesson plans for state-approved curriculums to their agriculture teachers. These lesson plans are typically professionally done and include illustrations, PowerPoint presentations, premade tests with answer keys, and suggested teaching–learning activities. These lesson plans can be a great help to both beginning and experienced teachers. Beginning teachers are saved the time of creating lesson plans from scratch. Experienced teachers may gain new ideas for teaching certain content.

These purchased lesson plans do not relieve the teacher of the responsibility of planning instruction. Commercial lesson plans may be generic in nature, requiring the agriculture teacher to make the content locally applicable. Often the plans consist of activities that center around a PowerPoint slide deck, with various rote memorization activities. A teacher wanting to provide more meaningful, student-centered experiences will need to modify those plans to meet the needs of their students. The plans may be written at the level of a typical student, requiring the agriculture teacher to modify them to differentiate instruction to reach all students in the classroom. In addition, each agriculture teacher has a style of teaching they develop over time. The commercial lesson plans are not written for a particular teacher's style, so modifications will need to be made in this area. Finally, commercial lesson plans, just like teacher-developed lesson plans, should be updated yearly with new content and teaching modifications.

USING INSTRUCTIONAL TECHNOLOGIES AND THE INTERNET

We live in a technological, information-rich society. Technologies seem to rise and become obsolete within a decade. The rate of information production continues to increase so rapidly that the time it takes for the volume of information to double is measured in years and months. Education is adopting instructional technologies at an ever-increasing rate. However, there are principles that apply, whether one is using a chalkboard/whiteboard, a PowerPoint presentation, or some other audiovisual aid.

The principles of interest discussed in Chapter 11 apply to using instructional technologies. Instructional technologies can be used to bring nature, humor, creativity, human interest, and the unusual into the classroom. When used effectively, instructional technologies can spread interest from interesting things to uninteresting ones. A caution is that the technology used should not be the focus but instead should enhance interest in the topic being studied.

Visuals

Well-designed visuals are important regardless of the instructional technology used. Visuals need to be visible and comprehensible to all students regardless of their location in the classroom. Instructional technologies need not be new to be valuable. The chalkboard/whiteboard is a useful tool in almost every agricultural education classroom. Because of its large size, ease of use, and ease of correction, this instructional tool is extremely functional.

Some instructional technologies do become obsolete. Filmstrips and 16 mm films gave way to first videotape and laserdiscs, then DVDs, and now to MP4 files that can be streamed from the internet. Reel-to-reel audiotapes were replaced by audiocassettes, which were then replaced by audio compact discs. Some of the instructional technologies discussed in this section will probably become obsolete within the near future.

Computers in the Classroom

The computer, in all of its forms and sizes, continues to be a valuable instructional tool in the classroom. Its uses cut across all areas of instruction. PowerPoint presentations are used to deliver notes, present procedures, and provide pictures to supplement lectures. Word processing, spreadsheet, and database programs are used in all aspects of teaching. Specialized computer programs are available to assist with agronomic, agribusiness, and management applications. The use of GPS and GIS has allowed agriculture teachers to teach about site-specific farming and other applications.

Some commercial curriculum services, such as iCEV Multimedia, have adapted their curricular materials to be accessible primarily through online classrooms such as Google Classroom. Students move through various learning modules and earn digital badges to illustrate their learning, which may result in industry-recognized certifications. More information on using computers and other digital technology in the agriculture classroom is described in Chapter 13.

REVIEWING SUMMARY

Today's learners bring unique experiences to the classroom and live in an information-rich world. Teachers need to design instruction that builds upon the assets students bring, while attending to their unique learning needs. Instruction that helps students develop deep understanding of agricultural concepts, provides opportunities for application, and weaves in a focus on metacognition of learning strategies is most effective for student learning. Teachers who organize their instruction using lesson plans, who carefully select teaching methods, and who appropriately utilize instructional technology have a positive impact on student learning. Teachers should strive to build continually upon their instructional repertoire and experiment with new techniques and technologies.

Bloom's taxonomy has guided teachers for more than 50 years in writing appropriate learning objectives. Although more recent research has shown students learn through their whole experience, Bloom's three domains of learning are still important for planning, assessment, and accountability purposes.

Teaching methods can be categorized as teacher-centered, social interaction, or student-centered. Effective teachers learn which methods work best in what situations, with what students, and for what purposes. Effective teachers also have a clear understanding of their teaching philosophy and how it impacts their instructional decision-making.

Lesson plans are critical for teachers to be effective and efficient. Although numerous styles of lesson plans exist, they all have certain similar components. Commercially or professionally developed lesson plans hold advantages for both beginning and experienced teachers; however, the local teacher must still personalize the plans.

Instructional technologies enhance the classroom environment. Through the use of these technologies, students are exposed to content and resources they would not have access to otherwise. Teachers need to take care that the subject matter, not the instructional technologies, is the focus of student attention. Computers, although extremely useful in the classroom, are merely another tool the teacher uses to enhance the student learning experience.

QUESTIONS FOR REVIEW AND DISCUSSION

1. Recall the principles that guide learning in the 21st century outlined by Darling-Hammond et al. (2008). How do agriculture teachers integrate these principles into their instruction?
2. What factors influence selection of a teaching method? Are there any other factors you can identify that would influence your selection of an appropriate teaching method?
3. How is Bloom's taxonomy used in the agricultural education classroom?
4. How are learning objectives written using the ABCD approach?
5. How do student-centered and teacher-centered instructional methods differ?
6. Discuss when it may be appropriate to utilize teacher-centered approaches such as lecture.

7. What is a lesson plan?

8. Why is it important for all agriculture teachers to use lesson plans?

9. What are the common components of all lesson plan styles?

10. What are the advantages and disadvantages of using commercially or professionally prepared lesson plans?

11. How do the principles of interest discussed in Chapter 11 relate to instructional technologies?

ACTIVITIES

1. Investigate the affective and psychomotor domains of learning. Why have these not been developed as fully or emphasized as much in education as the cognitive domain? How is this different or the same for agricultural education?

2. Investigate agricultural education resources on the internet. Develop and organize a sharable document (such as a Google Sheet) or create a Wakelet that serves as a collection of the resources you find.

3. Using the *Journal of Agricultural Education, The Agricultural Education Magazine,* and the proceedings of the National Conference of the American Association for Agricultural Education, write a report on block scheduling research in agricultural education. Current and past issues of these journals can be found in the university library. Selected past issues can be found online at www.aaaeonline.org or www.naae.org.

4. Explore the differences between behaviorist and constructivist teaching approaches. Which of these systems do you align more with? Revisit your teaching philosophy statement and consider how your teaching philosophy guides your instructional decision-making.

5. Explore culturally responsive teaching, culturally relevant pedagogy, and trauma-informed teaching. How do each of these approaches to instruction help students learn and grow?

REFERENCES

Areepattamannil, S., Cairns, D., & Dickson, M. (2020). Teacher-directed versus inquiry-based science instruction: Investigating links to adolescent students' science dispositions across 66 countries. *Journal of Science Teacher Education, 31*(6), 675–704. https://doi.org/10.1080/1046560X.2020.1753309

Bloom, B. S. (Ed.). (1956). *Taxonomy of educational objectives.* Longman, Green.

Darling-Hammond, L., Barron, B., Pearson, P. D., Schoenfeld, A. H., Stage, E. K., Zimmerman, T. D., Cervetti, G. N., & Tilson, J. L. (2008). *Powerful learning: What we know about teaching for understanding.* Wiley.

Easterly III, R. G., Warner, A. J., Myers, B. E., Lamm, A. J., & Telg, R. W. (2017). Skills students need in the real world: Competencies desired by agricultural and natural resources industry leaders. *Journal of Agricultural Education, 58*(4), 225–239. https://doi.org/10.5032/jae.2017.04225

Herreid, C. F. (1998). What makes a good case? Some basic rules of good storytelling help teachers generate student excitement in the classroom. *Journal of College Science Teaching, 27*(3), 163–165. https://sciencecases.lib.buffalo.edu/pdfs/What%20Makes%20a%20Good%20Case-XXVII-3.pdf

Krathwohl, D. R. (2002). A revision of Bloom's taxonomy: An overview. *Theory into Practice, 41*(4), 212–218. https://doi.org/10.1207/s15430421tip4104_2

National Council for Agricultural Education. (2016). *National quality program standards for agriculture, food, and natural resources education: A tool for secondary (grades 9–12) programs.* https://thecouncil.ffa.org/

Newcomb, L. H., McCracken, J. D., Warmbrod, J. R., & Whittington, M. S. (2004). *Methods of teaching agriculture* (3rd ed.). Prentice Hall.

Roberts, T. G., Stripling, C. T., & Estepp, C. M. (2010). A conceptual model of learning activities for college instructors [Abstract]. *NACTA Journal, 54*(supplement), 71. https://www.nactateachers.org/images/stories/nacta%20journal%20vol%2054%20supplement%201.pdf

ADDITIONAL SOURCES

Adams, S., & Leininger, G. (2019). *Next level: Classroom instruction for CTE teachers.* Self-published.

Blumberg, P. (2019). *Making learning-centered teaching work: Practical strategies for implementation.* Stylus.

Buck Institute for Education. (n.d.). What is PBL? *PBL Works.* Retrieved October 16, 2021 from https://www.pblworks.org/what-is-pbl

Burden, P. R., & Byrd, D. M. (2019). *Methods for effective teaching: Meeting the needs of all students* (8th ed.). Pearson.

Dean, C. B., Hubbell, E. R., Pitler, H., & Stone, B. (2012). *Classroom instruction that works: Research-based strategies for increasing student achievement.* ASCD.

Evans, D., & Taylor, J. (2005). The role of user scenarios as the central piece of the development jigsaw puzzle. In J. Attewell & C. Savill-Smith (Eds.), *Mobile learning anytime everywhere.* Learning Skills and Development Agency.

Garrett, T. (2008). Student-centered and teacher-centered classroom management: A case study of three elementary teachers. *Journal of Classroom Interaction, 43*(1), 34–47.

Hassad, R. A. (2011). Constructivist and behaviorist approaches: Development and initial evaluation of a teaching practice scale for introductory statistics at the college level. *Numeracy, 4*(2). https://doi.org/10.5038/1936-4660.4.2.7

Heinich, R., Molenda, M., Russell, J. D., & Smaldino, S. E. (1999*). Instructional media and technologies for learning* (6th ed.). Merrill/Prentice Hall.

Hill, J. D., & Miller, K. B. (2013). *Classroom instruction that works with English language learners* (2nd ed.). ASCD.

Marzano, R. J. (2017). *The new art and science of teaching.* Solutions Tree Press.

Mensah, F. M. (2021). Culturally relevant and culturally responsive: Two theories of practice for science teaching. *Science and Children, 58*(4), 10–13. https://www.nsta.org/science-and-children/science-and-children-marchapril-2021/culturally-relevant-and-culturally

Sorin, R., Errington, E., Ireland, L., Nickson, A., & Caltabiano, M. (2012). Embedding graduate attributes through scenario-based learning. *Journal of the NUS Teaching Academy, 2*(4), 192–205.

Wong, H. K., & Wong, R. T. (2009). *The first days of school: How to be an effective teacher.* Harry K. Wong Publications.

Digital Learning

Aja Burgess is an agriculture teacher at Sonoma High School. This year her school has introduced the use of one-to-one devices so students can have personalized access to digital tools and internet-based resources. Prior to this year, Ms. Burgess had primarily used paper-based resources, and she brought students to the computer lab in the library when needing to complete proficiency award applications or conducting research for a class project. In addition to the introduction of one-to-one devices, the school is also now using a learning management system (LMS) where teachers can post class assignments and resources, and students can submit their work. Although Ms. Burgess has been using an online gradebook for some time, she is rather unfamiliar with where to begin with her LMS and use of one-to-one devices.

Ayanna Ruth is an agriculture teacher at Santa Monica High School, where her students have been using one-to-one devices for several years. Her school administration requires all teachers to develop e-learning day lessons in advance of the school year so students can continue to engage in learning experiences from a distance while teachers are participating in professional development days. Ms. Ruth is very familiar with integrating the use of one-to-one devices in her teaching, and her students have come to expect using them most days.

Ayanna and Aja are good friends from their days in college together. Feeling overwhelmed as to where to start, Aja calls her friend Ayanna to seek her advice on where to begin. Where would you start to learn more about utilizing technology to accomplish instruction using digital learning tools?

TERMS

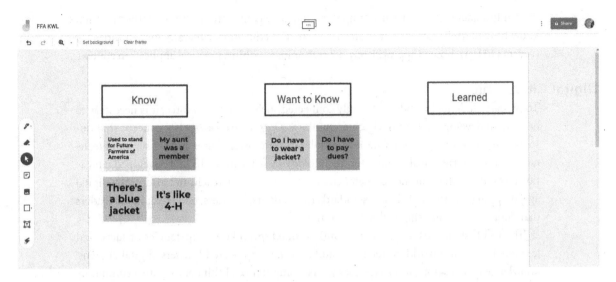

FIGURE 13.1 Google Jamboard is a free online tool that can be used in a variety of ways for a blended or virtual classroom. Here, students create digital sticky notes to capture their thoughts related to what they think they know and want to know about FFA.

THE 21ST-CENTURY CLASSROOM

Learning in the 21st century has drastically changed the way learners access and interact with information. Prior to the age of the internet, information was housed in libraries, encyclopedias, textbooks, and in the minds of educators. Now, knowledge and skills can be accessed through a quick internet search. Movies and music can be accessed through instant downloads instead of needing to make a trip to the store. As technology has increased our access to information, the role of a teacher has increasingly transitioned into one of a facilitator of learning, rather than a purveyor of knowledge.

Since technology is constantly evolving, it is likely certain technologies may become outdated very quickly. Consequently, this chapter will focus primarily on the guiding principles of teaching with digital technology instead of detailing how to use common educational software or tools. Examples of current technology will be provided, but readers should be cautioned to research the most up-to-date version of the technology to ensure they plan to use it appropriately. Technology access largely depends on the infrastructure to support digital and online technologies in local school districts and communities. Reliable high-speed internet access continues to be a challenge for many communities across the United States, limiting the full potential of many educational technologies.

Today's students are considered ***digital natives***; they have always known a world where internet access is omnipresent, social media plays a key role in how they experience the world, and technology is a necessary component of functioning in a global society. Many industries, including agriculture, are researching the development of the Internet of Things, or IoT. The ***Internet of Things*** describes a network of items that have digital sensors embedded within them to collect data that can be shared with other devices on a network to improve performance and efficiency. Today, many people wear digital watches that in addition to telling time, also track heart rate, breathing rate, sleep quality, and activity levels. These "smart" watches are an example of an IoT sensor. In agriculture, the IoT is particularly of interest in crop production to develop sensors and drones that can collect precise data to inform agronomic decisions.

While students may have grown up surrounded by connective technology, they may not know how to use it safely or ethically. It is essential educators integrate opportunities for students to learn how to appropriately use technology within a practical application context.

Digital Citizenship

To prepare students for today's digitally connected society and workforce, it is necessary to help them develop skills to navigate technology. *Digital citizenship* is a concept that describes the traits of an appropriate, responsible, and empowered user and consumer of technology. The International Society for Technology in Education (ISTE) developed educational standards that outline competencies necessary to learn, teach, and lead in the digital age, mapping out research-based standards for students, teachers, and educational leaders that define success in using technology in these spaces.

The ISTE standards for students (2016) outlined seven key competencies of successful learners in a digital world. Students should become empowered learners, digital citizens, knowledge constructors, innovative designers, computational thinkers, creative communicators, and global collaborators. Within each of these seven core competencies are specific standards that lead to overall competency. Students who demonstrate competency in digital citizenship "recognize the rights, responsibilities and opportunities of living, learning and working in an interconnected digital world, and they act and model in ways that are safe, legal, and ethical" (ISTE, 2016). Following are the four standards associated with developing this competency:

1. Students cultivate and manage their digital identity and reputation and are aware of the permanence of their actions in the digital world.
2. Students engage in positive, safe, legal, and ethical behavior when using technology, including social interactions online or when using networked devices.
3. Students demonstrate an understanding of and respect for the rights and obligations of using and sharing intellectual property.
4. Students manage their personal data to maintain digital privacy and security and are aware of data collection technology used to track their navigation online.

It is important that as agriculture teachers are planning instruction they consider how students will be using technology and plan to help students develop these competencies. ISTE offers several professional development opportunities and resources for educators to assist them in cultivating digital citizenship in their students. To learn more visit https://digcitcommit.org/.

Global Competency

Technology has allowed our world to become increasingly connected. Video chat platforms can connect students to industry professionals and classrooms across the country and around the world. While agricultural education programs do serve the needs of the local community, it is also necessary to prepare students to engage in a global society. Increasingly, employers want to hire workers who are globally competent. A person who demonstrates *global competence* has the disposition and knowledge to understand and act on issues of global significance. Educators are tasked with helping students develop their ability to compete, connect, and cooperate on a global scale to ensure they are prepared to meet the demands of our increasingly global world.

To help CTE teachers cultivate global competency in their students, the Association for Career and Technical Education (ACTE) collaborated with the Asia Society Center for Global Education to develop the Global CTE Toolkit (https://asiasociety.org/education

/global-cte-toolkit). Extensive resources are available to help teachers develop lessons that not only address global competence, but also cultivate students' ability to compete, connect, and cooperate through the appropriate use of technology.

Before considering how you might help your students develop their global competence, it is also important to understand your own global competence. The Globally Competent Learning Continuum (Tichnor-Wagner et al., 2019) was developed to help educators reflect on their own level of development of dispositions, knowledge, and skills necessary to teach diverse learners to be prepared to engage in a global society. This self-reflection tool can be a great place for educators to begin their journey of integrating global competence and responsible technology use into their instruction. The Globally Competent Learning Continuum and other educator resources is available at http://globallearning.ascd.org/lp/editions/global-continuum/home.html.

Specific resources in agricultural education to help students and educators develop their global competency within the context of agriculture, food, and natural resources are available through the Global Teach Ag Network. Each year a Global Learning in Agriculture (GLAG) conference is hosted online to connect with educators and students around the globe. For more information on the Global Teach Ag Network and the annual GLAG conference, visit https://sites.psu.edu/glag.

Ensuring Equity

When making any educational decisions it is essential to consider the needs of your students. Introducing the use of technology can definitely enhance your lessons, but may also have unintended inequitable consequences. In schools where students have been provided individual devices, it may be tempting for teachers to regularly assign homework that requires students to use their device to research issues using the internet. For students from low-income families or those who live in areas that do not have broadband internet access, completing these types of assignments would prove challenging, if not downright impossible. Additionally, technology should be accessible for students of all abilities. Students who are diagnosed with a learning disability may use assistive technology to help them interact with instructional materials. For example, a student who is diagnosed as having a visual impairment or blindness might use a screen reader that reads electronic documents aloud. To ensure educational equity in this space, it is essential the instructor develop educational materials that can be read aloud by such assistive technology.

If a school holds e-learning days or is conducting instruction in an online virtual setting, considering how you will work to ensure equity is paramount. Students completing work from their home environment may not always have access to a safe, comfortable, and ideal environment in which to focus on learning. Planning for ways to adjust instruction or assessment for students to allow for maximum student success is important. Be sure to check with your school's administration to determine how you can navigate issues of equity specific to your students and surrounding community.

SAMR MODEL OF TECHNOLOGY INTEGRATION

Technology should not be used for the sake of technology. It should be used to advance student learning and skill development. A useful framework for integrating technology into your instruction is called the SAMR model.

As technology has improved and changed over time, so have the types of technology utilized in education. The vast variety and options available in educational technology can feel

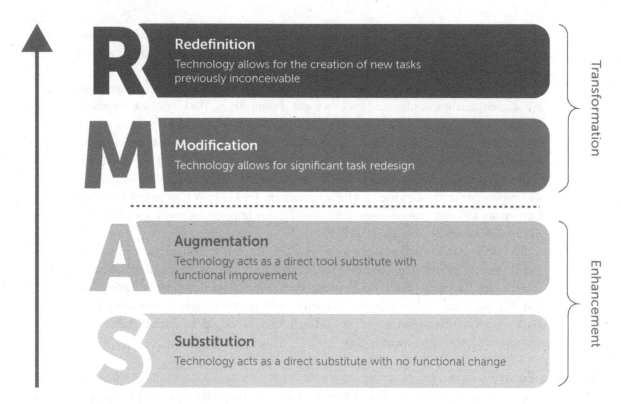

FIGURE 13.2 The SAMR model. (ADAPTED FROM PUENTEDURA'S [2006] SUBSTITUTION, AUGMENTATION, MODIFICATION, AND REDEFINITION [SAMR] MODEL.)

overwhelming at times when trying to decide how to best use technology in your instruction. When planning instruction, it is essential teachers continuously reflect upon how they can do a better job helping their students engage with the lesson content, which may mean deciding to use technology. To help teachers navigate making instructional decisions about integration of technology into their teaching, Dr. Ruben R. Puentedura (2006) developed the *SAMR model of technology integration*.

The SAMR model (Figure 13.2) illustrates how teachers can enhance existing instructional activities or even transform them into new learning experiences with the introduction of technology. Puentedura described many teachers as likely enhancing their lessons through substituting technology for previously existing approaches. For example, an agriculture teacher might decide to have their students research diseases affecting dairy cattle and assign students to create a poster presentation about the disease. Instead of having students create a physical poster with poster board, markers, and glue, they might have students create a digital poster using PowerPoint, Google Slides, or Canva. The assignment is still essentially the same, but substituting the type of technology used introduces an opportunity for students to practice digital citizenship skills.

Using the same assignment as an example, a teacher could choose to have students create infographics about dairy cattle diseases and then create a post about the diseases to share on the FFA chapter's social media accounts with the infographic. Students still create content about dairy cattle diseases, but the mode in which they are expressing their learning is different.

To transform a lesson using technology, a teacher would need to consider how they may redesign learning outcomes to better leverage student understanding of the content in tandem with the technology. Learning about dairy cattle diseases can occur in a variety of ways.

TABLE 13.1

Examples of How Technology Can Be Integrated Into Existing Instructional Activities Following the SAMR Model

	Substitution	Augmentation	Modification	Redefinition
Content	Unchanged	Unchanged	Unchanged	May be modified
Learning Tasks	Convert from hard copies to digital	Incorporate interactive digital features to enhance lesson content	Learning tasks completely redesigned to allow for a richer array of learning through digital tasks	Learning tasks fundamentally transformed to allow activities previously impossible in traditional brick and mortar classroom
Examples of Changes	Scan worksheets/ convert to PDFs Provide synchronous and asynchronous versions of lectures	Digital portfolios of learning Gamify quizzes with Kahoot, Quizizz, and GimKit Virtual bulletin boards like Padlet HyperDocs lessons	Use an LMS like Google Classroom, Moodle, Schoology, or Canvas Backchannel communication to engage quiet learners Asynchronous discussion threads	Virtual pen pals, field trips, panel speakers Students write their own wikis or blogs for public consumption and feedback

Instead of creating posters or infographics about a disease, a teacher might have students learn about disease diagnosis, management, and prevention through an online simulation game or through virtual field trips or conversations with dairy farmers and/or veterinary professionals. In these ways, the technology allows for new learning experiences that would not have been previously possible, transforming the type of learning occurring within the classroom. Table 13.1 provides examples of how technology can be integrated into existing instructional activities following the SAMR model.

Integration of technology can lead to more empowered students and is a natural fit for conducting student-centered instruction. To hear more from Dr. Puentedura about the SAMR model, go to https://youtu.be/ZQTx2UQQvbU.

DIGITAL LEARNING

Everyday life increasingly relies upon digital technology from the computer chips that regulate fuel consumption in the vehicles we drive to thermostats in our homes and robotic vacuums. Often without thinking about it, we put on smart watches, put smart phones in our pockets, and get into keyless entry cars. Digital technology is all around us! Education has begun to integrate digital technology into instructional activities, but digital learning is a concept that is difficult to define. Digital literally refers to technology that is coded using a series of ones and zeros, but it is often associated with technology that allows for connectivity to the internet. While digital learning does utilize educational technology, it is not the same thing. ***Digital learning*** leverages the connectivity of the internet along with educational technologies, allowing for the development of a wide range of approaches to learning to create individualized learning opportunities for students. There are several different approaches to digital learning that are discussed in the section that follows.

E-Learning

As technology and distance education have increasingly become a part of secondary and postsecondary education, the definition of e-learning has varied. Definitions have included a focus on technology, the delivery system in which knowledge is accessed, the communication technologies used, and on shifting the entire educational approach to teaching and learning. In this chapter we will focus on *e-learning* as a day of instruction that occurs either planned or unplanned, in which students learn from a distance. For example, students may have e-learning days when the school district has a planned teacher professional development in-service day, or they may have e-learning when a snowstorm unexpectedly cancels school. E-learning days may also be known as online learning days, cyber learning days, virtual days, or another similar term.

E-learning days typically count toward state-mandated attendance and instruction requirements, but the format of these days varies significantly from state to state and from district to district. Some schools may require teachers to develop e-learning plans at the beginning of the school year, much like emergency substitute teacher plans, while other districts allow teachers to develop e-learning plans as they go along, following the natural flow of instruction. Check with your school administration to identify the format and expectations of e-learning requirements in your community.

Virtual Learning

Learning that occurs from a distance through internet-based instruction is referred to as *virtual learning* or *online learning*, or *distance learning*. During the COVID-19 pandemic, schools around the world shifted to an emergency remote teaching model of instruction, in which all instructional activities were immediately moved into a virtual or distance learning format. In many cases, teachers who had never taught in an online format were asked to develop content and teach from a distance. As of the time of this book's printing, some schools are still teaching in a virtual format.

Prior to the COVID-19 pandemic, some states developed completely online virtual schools in which students attended all of their classes in an online format. Most of these schools do not include agricultural education as an offering, but some states such as Indiana and Florida do have agricultural education included as a virtual option. In states such as South Dakota, some students have accessed agricultural education coursework through distance learning, in which an agriculture teacher at a school in another part of the state would instruct students through a live video feed. Distance learning conducted in this manner is sometimes the only way students who live in remote areas can access agricultural education.

Depending on the expectations of your district and the needs of your students, educators may design virtual instruction to occur in synchronous or asynchronous formats. These approaches are discussed in more detail in the next sections.

Synchronous Learning

Synchronous instruction is also known as "live" or "real-time" instruction. In a traditional face-to-face classroom, students attend class all together at the same time each day, in a synchronous fashion. When teaching online at a distance, synchronous instruction occurs when students log in at the same time as the teacher and the teacher facilitates learning activities during that time. *Synchronous learning* sessions might be recorded and shared for students to access at a later point, but students would likely not be able to participate in any of the planned learning activities that occurred during a synchronous session. Teachers should be cautious about the potential effect of holding only synchronous instructional opportunities for students who do not have regular access to high-speed internet, might have other

---

BOX 13.1 *Examples of Learning Activities Appropriate for Synchronous and Asynchronous Learning Formats*

Synchronous Learning Activities

Provide real-time feedback on work in progress

Provide personalized 1:1 coaching

Facilitate real-time conversations

Incorporate guided practice and application

Build community and relationships

Lead interactive modeling sessions

Foster collaboration among students

Differentiate instruction for small groups

Asynchronous Learning Activities

Listen to podcasts

Explore teacher-curated resources

Reflect and document learning

Watch video-based instruction

Read and take notes

Engage in online discussions

Practice and review

Research and explore

Adapted from Tucker (2020b).

---

responsibilities at home (such as caring for siblings), or may have to share their device or internet bandwidth with others.

Since real-time instruction may be available at a premium, instructors should consider what learning activities will be of the greatest value to students during these sessions. These sessions are ideal for providing feedback and building relationships with students in an online instructional environment. Conducting a lecture in which students do not get to interact with the teacher will likely not be as valuable as getting individualized attention from the instructor, or an opportunity to connect with peers in a collaborative environment. Box 13.1 provides examples of learning activities to use in a synchronous online learning environment.

Asynchronous Learning

Unlike synchronous learning, *asynchronous learning* activities occur on demand at any time that works best for the student's schedule. Much of online coursework designed for adult learners who attend college part time is designed to be delivered in an asynchronous fashion. On-demand learning activities are available similarly to how movies are available to watch at any given time on a streaming platform like Netflix. Asynchronous activities have the benefit of being flexible and allowing students to work when they are most able to focus on or attend to their work. Conversely, students who may not be particularly motivated to engage in the learning, or who may have a difficult time managing their time independently may struggle to complete asynchronous learning activities. See Box 13.1 for examples of asynchronous learning activities.

Blended Learning

Blended learning, similar to many of the terms we have already discussed, is a broadly interpreted, difficult to define concept. *Blended learning* refers to a variety of modes of face-to-face instruction that integrate a component of online learning. Regardless of what type of approach is used, blended learning provides a large amount of flexibility for teachers and students. One particularly useful model for blended learning is the teacher-designed blend, a term coined by edtech expert Dr. Catlin Tucker. In the teacher-designed blend, the teacher is still the primary decision maker regarding curriculum content, organization, and edtech tools, and instruction still mainly occurs in face-to-face settings. This face-to-face instruction can be supplemented with online learning modules, and a learning management system is used to facilitate completion of assignments, share resources, and organize grades.

According to Tucker (2012), using a blended learning approach in the classroom should prioritize the teacher working directly with students to build positive relationships and promoting learning and discovery through an online network. Using a blended approach, teachers can leverage the power of technology not only to engage their students in learning in new ways, but also to save on preparation time, printing costs, and time spent grading.

At the heart of the success of the blended learning model is discussion. Without meaningful opportunities to discuss and connect new concepts, the various learning activities become rote tasks students complete in increasing isolation. Blended learning approaches allow for discussion to occur not just in a face-to-face, out-loud verbal conversation, but also to occur through online platforms, backchannel conversations, sticky notes, and in many more ways. Incorporating opportunities for students to practice discussing the content not only helps them become more connected to the instructor and their peers as an educational community, but also helps them practice the appropriate skills of communication using various technologies.

Flipped Classroom

In a *flipped classroom* teaching model, the teacher creates various forms of content delivery such as a recorded lecture, or curated readings for students to review outside of class time. Instead of focusing on content delivery during class time, the teacher prioritizes application and extension activities for students to clarify and deepen their understanding of the content. The teacher is also available for assistance to work with small groups of students who may need more support, while other students might begin working on other meaningful active learning activities. In this manner, students complete lower order learning activities on their own and engage in richer higher order learning activities together in class. For more information on higher order and lower order cognition, see the section on Bloom's taxonomy in Chapter 12.

EDUCATIONAL TECHNOLOGY TOOLS

Technology is constantly being developed, updated, or becoming obsolete. Advancements once thought innovative and cutting-edge may no longer become useful as improvements lead to replacement within months or years of release. Consequently, this section will not attempt to describe specific educational technologies in detail, but instead will provide a broad overview of currently existing resources to support implementation of agricultural education in the 21st-century classroom.

Regardless of what tool is selected, it is imperative that educators consider the accessibility of the tool to their students. *Accessibility* ensures that educational materials and technologies allow people with disabilities to learn the same content and experience the same educational experiences and benefits as those without disabilities (CAST, n.d.). In addition to considering the needs of students who have disabilities, it is helpful to consider resources to which your students have access. For example, some of your students may have high-speed internet at home, while others do not. Considering accessibility when planning your instruction is vital to ensuring that you are curating equitable learning experiences for all students.

Hardware

Physical equipment that allows users to interact with digital tools is referred to as *hardware*. For example, the laptop and its components (such as the keyboard, memory chip, and motherboard) would be hardware. The internal processors that allow for computing speed and quality have become smaller and smaller in size as chip technology has improved. As a result, many devices including laptops, phones, screens, and printers have become smaller or sleeker in design.

One-to-One Devices

Issuing each student a personal device to use for educational work is becoming increasingly the norm at schools around the United States. Schools "check out" a device to each student much like issuing a textbook for the year. When each student has been issued technology to use in this manner, the school has a *one-to-one device* program. Learning activities in a blended learning environment are most efficient when each student has access to their own fully charged device that can access the internet.

Examples of popular devices used in these programs include the Google Chromebook, Apple iPad, netbooks, and laptops. With the use of such devices, teachers will need to consider how students will learn appropriate care and use of the technology in the agricultural education classroom and lab.

Digital Lab Equipment

A variety of equipment used in the agricultural education laboratory is becoming digital. Soil probes, microscope lenses, wildlife trail cameras, and many more types of equipment are now able to connect wirelessly to other devices or upload data and/or pictures to the internet.

A popular device widely used in various CASE curriculums is the LabQuest, manufactured by Vernier. This small computer can connect to a variety of probes including those that measure temperature, pH, turbidity, dissolved oxygen, and many more. This device does need to be regularly charged and checked for updates. The LabQuest can organize and visualize data collected from these sensors and store the collected data on a cloud server. A *cloud*

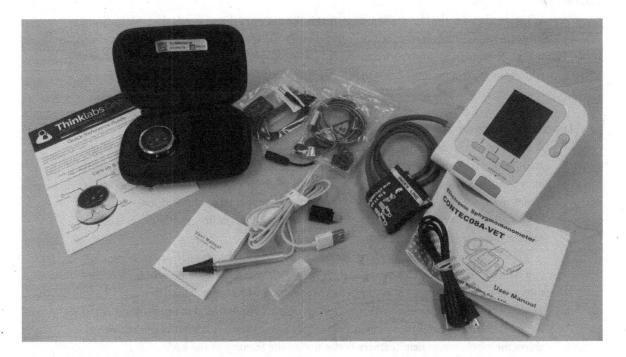

FIGURE 13.3 A variety of digital lab equipment exists including these tools used in a veterinary science class. From left to right: A digital stethoscope that can be connected to classroom speakers, a digital otoscope that connects to a laptop or computer to project a live video image to a screen, and a digital sphygmomanometer that can also connect to a laptop.

FIGURE 13.4 It is important to check that lab equipment is fully charged and has up-to-date software in advance of using it in the laboratory setting. Here, two agriculture teachers are ensuring that a Vernier LabQuest has up-to-date software.

server is a server that is accessible via a network, typically the internet. Data stored on an internet-based cloud server can be accessed from any approved devices that have an internet connection. Agriculture teachers should be sure to check their lab equipment in advance of needing them for a lab to ensure all devices are charged and up-to-date, as sometimes these updates significantly change the way the device operates.

Software

While hardware is the physical equipment, it is the *software* that controls the system. Some software needs to be downloaded onto a computer using a USB drive or CD, but increasingly software can be downloaded through a password-protected online portal. Agricultural education courses might use specific software to draft landscape designs or organize and track crop production records.

Learning Management System

A learning management system (LMS) is essential to digital learning. A *learning management system* is an online portal that serves as an access point for students and teachers to interact with class materials. Teachers can upload teaching materials for students to access, create spaces for students to submit work, post assessments, and facilitate discussion. Some learning management systems allow for integration with an online gradebook so students can see their grades in the same location as their classwork.

Popular learning management systems include Google Classroom, Canvas, Schoology, and Moodle. Each school district has their own designated LMS, so be sure to develop a thorough understanding of how to utilize the specific LMS at your school. If a school uses an LMS, it is likely they expect all teachers to regularly update and maintain their LMS. Each LMS software company usually posts extensive resource videos and training materials on their websites to help teachers better utilize the features of the LMS.

Mobile Apps

As smartphones and tablets have become increasingly a part of our everyday life, their functionality has been continuously improved to assist the user. Given the smaller dimensions and portable nature of smartphones and tablets, the user interface of the devices is different from that of a desktop or laptop computer. Software designed to operate on a mobile device is referred to as a *mobile application*, more commonly referred to as an *app*. An app is downloaded from an online source such as the Apple App Store, or Google Play Store. Apps may be available for free download or through a download fee.

Apps can be used for a variety of purposes including gaming, productivity, lifestyle/entertainment, mobile commerce, and education. Teachers might use certain apps to increase or regulate their own time management, or could use these same apps to assist their students in developing self-regulation skills. Educational games such as Farmers 2050 are accessible via an app, as are other farming simulation games like The Herd. Such games can prove useful for an agriculture teacher looking to develop opportunities for their students to practice making agricultural production decisions in a controlled environment. The popular agricultural education online record book Agricultural Experience Tracker (AET) has an app version that allows students to easily enter their records and upload photos from their phones. The ease and proximity of access to the AET may help students complete requirements of their SAE and agricultural learning experiences more fully. Apps might also be utilized in laboratory instruction to identify plant and animal specimens, such as the iNaturalist app developed as a joint initiative of National Geographic and the California Academy of Sciences.

While apps have the benefit of being transportable, they do need to be compatible with your school's technology requirements and restrictions. To maintain maximum functionality, apps may need to be regularly updated, requiring a clear and strong internet connection. Some apps may require users to create a username and password, while others may ask for access to personal information that may not be appropriate for the school setting. Be sure to appropriately screen use of apps before using them with your students and school equipment.

Learning Games

Capitalizing upon their ability to motivate learning, games can promote learning through apps, online environments, and downloadable software. Malone and Lepper (1987) identified four key characteristics of a learning game:

> **Challenge**—The player/learner works to attain realistic and relevant goals through a variety of difficulty levels. Activities completed to attain the goals provide opportunities for learners to receive feedback and create feelings of competence.
>
> **Curiosity**—Sensory and cognitive interest is promoted through the visual effects of the game as well as inherent challenges or problems that may surprise the learner.
>
> **Control**—The learner has the ability to make choices throughout the game that reinforce their sense of control over their learning/play experience.
>
> **Fantasy**—The learning game attends to both cognition and emotions of the learner in a way that is both appealing to the learner and directly related to the content to be learned.

Schaller (2005) suggested adding two more features of a learning game:

> **Iteration**—The game provides repeated opportunities for the learner to experiment, hypothesize, and synthesize their learning.
>
> **Reflection**—Opportunities for the learner to reflect upon their iterative learning experiences should be woven into the game.

Upon closer examination, one can see the elements of experiential learning (Kolb, 1984) present through these characteristics of a learning game. For more information about experiential learning, see Chapter 21.

Audiovisual Tools

Digital technology presents new opportunities for educators to incorporate immersive multimedia in the place of or extending real-life experiences. This section will describe how video, virtual reality, and podcasts can be utilized as audiovisual tools in agricultural education.

Video

Videos can be used for three primary purposes: delivering direct instruction, expanding upon concepts in a supplemental fashion, or assessing learning and skill development. Videos are a popular way to deliver instruction in a flipped classroom model and can be an efficient way to engage learners in an asynchronous learning environment. Videos can also be used to augment face-to-face direct instruction to illustrate a concept or skill that may not be possible to demonstrate with the available resources in the agriculture classroom or laboratory. Box 13.2 shows how videos can be used in a variety of ways in an agriculture classroom.

An agriculture teacher might choose to use preexisting videos or to create their own videos to accomplish a variety of purposes including demonstrating agricultural or computer-based skills, screen recording lecture notes, capturing skill development for instructional use in future classes, conducting a virtual field trip, interviewing potential guest speakers, or marketing their agriculture program for recruitment purposes. Keep in mind the agriculture teacher does not have to always bear sole responsibility for the creation of videos such as these; students might also create videos as part of an assignment or to assist in promoting the FFA chapter and agriculture program.

Selecting the right video for your students is an important decision. Previewing videos in their entirety prior to showing them to students is essential! Consider the following criteria when determining what video(s) to use during instruction:

- Does the video align with your educational standards and learning objectives?
- Is the vocabulary in the video appropriate and at an optimum comprehension level for your students?

BOX 13.2 *Examples of Video Use in an Agriculture Classroom*

Instructional

Can replace or serve as alternative to direct instruction

Useful in the flipped classroom model and asynchronous learning

May be created by the instructor or be previously existing from reputable source

Supplemental

Motivate student exploration and problem solving

Complement in-class work, piquing interest and driving inquiry

Expand on topics or offer an alternative explanation

Assessment

Students watch videos of themselves or others perform a skill and analyze for various items

Students create videos to demonstrate their learning and skill development

Students are assessed on their ability to evaluate specimens presented in video

---

Adapted from Tucker (2013).

- Will the video make sense in the context in which you plan to use it?
- Is the video up-to-date in language, production value, and examples?
- Does the video contain sensitive topics that may need parental or administrative approval prior to showing?
- Consider the video source, intended target audience, and potential bias of the video.
- Consider the length of the video; you may need to chunk the video into smaller segments.
- Is the resolution or pixilation appropriate for the size screen it will be viewed on?

Once you have settled upon the appropriate video for your instruction, consider how you will structure your students' viewing experience. Might it benefit students to pause the video at specific points to emphasize key concepts, check for understanding, or allow for discussion? Does the entire video need to be shown, or can you show excerpts? Should students view the video multiple times? The prior knowledge and learning experiences of your students can help guide you in making decisions on how to best scaffold the video viewing experience, whether that be in a teacher-led face-to-face classroom or in a virtual setting.

A useful video viewing platform for showing videos online in an asynchronous format is Edpuzzle (see Figure 13.5). Using this online platform, teachers can use online videos or upload their own video, and then embed questions within the video for their students to answer. Students can then view the video in the manner the teacher determines will best scaffold their understanding of the concepts presented.

Videos are useful not only for demonstrating skills and delivering content, but also for spurring critical thought. In an agriculture classroom, students might view a commercial that presents a particular perspective, and then analyze the video for evidence of that perspective. To enhance student understanding, it is important to teach students how to watch a video with purpose. To help students critique videos, Palmer (2014) suggested showing a short video clip multiple times. During each successive screening, students should be prompted to focus on different elements of the video:

> **Screening 1**—Look and listen for the main ideas. Write down the three that stood out to you the most.

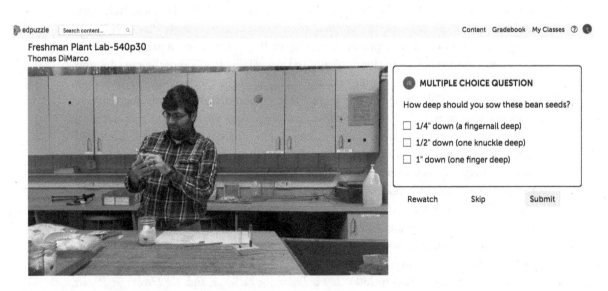

FIGURE 13.5 Edpuzzle (https://edpuzzle.com/) allows teachers to embed questions at key points within selected videos to check for student understanding. (COURTESY OF THOMAS DIMARCO.)

Screening 2—Look for evidence of support for these main idea.

Screening 3—(mute the video) Pay special attention to the images and selected scenes. What do they convey? How do they contribute to the main idea?

Screening 4—(unmute the video) Focus on the soundtrack. How does the music contribute to the message?

Screening 5—Pay special attention to the delivery technique of the speaker (if there is one). How does their delivery engage the listener/viewer?

Using the above technique can be helpful, particularly when showing videos of a prepared speech, debate, commercial, or a series of videos featuring individuals with varying perspectives on a controversial issue.

Virtual Reality

Virtual reality technology (or VR) allows students to explore a computer-generated three-dimensional virtual environment using multiple senses in an immersive fashion. Virtual environments present the participant with realistic, life-sized surroundings that are interactive, allowing for a free-flowing, seemingly natural experience. This technology is increasingly popular in the video gaming industry, but it is also used in various industries to train new employees in psychomotor skill development, including medicine, aviation, and construction trades such as welding. Wells and Miller (2020a, 2020b) determined school-based agricultural education teachers in Iowa generally view VR technology favorably, particularly as it relates to psychomotor skill development, but also believe the technology to be too costly to implement in their classrooms and laboratories. Virtual reality technology has been used to train skilled welders, but needs additional research in school-based agricultural education to determine the practical implementation of these technologies.

Virtual reality can also be used for giving virtual tours of your program as a program recruitment strategy. Tools such as ThingLink have the potential to be used for immersive tours, instructional aids, and helping students practice for a competitive event.

Podcasts

Podcasts are digital audio files created to share information with others. The word itself is a mashup of the words "iPod" and "broadcast," but podcasts are not limited to exclusively Apple devices. Podcasts are available to download from Apple iTunes, Google Play, Spotify, and many more online locations, typically free of charge. They can be listened to on a mobile device or on a computer, allowing for flexibility and accessibility that is appealing to educators.

There are several benefits to using podcasts in the classroom: they are portable, relevant to current events, available in a variety of lengths, help support multilingual learning, and meet academic standards. In some cases, podcasts can serve as an alternative or supplement to reading printed text to support reading proficiency. Many podcasts include rich storytelling components, featuring a variety of guests or characters, sound effects, and music. Listening to stories helps support brain development and cultivates psychological safety. Much like a thought-provoking video, podcasts can be used to kick off the start of a unit, deepen exploration of a topic, or listen to the perspectives of others. They can be used as the main resource for a lesson, or as a supplemental learning activity to support student learning.

In addition to utilizing podcasts for student learning, there are a variety of different podcasts that are great resources for educators looking to hone their craft. Some examples of popular education podcasts include *Cult of Pedagogy*, *EdSurge*, and *The Hechinger Report*. Agricultural education–specific podcasts are also available, including *Owl Pellets: Tips for*

Ag Teachers (2021) and NAAE's *Connect* (2021). *Owl Pellets* is a podcast that specifically explores practical tips for agriculture classrooms drawing from agricultural education research, while NAAE's *Connect* works to "educate listeners about NAAE resources, inform them of new and innovative practices, and connect current and future agricultural educators and supporters" (www.naae.org).

DIGITAL LEARNING ENGAGEMENT STRATEGIES

It is essential to design digital learning in a manner that builds on student motivation and interests, providing meaningful opportunities to interact with the material. Utilizing an active learning approach, educators can create meaningful opportunities for students to interact with the material and structure experiences they can reflect upon, which will drive future learning. Some of the major strategies utilized in digital learning are described below.

Game-Based Learning

Games can be used to educate and train students to develop skills and knowledge. While video games are quite popular among many students, video games tend to primarily focus on entertainment value, while game-based learning does not prioritize entertainment as a main focus. Instead, *game-based learning* uses games to educate.

There are two different categories of games leveraged in education: serious games and gamification. *Gamification* involves integrating elements of game play such as points, levels, and badges into learning activities, but in a manner that is not meant to entertain. *Serious games* provide non-entertainment-focused opportunities for students to wrestle with real-world problems and challenges, allowing for practice and training that is engaging.

Game-based learning can provide opportunities for students to directly participate in a role play or simulation experiences. These powerful learning experiences can lead to enhanced skill development, cultivation of critical thinking skills, problem-solving skills, information gathering, and team-building skills (Gros, 2007).

Project-Based Learning

As mentioned in Chapter 12, project-based learning can serve as a framework that guides units and courses. According to the Buck Institute for Education, *project-based learning* is a teaching method in which students tackle an "authentic, engaging, and complex question, problem, or challenge" (2021). This approach to teaching and learning can be useful in both face-to-face and online learning spaces. Entire courses can be structured around a particular challenge, with subsequent modules dedicated to elements of solving the messy problem. While some elements of project-based learning may require tactile manipulation of equipment or tools, some parts of the research and problem-solving process may be accomplished through independent or collaborative research in an online space. For more information on best practices in project-based learning, visit www.pblworks.org.

Scenario-Based Learning

A *scenario* is a type of story that focuses on a particular group of individuals and their goals within a described context. Scenarios can easily be integrated into e-learning through a text format, or shared in a short video prompt. Scenarios are commonly used in agricultural education to describe real-world situations that students might find themselves in. According to Sorin et al. (2012), there are five major types of scenarios, as follows.

Skill-based   Students would demonstrate skills they have already developed. For example, at the conclusion of an equine first aid unit, the agriculture teacher might draft scenarios that describe a scene in which the student would have to make decisions about the appropriate first aid measures to implement, such as bandaging the lower leg of a horse. Students would then actually demonstrate that skill.

Problem-based   Students are presented a problem to drive integration of their theoretical knowledge and practical knowledge. Through their investigation of the problem, students practice making decisions. In an agricultural mechanics course, students might reason through a scenario that describes a situation in which a piece of equipment failed to perform. Students would need to draw upon their knowledge of the equipment's functions and their practical experiences working with similar equipment to troubleshoot the source of the problem and brainstorm possible solutions.

Issue-based   Students take a stand on a particular issue. Considering the wide array of controversial issues related to agriculture, the possibilities are endless in drafting issue-based scenarios. Students might be prompted to form their own stance based on the described situation, or they may be assigned a perspective to illustrate. For example, when learning about animal rights versus animal welfare, students may be asked to take an initial stand on issues that emerge through a scenario describing the conditions of animal care. After exploring the differences and similarities between rights and welfare, students might revisit those same scenarios and take the stance of someone who advocates for one or the other perspectives.

Speculative   Students make hypotheses or predictions about the outcome of a described situation, based on their current knowledge and context clues. These types of scenarios are helpful for kicking off a unit and uncovering prior knowledge and misconceptions. They can also be helpful when setting up an inquiry-based approach to learning. For example, scenes that illustrate animal health might ask students to explore questions such as "What is wrong with this animal?" and "How do I know what is wrong with this animal?"

Gaming   These scenarios use games as learning tools. Through the use of games such as Farmers 2050, students can explore different agricultural decisions and concepts by playing a game that simulates real-world settings.

CLASSROOM MANAGEMENT IN A DIGITAL ENVIRONMENT

Managing a classroom is comprised of all the things that a teacher does to organize students, space, time, and materials to maximize student learning (Wong & Wong, 2009). Teaching in a face-to-face setting with devices or teaching in an online setting might look a bit different from teaching face-to-face without any digital devices, but teachers can still be proactive in managing the learning environment to promote learning. In this section, we will discuss considerations for teaching in digital or blended spaces, but will not address discipline. Effective discipline usually begins at the school-wide level, with consistent expectations and consequences upheld across the whole school rather than reactive individual classroom-level discipline procedures. A word of caution: When teaching with individual devices, it is inadvisable to take away a device from a student as a consequence of inappropriate behavior; that should only be used as a last resort. Getting at the root of the behavior, but providing students with the tools to complete a learning task is important to help students become good digital citizens (Dowd & Green, 2016).

The following sections will outline building positive relationships with students, establishing expectations, considering organizational procedures, and managing the learning management system. More detailed information on classroom management in digital or virtual learning spaces can be found in the references list at the end of this chapter.

Building Positive Relationships

The foundation of any learning experience rests upon the relationship (sometimes referred to as rapport) between the student and the educator, and the students with each other. It is essential that teachers dedicate intentional time to cultivating positive relationships with their students, regardless of the learning environment. Cornelius-White (2007, p. 113) identified five elements of teacher-student relationships that highlight a student-centered approach to teaching and relationship building:

1. The teacher is empathetic and understanding.
2. The teacher has unconditional positive regard for students.
3. The teacher is genuine and authentic.
4. The teacher cultivates student-led and -regulated activities; the teacher is nondirective.
5. The teacher prioritizes critical thinking as opposed to rote memorization.

When teachers and students have a strong, positive relationship, student mistakes are seen as learning opportunities rather than punitive, negative experiences. How teachers handle situations in which students stumble in their behavior or academic performance can strengthen or destroy a student's trust in the teacher. Clear communication between teachers and students enhances the development of positive, trusting relationships.

Building relationships with students can begin before they even step foot into your physical or virtual classroom. Sending home introductory letters/postcards/emails to share about yourself can be one way to make students feel more comfortable with you when they arrive in your class on the first day of school. You might also conduct student and family questionnaires to obtain baseline information about your students, their home life, what they value, and what motivates them. While a face-to-face setting allows plenty of informal opportunities for teachers to use humor, body language, and facial expressions to interact with students as they walk in the door, while they are working in groups, or see each other in the hallway, teachers need to be more intentional about how they build a relationship of trust and empathy in a virtual setting, particularly if it is in an asynchronous format.

Research indicates the more regular contact a teacher has with students, the more likely they will be successful in building positive relationships from a distance. In virtual settings, six key indicators contribute to building positive rapport at a distance (Murphy & Rodriguez-Manzanares, 2012):

1. The teacher makes an effort to recognize each individual student.
2. The teacher provides ongoing support and monitoring of learning through feedback and praise.
3. The teacher is available, accessible, and responsive within a timely manner.
4. The teacher and student are able to engage through other means beyond written text, such as phone calls, video chats, or audio recordings.
5. When the teacher does interact with students, the tone of the interaction is positive, respectful, friendly, and honest.

6. The teacher initiates conversations with students that are more social than academic, initiating small talk and demonstrating care and concern for their well-being as individuals beyond being a student in their class.

As you can see, many of these features of building rapport with students in online settings mirror what an agriculture teacher might do in an exclusively face-to-face setting. Should you find yourself needing to move from a face-to-face setting to an exclusively online setting, make sure to continue to cultivate positive relationships with your students in this new space.

Establishing Expectations

Many agriculture teachers use activities during the first week of school to get to know their students and to develop course expectations as a group together. Expectations are also important in a virtual or blended space. As with managing a face-to-face classroom, it is important to approach development of your expectations by evaluating your own beliefs about teaching and learning as outlined in your teaching philosophy statement. What type of learning environment do you hope to cultivate? What are your goals for your students upon completion of your course or graduation from the agriculture program? How do you believe the diversity of your students can be an essential component of cultivating classroom community? Your beliefs undergird all other decisions you make as an educator. Consider drafting a condensed version (one paragraph or so) of your teaching philosophy to share with students and their families at the beginning of the year.

Working together with your students, develop a list of norms or expectations of behavior and learning in your class. When students have the opportunity to contribute to the creation of these expectations, there is a greater likelihood of student buy-in and self-regulation in adherence to these community agreements. It is also beneficial to identify expectations for learning behaviors in the following settings: face-to-face with one-to-one or shared devices, asynchronous distance learning, and synchronous distance learning. Will students be expected to come to class with a fully charged device? How will you gain students' attention while they have their devices out? What is the appropriate time (if any) for food and drink to be present in the face-to-face classroom, especially when devices are being used? When is the appropriate time for students to be using their devices, and how will the teacher communicate this? How will earbuds/headphones be managed to ensure on-task behavior? Obviously, there are many more questions to be considered than these. How might you manage students, space, time, and materials in a blended or virtual setting?

Organizational Procedures

Keeping students organized and aware of the learning activities and upcoming assessments is important in any classroom. In a face-to-face setting, teachers often write these things on the whiteboard or chalkboard, but in a virtual or blended setting, they may post them digitally on a slide deck or learning management system home page. Posting weekly and daily schedules can be a helpful tool to communicate with students and their parents to ensure clarity and efficiency. Using online timers during synchronous online or face-to-face instruction can help keep both the teacher and the students on track to make maximum use of class time together.

In addition to organizing the structure of the class, it will be important for the agriculture teacher to consider how they will distribute assignments, collect student work, and any particular ways they want students to label their work upon submission. Will you have hard copies or digital copies of work for in-class assignments? What is considered "on time" in a virtual learning environment? Just as you may have organizational or clean-up procedures in your agriculture laboratories, you should consider how students should organize or clean up equipment used in a blended setting. Communicating these procedures and having students practice them can ensure a smoothly run classroom, regardless of the learning environment.

Managing the Learning Management System

The layout and organization of your online presence can do much to support student learning and communicate the expectations of performance in your class to their families. Most learning management systems have customizable pages within the course shell. It is important to keep your classroom page up-to-date, organized, and uncluttered. Key questions to consider as you are organizing your classroom page include the following (Fisher et al., 2021):

- Where are the weekly/monthly schedules posted?
- Where are assignments and learning materials posted? Are they consistently available in the same space?
- How do students submit work?
- How do students find their graded work and read feedback?
- Is it apparent where students can contact the teacher for help with learning activities?
- How does the student get technical support?

To ensure maximum accessibility for all students, but especially those with disabilities, make sure to follow Web Content Accessibility Guidelines 2.0 (available at https://www.w3.org/TR/WCAG20/). This includes making sure websites are perceivable (color, background, text), operable (page can be navigated using a keyboard), understandable (text is readable and developmentally appropriate), and robust (the website is compatible with other assistive technologies). For more assistance with ensuring the accessibility of your online classroom portal, connect with your school district's IT staff.

ASSESSMENT OF LEARNING IN DIGITAL SETTINGS

As discussed in Chapters 19 and 21, assessment is an integral part of instruction. When planning instruction, the agriculture teacher should begin with identifying what they want students to know, do, and understand by the end of an instructional unit, and then work backward from there, determining what assessments will provide the best evidence students have indeed met the identified target goals. These goals are informed by educational standards as well as the context of the local community and the students themselves. Teachers can use a variety of digital tools to enhance their ability to conduct formative and summative assessments of student learning. Each of these tools can be used in a variety of ways; a video recording tool such as FlipGrid could easily be used to assess prior knowledge, capture student learning during the learning activities, and be used to illustrate student learning at the end of a unit of instruction. The tools themselves are not forms of assessment, rather the way the teacher chooses to use that tool defines that type of assessment occurring.

Assessing Prior Knowledge

Prior to beginning a unit of instruction, it is essential the agriculture teacher identify what students already know or believe they know about the unit content. Conducting a preassessment or diagnostic assessment allows the teacher to collect such evidence to inform their planning of the unit, and also allows students to gauge their current level of understanding of the content. Many of the strategies utilized in a face-to-face setting can be accomplished using digital tools. For example, a common activating strategy is a KWL chart. Students fill in what they think they know about a topic in the "K" column, what they want to know in the "W" column, and then at the end of the unit or a portion of instruction, fill in the "L" column with what they learned. Without digital devices, this typically occurs on paper and the teacher captures student responses on a whiteboard or chalkboard. In a digital setting, the teacher could use a variety of tools such as Google Jamboard, Padlet, or Google Slides to capture student responses. Any of these collaboration tools allow participants to directly edit the KWL chart by adding their responses.

Formative Assessment

Formative assessment provides ongoing feedback to the instructor and students as to their progress toward the established learning goals. For quick checks of understanding, an agriculture teacher might choose to integrate questions throughout their lecture using programs like Nearpod or Peardeck. Each of these programs has a variety of options for students to respond to prompts including multiple choice, short answer, click and drag, freehand drawing, and many more. The teacher might also choose to use an online quiz tool such as Kahoot!, Quizizz, or GimKit to integrate an element of competition into checking for understanding. Videos can be used to assess student learning using an online tool like Edpuzzle, in which teachers craft questions at key points throughout the video for students to respond to as they move through the video. Paper exit slips can easily be converted into digital exit slips by using Plickers or Google Forms. Often, the type of technology chosen by the instructor is influenced by compatibility with the operating system and technology requirements of their school.

Summative Assessment

Summative assessment of learning typically takes place at the end of a unit of instruction as the final opportunity for students to demonstrate their learning on a given topic. In a brick-and-mortar classroom, this often is a written test, but it can also include performance-based assessments such as a speech or presentation, demonstration of skill acquisition, case study analysis, or a project in which students apply their knowledge from across the unit to solve a problem. Many schools use test proctoring software when conducting summative assessments from a distance in an attempt to discourage cheating. When administering a digital test in person, teachers may have access to screen locking software that prevents students from clicking to other windows while completing their assessment.

To facilitate performance-based assessment, digital tools open up a plethora of options. Students might create digital portfolios of their work using Google Sites. The portfolio can be easily updated and shared throughout the year. This might be particularly useful for documenting flower arrangements, welds, or speeches. Digital cameras are often embedded in mobile devices such as smartphones and tablets. Students can use these cameras to document video evidence of their performing a specific skill and upload the video to the course LMS as a component of a summative assessment. Students might also create pamphlets or books using Book Creator, record a podcast, or write a blog to demonstrate their learning across a unit. The possibilities are endless!

REVIEWING SUMMARY

Advancements in technology and communication have drastically changed the way we live and learn. Digital technologies have opened up new possibilities for educators to connect students with their local and global communities, but they also need to be introduced to students as a powerful tool that they need to learn how to use responsibly and ethically.

Although the students of today have always known a world where internet access and digital devices are ubiquitous, teachers should integrate opportunities for students to develop digital citizenship across their coursework. The ability to quickly connect with others around the world using these technologies serves to emphasize the need for students to also demonstrate global competence. Employers are increasingly desiring employees who are digitally and globally competent in addition to being well versed in subject-matter knowledge.

While digital tools have the ability to connect students with others, they also may be a source of inequity. When making decisions about which tools and strategies to use in classrooms, it is essential that agriculture teachers consider how they will ensure that the learning activities are accessible to students with disabilities and those who may not have regular access to the internet outside of school.

Technology should be purposefully integrated into instruction to advance student learning and skill development. The SAMR model is a useful tool for helping teachers determine how they can enhance or transform existing learning activities into those that utilize more digital technologies.

There are several terms used to describe learning with technology and internet connectivity. Digital learning uses the combination of internet-based technologies and other educational tools to broaden the ways that educators can customize instruction for students. Learning might take place from a distance through virtual instruction, in real time (synchronous) or on demand (asynchronous). Schools might plan for students to have e-learning days in which students complete planned learning activities from home using a one-to-one device. Blended learning approaches typically occur in person in a classroom, but elements of instruction may be self-guided using online resources. A flipped classroom model presents content to students outside of class time and leverages face-to-face time with the instructor to conduct application and extension activities to enhance student comprehension.

Teaching in a digital world requires use of hardware and software designed to engage learners. Hardware such as one-to-one devices, cameras, and interactive whiteboards run software programs that help facilitate learning. More discussion of hardware found in agriculture classrooms can be found in Chapter 10.

Many of the same instructional and assessment strategies used in a traditional classroom can be converted or enhanced by digital technologies. Game-based learning, project-based learning, and scenario-based learning are all common approaches to engaging students in the content. Diagnostic, formative, and summative assessment can be accomplished using an array of online tools, many of which are available at free or reduced cost to educators.

Regardless of the learning environment, building positive relationships with students is essential to their learning experience. Frequent, positive, and honest contact with students in a virtual learning environment helps to foster positive rapport between instructors and students. Establishing norms and expectations in an online environment ensures clear communication between the instructor and students to enhance the flow of learning.

Digital technologies present novel opportunities for agriculture teachers to design meaningful, up-to-date instructional experiences that prepare students for our digital, globalized society. Not every technology needs to be used simultaneously, but those that are used by teachers should be carefully selected to enhance student understanding and learning performance.

QUESTIONS FOR REVIEW AND DISCUSSION

1. What is the role of technology in instruction? Should technology be used in every lesson?
2. How can agriculture teachers ensure the technology used in their courses is accessible to all students?

3. What are some examples of instructional technology that can be used in the classroom setting in agricultural education? The laboratory setting? When facilitating SAE projects?

4. What role do agriculture teachers play in developing digital citizenship and global competency skills in their students?

5. What is the difference between synchronous and asynchronous instruction? What are the advantages and disadvantages of each?

6. How can agriculture teachers ensure virtual learning is engaging from a distance?

7. How does your teaching philosophy influence the management of a classroom in a digital environment? How can you translate this into action?

8. How can an agriculture teacher ensure their course page in their learning management system is accessible to all?

9. How might a single digital assessment tool be used to diagnose prior learning, formatively assess learning in the moment, and be used as a summative assessment?

ACTIVITIES

1. Interview an agriculture teacher about their use of technology. What types of digital tools do they use to assess student learning? Do they have any particular tools they use in the laboratory setting that have enhanced student learning?

2. Explore various LMS companies such as Google Classroom, Canvas, Blackboard, Moodle, and Schoology. Compare and contrast the options available.

3. Become a Google Certified Educator or complete the Apple Teacher professional learning program.

4. Explore various learning technologies and create a chart that organizes them by use. Consider using categories such as discussion, collaboration, demonstration, content delivery, and assessment.

5. Using the *Journal of Agricultural Education*, *The Agricultural Education Magazine*, and the proceedings of the National Conference of the American Association for Agricultural Education, write a report on technology use in agricultural education. Current and past issues of these journals can be found in the university library. Selected past issues can be found online at www.aaaeonline.org or www.naae.org.

6. Using the SAMR model as a framework, retool an existing lesson or unit to integrate the use of technology.

7. Research how HyperDocs can be utilized in a blended or virtual learning environment. Convert student handouts into HyperDocs for future use.

REFERENCES

Buck Institute for Education. (2021, September 28). *What is PBL?* PBL Works. https://www.pbl works.org/what-is-pbl

CAST. (n.d.). *What is accessibility?* National Center on Accessible Educational Materials. https://aem.cast.org/get-started/defining-accessibility

Common Sense Education. (2016, July 12). *How to apply the SAMR model with Ruben Puentedura* [Video]. YouTube. https://www.youtube.com/watch?v=ZQTx2UQQvbU

Cornelius-White, J. (2007). Learner-centered teacher-student relationships are effective: A meta-analysis. *Review of Educational Research, 77*(1), 113–143. https://doi.org/10.3102/00346543 0298563

Dowd, H., & Green, P. (2016). *Classroom management in the digital age: Effective practices for technology-rich learning spaces.* Dowd Green EDU.

Fisher, D., Frey, N., & Hattie, J. (2021). *The distance learning playbook, grades K–12: Teaching for engagement and impact in any setting.* Corwin Press.

Gros, B. (2007). Digital games in education: The design of games-based learning. *Environments—Journal of Research on Technology in Education, 40*(1), 23–38. https://doi.org/10.1080/15391523.2007 .10782494

ISTE. (2016). *ISTE standards for students.* International Society for Technology in Education. www .iste.org/standards

Kolb, D. A. (1984). *Experiential learning: Experience as the source of learning and development* (Vol. 1). Prentice-Hall.

Malone, T. W., & Lepper, M. R. (1987). *Making learning fun: A taxonomy of intrinsic motivations for learning.* In R. E. Snow and M. J. Farr (Eds.), *Aptitude, learning and instruction III: Conative and affective process analyses.* Erlbaum.

Murphy, E., & Rodriguez-Manzanares, M. A. (2012). Rapport in distance education. *International Review of Research in Open and Distance Education, 13*(1), 167–190. https://doi.org/10.19173/irrodl .v13i1.1057

NAAE's *Connect* Podcast. (2021). National Association of Agricultural Educators. https://www .naae.org/profdevelopment/podcast.cfm

National Council for Agricultural Education. (2016). *National quality program standards for agriculture, food, and natural resources education.* https://thecouncil.ffa.org/

Owl Pellets: Tips for Ag Teachers. (2021). https://owlpelletsforag.podbean.com/

Palmer, E. (2014). *Teaching the core skills of listening and speaking.* ASCD.

Puentedura, R. (2006). *Transformation, technology, and education* [Blog post]. Retrieved from http:// hippasus.com/resources/tte/

Schaller, D. (2005). *What makes a learning game?* Retrieved from https://eduweb.com/schaller -games.pdf

Sorin, R., Errington, E., Ireland, L., Nickson, A., & Caltabiano, M. (2012). Embedding graduate attributes through scenario-based learning. *Journal of the NUS Teaching Academy, 2*(4), 192–205.

Tichnor-Wagner, A., Parkhouse, H., Glazier, J., & Cain, J. M. (2019). *Becoming a globally competent teacher.* ASCD.

Tucker, C. (2013). Teachers' guide to using videos. *MindShift.* https://cdn.kqed.org/wp-content /uploads/sites/23/2013/03/MindShift-Guide-to-Videos.pdf

Tucker, C. R. (2012). *Blended learning in grades 4–12: Leveraging the power of technology to create student-centered classrooms.* Corwin Press.

Tucker, C. R. (2020b, August 19). Asynchronous vs. synchronous: How to design for each type of learning. *Dr. Catlin Tucker.* https://catlintucker.com/2020/08/asynchronous-vs-synchronous/

VMware, Inc. (2022). *What is a cloud server?* Retrieved from https://www.vmware.com/topics/glossary /content/cloud-server.html

Wells, T., & Miller, G. (2020a). Teachers' opinions about virtual reality technology in school-based agricultural education. *Journal of Agricultural Education, 61*(1), 92–109. https://doi.org/10.5032 /jae.2020.01092

Wells, T., & Miller, G. (2020b). The effect of virtual reality technology on welding skill performance. *Journal of Agricultural Education, 61*(1), 152–171. https://doi.org/10.5032/jae.2020.01152

Wong, H. K., & Wong, R. T. (2009). *The first days of school: How to be an effective teacher.* Harry K. Wong Publications.

ADDITIONAL SOURCES

Alber, R. (2019, March 19). *Using video content to amplify learning*. Edutopia. https://www.edutopia.org/article/using-video-content-amplify-learning

Budhai, S. S., & Skipwith, K. B. (2017). *Best practices in engaging online learners through active and experiential learning strategies*. Routledge.

Clark, H., & Avrith, T. (2017). *The Google infused classroom*. EdTech Team Press.

Digital Learning Collaborative. (2019). *eLearning Days: A scan of policy and guidance*. https://www.digitallearningcollab.com

Highfill, L., Hilton, K., & Landis, S. (2016). *The HyperDoc handbook: Digital lesson design using Google Apps*. Elevate Books EDU.

McKibben, S. (2014). *Showing videos in the classroom: What's the purpose?* Retrieved from https://www.ascd.org/el/articles/showing-videos-in-the-classroom-whats-the-purpose

Mobile Application (Mobile App). (August 7, 2020). Techopedia. https://www.techopedia.com/definition/2953/mobile-application-mobile-app

Novak, K., & Tucker, C. R. (2021). *UDL and blended learning: Thriving in flexible learning landscapes*. Impress.

Patterson, L. (2021, August 8). Why you should bring podcasts into your classroom. *Cult of Pedagogy*. https://www.cultofpedagogy.com/podcasts-in-the-classroom/

Sangrà, A., Vlachopoulos, D., & Cabrera, N. (2012). Building an inclusive definition of e-learning: An approach to the conceptual framework. *International Review of Research in Open and Distributed Learning, 13*(2), 145–159. https://doi.org/10.19173/irrodl.v13i2.1161

Terada, Y. (2020). A powerful model for understanding good tech integration. *Edutopia*. https://www.edutopia.org/article/powerful-model-understanding-good-tech-integration

Tucker, C. R. (2020a, March 8). Tips for designing an online learning experience using the 5Es instructional model. *Dr. Catlin Tucker*. https://catlintucker.com/2020/03/designing-an-online-lesson/

Virtual Reality Society. (2017). *What is virtual reality?* Retrieved from https://www.vrs.org.uk/virtual-reality/what-is-virtual-reality.html

Williams Jr., K. (2018, November 20). *Digital learning: What exactly do you mean?* Michigan State University Hub for Innovation in Learning and Technology. https://hub.msu.edu/digital-learning-what-exactly-do-you-mean/

14

Classroom Management

The stories about Mr. "Hubcap" Williams are true and legendary (Croom & Moore, 2004). During a welding class, students in his class welded the tines of the garden tiller to a steel support post in the agriculture shop. When Mr. Williams tried to take it outside for use by his horticulture classes, he found he could not budge it. On another occasion, students took the agricultural mechanics textbooks in his classroom and sawed them in half using the table saw in the agricultural mechanics shop. On another occasion, students locked him in the horticulture storage shed. The school custodian heard him banging on the shed's doors and released him after he had been locked inside for two hours.

Hubcap Williams acquired his nickname after some students put rocks in the hubcaps of his automobile. You can imagine the sound those stones made as Mr. Williams drove the car. He eventually sold the automobile after being unable to cure this apparent mechanical defect. The misbehavior of his students was a source of incredible frustration to Hubcap and led to many miserable days for him in the classroom.

Hubcap had an impressive reputation at his school. Students enrolled in his classes because they found it an easy period during the day when they could kick back and avoid strenuous mental activity. Hubcap did not put much effort into teaching. He would instead work on projects that struck his fancy. Frequently, class projects involved a few of the many students in his classes, leaving most students to find things to do to entertain themselves. Other students avoided his classes because they did not offer the challenge and rigor they desired.

Hubcap Williams was a real agriculture teacher and has long since retired from teaching. His name has been changed to protect his identity. Would you prefer to be a student in his classroom? Would you like to be his colleague, working in the same agriculture program?

OBJECTIVES

This chapter addresses the National Quality Program Standards for Agriculture, Food, and Natural Resources Education (National Council for Agricultural Education, 2016), specifically Standard 1B: Program Design and Instruction. It has the following objectives:

1. Discuss unique attributes of agricultural education as related to establishing a quality learning environment.
2. Describe the characteristics of student behavior in a learning environment.
3. Identify practical approaches in student discipline.
4. Describe how to handle misbehavior when it occurs.
5. Identify student behavior trends in agricultural education.

TERMS

discipline
engagement
misbehavior
proximity control

punishment
restorative practice
reward
trauma-informed teaching

CLASSROOM MANAGEMENT, AN OVERVIEW

The American people like their local schools, and they trust and have confidence in their children's teachers. But there are concerns. Bullying is becoming an issue of importance to parents who send their children to public schools, and there is increased concern about the use of tobacco, vaping, and drug use among students (Heller et al., 2020). These concerns demonstrate that agricultural education teachers must have the public's support necessary to create and maintain a positive classroom environment.

Classroom management starts with the environment the teacher creates in the classroom. Students require clear direction toward learning goals. The mechanism for achieving those goals is **engagement**. Engagement occurs when students make a psychological investment in learning. The lessons must be engaging and relevant to the interests of the students. Often, the instructor needs to help students see the connection between the lesson content and students' interests and experiences.

Students typically respond positively to routines and established norms for behavior in the classroom. Teachers need to develop those norms on the first day of classes and reinforce them throughout the academic year. The teacher communicates expectations to students through direct communication, such as "I appreciate how you went about starting on your lab assignment today" or "Thanks for cleaning up the lab area." The teacher sets the standard for behavior in the classroom, and their behavior should encourage students to perform their best (Marzano, 2017).

LEARNING ENVIRONMENTS IN AGRICULTURAL EDUCATION

The learning environment is large, complex, and reflects the society from which students come. School violence, the use of alcohol and illegal drugs, and sexual and racial harassment are widely discussed today. These and other student behaviors are not just in agricultural education classrooms but are school-wide. Educators must be aware of these issues as they go about establishing a learning environment.

The agricultural education classroom generally has multiple learning environments, including, but not limited to, standard classroom with desks, greenhouse, land laboratory, mechanics laboratory, and biotechnology laboratory. In addition, it may have storage areas, an instructor's office, and other facilities. These learning environments contain numerous industrial tools and equipment. They may contain hazardous materials such as compressed gases, fuels, paints, and pesticides. The key to managing these facilities is found in a sound classroom management plan. Box 14.1 identifies the elements of a positive classroom environment.

Student performance depends upon the level of expectation held by the teacher. A key in managing agricultural education is to establish and maintain a quality learning environment. The learning environment includes all the conditions that influence teaching and learning. Some environments promote learning; others may impair learning. The teacher's responsibility is to be in charge of the situation and establish a quality learning environment. A teacher may use a classroom management plan based on the best practices of a quality learning environment.

Agricultural education may involve the use of power tools, chemicals, lab equipment, and many other potentially hazardous devices and materials. Misbehaving students could seriously injure themselves and others around them. Teachers should develop organizational skills and lesson plans that minimize the risk of injury to students.

BOX 14.1 *Positive Classroom Environment at Multiple System Levels*

The School
School-wide rules are clear and developed with student and teacher input.

School administrators support teachers in the application of school-wide policies.

School officials conduct in-service training and staff development that help teachers develop positive classroom environment strategies.

In the Agricultural Education Classroom and Laboratory
Classroom and laboratory rules are posted and referred to frequently.

Exceptions to classroom and laboratory rules are clearly explained.

The room is organized and arranged to allow for efficiencies in student movement and learning.

Lessons should be well planned, interesting, and engaging to learners.

Support structures should be in place to allow extra assistance for students experiencing difficulties concerning learning.

The Agriculture Teacher
A teacher's language should be clear, and positive, and free of labels.

A teacher's feedback should be based on performance rather than student characteristics.

The teacher should provide a clear explanation of tasks and assignments and behavioral and learning expectations.

The teacher should develop a system for creating a meaningful dialogue with students on items of mutual importance.

Teachers should practice empathy and work to develop positive relationships with students based upon respect and trust.

Teachers should look for opportunities to give praise.

System-Wide Responses to Student Behavior
Appropriate behavior should be rewarded or reinforced through effective communication strategies between the student and teacher.

Teachers should use low-level and discreet strategies for dealing with inappropriate behavior before it reaches the stage of disrupting the whole class.

Serious misbehavior is dealt with seriously and promptly.

Furthermore, students come into agriculture classes with varying degrees of skill in the agricultural sciences. Some students have experience with the proper use of tools, equipment, and chemicals, while others have no experience in these areas. Some students have in-depth knowledge of some agricultural subjects but very little knowledge of others. For instance, the student who is an expert welder and machinist may not have much skill in handling livestock.

Laboratory work involves a significant amount of psychomotor ability, and students must be taught how to work safely and follow procedures. The hands-on nature of agricultural education demands much attention to the development of new psychomotor skills in addition to cognitive and critical-thinking skills.

Agriculture classes are elective courses. Students may enroll in these courses because they are personally interested in the subject matter. However, most experienced teachers recognize not every student who enrolls does so because of a deep personal interest in an agricultural career. Some students take the courses because they have a nominal interest in agriculture and wish to explore further. These students can either become more interested in the subject matter or lose interest as the courses progress.

Some students take agriculture courses to avoid other required subjects or because they need additional electives to graduate. Sometimes, administrators and guidance counselors place students in agriculture classes without regard to their career interests. Whatever the method by which students arrive in the classroom, the possibility exists that student behaviors could disrupt the learning experiences of other students.

It would be an error to assume that discipline and classroom management are what teachers do to get the lesson accomplished and the content delivered. Maintaining a positive learning environment is more than an ancillary duty; it is part of the learning process itself. The teacher who understands the psychology of learning and human growth and development processes at the adolescent and preadult stages can effectively help students make the right choices regarding their behavior.

MISBEHAVIOR IN THE CLASSROOM

Misbehavior is the condition that results from a person's inappropriate attempt to solve a problem or adapt to a situation (Maslow & Frager, 2003). This inappropriate coping behavior is purposeful and motivated. It is often a learned response to conditions in the student's environment. In Maslow's words, "It is an attempt to make up for internal deficiencies by external satisfiers" (p. 132).

Students express coping behavior in several ways. For instance, a student coping with a need to be part of a group might misbehave to gain acceptance from their classmates. Another student who works a part-time job well into the late evening hours may cope with a physiological need for rest through frequent absences from school (Dahl, 1999). When the coping mechanism violates social norms within the classroom and school, it becomes misbehavior (Croom & Moore, 2003).

Working through the problems associated with misbehavior in the classroom is an essential part of teaching. However, teachers rarely communicate among themselves in any depth about the subject of student misbehavior even though the stress generated by misconduct is a more significant concern than problems with other working conditions (Abel & Sewell, 1999).

FIGURE 14.1 What features do you see in this learning environment? Are students demonstrating on-task behavior?

How would you respond?

Ms. Mary Jackson sat at her desk grading the last of the animal science unit tests. The last test she picked up to grade belonged to Jeff Wright. Immediately, she noticed most of the questions on the test were unanswered. Ms. Jackson put down the paper, picked up the grade book, and found Jeff's name in her second-period class. Except for a few directed lab exercises where he worked with a partner, Jeff had few grades to show for his efforts. She picked up her test and walked to the next-door classroom to speak to her teaching partner, Tom Davis.

"Tom, did you have Jeff Wright in class last semester?"

Tom looked up from his paperwork and replied, "I sure did. Do you have him this term? I'll bet I know why you are asking about him."

"Yep, I have him in the animal science class this semester. How well did he do in your class?" asked Ms. Jackson.

"Not good at all. It was the introduction to agricultural science class, and Jeff barely made a D. He's not a great student," said Mr. Davis.

"What's the deal with Jeff, then?" asked Ms. Jackson. "How can I get him engaged in this class? Every day he falls farther and farther behind. He doesn't bring a notebook to class, asks no questions, and often tries to sleep while I am teaching. I have to stay on him to keep his head up off the desk."

"How do the other students act around him?" asked Mr. Davis.

"See, that's the interesting thing. When the other students are engaged in what we are doing, Jeff just sits there like a bump on a log during labs. He doesn't communicate well with the other students," explained Ms. Jackson.

"What are you going to do?" asked Mr. Davis.

"I don't know, but if I don't do something soon, he is going to fail my class," replied Ms. Jackson.

• • •

This misbehavior is a common experience for teachers. How would you address the problem Jeff is having in this class?

Questions for discussion

What misbehaviors do you notice in this scenario?

What do you think is the origin or cause of Jeff's behavior?

How should the instructor respond to the misbehavior?

Because teachers spend much of their workday almost exclusively with pupils, they tend to formulate their own definition of misbehavior and handle misbehavior accordingly (Borg & Riding, 1991). Thus, there is a wide range of behavioral modification techniques used to redirect or repress misbehavior.

Several studies have addressed student behavior in agricultural education. Most found student attitude and disposition toward schooling was a significant factor in misbehavior. Furthermore, the ambivalent attitude of students, students being poorly prepared for class, and a general negative student attitude were the most frequently cited behavioral problems (Burnett & Moore, 1988; Camp & Garrison, 1984; Croom & Moore, 2003). Some students develop a dislike for schooling. Their attitude carries over into the thoughts, feelings, and reactions they have related to the efforts of instructors to teach them. In the agricultural education program, the task is to provide conditions under which student interest remains high. The effective agriculture teacher constantly seeks effective methods to keep student learning at the highest possible level. Box 14.2 identifies environmental variables that may influence student behavior.

BOX 14.2 *Environmental Variables Influencing Student Behavior*

The Student
Socioeconomic status and family structure
A disposition toward school and learning
General physical and mental health
Dietary habits and sleep patterns
Level of interest in the class or subject matter
Learning style
Effect of peer pressure

The Teacher
Teaching ability
Knowledge and expertise in student behavior management
General physical and mental health
Dietary habits and sleep patterns
Personality
Quantity and quality of student supervision
Organization of instruction and quality of curricular planning
Teaching methods employed

The Rest of the Class
Time of day
School activities (prom day, last day of school, other special activities/events)
Size of class
Peer pressure
Subject matter or teaching method employed

The Teaching and Learning Facility
Cleanliness of the facility
Size of classroom
Student desk style, arrangement, and quantity
Adequate lighting in the classroom and lab environments
Safe use and maintenance of lab equipment
Space for students to sit and move about the classroom
Ambient air temperature of the classroom and labs
Ambient noise in the classroom
Organization of learning materials and resources
Adequacy of personal protective equipment
Frequency and intensity of external interruptions of a lesson in progress by announcements
   from the public address system, classroom visitors, classroom telephone interruptions

---

Sources: DeBruyn (2011); DeBruyn & Larson (2009).

DISCIPLINE VERSUS PUNISHMENT

The word *discipline* is derived from the Latin *disciplina*, which means "to teach," and the Latin *discipulus*, which means "pupil" (Oxford University Press, 2000). Educators view discipline as authoritative control of behavior through punishment and rewards. It is related to the kind of behavior students need to express to make learning meaningful.

Punishment and rewards are discipline functions and are often used to encourage students to learn new choices in personal behavior management. ***Punishment*** is the administration of a penalty or undesirable action designed to discourage socially unacceptable behavior.

A ***reward*** is a pleasant or satisfying consequence of an action. The goal of giving a reward is to gain desired behavior. Rewards include allowing a student to be first to perform a desirable task, awarding a small prize, placing an individual's name on the honor roll, and using other means viewed as favorable outcomes.

Perhaps, at this point, it is necessary to define misbehavior further. Misbehavior is any behavior that is inappropriate for the moment. For instance, certain classroom activities may require students to stand up and move about the classroom, and that might be acceptable behavior at the moment. However, at other times, it is the instructor's wish that the students remain seated and attentive to a lecture. Standing up and moving about the classroom then becomes misbehavior at that moment.

It is crucial to see misbehavior as an individual choice. Teachers perceive misbehavior as something that a student does that directly and adversely affects others in the classroom. Misbehavior is a choice individual students make, and teachers should view it as such. For instance, the student who has her head down on her desk and is sleeping during a lecture period may not be disturbing other students, but this behavior is inappropriate. It should receive some corrective attention from the teacher.

FIGURE 14.2 Students in agricultural education programs use tools and equipment that could cause injury if misused. There is little room for misbehavior in the agricultural lab setting.

THREE VARIABLES OF A DISCIPLINE PROBLEM

DeBruyn (2011) found three variables (the student, the teacher, and the class) are present in almost every discipline problem. These variables determine the frequency, intensity, and duration of misbehavior. Further, these three variables influence the type of misbehavior that occurs.

The Student

A student's behavior could change from socially acceptable for the classroom into misbehavior on any given day and in any given situation. As human beings, students are often influenced or motivated by certain external factors in their environment. A student who did not get enough sleep the night before is going to be sleepy in class. The student who failed to eat a nutritious breakfast might be passive in class, or the student who consumed too many caffeinated soft drinks and energy drinks might be restless and excited. Student behavior is likely to be different on the day of the school prom than on an ordinary school day. The experienced teacher can frequently determine the type of behavior students will exhibit by observing them as they come into the classroom. A primary goal should be to help students practice self-discipline and exercise self-control.

The Teacher

Misbehavior is an individual choice, but the teacher may play a role in whether or not misbehavior occurs. The teacher sets the tone and pace of the class and exerts a significant influence over students. Therefore, the teacher's actions may influence student behavior either positively or negatively. The following is an example of how a teacher can negatively affect student behavior. A student puts his head down on his desk and falls asleep during class, missing important content and concepts. The teacher notices the behavior but fails to wake the student or somehow intervene in the behavior. In this case, the teacher is contributing to the misbehavior.

FIGURE 14.3 The dynamic presence of the teacher in the classroom mitigates many misbehavior problems.

Classroom management is an integral part of having a good learning environment. The teacher's presence in the classroom or the supervision of students during lab activities is a significant factor in student behavior. Teachers who are constantly supervising students and directing their behavior generally have fewer behavioral problems.

The Class of Students

The third factor in student misbehavior is found in the actions of the class as a whole. Students need to be accepted by their classmates. Moreover, they have a strong need for belonging in the sense of acceptance by the group members. Therefore, students will often act in a manner that brings admiration and attention to themselves, even though their actions are misbehavior.

Furthermore, this need for belonging and safety is so strong in students that their response to corrective action by the teacher often leads to an escalation of misbehavior. For instance, the disciplined student often becomes defiant and argumentative with the teacher. That very same student might be altogether different and exhibit more socially acceptable behavior in a private student–teacher conference. Effective teachers recognize they greatly influence student behavior.

The easiest variable to control is the teacher's behavior. A teacher is responsible for attempting to diagnose the problem and adjusting instruction and interaction with students to help students make the right choices about socially acceptable behavior. Recognizing and coping with student misbehavior is an essential part of instruction. The beginning teacher must develop tools for effectively dealing with behavior.

CONDITIONS THAT PREVENT MISBEHAVIOR

One of the best methods of managing student misbehavior is to minimize opportunities for it to happen. The effective agriculture teacher recognizes the conditions that may cause mischief and modifies the instructional environment so misbehavior cannot flourish. For instance, the teacher can often improve student behavior simply by reorganizing the seating arrangement in a classroom. To prevent students from mishandling shop tools, the teacher can establish procedures by which the tools are locked in the tool room when not in use. If students appear sleepy and unfocused during class on a warm day, the instructor can modify instruction to engage the students in activities or movement. In some cases, simply changing the air temperature in the classroom can be conducive to good student behavior. A classroom that is too warm or too cool can discourage student participation.

Effective teachers know their students well enough to identify uncharacteristic behavior. It should be easy for a teacher to spot the behavior of students who come to class angry, worried, or excited about some prior experience. By adjusting the lesson, the teacher can minimize inappropriate student behaviors in the class. The angry student could be pulled aside for a private conference with the teacher. By listening to the student's problem and offering constructive solutions in a nonthreatening manner, the teacher might defuse that behavior before it affects the whole class. The key is to remove as many distractions from the learning environment as possible.

Have a Well-Prepared Lesson

Good preparation by the teacher helps prevent many misbehavior problems. With good planning, the learners are engaged and active. Busy learners do not have the time to contemplate misbehavior.

A well-prepared lesson plan promotes student learning. The single most effective method for preventing misbehavior is to have a well-prepared, well-executed lesson delivered enthusiastically. Realistic, relevant lessons, presented interestingly, will engage the students inclined not to misbehave. When planning a lesson, the teacher needs to think like a student but act like a teacher. Answer the question, "How can I plan this lesson so it is interesting to the students and meets their needs?" The same things that motivate adults also motivate students. Attractive visuals, exciting demonstrations, and real-life applications of learned skills are all things that encourage adults. These same methods also engage student interest.

Through careful planning, the agriculture teacher can usually find a way to engage students in the lesson. Teachers can find unique ways to keep students involved in a class by using their creative talents, even when resources are limited. For example, a teacher is planning a lesson on taking a soil sample. The least preferred method is for the teacher to teach the use of the soil probe to students one at a time. Invariably, while the teacher is showing one student how to collect a soil sample effectively using the soil tube, the remainder of the class is talking, joking, or wasting time.

A more effective method for teaching this procedure would be this: The teacher gives each student a shovel to collect a soil sample. Most homeowners do not own soil tubes for taking soil samples, but they would have shovels to help complete the task. The teacher demonstrates the procedure to all students and then assigns them to practice the skill under supervision in a suitable location. Thus, students learn the technique for taking soil samples promptly.

By developing instructional activities that engage all students in a lesson, the instructor can prevent misbehavior while using instructional time more judiciously. The best defense against student misbehavior is a well-planned and well-executed lesson.

Teach the Entire Period

Making full use of class time is a companion to good instructional planning. Student teachers are often surprised at how quickly they execute their prepared lessons. They are at a loss about what to do with the extra time remaining at the end of the class period.

Do not allow dead time in a lesson. The students should be working and thinking throughout the entire class time. It takes less effort and causes less aggravation to prepare and deliver a lesson that engages the students for the entire class period than it does to constantly chase down and discipline students who have been given too much free time. Students dislike being bored. Lessons that are irrelevant to students' interests and needs, not well planned, or tedious lead to student disengagement with the lesson. When this occurs, learning takes a backseat to other things students might do to amuse themselves. By engaging students in learning for the entire class period, the teacher can control student behavior in the best possible manner.

A matter of time

If a teacher allows a class to have 10 minutes of free time at the end of a class period, that adds up to approximately 15 hours of free time by the end of the semester and 30 hours of free time by the end of the school year. If the average instructional day is 6 hours long, that equals five school days of free time for students.

Consider the amount of instruction lost in those five school days, and the opportunity cost of this lost instructional time.

BOX 14.3 *Sample Bell Ringer Questions*

- What are the six classes of nutrients in livestock feed?
- Of the four pines we identified yesterday in class, which ones are naturally adapted to drier sites and naturally found to grow in wet soils?
- How many uses of the peanut can you list? How about cotton?

Use Interest Approaches

Teachers should use a focusing activity at the beginning of a lesson. The activity should direct the attention of the students to the lesson at hand. Before that, "bell ringer" activities should be available for students as they enter the classroom (see Box 14.3). This activity usually consists of a question or statement written on the chalkboard/whiteboard at the front of the classroom. Students should enter the classroom, get their materials ready, and respond to the question or statement on the board. The bell ringer is an excellent method for reviewing at the beginning of a class, and it folds nicely into a more elaborate interest approach as the teacher begins to teach. Students should have something positive and constructive to do the moment they enter the classroom.

Box 14.4 provides three examples of interest approaches to lessons to gain students' attention and develop a genuine interest in the subject.

BOX 14.4 *Three Examples of Interest Approaches*

Interest Approach #1

Lesson: "Grafting With Skill"

Class: This is a Horticulture 2 class with 20 students.

Interest approach:

1. Bring in a bag of white seedless grapes and a few cut-up Red Delicious apples.
2. Share these with the students and ask them to identify what they are eating. Ask if they have ever eaten these items before.
3. Draw on their previous horticulture experience by asking them to find the seeds. (*Note:* They may or may not be able to find seeds in the Red Delicious apples.)
4. Ask them how many have apple trees at their houses or their grandparents' houses. Explain that the seeds in the apples will not produce more Red Delicious apples.
   a. Ask: "Since there are no seeds, how do we get more seedless grapes?"
   b. Ask: "If we don't get more Red Delicious apples from the seeds, how do we get more?"
   c. Ask: "What DO we get if we plant the Red Delicious apple seeds?"
5. Begin the discussion of grafting.

Interest Approach #2

Lesson: "Understanding Plant Terminology"

Class: This is a Horticulture 1 class with 25 students.

Interest approach:

1. Give each student a flower and a large piece of paper. Tell the student to sketch the flower and write down the name of each part they recognize.

Continued

BOX 14.4 *Continued*

2. Ask each student to identify their flower. Allow the student to identify someone else's flower with which the student might be familiar. Have a few students share their specified drawings.
3. Display a list of plant terms.
   a. Ask: "Who can identify a term from the screen?"
   b. Ask: "What is a calyx, what is a corolla, and how do you tell the difference?"
   c. Ask: "How many of these terms apply to the flower in front of you?"
4. Begin the discussion and definition of the plant terms.

Interest Approach #3

Lesson: "Identifying Plant Pests"

Class: This is a Horticulture 1 class with 25 students.

Interest approach:

Bring in a display of plant pests that include leaves with fungus and disease, plants damaged by whiteflies and aphids, and weeds such as morning glory or stinging nettle. You may also include some pictures of mammals, such as whitetail deer. Altogether, there should be about 10 specimens. Have students observe the specimens and describe what they see. Lead from this discussion into the presentation of the objectives for the lesson. Begin covering the content for the first objective.

Develop Classroom Routines

Routines are force multipliers. Routines operate in the background of a person's mind without much conscious thought. This frees the brain to focus on other things. On the first day of class, the teacher should explain the classroom rules and procedures for students to follow. Then, the teacher should firmly adhere to those policies to establish a routine for the

FIGURE 14.4 The table arrangement facilitates student discussion, as opposed to lecture. Notice there is room for the teacher to circulate between tables to monitor student performance.

students. For example, some teachers have a station set up with typical classroom materials such as scissors, tape, markers, colored pencils, glue, and paper. Students who need those materials go directly to the materials station without delay, preventing the teacher from being inundated with requests for those same materials during a class activity.

Teachers should also post their classroom rules in all instructional areas for students to see. Here is a sample of classroom rules:

- Be seated and have materials ready for class before the tardy bell rings.
- Respect the responsibility of the instructors to teach.
- Respect the rights of your fellow students to learn.
- Be kind and courteous to visitors.
- Make up work after an absence.
- Keep work areas, desks, and tables clean.

WHEN MISBEHAVIOR HAPPENS: SOME GENERAL GUIDELINES

What should be the teacher's response to misbehavior when it happens? Invariably, teachers will be confronted with some type of student misbehavior during the school day. All teachers are responsible for all of the students all of the time.

During the Lesson

Misbehavior during a lesson is particularly disruptive to both teaching and learning. One mistake a new teacher often makes is attempting to ignore misconduct by continuing to teach. If students are talking, the teacher will try to speak louder to drown out the students. This procedure is both ineffective and exhausting for the teacher. The teacher can use their silence, posture, positioning in the room, and "teacher look" to quiet students and then resume instruction. Again, the use of interest approaches and high-quality lessons will negate a large portion of the discipline problems in a class.

Handle class disruption and misbehavior quickly and decisively. Keep teaching, but use gestures and eye contact to communicate messages to the misbehaving student. If the misbehavior continues, try to deal with the student privately. A public reprimand will often result in a public spectacle. If the student fails to heed the teacher's warnings, the teacher should follow the procedure outlined in the school handbook such as assigning detention or sending the student to the school disciplinarian. The key is to be consistent, decisive, and task-oriented.

Proximity Control

One effective method for encouraging good student behavior is moving around the classroom, observing students working. Whenever a student is misbehaving, the misbehavior is often curtailed by the proximity or presence of the teacher. *Proximity control* can shape student behavior by being close to the student in the classroom or laboratory area. The teacher need not say anything to the student, and often the simple presence of the teacher will cause the student's behavior to improve.

An additional benefit of moving about the class is that the variation of the teacher's location relieves student boredom. If the instructor stands for long periods in the same spot, habituation sets in, and the students lose focus and attention. By moving around the classroom, the teacher breaks the monotony of the environment and causes students to remain attentive.

FIGURE 14.5 Brief teacher–student conferences out of earshot of other students can identify and resolve minor misbehavior problems.

Setting an Example

Teachers must be good, professional role models for students. They must refrain from acting in a manner that encourages misbehavior. Teachers who fail to follow classroom rules are setting a poor example for the students to follow. Middle school and high school students have a keen sense of fairness, and they resent actions by the teacher that show a disregard for the rules. For example, the teacher who sends students to the school office for carrying knives on campus will be looked upon poorly by students if that teacher continues to carry a pocket knife. Teachers who are following their own rules are good role models of self-discipline for students.

Use Restorative Practices

Some school systems still operate under the practice of punishing student misbehavior and then moving on without dealing with the causes of the misconduct. However, this practice has come under increasing public scrutiny.

Students who are misbehaving may be doing so as a reaction to poor academic performance. Students who are routinely absent from school feel alienated from their peers. Zero-tolerance discipline procedures are draconian at best and cause emotional stress and frustration for all involved. At their worst, zero-tolerance discipline procedures exacerbate existing racial, ethnic, and socioeconomic inequalities. Students who misbehave in a zero-tolerance environment may escalate their behavior out of frustration or embarrassment. Severe punishment that removes students from the learning environment without restorative practice damages the student–teacher relationship.

The 2000 report of the American Psychological Association Task Force on Zero Tolerance found zero tolerance for misbehavior that does not present a threat to safety was indeed harmful to the adolescent learner's social, emotional, and cognitive development (American Psychological Association Zero Tolerance Task Force, 2008). A better option is *restorative practice*, which is restoring the school or classroom community by performing service.

The School Community and Restorative Practices

One method for dealing with misbehavior is to involve the school community in the righting of the wrongs committed. Whenever a wrong is committed in a school's social environment, everyone affected by the misbehavior contributes to acknowledging the problem and has a stake in its remedy. Students who misbehave must restore the community to the status quo by performing helpful community service. Some schools use peer juries to recommend the best course of restoration.

Prevention of Misbehavior in a Restorative Practices Model

An ounce of prevention is worth a pound of cure. Misbehavior can be prevented through several strategies. These include

- Teaching students how to resolve conflict proactively
- A peer mediation program that helps students work through conflict
- Utilizing a multiprong approach that incorporates a variety of restorative practices consistently and purposely

Restorative Practices in the Agriculture Class

One form of restorative practice is restorative justice—the process of righting wrong behavior and repairing the harm done. An example of restorative justice for a student with excessive absences from school might be

- The student spends time after school catching up on work missed.
- The teacher connects with the student's parents to develop solutions to mitigate excessive absences in future.
- Periodic student–teacher conferences are scheduled to measure progress and academic achievement.

The focus of restorative justice is less about stopping the behavior and more about addressing the harm done by the behavior.

The First Class Meeting

A teacher's expectations should be made known at the first meeting of a class. By doing so, students understand the kind of behavior they must exhibit. It is far easier to have a good learning environment from the beginning than it is to try to impose it later. Here are some things to do at the first meeting of the class:

1. Inform the students of your philosophy of teaching and learning. Take the time necessary to explain how you plan to teach, what you plan to teach, and why you will be teaching it. Students will understand very clearly the teacher's commitment to teaching.
2. Explain the general rules of conduct for students in the class. Take the time to explain why these rules are essential.
3. On the first day of class, teach something from the content area. Do more than just distribute the syllabus and go over classroom rules. Set the tone and pace of the course the very first day by committing to teach some agricultural subject matter.

Avoid Disruptions

The purpose of misbehavior with some students is to cause the teacher to stop teaching. It is in the teacher's best interest to thwart those efforts to stop instruction. Continue to teach if at all possible. Use proximity control to manage student behavior while continuing the forward motion of the lesson. Handle discipline issues quickly and quietly if possible so the lesson keeps moving. Do not give the misbehavior more attention than necessary.

Under no circumstances should the teacher shout or lose their temper in class. When teachers lose their tempers, they may resort to bullying to get the students to behave. Students may consider the ranting and raving teacher to be a form of amusement. If a student chooses to argue with you over a discipline issue in the classroom, defer it until after class or meet briefly in private outside of class.

Follow School Policies

A teacher should know and follow the policies of the school on student misbehavior. Use only school-sanctioned means to correct student behavior. To do otherwise could cause harm to the student and, subsequently, lead to a teacher's dismissal for administering inappropriate discipline measures.

Make sure administrators are familiar with and approve of the classroom behavior plan. Ask administrators to help you find ways to engage problem students productively in the classroom. It is better to ask administrators to assist you in preventing misbehavior than to send students to the office for the school disciplinarian to handle. Good administrators appreciate a proactive approach to discipline and will applaud the teacher's efforts in keeping students in class and on task.

Communicate With Parents

Teachers should contact parents when student misbehavior reaches the point where some intervention is necessary. Teachers should seek to balance their communication with parents by reporting student success when it occurs. Otherwise, contacting parents only when the students misbehave creates an antagonistic relationship between parents and teachers. Ask parents to help you with discipline issues and keep them informed of their children's academic performance. Try to solve problems before they become serious issues.

Handle Misbehavior With Finesse

If a teacher notices misbehavior, they should confront the student about the behavior and appropriately handle it. Under no circumstances should a teacher ignore misbehavior simply because it is not happening in their assigned work area. Student behavior is a shared

FIGURE 14.6 On-task behavior is an integral part of the learning environment. Do you consider this to show misbehavior?

responsibility between parents, students, and all teachers. Effective schools create the mindset that there is no safe place on campus for misbehavior.

Organize the Learning Environment

The physical environment in which students learn is an essential factor in preventing or encouraging student misbehavior. The arrangement of student desks can either alleviate or contribute to student misbehavior. Where the teacher is the primary focus, the students' desks should face the teacher. Students should have adequate desk space for writing in their notebooks. If the primary method for the lesson requires student discussion, then the desks should be arranged to allow student groups to discuss the topic comfortably.

It is good to modify instruction to use various teacher-centered, social interaction, and student-centered instructional methods. To help ensure the success of these methods, the instructor will need to modify the seating arrangement. Simply rearranging student desks can clear up minor misbehavior problems. (See Figure 14.7 for two arrangements.)

What would you do in this situation?

Joanne entered the classroom just as the tardy bell rang, engaged in a loud argument with another student. Upon seeing Mr. Williams, the agriculture teacher, she immediately stopped arguing and loudly proclaimed, "Hi, Mr. Williams. My favorite teacher in the whole world!"

Joanne was a junior in Mr. Williams's horticulture class. Joanne has been a challenging student to teach because she often argues with her teachers and other students. She is frequently late for class, rarely turns in assignments on time, and usually asks inappropriate questions or uses crude or unsuitable language. Her relationships with other students are strained at best.

"Hello, Joanne, please take a seat so we can get started," replied Mr. Williams.

"I would, but Tom is in my seat again! Boy, you had better move before I—" said Joanne, getting ready to take a swing at the student.

"Hold on, Joanne, take any open seat. The late bell has rung, and class has started."

"Okay, I'll take your seat then," replied Joanne, as she sat down at the teacher's desk.

"Nope. Not that one, a regular student desk," replied an exasperated Mr. Williams.

"You said to take any seat," said Joanne. Several students laughed at Joanne's response.

"You know which seats you are supposed to sit in, so let's get moving," said Mr. Williams as he moved to the front of the class to begin the lesson.

"Today, we are examining the growth and development of . . ." began Mr. Williams.

While moving to a student desk in the back of the room, Joanne loudly interrupted her teacher again. "Are we going out to the greenhouses today?"

"Joanne, you know the rules. Raise your hand if you have a question," replied Mr. Williams.

Joanne, still not in her seat at the back of the room, scowled and crossed her arms. "It was just a simple question," she responded, throwing down her school backpack and falling into a student desk at the rear of the class.

• • •

This scenario plays out every day in schools. Your response in this situation can mitigate student misbehavior or cause it to escalate into more severe misconduct.

Questions for discussion

What misbehaviors do you notice in this scenario?

What do you think is the origin or cause of Joanne's behavior?

Which responses by Mr. Williams have been successful in helping Joanne manage her behavior?

How might the instructor respond more effectively to the misbehavior?

How might you employ restorative justice measures in this situation?

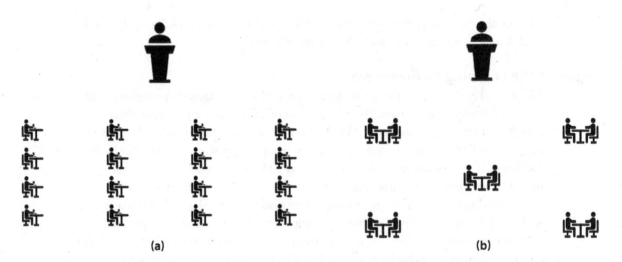

FIGURE 14.7 Classroom desk arrangement for (a) a presentation (lecture) and (b) small group discussion and activity.

Develop a Laboratory Safety Plan

Because an agriculture teacher usually manages a laboratory, a lab management plan is necessary. There are three essentials for effective shop or laboratory management.

1. Position yourself in the laboratory so you can observe all students working. Be where you can visually locate each student in the class with a minimal degree of effort. If the students are aware you monitor their actions, they should be less inclined to misbehave. Be highly visible in a shop or laboratory setting so students can get answers to questions and assistance on projects.

2. Securely store unused tools and materials out of the reach of students. Securely store agricultural chemicals, fuels, paints, glues, and other toxic substances to prevent accidental exposure. By following the same procedure for infrequently used tools, equipment, and supplies, you will ensure these will be available for use the next time.

3. Keep all tools and equipment in safe working order, with all guards and shields in place. Broken equipment or equipment without the necessary safety guards is a severe safety hazard. This equipment should be tagged and locked out of service and repaired at the earliest possible opportunity.

TRAUMA-INFORMED TEACHING

Trauma is one of the most prominent public health issues in the United States (Felitti et al., 1998). Students cannot learn well if they experience concerns about their safety. Traumatized students have suffered through an alarming, inadequately resolved experience, such as experiencing violence, neglect, natural disaster, or abuse. Traumatic experiences are difficult for adults to cope with, but trauma is keenly felt in youth. They do not yet have the positive coping mechanisms necessary to deal with trauma.

Trauma-informed teaching requires teachers to anticipate unexpected behavior from students and respond to reduce the student's anxiety and stress. A student's sudden spontaneous emotional outburst in the middle of classroom activity can be problematic to teachers and classmates. Teachers should prepare to deploy strategies that help defuse traumatic responses in class. This may include giving students extra time to comply with requests, handling misbehavior issues discreetly and out of earshot of other students, and maintaining a calm and poised demeanor.

Students suffering from trauma often dwell on the negative. Teachers can employ strategies that disrupt negative self-talk by distracting students by transitioning from one activity to the next. Teachers can also provide supportive feedback designed to counter negative thinking, such as using praise, maintaining a calm demeanor, and refusing to argue with the student.

Overall, the teacher can defuse some of the stress students experience by using positive reinforcement, using praise to recognize performance, and maintaining an inviting and cheerful classroom (Minahan, 2019).

REVIEWING SUMMARY

The learning environment is the nature and quality of the conditions in which teaching and learning occur. Some environments promote learning; others may impair learning. The teacher's responsibility is to be in charge of the situation and establish a quality learning environment.

Misbehavior may occur in several forms. Sometimes it does not disrupt the learning environment beyond the student involved in something that detracts from learning. Misbehavior is the condition that results from a person's inappropriate attempt to solve a problem or adapt to a situation.

Discipline is the authoritative control of behavior by using punishment and rewards. Some individuals feel it is more critical to stress establishing and maintaining a quality learning environment. Discipline tends to create negative emotions about education and the processes involved.

Three variables in a behavior problem are the student, the teacher, and the class of students. These three create an environment in which teachers and students carry out the learning process.

Prevention of misbehavior is by far the preferred approach. Teachers can use practices that promote desired behavior. Having well-prepared lessons that engage students is among the first steps. The furniture arrangement in the classroom can enable students to be more engaged with the lesson. A teacher should teach the entire period and not have unused time at the end. Robust interest approaches help engage students and motivate them to learn what is being taught. Having classroom routines and safety plans also promotes a good learning environment.

Teachers must appropriately respond to misbehavior situations. It is best to avoid disrupting a lesson and the learning that is occurring. Proximity control is a valuable strategy. Present behavioral expectations during the first meeting of the class. Once learners know expectations, they are more likely to follow them. Always follow school policies. Get to know parents and involve them in your program.

QUESTIONS FOR REVIEW AND DISCUSSION

1. What is the "learning environment"?
2. What are the features of learning environments in agricultural education?
3. What is misbehavior? Why is misbehavior said to be a "coping mechanism"?
4. What are the three variables in a discipline problem? Briefly describe each.

5. What is "restorative practice" in education?
6. How do well-prepared lessons minimize student misbehavior?
7. How should a teacher organize a classroom to minimize misbehavior?
8. What is trauma-informed teaching?
9. What classroom rules may help establish a good learning environment?
10. What should be a teacher's response when misbehavior happens?
11. What is proximity control? How is it useful?
12. Why is it beneficial to make expectations known during the first meeting of a class?
13. How can working with parents help minimize misbehavior?

ACTIVITIES

1. This activity focuses on quality learning environments as related to student achievement.
   a. Goal: Identify principles and teaching strategies that yield good classroom and laboratory management results.
   b. Procedure: Observe a teacher at work in a local agriculture program. Do not participate in class discussions or activities. Observe how the teacher manages the classroom or laboratory during the class period and complete the following assignments:
      i. Draw a diagram of the classroom or laboratory setting in which you observe the teacher. Write a paragraph describing positive and negative attributes of how the classroom or laboratory is organized. Determine the mode for which the furniture is arranged: lecture mode, small-group mode, demonstration mode.
      ii. Observation: Managing student work—Describe how the instructor grades student work. What are the procedures performed by the instructor to account for absent students? How does the instructor arrange for student work to be made up after an absence?
      iii. Observation: Managing instruction—Describe how the instructor keeps students focused on instructional activities and communicates what is expected. Also, include an analysis of the efficiency of the lesson (how time and other resources are managed).
      iv. Observation: Managing special groups—Describe how the instructor manages instruction for students of varying ability levels. Evaluate the effectiveness of these methods.
      v. Synthesis: A final question—Who worked harder today, the students or the instructor? Why?
2. Read peer-reviewed articles on trauma-informed teaching. What elements of the reading can you employ in the classroom when you begin your career as a teacher?

REFERENCES

Abel, M. H., & Sewell, J. (1999). Stress and burnout in rural and urban secondary school teachers. *Journal of Educational Research, 92,* 287–294.

American Psychological Association Zero Tolerance Task Force. (2008). Are zero tolerance policies effective in the schools?: An evidentiary review and recommendations. *American Psychologist, 63*(9), 852–862. https://doi.org/10.1037/0003-066X.63.9.852

Borg, M. G., & Riding, R. J. (1991). Stress in teaching: A study of occupational stress and its determinants, job satisfaction and career commitment among primary schoolteachers. *Educational Psychology, 11*(1), 59–77.

Burnett, M. F., & Moore, G. E. (1988). Student misbehavior in vocational agriculture and other vocational programs: A comparison. *Proceedings of the National Agricultural Education Research Meeting, 15,* 42–47.

Camp, W. G., & Garrison, J. M. (1984). The seriousness of student misbehavior in vocational agriculture. *Journal of the American Association of Teacher Educators in Agriculture, 25,* 42–47.

Croom, D. B., & Moore, G. E. (2003). Student misbehavior in agricultural education: A comparative study. *Journal of Agricultural Education, 44*(2), 14–26. https://doi.org/10.5032/jae.2003.02014

Croom, D. B., & Moore, G. E. (2004, August). Dumping grounds. *The Agricultural Education Magazine, 77*(1), 14–18.

Dahl, R. E. (1999). Consequences of insufficient sleep for adolescents. *Phi Delta Kappan, 80*(5), 354–359.

DeBruyn, R. L. (2011). *Before you can discipline: Vital professional foundations for classroom management.* Master Teacher.

DeBruyn, R. L., & Larson, J. L. (2009). *You can handle them all: A discipline model for handling 124 student behaviors at school and at home.* Master Teacher.

Felitti, V. J., Anda, R. F., Nordenberg, D., Williamson, D. F., Spitz, A. M., Edwards, V., Koss, M. P., & Marks, J. S. (1998). Relationship of childhood abuse and household dysfunction to many of the leading causes of death in adults: The adverse childhood experiences (ACE) study. *American Journal of Preventive Medicine, 14*(4), 245–258. https://doi.org/10.1016/S0749-3797(98)00017-8

Heller, R., Preston, T., Hirshfeld, J., Jacques, M. S., Langer, G., Filer, C., Sinozich, S., Jong, A. D., Barron, B., Cardichon, J., Fagell, P. L., Gangone, L., & Johnston, R. (2020). *The 52nd annual PDK poll of the public's attitudes toward the public schools.* 16.

Marzano, R. J. (2017). *The new art and science of teaching.* Solution Tree Press.

Maslow, A. H., & Frager, R. (2003). *Motivation and personality.* Longman.

Minahan, J. (2019). Trauma-informed teaching strategies. *Educational Leadership, 77*(2), 30–35.

National Council for Agricultural Education. (2016). *National quality program standards for agriculture, food, and natural resources education: A tool for secondary (grades 9–12) programs.* https://thecouncil.ffa.org/

Oxford University Press. (2000). *Oxford English dictionary.* http://dictionary.oed.com/entrance.dtl

15

Agricultural Literacy

Barbara Green teaches a variety of agriculture courses for students in grades 8–12. At the National Association of Agricultural Educators (NAAE) Convention she participated in a workshop advocating the need for all students to be agriculturally literate. Barbara has always thought her mission with the eighth graders was to get them interested in agriculture so they would enroll in her high school courses. Based on the workshop presentation, she is now asking herself if her eighth-grade agriculture class should have more of an agricultural literacy focus. She is also wondering what it means for her high school students to become agriculturally literate. Barbara is asking very important questions.

What exactly is agricultural literacy? What content and experiences do all students need to become agriculturally literature? Why do students need to be agriculturally literate?

Just where do we get our food? The answer is not easy and is far more involved than "at the local grocery store." In our modern society, an increasing number of people do not know what is involved in producing wholesome, nutritious food and making it readily available. This is one rationale for why we need to provide education for agricultural literacy.

With only about 1% of the nation's workforce currently involved in producing food and fiber, a large part of our society has limited exposure to agricultural production. Today, many people are unaware that agricultural producers are the primary source of the nation's food and fiber products. Agricultural educators can play a key role in educating about the importance of the plants, animals, and natural resources systems.

OBJECTIVES

This chapter addresses the National Quality Program Standards for Agriculture, Food, and Natural Resources Education (National Council for Agricultural Education, 2016), specifically Standard 1A: Program Design and Instruction—Curriculum and Program Design. It has the following objectives:

1. Explain the meaning and importance of agricultural literacy.
2. Explain National Agricultural Literacy Standards and how they are used.
3. Describe strategies for developing agricultural literacy.

TERMS

agricultural awareness
agricultural literacy
agriculture
Agriculture in the Classroom

American Farm Bureau Foundation for Agriculture
infuse
literacy

FIGURE 15.1 Children are enjoying an agricultural literacy presentation from a high school student and FFA member. (COURTESY OF NATIONAL FFA ORGANIZATION.)

THE MEANING AND IMPORTANCE OF AGRICULTURAL LITERACY

Agriculture is the science, business, and technology of plant and animal production. It includes the systems for producing food and fiber and managing the environment and natural resources used by society. Today, we often refer to all of agriculture as the agricultural industry. This is because it includes agricultural supplies and services, marketing and processing, horticulture, forestry, wildlife, aquaculture, agricultural power, and other areas. The functions carried out that support production agriculture are called agribusiness. *Agricultural awareness* is the realization, perception, or knowledge of the agricultural industry, which is sometimes referred to as the food, fiber, and natural resources system.

Agricultural education is a systematic program of instruction available to students wanting to learn about the food, fiber, and natural resources system. Approximately 6% of the nation's school population is enrolled in agricultural education programs. This leaves about 94% of public-school students with no formal in-school instruction regarding the nation's food, fiber, and natural resources system.

FIGURE 15.2 Using an agricultural literacy lesson plan library to prepare for teaching elementary school children.

Literacy

Literacy is a term used to describe the ability to read and write. It can also refer to having more than average knowledge. A person who is literate is someone who is educated. An educated individual should have at least a basic understanding of the food, fiber, and natural resources system and the role it plays in the world. However, many "well-educated" individuals in the United States have no knowledge of the agricultural industry, which has an impact on their daily lives and is essential for their health and well-being.

In 1988, the National Research Council of the National Academy of Sciences released a study of agricultural education that made a distinction between education *in* agriculture and education *about* agriculture. This study defined "education in agriculture" as the vocational or career preparation component of agricultural education. It also surfaced the idea of ***agricultural literacy*** and defined it as "education about agriculture." In this chapter, agricultural literacy describes the attainment of knowledge and understanding of the food, fiber, and natural resources system. The purpose of agricultural literacy is for youth and adults to use their knowledge and understanding to make informed purchasing, land use, and political decisions.

Importance

The National Academy of Sciences study stated the topic of agriculture was so important it should be taught to more than the small percentage of students enrolled in agricultural education programs. It concluded all individuals should have a basic understanding of the nation's agricultural systems and everyone should have some knowledge of where and how food and fiber are produced, processed, and distributed. The study also indicated all individuals

FIGURE 15.3 All people need an understanding of the importance of agriculture in feeding the world. (COURTESY OF U.S. DEPARTMENT OF AGRICULTURE.)

should be able to make informed choices about their own diet and health and possess the knowledge needed to understand the importance of caring for the environment.

The National Academy of Sciences study generated considerable interest among agricultural educators for agricultural literacy and served as a catalyst for numerous additional studies on this topic. In 1999 the National Council for Agricultural Education released *Reinventing Agricultural Education for the Year 2020*. The plan's stated vision was that all people will value and understand the important role agriculture plays in individual and world well-being. One goal of this strategic plan was for all students to attain agricultural literacy. Although the agricultural education profession does not currently have a new strategic plan, efforts such as SAE for All continue to emphasize the importance of an agriculturally literate population.

Few people realize approximately 20% of the U.S. workforce is involved in the agricultural systems and, worldwide, more people work in agriculture than in any other occupation. Broadly defined, the food, fiber, and natural resources system includes the science and business of plant and animal production, along with management of wildlife, forests, rangelands, rivers, oceans, and renewable natural resources. The system also includes horticultural crops, such as lawns, trees, gardens, landscaping, and turf for sports fields and golf courses.

The agricultural system impacts many aspects of each individual's everyday life and should not be taken lightly. The world's population is continuing to increase, which means the demand for agricultural products used for food, shelter, and clothing will also increase. It is important for all citizens to have an awareness of the importance of agriculture in the world. This is a tremendous undertaking and requires the commitment and cooperation of many agencies, organizations, and groups, including agricultural education teachers and their students. Agricultural education teachers and students should be primary providers and facilitators of opportunities for increasing agricultural literacy. According to McDermott and Knobloch (2005), national leaders in agricultural education have identified literacy and awareness of agricultural education as major issues of importance to the future of agricultural education. The American Association for Agricultural Education (AAAE) National Research Agenda 2016–2020 has as its first research priority that the public and policy makers understand agriculture and natural resources (Roberts et al., 2016).

NATIONAL AGRICULTURAL LITERACY STANDARDS

Achieving agricultural literacy for all students requires classroom teachers to infuse agriculture into their curriculums. To *infuse* is to introduce one thing into another so as to affect it. In relation to agricultural literacy, it means to include agricultural lessons, examples, and topics in courses for other subjects. Leising et al. (1998) presented a framework for infusing agricultural literacy into core academic subjects across grade levels and described what young people should know about agriculture when they graduate from high school. A set of sample lesson units and a literacy test were included in the guide they developed.

Agricultural literacy efforts are organized by multiple organizations on the state and federal levels. The United States Department of Agriculture–National Institute of Food and Agriculture, the National Agriculture in the Classroom Organization, and Utah State University collaborated to create a National Center for Agricultural Literacy. The center's website is a hub for resources, research reports, and learning opportunities. The website can be accessed at www.agliteracy.org.

In 2013, Spielmaker and Leising authored the publication *National Agricultural Literacy Outcomes*. These outcomes are organized by theme and have grade-level benchmarks for kindergarten through grade 12. Academic subjects recommended for integration of agricultural literacy outcomes include social studies, science, and health.

Role of Secondary Agricultural Educators

As recognized professionals, agricultural education teachers have a large role to play in creating an agriculturally literate society. The major deterrent to agricultural education teachers advancing agricultural literacy to students other than those enrolled in their programs is finding sufficient time to address this issue. Agriculture teachers generally view assisting other students as a good thing but outside the primary focus of the agricultural education program. However, many agricultural education teachers do realize there is value in spending time on agricultural literacy. For example, they understand educating younger students about agriculture and agricultural education is a prime recruitment opportunity for attracting more students into the local agricultural education program.

Agricultural education teachers can facilitate efforts within their school system to provide agricultural literacy through elementary and other secondary teachers. Agricultural education teachers can use the many resources and strategies available for infusing agriculture into the public school curriculum.

Role of Agencies, Organizations, and Businesses

A number of agencies, organizations, and businesses have developed materials for use in teaching about many topics related to agriculture. However, these groups are not directly connected to the schools and often face barriers getting their materials into the hands of teachers most likely to use them. Most teachers, on the other hand, do not understand where to find good agricultural instructional materials and do not have the time to search for materials outside their subject areas. Agricultural education teachers can become links between teachers desiring to teach agriculture and groups desiring to have their materials utilized by classroom teachers.

FIGURE 15.4
Young MacDonald's Farm exhibits promote agricultural literacy. (COURTESY OF LEDYARD AGRI-SCIENCE & TECHNOLOGY PROGRAM.)

USDA and State Agriculture Offices

The U.S. Department of Agriculture (USDA) and the state agriculture offices are good sources of information. Websites, personnel, materials, and other resources related to agricultural literacy are available. For example, the USDA website includes information for children, youth, and other groups. The USDA web address is www.usda.gov/.

Agriculture in the Classroom

Agriculture in the Classroom (AITC) is among the best-known examples of an agricultural literacy initiative. AITC exists in every state and has had a federal presence within the U.S. Department of Agriculture. By merely introducing an AITC resource person to the faculty in an elementary school, the agriculture teacher can help open an entire school to agricultural literacy instructional materials, human resources, and teacher training opportunities.

Many resources are available through AITC. A good place to begin is the Agriculture in the Classroom website: www.agclassroom.org. This site features a wide range of materials at the beginning, intermediate, and advanced levels. Teachers can identify key contacts in their states who have resources available and who serve as resource persons for instruction. The resources often focus on the early childhood years as well as the middle school grades.

The American Farm Bureau Federation and its state affiliates have actively pursued AITC. Materials about agriculture have been provided to elementary school teachers. Workshops, field trips, and other means have been used to train teachers in AITC. Resource persons are available through local farm bureaus and other organizations.

American Farm Bureau Foundation for Agriculture

The *American Farm Bureau Foundation for Agriculture* is a source for materials, programs, and links to promote agricultural literacy. Its website, www.agfoundation.org, has free resources including agricultural statistics, games, publications, and resource guides.

FIGURE 15.5 A community garden is a popular setting for agricultural literacy education.

Displays and Museums

Government agencies, organizations, and businesses often promote or establish exhibits. Most states have agricultural museums that help promote an understanding of agricultural history. Many states have working farms open to the public for tours and education. Land-grant colleges and universities have experiment stations that are excellent for field trips and other activities.

National FFA Organization

Many agricultural education teachers teach middle school/junior high courses. One of the more visible results of agricultural education's recent attempts to develop agricultural literacy programs has been the increasing number of middle school/junior high agricultural education programs. The National FFA Organization made changes to its constitution and bylaws that have promoted middle school participation. The change officially allowed agricultural education students in grades 7 and 8 to become FFA members (National FFA Organization, 2021).

The middle school/junior high curriculums are generally exploratory and designed to increase awareness and knowledge of agriculture, whereas high school programs are more focused on career preparation. The middle school/junior high programs provide the most effective agricultural literacy education in the nation; however, they reach only a very small percentage of the total school population.

Teacher Education

Teacher training programs for agricultural education focus primarily on working with middle school and high school students. Thus, agriculture teachers sometimes fail to realize children in the lower grades are the ones most receptive to accepting and applying new concepts. This means agricultural awareness should be an important part of public education at the very beginning of a child's formal schooling.

Currently the agricultural literacy efforts for the lower grades are less structured than those for the middle school/junior high levels. Agriculture teachers desiring to make an impact on increasing all students' awareness of agriculture should devote time to assisting with agricultural literacy efforts for students in the lower grades.

Most elementary and high school teachers have no background in agriculture and are, therefore, not inclined to try an area that is of little interest to them. A teacher's background and experience are important in determining the topics they are willing to teach. Teachers need an awareness of agriculture before they can be successful teaching it.

Universities that prepare agriculture teachers can assist in agricultural literacy efforts by helping teacher candidates in other subject-matter areas increase their understanding of agriculture and how it can be infused into academic courses. State education agencies and teacher preparation programs can help prepare for and support the teaching content of junior high/middle school agricultural education programs and provide leadership for articulating these programs with those offered at the high school level.

Food for America

Agricultural education teachers in many schools provide agricultural awareness programs for elementary students through the FFA Food for America program. This program offers opportunities for leadership development to agricultural education students while increasing the agricultural awareness of elementary students and teachers. Joint activities with the FFA Alumni and Supporters, Young Farmers, and parent booster clubs are very effective strategies in which the teacher is the facilitator and not the primary deliverer of instruction. These joint efforts may include tours of school farms, Young MacDonald's Farms, school assembly programs, and/or demonstrations presented to individual classes. The activities are often conducted during National FFA Week. The Food for America curriculum can be found on the National FFA website, www.ffa.org.

Activity Orientation

Many students are very interested in animals and plants. They will respond positively to instruction that is hands-on and actively involves them. With the tremendous pressures on teachers and students to increase performance on state and national assessments, instructional materials that help teachers achieve the objective of higher student academic performance would be welcomed. Providing agricultural lessons that help students gain academic knowledge and skills measured by standardized assessment exams is critical for enticing other teachers to utilize agricultural instructional materials. Agriculture teachers can demonstrate how agriculture can be used as a motivational tool to increase student interest and enthusiasm for learning. Powell, Agnew, and Trexler (2008) recommended multiple approaches to delivering agricultural literacy.

Shared Resources and Facilities

Agriculture teachers can team teach with other teachers to present agricultural information, and/or they can help other teachers find materials that focus on the subject of agriculture and on an increase in academic achievement in core subject-matter areas. The internet enables agriculture teachers and students to more readily find high-quality instructional resources for increasing agricultural literacy. Agriculture teachers can also provide educational programs directed at students in other courses and programs at various grade levels by using today's technology. This technology allows lessons to be delivered synchronously or asynchronously to multiple classrooms.

Another important strategy often overlooked is for the agriculture teacher to provide other teachers with access to facilities, resource people, equipment, and the local agricultural community. Other teachers do not have the connections with the agricultural community that the agricultural education teacher possesses. By connecting other teachers to key agricultural resources, the agricultural education teacher can facilitate the teaching of agriculture by numerous other teachers.

REVIEWING SUMMARY

All people need to value and understand the importance of agricultural systems and what they contribute to world and individual well-being. Agricultural literacy is a major goal of agricultural education; however, widespread cooperation and commitment are required to achieve nationwide knowledge and understanding of agriculture. Of necessity, agricultural literacy involves many agencies, organizations, and groups interested in creating awareness and understanding of the agricultural systems in this country.

Agriculture teachers and their students play an important role in educating all students about the plants, animals, and natural resources systems. In some instances, they are the primary deliverers of instruction about agriculture, and in many other endeavors, they serve as facilitators of the process.

QUESTIONS FOR REVIEW AND DISCUSSION

1. What is agricultural literacy? Why is it important to society?
2. What are some strategies agricultural educators use to increase agricultural literacy?

3. What should be the role of the agricultural education teacher in providing agricultural literacy?
4. What is Agriculture in the Classroom?
5. What are the ways an agriculture teacher can collaborate with other teachers to support agricultural literacy?

ACTIVITIES

1. Develop an agricultural literacy lesson appropriate for infusion into a middle school course. Identify the course for which the lesson is designed, the estimated time required to teach the lesson, and the best time of year to teach it.
2. Use the internet to investigate agricultural literacy instructional materials. Develop a list of sources for the materials and identify the grade level(s) for which the materials are recommended. A site for starting your search is www.agclassroom.org.
3. Investigate the agricultural literacy structure, providers, and training opportunities within your state and prepare an infographic you could share with your future teaching colleagues.

REFERENCES

Leising, J. G., Igo, C. G., Heald, A., Hubert, D., & Yamamoto J. (1998). *A guide to food & fiber systems literacy.* W. K. Kellogg Foundation and Oklahoma State University.

McDermott, T. J., & Knobloch, N. A. (2005). A comparison of national leaders' strategic thinking to the strategic intentions of the agricultural education profession. *Journal of Agricultural Education, 46*(1), 55–67. https://doi.org/10.5032/jae.2005.01055

National Council for Agricultural Education. (1999). *A new era in agriculture: Reinventing agricultural education for the year 2020.*

National Council for Agricultural Education. (2016). *National quality program standards for agriculture, food, and natural resources education: A tool for secondary (grades 9–12) programs.* https://thecouncil.ffa.org/

National FFA Organization. (2021). *Official FFA manual.* https://www.ffa.org

National Research Council. (1988). *Understanding agriculture: New directions for education.* National Academy Press.

Powell, D., Agnew, D., & Trexler, C. (2008). Agricultural literacy: Clarifying a vision for practical application. *Journal of Agricultural Education, 49*(1), 85–98. https://doi.org/10.5032/jae.2008.01085

Roberts, T. G., Harder, A., & Brashears, M. T. (Eds). (2016). *American Association for Agricultural Education national research agenda: 2016–2020.* University of Florida, Department of Agricultural Education and Communication.

Spielmaker, D. M., & Leising, J. G. (2013). *National agricultural literacy outcomes.* Utah State University, School of Applied Sciences and Technology. https://www.agliteracy.org/

16

Middle School Agricultural Education

Thomas Watson, a college senior in agricultural education, was having some doubts about whether or not he should begin his teaching career in a high school or a middle school. He called his former agriculture teacher, Bob Kendall, for advice. Bob, now retired, agreed to meet Thomas for lunch at a local café.

"So what questions do you have about your teaching career? You aren't getting cold feet, are you?" asked Mr. Kendall.

"No sir, I am looking forward to teaching. But where? Milltown Middle School has a position open, and the principal there contacted me about it. But I don't know ..." Thomas's voice trailed off.

"But you aren't sure if teaching middle school is for you?" Mr. Kendall interjected.

"That's right," said Thomas. "I always imagined teaching at the high school and never thought much about middle school. I don't know if I can chase a bunch of little kids around the classroom all day."

Mr. Kendall thought for a moment. "Have you been in a middle school lately?"

"Not since I was a student in one," Thomas replied.

Mr. Kendall looked at his watch. "When you finish your coffee, let's ride over to Brookfield Middle School. I know the principal and teacher there, and maybe they would show us around."

Thomas agreed. "Anything that will help me make a decision is welcome."

Mr. Kendall and Thomas climbed into Mr. Kendall's car and drove to Brookfield Middle School. After meeting the principal, Mr. Kendall and Thomas walked through the hallways on the way to the agricultural education department. After meeting with Ms. Welborn, the middle school agricultural education instructor, Thomas and Mr. Kendall were treated to a facility tour by two students. They observed students in the greenhouse conducting germination experiments and noted several agricultural science projects in progress in the classroom.

OBJECTIVES

This chapter addresses the National Quality Program Standards for Agriculture, Food, and Natural Resources Education (National Council for Agricultural Education, 2016), specifically Standard 6: Certified Agriculture Teachers and Professional Growth. It has the following objectives:

1. Explain the concept of middle schools in American education.
2. Discuss the role of middle school agricultural education.
3. Identify stages of student maturation and other developmental factors influencing methods of instruction in middle schools.
4. Describe instructional strategies in the middle school environment.
5. Describe the function of FFA and supervised experience.

TERMS

Association for Middle Level Education
concrete operations
Erik Erikson
Food for America
Jean Piaget
preoperational
student-centered instruction

On the drive back to the café Mr. Kendall asked Thomas what he thought of the school. Thomas could scarcely contain his enthusiasm for what he had just seen. "I thought I would see loud, rude children running in the hallways. I thought I would see hyperactive young people trying to figure out what life was all about. And the teachers . . . I thought they would be frantically chasing about trying to establish order and they would have few instructional resources. Was I surprised! The hallways were quiet. The students were pleasant and cooperative. When the class period was over, they changed classes in an orderly manner and without lagging. The teachers were businesslike and professional. They appeared to have an abundance of instructional resources. There was no downtime—everyone was engaged in the teaching–learning process. This settles it for me. I am going to apply for the job at Milltown Middle School."

MIDDLE SCHOOLS IN AMERICAN EDUCATION

Junior High Schools

Junior high schools came into being in the early 1900s. These schools carved away the top two to three grades from the elementary school and often the lowest high school grade and placed them in a separate school. Junior high school curriculum followed a two-pronged approach—the college-bound program and the vocational program (Manning, 2000). School administration funneled students into one or the other through educational and aptitude assessments. This method met the needs of the established high school curriculum by preselecting students into the college or vocational categories.

The name "junior high school" implies that the purpose of these schools is to acclimate students to the high school model of education, with the sounds of ringing bells and slamming lockers echoing down the hallways as students rush from one classroom to another. Gruhn and Douglass saw junior high schools differently and expressed their views in the seminal work, *The Modern Junior High School* (Gruhn & Douglass, 1956). From their inception, junior high schools should have done the following:

- Integrated the school curriculum, blending math, science, language, and other subjects horizontally across grades and vertically across grade levels
- Provided for exploration in all aspects of the curriculum, including career exploration through career and technical education
- Introduced students to academic and career guidance
- Differentiated instruction
- Provided for the continued development of social skills

Many junior high schools were scaled-down versions of high school. Teachers, administrators, educational policy makers, and researchers concluded junior high schools failed to meet students' unique social and academic needs. In the early years of these junior high schools, there was a concern that the instructional program did not meet the developmental needs of students. This model failed to serve the needs of adolescent learners because it was unable to account for cognitive, social, and emotional development. Students grapple with the transition from adolescence to the early stages of adulthood. The students in the middle school needed an educational program that was appropriate for their cognitive needs.

The Modern Middle School

A middle school is an administrative unit between the primary or elementary grades and the high school grades. The grades included are typically 6 through 8, though other variations

FIGURE 16.1 Differentiated instruction is an essential ingredient in the middle school agricultural education program.

exist. To meet the needs of young adolescents, the Association for Middle Level Education (Bishop & Harrison, 2020) determined there are essential elements the instructional program provides at the middle school level. These are described in Box 16.1.

The modern middle school movement began to develop in the early 1960s. In the 1970s, junior high schools started to change their names to middle schools. Coupled with this name change was a new identity and role for these schools. This new structure for middle schools changed the curriculum, teacher training and development, and school leadership. Research on adolescent learners in the middle grades began to inform teaching practice and educational policy. The social and emotional well-being of students became a priority. The

---

BOX 16.1 *Essential Elements of Middle School Education*

- Student-centered and teacher-centered instructional methods encourage and develop critical thinking
- A rigorous, age-appropriate curriculum that uses technology to bring the world into the mind of the learner
- Learning experiences that allow for intellectual exploration, including education in the arts and humanities
- Instruction is organized for maximum benefit to students
- Learning experiences foster the development of the whole person
- Academic and career guidance opportunities encourage personal wellness and stewardship of a sustainable environment and economy
- Appropriate assessment strategies that adequately measure student progress
- Learning experiences that foster lifelong learning along with a desire to contribute to the growth of the community
- Educational experiences that demonstrate citizenship and encourage students to be civic-minded learners

Source: Bishop & Harrison (2020).

curriculum, in the 2000s, started to adopt elements that address issues of diversity and inclusion. Cooperative learning became a much-used strategy for teaching. Interdisciplinary teams of teachers have a positive effect on student learning.

Throughout the 1990s, research and best practices guided middle grades education. With the passage of the No Child Left Behind Act of 2001, the focus shifted toward testing and accountability measures (Elementary and Secondary Education Act, 2002). The overall sense of what middle grades education should "look like" was driven by accountability measures.

Successful Middle School Attributes

In *The Successful Middle School: This We Believe*, Bishop and Harrison (2020) described the essential attributes of successful middle schools as defined by the *Association for Middle Level Education*—the primary professional organization for middle grades instructors and administrators. These are displayed in Box 16.2, along with a description of how agricultural education could provide these essential attributes in middle school.

LEARNING THEORIES ASSOCIATED WITH MIDDLE GRADES EDUCATION

Teachers through the years have wondered what goes on the minds of middle grades students. The physical, social, emotional, and cognitive changes their bodies and minds are going through during adolescence have a profound and visible effect. These changes unfairly depict middle grades children as "unguided missiles" bouncing around in the classroom, unable to contain themselves. But what goes on in the minds of middle graders? The answer may exist in the theoretical work of educational psychologists and educators. Theories attempt to explain natural and observed phenomena. In the context of middle schools, theories help define what teachers are observing and experiencing in their relationship with their students. These theories can guide the selection and use of teaching strategies and even educational policy.

Piaget's Cognitive Development Theory

Jean Piaget (1896–1980) was a Swiss developmental psychologist. Although Piaget's doctorate was in natural sciences, he had a strong interest in philosophy, sociology, and psychology. Upon receiving his doctorate at the University of Neuchatel in 1918, he pursued additional training in psychology. Through his work in school laboratories, Piaget began to develop his model of adolescent development. Piaget studied the language use and reasoning abilities of children and the intellectual development of infants.

Piaget proposed that children develop socially and intellectually in stages. Piaget defined the first stage of development as the sensorimotor intelligence stage. This stage begins at birth and continues through the second year of life. This stage involves learning basic psychomotor skills and the simple perception of things. The second stage is the preoperational stage. It begins at age 2 and progresses through age 7. In this stage, the child begins to understand the symbolic nature of things. A child's intuition begins to develop. While the child cannot perform complex mental operations, they develop logical thought (Flavell, 1963). For instance, the child can determine whether two objects are tall or short compared to each other. However, the child may not correctly select from containers of various shapes and sizes the one that likely holds the most liquid.

The *preoperational* stage occurs when children begin to talk. Until about age 4, children only think in terms of themselves. By the late preoperational stage, they will have developed to the point where they have a limited ability to see another person's point of view. The child's imagination begins to come alive, and they learn how to play alongside other children.

BOX 16.2 *What Successful Middle School Programs Are*

Responsive to the unique characteristics of the adolescent learner—The SAE program is designed to provide individualized learning experiences based upon a student's interests and preferred career path.

Challenging by creating an environment of high-performance expectations—The FFA awards program and the financial incentives and awards provided by the SAE program set a high bar for performance.

Empowering students to take charge of their learning—Students develop individualized learning plans through their SAE.

Equitable learning opportunities for all students—The FFA provides leadership and career development opportunities for all members.

Engaging students in the learning process involves them in decision-making and motivates them through personal ownership of the content—The FFA elects student officers who provide leadership for the FFA chapter.

Educators respect young learners—Agriculture teachers mentor students by coaching them in FFA Career Development Events and meet the student and their parents during SAE program visits at the student's home or place of work.

The school welcomes all students—The agriculture teacher uses teaching materials that are free from bias.

Every student has an adult advocate—The agriculture teacher advocates for students through mentorship of the students' SAE project and serving in the advisor role for the FFA chapter.

Schools are safe without compromising student rights—Agriculture teachers manage labs and all learning activities with student safety as the primary consideration.

Counseling services meet the needs of young learners—The agriculture teacher provides mentoring and advice for all students through their SAE program.

Families are partners in learning—The agricultural education program involves parents and families in program advisory committees through the FFA Alumni and Supporters Organization and SAE programs.

The school utilizes community and business partners—Students are placed with community and business partners in the SAE program.

Educators are well prepared and skilled at teaching—Agriculture teachers must be certified to teach middle grades in agricultural education.

The curriculum is relevant to student needs and rigorous—The agricultural education curriculum is devised to increase student knowledge about modern best practices in the agricultural sciences.

The school and curriculum support the well-being of the whole child—The FFA program provides awards and recognition for student achievement at all grade levels.

Active learning opportunities abound and assessments measure learning—Standards and metrics devised by the FFA provide a measure of progress for students involved in FFA activities. The SAE program allows students to measure progress toward career goals through extensive record keeping and journaling.

Stakeholders have a shared vision of the school—Stakeholders participate in program advisory committees, advising school leaders on areas for growth and development.

Policies are student-centered and unbiased—The FFA has established a clear structure for student leadership in FFA programming, and the SAE program is focused solely on a student's career interests.

School leadership is skilled and applies research-based strategies for instruction—Agriculture teachers differentiate instruction through hands-on learning activities designed for students to progress at their own pace.

School leaders build a culture of collaboration—Students participate in team-building activities in the FFA and group learning activities in hands-on lab activities.

Teachers adopt a model of continuous improvement—Agriculture teachers participate in regional and state in-service training opportunities. The National Association of Agricultural Educators provides opportunities for collaboration across state boundaries.

The school is designed for the benefit of students—The agricultural education program follows the career and technical education model, which provides students with the knowledge and skill to pursue their preferred career choices.

Children enter the third stage, concrete operations, at around age 7. Children in this stage are able to logically organize their thoughts. Their thinking is very concrete, with definite ideas of right and wrong. They can develop ideas and reach conclusions based on inductive reasoning, or rather, can take specific information and apply it to a general principle. In this stage, a child begins to think about how other people might view a situation.

The formal operations stage begins around age 12, or when a young person enters middle school. This stage is the threshold of adulthood in terms of cognitive development, according to Piaget. Children can begin to consider moral, philosophical, and ethical issues. They begin to exercise their ability for abstract thought and reasoning. And, they begin to use deductive reasoning to solve a problem or answer a question. A young person can start to apply the scientific method to solve problems and answer questions. Piaget's theory provides insight into the development of adolescents, but other theories add to the body of knowledge.

Erikson's Stages of Social-Emotional Development

Erik Erikson, a developmental psychologist in the late 20th century, proposed a psychological theory of development. He described human socialization as an eight-stage process. Each stage results from a felt internal need that must be resolved to reach the intended outcome—an average, well-balanced, emotionally stable child. The first stage encompasses the period from birth to age 1 when children learn to trust others. If a nurturing environment embraces the child, they develop a fundamental optimism and trust of others. Jean Piaget identified this stage as a child's sensorimotor period, during which the child develops physical coordination of limbs and moves from reflexive action to synthesized action.

In the second stage, the 2 to 3 age bracket, the individual develops a need for autonomy. Many parents know this stage as the "terrible twos," characterized by a child's need to be assertive.

Erikson identified the third stage as the "play" stage. In this stage, the child learns to imagine and plays cooperatively with other children. It is in this stage that the child first begins to develop skills in leading and following. Piaget's preoperational period coincides with Erikson's third and fourth stages, in that the child develops reasoning skills and more complex speech patterns.

Erikson's fourth stage begins sometime around the age at which the child goes to school. In this stage, the child starts mastering peer relations, cognitive skills in reading and math, and the complex rules associated with formal play and organized recreation. The child is likely to enter middle school during this stage and progress from it to the next phase before going to high school.

Erikson's fifth stage is just beginning upon the child's departure from middle school, when the child needs to develop a unique identity. Piaget identified this stage as the *concrete operations* period, in which the child develops organized and logical thinking skills. Before leaving middle school, the child will have entered Piaget's period of formal operations and begun to develop abstract thought and prepositional logic ("if-then" processes). Furthermore, the middle-grade years see the beginning of the physical maturation process and puberty.

The transition from middle school to high school is a challenge for most adolescents. The high school learning environment may seem impersonal and grade-oriented. The competition for academic scholarships begins in high school, along with physical competition in organized athletics. The high school years are the beginning of part-time jobs and serious talk about colleges and careers. In agricultural education, the curriculum narrows to specialized career pathways, and students are required to choose the classes that most nearly match their career goals. The high school environment is very different from the middle school environment, and preparing for the transition from one to the other begins in middle school.

FIGURE 16.2 Middle grades students practice academic skills in a safe and welcoming social environment.

The first five stages of Erikson's model provide the foundation for the three remaining phases of human development. Students who have successfully resolved conflicts in the first five stages are prepared for young adulthood, which is stage 6. In stage 6, young adults hone their sense of identity in preparation for developing love relationships with others. Young adults who find it difficult to trust and commit to others will experience difficulty developing intimate relationships and may feel socially isolated.

As individuals move into middle adulthood, they experience the seventh stage by building upon intimate relationships and become caregivers through parenting and mentoring. In Erikson's final and eighth stage of human growth and development, adults come to terms with their existence by reflecting positively on their lives. Adults in this stage of maturity understand they cannot change their past and accept responsibility for what their lives have become. In the last stages of life, the maturation process is complete.

Bandura's Social Learning Theory

Bandura proposed that learning occurs through observation and modeling. The learner gathers new knowledge by observing the behavior of others. The learner stores the information about the behavior for retrieval later. This information is tagged and coded in the learner's mind. In a future circumstance, the learner retrieves the data and applies it. An example of Bandura's model in action is in the safe use of power tools. The learner observes the experienced professional teacher using the radial arm saw in an agriculture shop. The student stores information in memory about the adjustment levers, start/stop switch, shields, and guards; and how the teacher places her hands on the saw controls. When the learner has the opportunity to use the saw for the first time, he will likely have stored enough knowledge

FIGURE 16.3 Teaching strategies that encourage positive social interaction are essential in the middle school agricultural education classroom.

to know where to place his hands to operate the saw. Perhaps a more elementary example would be using a hammer. The learner would only need to see someone using a hammer to drive a nail to know the hammer's essential function. Bandura further proposed that learning depends upon the learner's mental state. The learner's preferences, personal interests, and preferred methods of motivation can determine whether or not the behavior is learned. Finally, Bandura proposed that observing behavior in the right frame of mind may not be enough to cause a behavior change.

The Rapid Development of the Adolescent Mind

In Piaget's model of cognitive development and the theories of Bandura and Erikson, the adolescent brain has gone through tremendous changes in the first 12 years of life. Remarkably, a child can learn to speak at around age 2 and discuss basic tenets of philosophical thought within 10 years. By the time a child reaches middle school, their brain has begun to develop at "warp speed."

Transitioning From Elementary School to Middle School

Belonging is a significant concern to middle school students. They have to feel they belong in the school community. The most effective teachers empathize with middle school students and help students respond to the pressures of homework, middle school culture, and social relationships. Middle school students are developing but have not fully developed planning, decision-making, and ethical reasoning skills. They are learning how to control impulsive behavior. It may prove helpful for the agriculture teacher to employ techniques

> ## BOX 16.3 *Helping Students Transition to the Middle Grades*
>
> - Arrange a school visit day for elementary school students to the agricultural program at the middle school. Have tours of the labs and greenhouse facilities, and teach a mini-unit of instruction on an agricultural subject.
> - Meet with elementary school teachers of rising middle grades students to gather information on new students in your program. Information about particular student strengths and their personal or career interests is helpful.
> - Send a letter of congratulations to rising middle grades students on their graduation from elementary school.
> - Continue to make connections with parents. Invite them to open house programs and provide regular feedback on student progress.
> - Remember, the teacher sets the tone and morale in the classroom. A healthy dose of humor keeps the stress levels down and encourages students to develop an excellent professional working relationship with the teacher.

that help middle school students survive the transition from elementary school. Box 16.3 provides tips for helping students transition to middle school.

MIDDLE SCHOOL AGRICULTURAL EDUCATION

There are approximately 13,300 middle schools in the United States, a 537% increase from 2,100 schools in 1971. The number of schools designated as junior high schools decreased to 2,500 by 2017 (National Center for Educational Statistics, 2021). The number of students enrolled in agricultural education coursework is not well known. The National Center for Educational Statistics, which collects data on curriculum and instruction, defines middle schools as a subset of elementary schools. This makes it difficult to find or select data about agricultural education in middle schools. While there is no central statistical database to draw from, an examination of state education departments finds curriculum for middle grades in agricultural education.

One significant advantage of the middle school agricultural education program is that it exposes students to various learning experiences related to the agricultural sciences. The learning experiences are coupled with opportunities for career exploration, thus enabling students to begin formulating career goals based upon their occupational interests. The middle school mission of literacy and exploration in the food, fiber, and natural resources systems aligns with the strategic plan for agricultural education (National Council for Agricultural Education, 2021).

Middle school agricultural education programs assist students in developing necessary social skills and decision-making skills while providing basic knowledge about agriculture and natural resources. This allows for the practical application of academic disciplines. Furthermore, middle school programs also allow for the developmental needs of youth. To help students develop a sense of belonging and self-worth, the agriculture teacher arranges career interest surveys to help students identify their strengths. Learning activities that develop citizenship and teamwork skills provide outlets for the growth of social skills. Lessons that focus on international agriculture build a sense of community in the classroom among students of differing cultures. To promote problem-solving and critical thinking, the teacher provides instructional activities that require these skills to solve authentic problems.

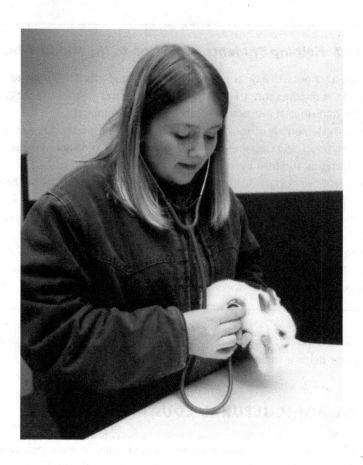

FIGURE 16.4 Middle grades students need opportunities to practice critical thinking and problem-solving. Agricultural education programs provide plenty of options.

The Middle School Agricultural Education Curriculum

The agricultural education program in middle school focuses on two significant areas: agricultural literacy and career exploration. The program aims to provide the fuel that ignites the spark of interest in a student for agricultural occupations. Because many agricultural disciplines are available to high school and college graduates, it is essential students receive a concentrated and directed study of agricultural careers at some point before graduating from high school.

Career exploration is the use of career-related investigative activities conducted in and outside the classroom. The goal is to help students identify future potential career interests and gain information to help plan their higher levels of education. Career exploration may involve reading about careers, watching videos, participating in job shadowing, volunteering, and gaining information in other ways. Overall, the best career exploration involves "hands-on" experiences that provide vivid real-world opportunities.

Through *student-centered* instructional methods, the middle school agricultural education program offers students the opportunity to learn about agricultural careers and some of the skills needed in those careers. Following are some basic things a student might learn about a given agricultural job:

- Vocabulary and terminology specific to an agricultural occupation
- The impact of agriculture on the local, state, national, and global economy
- Basic skills in agricultural sciences, biotechnology, and engineering
- Leadership and human relations skills
- The overall importance of agriculture to humanity

(National FFA Organization, 2009)

Since middle school is an excellent place for students to begin developing personal and social skills for life after high school, the middle school agricultural education program should incorporate experiences that expose students to the world they will soon enter as adults. Through exploratory activities, students can learn firsthand about the technological and intellectual demands placed on workers in the industry.

Agricultural education has always gone beyond learning about "cows and plows." With the exponential increase in technology in the agricultural sciences, there is a greater need for highly skilled workers. Career preparation has a place in the middle school agricultural education program.

Because adolescents are changing socially, emotionally, and physiologically in the middle school period, the agriculture teacher must design the curriculum and instruction to recognize the specific needs of these students. Adolescent students may have difficulty engaging in the lesson from time to time. Disengaged students need more exposure to differentiated instruction, as delivered by a competent and caring teacher. In middle school agricultural education, the teacher can engage reluctant learners by

- Focusing student attention on the lesson and eliminating or curtailing distractions
- Frequently explaining the importance of the lesson
- Using the personal experiences of students as part of the instructional process
- Providing several methods or opportunities for students to learn the lesson content
- Conducting diagnostic tests to determine how students are constructing meaning from the lesson content
- Paying careful attention to students' cultural backgrounds and providing an atmosphere of acceptance in the learning environment

(Turner, 2011)

Interdisciplinary Methods of Instruction

A unique advantage of middle school education is the opportunity for interdisciplinary methods of instruction. Agricultural education teachers should grasp the chance to work with other teachers in developing meaningful learning experiences for students. For instance, an agriculture teacher could partner with a science teacher to study aquatic life in a pond near the school. The students would meet the science teacher's goals through an exciting lab activity. In contrast, the students would complete the agriculture teacher's goals through a realistic exercise that demonstrates the tasks associated with an agricultural career area, such as wildlife biology.

An essential phase of planning a middle school agricultural education curriculum is to articulate it with the curriculum at the high school level. Articulation is achieved when educational activities are organized to facilitate students' continuous and efficient progress from one grade or class to another. Courses of study are sequenced so prerequisites are developed in the lower grades. For example, students in grade 7 learn skills that prepare them for grade 8, those in grade 8 learn skills that prepare them for grade 9, and so on.

The agricultural education curriculum may also be articulated within the same grade level among the various courses a student takes. For example, the science and language arts curriculums may be articulated and demonstrate continuity from one class to another.

A prime example of how articulation plays out in school is through school gardens and outdoor learning labs. For instance, research studies have demonstrated middle school outdoor learning labs in plant sciences can

- Increase a student's knowledge and interest in plant science
- Provide awareness about the environment

- Provide knowledge about food production and food systems
- Encourage teamwork and social skill development

There are more than 10,000 school gardens and outdoor learning labs on middle grades school campuses in the United States. Students who are actively engaged in gardening on the school campus are more likely to be involved in science education and participate more effectively in other school disciplines. The excitement of learning in the garden often transfers into learning in the science classroom (Skinner et al., 2012).

FFA in the Middle Grades

Career Development Events

The National FFA Organization offers a variety of Career Development Events for middle school students. Some of these, such as the floriculture event, may have junior divisions for middle school students at the state level. Still others, such as the conduct of meetings CDE, are designed specifically for middle school FFA members. The FFA offers leadership programming at the local, state, and national levels. These conferences build leadership and communication skills for members of all grade levels.

Leadership Programs

FFA offers several leadership programs at the state and national levels for middle-grade students. Participants explore shared leadership and career interests with others, and the conference provides introductory training in service-learning and community development.

FIGURE 16.5 Middle grades agricultural education programs support other academic disciplines through contextual learning experiences for students.

FFA Membership and Member Degrees

FFA membership is open to students in an agriculture class at the middle school level. The Discovery FFA Degree is the first degree a student can obtain as a member of FFA. This degree encourages middle-grade students to learn more about FFA and the agricultural industry. The requirements are consistent with the career exploration elements of middle grades agricultural education. This degree is open to agriculture students in grades 7 and 8. Additional requirements for this recognition include basic knowledge of the purpose of the FFA and involvement in FFA activities.

SAE in the Middle Grades

Middle school is a great place to start an SAE project. Middle-grade students can create exploratory experiences based upon their own learning experiences in the agricultural education classroom. Exploratory supervised experience can lead to a more sophisticated understanding of the agricultural industry (Moore & Flowers, 2012). Suppose the middle-grade student stops taking agricultural education courses in the eighth grade. In that case, they should have learned enough about the agricultural industry to have a significant measure of appreciation for it. If the student chooses to continue to take agriculture courses in high school, the exploratory experiences they began in middle school can blend into foundational and immersion supervised agricultural experiences. (See Chapter 22 for more information on SAE.)

If the student was active in the FFA Agriscience Fair program, they might continue exploratory work by developing an experimental or analytical supervised experience. The middle-grade student need not wait until enrollment in a high school agricultural education program to begin work on an entrepreneurial, placement, or research-based SAE.

For some students, their first significant experience with agriculture happens during a *Food for America* program at their school (see Chapter 15). Students interested in the agricultural sciences can investigate them more thoroughly when they enter middle school and enroll in an occupational exploration course. The middle school agricultural education course can provide appropriate learning experiences that encourage students to seek more knowledge about agriculture and perhaps become valuable and productive in the agricultural industry.

REVIEWING SUMMARY

A middle school is a school between the primary or elementary grades and the high school grades. With grades 6 through 8 typically included, the students are at a time of significant personal development. Middle school agricultural education is an important component of the curriculum as it provides skill development and career exploration. Agricultural education in middle school meets two significant goals: agricultural literacy and career exploration. Agricultural literacy focuses on developing a proper understanding of food, fiber, and natural resources. Career exploration is the investigation of various career areas using a wide range of approaches.

Piaget's cognitive development model and Bandura's model are foundational works in understanding adolescent development. Adolescent development is a stage-based process. Erik Erikson is well known for his eight-stage human socialization process as related to middle school students.

Instructional strategies may be student-centered and interdisciplinary. Student-centered instruction matches the maturation levels of the learners. Teachers determine and use a range of approaches in teaching.

FFA has an essential role in middle schools. Career Development Events, leadership programs, and related activities promote student development at the middle school level. Supervised experience beginning in middle school can lead to a more sophisticated understanding of the agricultural industry.

QUESTIONS FOR REVIEW AND DISCUSSION

1. What types of supervised experiences can a student have in a middle school agricultural education program?
2. What is a middle school?
3. What is the name of the association that serves middle schools? What is the central thrust of this organization?
4. What is career exploration? What is the goal of career exploration?
5. What are the eight stages in the human socialization process developed by Erik Erikson?

ACTIVITIES

1. Visit a middle school agricultural education program. Ask the agriscience teachers the following questions. Then, write a summary of your interview and use it as a basis for a classroom discussion.
   a. Why did you choose to become a middle school teacher? Why did you choose agricultural education as your field of expertise?
   b. What do you enjoy most about teaching in middle school?
   c. What do you enjoy least about teaching in middle school?
   d. How have middle schools changed since you first entered the teaching profession? How has agricultural education changed as well?
   e. How do you handle the demands of teaching?
   f. How is your program funded?
   g. What effect does FFA have on your program? Does it help or hinder your instruction?
   h. Generally speaking, what teaching methods work best for your students?
   i. How do you manage your time as a teacher? (Alternate question: How do you accomplish all that you must do to be an effective teacher? What are some methods you use to utilize your time better?)
   j. What types of assistance do you get from school administrators and parents?
2. Investigate the Association for Middle Level Education (AMLE). Begin with the association's website: www.amle.org. Determine the mission and goals of AMLE, how it supports the professional development of middle school teachers, and its advocacy role with policymakers. Prepare a report on your findings.

REFERENCES

Bishop, P., & Harrison, L. (2020). *The successful middle school: This we believe*. Association for Middle Level Education.

Elementary and Secondary Education Act. (2002). Public Law 107-110, 107th Congress.

Flavell, J. H. (1963). *The developmental psychology of Jean Piaget*. Van Nostrand Reinhold.

Gruhn, W. T., & Douglass, H. R. (1956). *The modern junior high school* (2nd ed.). Ronald Press.

Manning, M. L. (2000). A brief history of the middle school. *Clearing House: A Journal of Educational Strategies, Issues and Ideas, 73*(4), 192–192. https://doi.org/10.1080/00098650009600946

Moore, G. E., & Flowers, J. L. (2012). *The North Carolina SAE model.* North Carolina State University.

National Center for Educational Statistics. (2021, May 13). *NCES fast facts.* National Center for Education Statistics. https://nces.ed.gov/fastfacts/display.asp?id=84

National Council for Agricultural Education. (2016). *National quality program standards for agriculture, food, and natural resources education: A tool for secondary (grades 9–12) programs.* https://thecouncil.ffa.org/

National Council for Agricultural Education. (2021). *National Council for Agricultural Education strategic plan 2012–2015.* National Council for Agricultural Education.

National FFA Organization. (2009). *A guide for middle school agricultural science teachers.* National FFA Organization.

Skinner, E. A., Chi, U., & The Learning Gardens Educational Assessment Group 1. (2012). Intrinsic motivation and engagement as "active ingredients" in garden-based education: Examining models and measures derived from self-determination theory. *Journal of Environmental Education, 43*(1), 16–36. https://doi.org/10.1080/00958964.2011.596856

Turner, S. L. (2011). Student-centered instruction: Integrating the learning sciences to support elementary and middle school learners. *Preventing School Failure: Alternative Education for Children and Youth, 55*(3), 123–131. https://doi.org/10.1080/10459880903472884

17

High School Agricultural Education

What is high school agricultural education like today? Consider these examples:

- Lauren Wise teaches in a one-teacher agriculture department at a small school with grades kindergarten through 12 all in the same building.
- Jennifer Smart teaches in a five-teacher agriculture department in a large 9th- through 12th-grade comprehensive high school.
- Octavius Brown teaches horticulture and landscape management at a 10th- through 12th-grade area career and technical education center.
- Cato Cantrell rides the bus to school, has four classes on block scheduling (one is veterinary science), and gets off the bus each afternoon for supervised experience at a veterinary clinic near his home.
- Jasmine Robles goes to her local high school for first-period agriculture, then returns to her house where her mother home schools her.
- Santiago Lopez attends a virtual school online. He is excited as the genetics lab kit arrived at his home last week and today he performs the experiments.

OBJECTIVES

This chapter addresses the National Quality Program Standards for Agriculture, Food, and Natural Resources Education (National Council for Agricultural Education, 2016), specifically Standard 6: Certified Agriculture Teachers and Professional Growth. It has the following objectives:

1. Explain high schools in American education.
2. Discuss the role of high school agricultural education.
3. Identify student maturation and developmental factors influencing instruction.
4. Describe instructional strategies appropriate for high school students.

TERMS

area career and technical education center
charter school
cognitive development
comprehensive high school
development
elementary school
home school
magnet school
Jean Piaget

scaffolding
scheme
secondary school
Seven Cardinal Principles
tertiary system
virtual high school
voucher
Lev Vygotsky
zone of proximal development

FIGURE 17.1 A current high school agriculture facility in Connecticut bears little resemblance to a high school agriculture facility from the mid-1900s.

HIGH SCHOOLS IN AMERICAN EDUCATION

Formal schools have existed in the United States since its founding. The majority of these schools in the 1600s and 1700s were religious in nature, whether publicly or privately funded.

The 1800s brought structure to education, and eventually a laddered system of elementary, secondary, and tertiary schools took form. An *elementary school* was typically a school with grades 1 through 6. A *secondary school* was usually a school with grades 7 through 12. The *tertiary system* was the colleges and other institutions after high school. This structure served as a good foundation on which today's education was built.

The 20th century saw an ever-increasing percentage of American adolescents attending at least some high school. In the early 1900s, the junior high school developed as an entity separate from the high school. The junior high school contained grades 7 through 9, and the high school contained grades 10 through 12. Although in its beginnings the junior high school was designed to meet the unique needs of preadolescents, it quickly came to follow the high school in structure and organization.

Beginning in the 1960s and continuing throughout the remainder of the century, the middle school concept took hold and spread. The middle school was designed for students in grades 5 or 6 through grade 8. Ninth-grade students were moved to the high school so the curricular, physiological, and social needs of middle school students could be better met. (Middle school agricultural education was discussed in Chapter 16.)

The first decades of the twenty-first century have seen continued changes. The number of students who are being home schooled has increased. *Home school* consists of teaching by parents and others outside of traditional school buildings and typically in the home. Home school may include face-to-face instruction, online learning, or a combination of the two. Students may be home schooled until the start of middle or high school or until completion of high school.

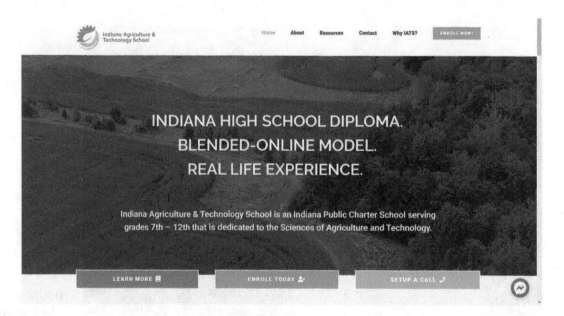

FIGURE 17.2 Virtual and charter schools may offer agricultural education. This Indiana school uses a blended instructional model for agriculture students.

Another kind of school is the charter school. A *charter school* is a publicly supported school given freedom to operate outside of state and local bureaucratic control. Charter schools are able to experiment and try educational innovations. However, they are still required to meet local and state performance standards for their students.

A *virtual high school* is staffed by teachers with instruction delivered primarily online. Virtual high schools are managed in various ways including local school corporation, charter school management company, or state institution. Many virtual high schools offer a flexible curriculum allowing students to complete a course in an individualized time frame.

Early college high schools are another form of secondary education. High school students earn both high school graduation credit and college credit for their coursework. Students typically graduate with an associate's degree or completion of two years of college credits. Early college high schools may be aligned with a community college, another two-year college, or a four-year university. Students may take dual credit courses from their high school teachers, travel to the college campus, or take online college courses.

A voucher approach has received considerable debate. With *vouchers*, parents can choose the school their children attend, whether public or private, and receive public funds through a direct payment or tax credit arrangement. A voucher system has the potential to erode financial and community support for the public schools.

Purposes of a Secondary Education

Education in the United States has been organized according to three guiding principles: local control, federalism, and professionalism (Chubb, 2001).

In the principle of local control, public schools are controlled and administered by local boards of education. These entities make curricular, staffing, and other educational decisions. A local body may control the collecting of taxes for school use and typically must approve a budget of expenditures.

The federalism principle gives the states the responsibility for education. This ensures that no one group can exercise its force on a national level. At the same time, it provides flexibility among the states for innovation and experimentation.

The principle of professionalism states education is best delivered by professional educators who make decisions based on the best interests of the students. Among others, this is one reason teachers are held to high standards of credentialing.

Committee of Ten Report

Over the years, groups, committees, and commissions have been formed to determine the purposes of a secondary education. In 1893 the National Education Association (NEA) funded a study that came to be called the Committee of Ten Report. The committee recognized a secondary education should be for all students, not just those going to college. However, the nine subjects recommended by the Committee of Ten were traditional liberal arts in nature, thus continuing the focus of secondary education as primarily preparation for college. The nine subjects were Latin, Greek, English, modern languages, physics, astronomy and chemistry, natural history, history, and geography. The work of the Committee of Ten influenced other committees that established a standard unit of credit for high school subjects called the Carnegie unit. The Carnegie unit, which is defined in Chapter 1, is still used today to determine high school graduation and college entrance requirements. The Committee of Ten Report directed secondary education for the next 25 years and continues to influence educational practices to this day.

Cardinal Principles Report

Secondary education was dramatically shaped by another NEA committee. The Cardinal Principles Report of 1918 changed the focus of secondary education and shaped it for the remainder of the 1900s. Instead of making subject area recommendations, the Cardinal Principles Report gave objectives to be met. These objectives are known as the *Seven Cardinal Principles.* They include command of the fundamental processes (basic skills, such as reading, writing, and mathematics), worthy home membership, health, vocation, citizenship, worthy

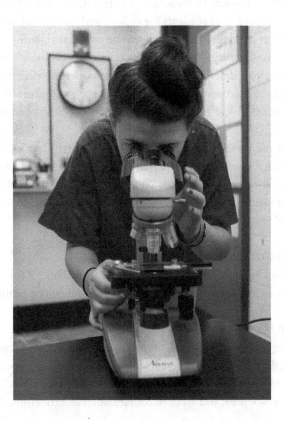

FIGURE 17.3 The Committee of Ten Report included science as part of a high school education. (COURTESY OF LEDYARD AGRI-SCIENCE & TECHNOLOGY PROGRAM.)

use of leisure time, and ethical character. For the first time, vocational, citizenship, and personal development areas were to be a part of secondary education for everyone. To meet this broadened purpose, the comprehensive high school was developed.

Reports From the 1980s to the 2000s

In 1987 the Southern Regional Education Board (SREB) and several state partners began an initiative called High Schools That Work (SREB, 2020). The initiative had 10 key practices including high expectations and continuous improvement to prepare students for the world of work and further education. The philosophy of the program was to create a learning environment that allows students to master a rigorous academic and technical curriculum. The initiative has broadened to all grades and is now called Making Schools Work.

In the 1990s the U.S. secretary of labor appointed a commission to find the skills students need to be successful in the workforce. The Secretary's Commission on Achieving Necessary Skills (SCANS) concluded that all high school students must be competent, all companies/employers must be characterized by high performance, and all schools must become high performing. The SCANS report identified five competencies (resources, interpersonal skills, information, systems, and technology) and three foundational skills (basic skills, thinking skills, and personal qualities) essential for job performance (SCANS, 1991).

The National Governors Association (NGA) was formed in 1908 to provide the nation's governors a vehicle to promote leadership, share best practices, and speak with a unified voice on public policy. Over the past century, the NGA has issued numerous reports regarding education. The 1989 National Education Summit led to *Goals 2000*, which was signed into law by President Clinton. The 2005 National Education Summit had as its goal to redesign high schools, particularly to improve graduation rates (NGA, 2021).

The NGA Center for Best Practices and the Council of Chief State School Officers (CCSSO) released *Common Core State Standards* in 2010 (Common Core State Standards Initiative, 2021). Two categories of standards, English–Language Arts and Mathematics, were developed to prepare K–12 students for college and career readiness. Most states adopted the standards or developed a modified version of the standards.

Partnership for 21st Century Learning (2009 document; online at https://www.battelle forkids.org/networks/p21) developed a framework of skills, knowledge, and expertise students must master to be successful for life and work in the 21st century. The partnership also advocates for the 4Cs (Critical Thinking, Communication, Collaboration, and Creativity) of learning.

Starting in 2011 and completed in 2013 (National Research Council, 2012; NGSS Lead States, 2013), the Next Generation Science Standards (NGSS) revised how science is taught and what is taught. The standards are formed around the three dimensions of crosscutting concepts, science and engineering practices, and disciplinary core ideas. Agricultural education research and practice has been influenced by NGSS.

Organization of American High Schools

High schools in the United States are organized in various ways. All are intended to provide the amount of education associated with a specific diploma or certificate.

The Comprehensive High School

A *comprehensive high school* encompasses the full range of subjects and activities in academic, career and technical, citizenship, and personal development areas. The curriculum includes courses that all students take, such as English; courses that students select based on their career goals, such as sciences; and elective courses, such as agriculture. In addition to the classroom subjects, a comprehensive high school has citizenship development activities, such as clubs and organizations, and personal development activities, such as sports and the arts.

The number of career and technical education areas taught in comprehensive high schools is dependent upon their size and communities. Some may offer a minimum of technology education, business education, and consumer and family sciences. Others may offer a wide array, including automotive technology, building trades, metalworking, electronics, and health sciences. Agricultural education is offered in about 8,500 U.S. public high schools.

Specialized High Schools

An ***area career and technical education center*** is a specialized type of high school designed to prepare students for careers. Students may enter the workforce directly upon high school graduation or continue on to an apprenticeship or a trade, technical, community, junior, or four-year college. Area centers may draw from comprehensive high schools and teach only career and technical subjects, or they may be self-contained, teaching academic as well as career and technical subjects.

Area centers tend to offer a great variety of career and technical education areas. Teachers in these schools typically must document work experience in the areas in which they teach. Courses may be in two- or three-hour blocks, allowing students to complete detailed, complex assignments and more closely simulating the real world of work. These longer blocks of time also allow students to put classroom theory immediately into practice in the laboratory.

A ***magnet school*** is a specialized high school designed to attract students for a specific purpose. A magnet school is organized around a theme, such as the arts, science, mathematics, or even agriculture. To parents, magnet schools may be viewed as safer and more academically challenging. In most instances, students must apply to attend a magnet school, and competition for available spaces is intense. Magnet schools are most prevalent in urban areas. The Chicago High School for the Agricultural Sciences in Chicago, Illinois, and the STAR (Science and Technology of Agriculture and its Resources) Academy in Indianapolis, Indiana, are two examples of agricultural magnet schools.

FIGURE 17.4
Georgia students stand beside the sign of their bedding plant facility.

HIGH SCHOOL AGRICULTURAL EDUCATION

Agricultural education emerged along with American education to serve an important role. The history of education and the history of agricultural education were covered earlier in this book. This section reviews and expands upon agricultural education in high schools.

Before Smith-Hughes

For the first 75 years of the 19th century, agricultural education was not a part of public secondary education. However, during the period 1875 to 1900, many states established residential agricultural schools. Passage of the Hatch Act of 1887 contributed greatly to the increase of high school agricultural education. Beginning in 1889, Alabama established secondary agricultural schools in each of its congressional districts. Many state agriculture colleges offered agricultural instruction of less than college grade either as separate secondary schools or as short, intensive courses.

The first decade of the 1900s brought efforts to establish secondary agricultural education as a course of study in county high schools. Many states provided funding for counties that taught and maintained agriculture courses. Other high schools added agricultural education even without the benefit of state monetary aid. Some high schools teaching agriculture had land and facilities available for experiments, demonstrations, and student work.

By 1915, agricultural education was taught in almost 4,000 secondary schools in the United States. The level and quality of instruction varied greatly. In some schools, agriculture was taught as a science by teachers who were not always trained at a state agricultural college. In other schools, the instruction was based on textbooks or covered the sociological aspects of the subject and was taught by teachers not trained in agriculture. In still other secondary schools, agriculture had a more vocational focus, included laboratory and hands-on instruction, and was taught by trained agriculture teachers.

From Smith-Hughes to the 1970s

In 1917 the Smith-Hughes Vocational Education Act provided federal funds for instruction in agriculture, home economics, and trades and industry. This act dramatically changed the scope and focus of secondary agricultural education. The number of agriculture teachers and secondary students of agriculture greatly increased.

Teacher education for agriculture teachers, although occurring before 1917, became more important to assure a supply of qualified teachers for all the new secondary agricultural education programs.

State supervisors of agricultural education were hired. Home projects became a required part of secondary agricultural education. Local extension agents, as a part of the 1914 Smith-Lever Act, and secondary agriculture teachers, as a part of the 1917 Smith-Hughes Act, had to work out agreements on how they were to work with both children and adults.

From 1917 to the early 1960s, agricultural education on the secondary level grew and expanded. Although certain aspects changed, such as instructional technologies and subject-matter content, the focus remained vocational, oriented on production agriculture primarily for males. Both societal and legislative changes of the 1960s affected agricultural education.

The Vocational Education Act of 1963 expanded agricultural education instruction to off-farm agricultural occupations and subjects. This opened the door for instruction in areas such as horticulture, natural resources, and agricultural business. The combination of the addition of courses in off-farm agricultural subjects and the acceptance, in 1969, of females into FFA greatly increased the number of females enrolling in secondary agricultural education.

The civil rights movement of the 1960s and concurrent civil rights legislation led to the desegregation of public schools and therefore of agricultural education. This change was epitomized by the absorption of the New Farmers of America (NFA) into the Future Farmers of America (FFA) in 1965. Desegregation led to a decline in the number of African American agriculture teachers and over the years to a decline in the number of African American secondary agriculture students. (More information about the NFA is in Chapter 23.)

From the 1970s to 2000

The 1970s saw increasing enrollments in secondary agricultural education, peaking in 1976 at almost three quarters of a million students (Camp et al., 2002). The Vocational Education Amendments of 1968 and 1976 brought about an increased emphasis on overcoming gender discrimination and gender stereotyping, which further enhanced the effort to increase enrollment of females in agricultural education. These amendments, along with other federal legislation, encouraged the inclusion of special needs students in agricultural education.

From the mid-1970s to the early 1990s, enrollment in secondary agricultural education declined. This decline has been attributed to numerous factors. In the 1980s a major farm crisis accelerated the decline of families earning their living from farming. This was coupled with the long-term trend of overall declining farm numbers. The pendulum of educational reform swung toward increasing requirements in core academic subjects to the detriment of electives. In addition, the generation of baby boomers had worked its way through the secondary school system by the mid-1980s, and a smaller population of Generation Xers followed. The publication of *Understanding Agriculture: New Directions for Education* (commonly referred to as the *Green Book*) in 1988 by the National Research Council set the course for numerous changes in agricultural education over the next decade.

The 1990s saw increasing enrollments, greater emphasis on the recruitment and retention of female and minority students, and a greater diversity of curriculum offerings. This decade also saw the rapid increase in the use of microcomputers in the agricultural education classroom and eventually access to the internet. During this time, many states greatly expanded the offering of agricultural education courses, especially nonsequenced, agricultural literacy–type courses. Although in existence in a few cities before this time, urban agricultural education received a renewed emphasis.

Beginning of the 21st Century and Beyond

Agricultural education at the beginning of the 21st century is taught in approximately 8,500 U.S. middle and high schools to more than 1 million students in grades 6 through 12. Classes include a diversity of students with regard to ethnicity, gender, geographic area, and ability level. This diversity extends to the reasons students enroll in agricultural education. Some still enroll for vocational reasons, whether they are planning to enter agricultural occupations immediately or continue on to further education after high school. Others enroll for avocational or personal-interest reasons. In some high schools, certain agriculture classes count toward science graduation requirements. Finally, the extension of agricultural education classes into middle schools has greatly expanded.

In 2000, the *National Strategic Plan and Action Agenda for Agricultural Education: Reinventing Agricultural Education for the Year 2020* was released. This document set out the vision, mission, and goals for agricultural education for the next 20 years. The vision of agricultural education was "a world where all people value and understand the vital role of agriculture, food, fiber, and natural resources systems in advancing personal and global well-being" (National Council for Agricultural Education, 2000, p. 3). The mission of agricultural education was set as preparing "students for successful careers and a lifetime of informed choices

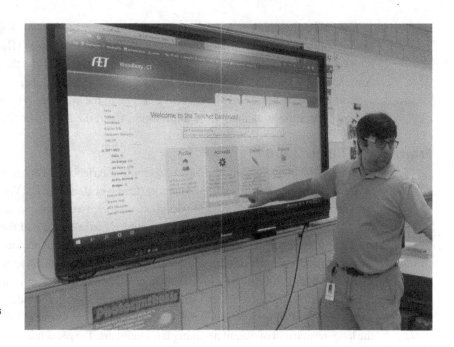

FIGURE 17.5 Information technology has changed many practices in high schools.

in the global agriculture, food, fiber, and natural resources systems" (National Council for Agricultural Education, 2000, p. 3). This mission envisioned an expansion of agricultural education in both reach and impact. The National Council for Agricultural Education worked to accomplish the vision, mission, and goals set forth in this plan.

National AFNR Content Standards were revised in 2015 putting more emphasis on career preparation in AFNR pathways, up-to-date knowledge and skills on which to build agricultural education courses, and measuring student attainment of knowledge and skills. The standards were cross-walked to NGSS and Common Core, which were described earlier in this chapter. Standards cutting across all pathways, cluster skills, are included as well as career-ready practices.

In 2017, the SAE component of agricultural education changed with the publication of SAE for All teacher and student guides (National Council for Agricultural Education, 2017). SAE for All places a greater emphasis on those aspects of experiential learning that all students should experience in their agricultural education courses. SAE for All is discussed in Chapter 22.

The start of the 2020s saw a renewed national focus on diversity, equity, and inclusion. This is further discussed in Chapters 20 and 23. The agricultural education profession also started efforts to be more inclusive in its course enrollments and FFA participation. The Agricultural Education for All efforts will need to be purposeful and ongoing for lasting change.

ADOLESCENT DEVELOPMENT

Secondary students are going through mental, emotional, physical, hormonal, and social changes. This adolescent development affects what is taught and how it is taught. *Development* is the orderly, lasting change that occurs in humans as a result of maturation, learning, and life experiences. Many scientists have studied and written about adolescent development across biological, cognitive, and socioemotional domains. Recently, these have been explored through a cultural lens as development varies across groups and societies in the world (Gibbons & Poelker, 2019). For example, biologically the beginning of adolescence is typically defined as the start of puberty. However, different cultures acknowledge this transition in various ways, which can impact how students view themselves and their behaviors inside and outside of school. Cognitive and socioemotional development in adolescents is further discussed later in this chapter.

Woolfolk (1998) gave the following three general principles of development:

1. People develop at different rates. Adolescents of the same age will be at varying stages of physical, mental, and social development.
2. Development is relatively orderly. We can predict that most people will go through the same stages of development in about the same order.
3. Development takes place gradually. Change and growth take time and, at least in mental and social development, may include periods of decline before advancement.

Cognitive Development

Cognitive development is the development of an individual's thought processes. It includes adaptation to the environment and the assimilation of information.

Piaget

In the mid-1900s, *Jean Piaget*, a Swiss psychologist, developed a theory of cognitive development that is still in use today (Piaget's theory is detailed in Chapter 16). He proposed that we all have a tendency toward organization and adaptation. We organize our thinking into structures called schemes. *Schemes* are the mental representations of objects and events of the world. We then adapt our schemes as new information and experiences cause us to adjust our thinking.

Piaget is most famous for his "four stages of cognitive development" theory. Although psychologists have questioned whether the number and order of stages is correct, most agree that Piaget's observations about cognitive development are accurate. The first two stages, sensorimotor and preoperational, occur in children from birth to about 7 years of age. Although agriculture teachers rarely teach students of this age range, certain characteristics of the preoperational stage are important for teachers to consider. Students in this stage need visual aids to help them learn concepts. Demonstrations and student practice are important for student understanding. Also, students need real-world experiences, such as those provided by field trips, to build a foundation for learning concepts.

Piaget's third stage, concrete operational, ages 7 to 11, is when students begin to think logically. They still need concrete examples, but these can increasingly take the form of symbols as opposed to real objects. Students continue to need clear instructions and opportunities to experiment. Students can now understand more complex ideas but benefit when the concept presented is connected to a familiar idea.

The fourth stage in Piaget's theory, formal operational, age 11 to adult, is when students are able to think hypothetically and abstractly and to combine information into new concepts. Teachers should involve students in solving problems, in developing hypotheses, and in testing and exploring complex concepts. Teachers should not expect all high school students to be in Piaget's fourth stage. In fact, many psychologists debate whether adults operate at the formal operational stage in all areas.

Vygotsky

Lev Vygotsky, a Russian psychologist, had a view of cognitive development different from Piaget's view. The sociocultural theory emphasizes the role people play in a child's cognitive development. This is particularly true with regard to the child's culture and the child's

FIGURE 17.6 High school students practice social skills they began developing during the middle school years.

tools, such as language. Children develop through their interactions with adults, especially as adults provide assistance in the form of scaffolding. *Scaffolding* is the help, prompts, and questions an adult gives a child to assist the child's understanding.

Vygotsky proposed the concept of the "zone of proximal development." For any particular range of tasks, the *zone of proximal development* is that level of proficiency in which students cannot accomplish the tasks alone but can accomplish them with assistance (Woolfolk, 1998). Below the zone, students are capable of accomplishing the tasks by themselves. Above the zone, students cannot accomplish the tasks even with assistance. Only within the zone can learning take place and instruction be successful.

Personal, Social, and Emotional Development

Erikson's psychosocial theory of human development was introduced in Chapter 16. It is expanded here as a base for high school students.

Erikson based his theory on his observation that people have the same basic needs and that a purpose of society is to provide for those needs. He proposed that all people go through eight stages of psychosocial development. Table 17.1 provides a summary of these stages. Each stage is characterized by a crisis or conflict between a positive alternative and a potentially unhealthy alternative. How a person resolves each stage has an impact on later stages. The first three stages cover birth to approximately 6 years of age. The final two stages cover middle and late adulthood. The middle three stages occur during the time a person is in school or just out of school.

Erikson's fourth stage occurs during middle childhood and involves the conflict of industry versus inferiority. This conflict is characterized by the learning of new skills. Children who master the skills and become competent develop a sense of accomplishment. Those who do not master the skills risk developing a sense of failure and inferiority. Teachers can encourage industry by making sure students can set and work toward realistic goals, show independence and responsibility, and recognize their progress and improvements (Woolfolk, 1998).

Erikson's fifth stage occurs during adolescence and involves the conflict of identity versus confusion. At this stage, roughly ages 12 to 18, peers become much more important.

TABLE 17.1

Erikson's Stages of Human Growth and Development Summarized

Stage	Description
1–4	Life stage: Infancy, Childhood Approximate ages: Birth–12 Conflicts: Trust, Autonomy, Initiative, Industry
5	Life stage: Adolescence Approximate ages: 13–18 Conflict: Identity vs. Role confusion
6	Life stage: Young adulthood Approximate ages: 19–25 Conflict: Intimacy vs. Isolation
7–8	Life stage: Middle adulthood, Late adulthood Approximate ages: 26–Death Conflicts: Generativity, Integrity

Adolescents must develop their identity in terms of personal, social, occupational, and sexual roles. This stage involves experimentation and clarification of roles. Teachers can support identity formation by providing role models, resources to help work out personal problems, acceptance of experimentations that are not harmful and do not interfere with learning, and realistic feedback (Woolfolk, 1998). Students whose identity formation is impacted by being a member of a racial/ethnic, gender, cultural, or cognitive developmental minority group

FIGURE 17.7 An agriculture teacher uses an animal in the school's lab to provide hands-on skill development.

may experience the conflict earlier in life and may have difference experiences from those from a majority group (Martin & Torok-Gerard, 2019).

Erikson's sixth stage typically occurs during young adulthood but may have an impact on some high school students. At this stage the conflict is between intimacy and isolation. Young adults need to be willing to relate to others and to develop intimate relationships. Teachers need to watch for students who isolate themselves and have few friendships.

Just before high school, students in grades 5 through 8 (middle school) have gone through rapid developmental changes. This age group demonstrates a great range of individual developmental levels. Students of the same chronological age will exhibit vastly different levels of physical, cognitive, and social development.

Physically, middle school students are developing in weight, height, strength, and endurance. Many middle school students will be awkward and uncoordinated as their bones grow faster than their muscles. Students become self-conscious about how they look as their bodies change in size and structure. Students at this age, especially females, will begin puberty and may not understand all the changes their bodies are going through. Students may grow tired more easily, have periods of restlessness, and eat more in both frequency and quantity.

Emotionally, middle school students are developing their self-identity. They are beginning to define themselves as being independent from their families. They may have exaggerated emotions, experiencing high "highs" as well as low "lows." Middle school students may have unwarranted fears and at the same time have feelings of invulnerability.

Socially, peer groups become increasingly important for middle school students. They compare themselves with peers and follow group norms in dress, speech, and behavior. A middle school peer group may become exclusive, territorial, and even cruel to those not in the group. On the other hand, middle school students are becoming increasingly conscious socially and may take on causes of justice and assist those who are less fortunate. Parents and authority are still important, but this age group may also seek other adults as mentors and role models.

Chapter 11 introduced mindset and its effect on learning and growth. High school students are at a developmental stage where a fixed mindset can limit current successes and future opportunities. High school adolescents are further forming their identity, especially in sexual orientation/gender identity, ideological stance, and career direction (Marcia, 1980). Bullying against those seen as nonconforming can take many forms and must be addressed.

INSTRUCTIONAL STRATEGIES APPROPRIATE FOR SECONDARY STUDENTS

Agriculture teachers should organize their classrooms according to the developmental levels of students. For example, with middle school students, some will be able to think and work abstractly, whereas many others will still be in the concrete operational stage. For the benefit of both groups of students, agriculture teachers should include problem-solving experiences but make sure to give clear instructions and expectations.

Agriculture teachers of middle school students must realize these students need social opportunities to learn effectively. Teachers should plan activities that allow students to talk and move while accomplishing educational objectives. Problem-solving exercises that involve justice, righting wrongs, or helping those less fortunate can actively engage middle school students in the learning situation.

High school students, more so than middle school students, will be on similar developmental levels. However, students may have trouble transferring learning from one situation to another. For example, they may have learned concepts in geometry and biology that are applicable in agriculture but not be able to use the learning in agriculture. If possible, the agriculture teacher should individualize scaffolding so each student receives only the help and prompts needed for advancement to the next step. The teacher should design tasks and problems so each student can operate within their own zone of proximal development.

High school students are very much in Erikson's fifth stage of identity development. Agriculture teachers should use SAE and FFA activities to help students in this task. Assistance in occupational identity development should be provided to all agricultural education students. The opportunities within FFA for social and personal identity development are numerous. Individualizing instruction within the classroom and laboratory can help students safely experiment with different roles, especially in the occupational area.

REVIEWING SUMMARY

Secondary education in the United States has gone through numerous transformations during the past 300-plus years. What was once reserved for the rich and elite is now a universal right for all American children. In the transformations, secondary education has included everything from purely academics for the training of the mind, to purely vocational for the working classes, to a mixture of academic and vocational designed to meet individual student needs.

In recent years, greater emphasis has been placed on early secondary education. The middle school concept recognizes students in grades 5 through 8 are in a unique developmental stage and should be educated accordingly. This concept also takes into account that middle school students are developing physically, emotionally, and socially in addition to intellectually.

Agricultural education has been a part of secondary education since the late 1800s. With the passage of the Smith-Hughes Act in 1917 and the provision of federal support, agricultural education gained a larger role in public secondary education. The 1960s brought social and legislative changes to agricultural education. The 1980s saw declining enrollments and a reexamination of the role of agricultural education. At the beginning of the 21st century, agricultural education is working to set its own vision and chart its own course.

Our understanding of adolescent cognitive development has been greatly influenced by Jean Piaget and Lev Vygotsky. Piaget's theory has had substantial impact on what we do in the classroom and how we do it. Vygotsky's theory emphasizes the role language plays in cognitive development.

Erikson's psychosocial theory of development explains personal, social, and emotional development. Three of Erikson's eight stages occur during the years within secondary education. Erikson's theory has implications for teaching, especially middle school students.

QUESTIONS FOR REVIEW AND DISCUSSION

1. What is the difference between a comprehensive high school and an area career and technical education center?
2. Explain the rationale for charter schools.
3. What are the Seven Cardinal Principles?
4. What was the impact of the Smith-Hughes Act on secondary agricultural education?
5. Describe the legislative and social forces that shaped agricultural education during periods of significant change such as the 1960s and 2000s.

6. What are the characteristics of agricultural education in the 2020s?
7. How can an agriculture teacher's knowledge of Piaget's four stages of cognitive development enhance the agricultural education learning experience for students?
8. What are examples of how the zone of proximal development can be used in agricultural education instruction?
9. How can Erikson's psychosocial theory of development influence the total agricultural education program?
10. What variations may be made between teaching middle school students and high school students?

ACTIVITIES

1. Investigate what the Seven Cardinal Principles for a secondary education should be for the 21st century.
2. Investigate current legislation from your state and the federal government regarding secondary education. On the federal level, you will want to investigate the latest Perkins Act, the Individuals with Disabilities Act, the Elementary and Secondary Education Act, and the Higher Education Act, among others. On the state level, you will want to search for information regarding home schooling, charter schools, school vouchers, and other alternatives to the traditional public secondary education.
3. Investigate current research and theory on how adolescents' use of social media influences their identity development.
4. Using the *Journal of Agricultural Education*, *The Agricultural Education Magazine*, and the proceedings of the National Conference of the American Association for Agricultural Education, report on research about an aspect of secondary agricultural education. Current and past issues of these journals can be found in the university library. Selected past issues can be found online at www.aaaeonline.org or www.naae.org.

REFERENCES

Camp, W. G., Broyles, T., & Skelton, N. S. (2002). *A national study of the supply and demand for teachers of agricultural education in 1999–2001.* http://www.aaaeonline.org

Chubb, J. E. (2001). The system. In T. M. Moe (Ed.), *A primer on America's schools* (pp. 15–42). Hoover Institution Press.

Common Core States Standards Initiative. (2021). *Common core states standards initiative.* http://www.corestandards.org

Gibbons, J. L., & Poelker, K. E. (2019). Adolescent development in a cross-cultural perspective. In K. D. Keith (Ed.), *Cross-cultural psychology: Contemporary themes and perspectives* (2nd ed.). Wiley.

Marcia, J. E. (1980). Identity in adolescence. In J. Adelson (Ed.), *Handbook of adolescent psychology.* Wiley.

Martin, J. L., & Torok-Gerard, S. (Eds.). (2019). *Educational psychology: History, practice, research, and the future.* ProQuest Ebook Central. https://ebookcentral.proquest.com

National Council for Agricultural Education. (2000). *National strategic plan and action agenda for agricultural education: Reinventing agricultural education for the year 2020.*

National Council for Agricultural Education. (2016). *National quality program standards for agriculture, food, and natural resources education: A tool for secondary (grades 9–12) programs.* https://thecouncil.ffa.org/

National Council for Agricultural Education. (2017). *SAE for all: Teacher guide.*

National Governors Association (NGA). (2021). *National governors association.* www.nga.org

National Research Council. (1988). *Understanding agriculture: New directions for education.* National Academies Press. https://doi.org/10.17226/766

National Research Council. (2012). *A framework for K–12 science education: Practices, crosscutting concepts, and core ideas.* National Academies Press. https://doi.org/10.17226/13165

NGSS Lead States. (2013). *Next generation science standards: For states, by states.* https://www.next genscience.org/

Secretary's Commission on Achieving Necessary Skills (SCANS). (1991, June). *What work requires of schools: A SCANS report for America 2000.* U.S. Department of Labor. https://wdr.doleta .gov/SCANS/

Southern Regional Education Board (SREB). (2020). *School improvement process.* https://www.sreb .org/school-improvement-process

Woolfolk, A. E. (1998). *Educational psychology* (7th ed.). Allyn & Bacon.

18

Adult and Postsecondary Education

Beyond secondary school classes, agricultural education takes on different forms. Two examples are described here. How do you feel these appeal to adults and meet their needs?

1. The Willis River Young Farmers chapter will conduct machinery and grain bin safety demonstrations at the high school agriculture facilities next Saturday from 1:00 to 4:00 p.m. Demonstrations include safety with PTO, augers, and grain heads, the prevention of tractor rollover, and the dangers of grain bin entrapment. The public is invited free of charge.

2. The Agriculture Department at Slate Valley Community College announces a new Technical Certificate in Controlled Environment Agriculture. This 34-credit program can be completed in three semesters. Graduates can enter directly into exciting, competitive salary careers in agriculture. Call (555) 555-1234 between 8:00 a.m. and 4:00 p.m. or visit the college's website for more information.

OBJECTIVES

This chapter addresses the National Quality Program Standards for Agriculture, Food, and Natural Resources Education (National Council for Agricultural Education, 2016), specifically Standard 6: Certified Agriculture Teachers and Professional Growth. It has the following objectives:

1. Explain the meaning of adult, adult education, postsecondary education, and continuing education.
2. Describe the nature of adults as learners.
3. Describe the organization of adult/young adult agricultural education.
4. Discuss the role of postsecondary education.
5. Discuss adult education in agriculture offered through secondary schools.
6. Explain the laws of adult learning.
7. Discuss teaching approaches for adults and young adults.
8. Explain the role of student organizations with adult and postsecondary students.

TERMS

adult
adult agricultural education
adult basic education
adult education
andragogy
Chautauqua movement

lyceum
National Professional Agricultural Student Organization
National Young Farmer Education Association
postsecondary education

FIGURE 18.1
Adult instruction in agricultural education covers a variety of topics and includes a wide range of audiences.

ADULTS AND ADULT EDUCATION

Who is an adult? Is the definition based on age, life status, independence from parents, or some other condition? What is adult education? This section clarifies these two concepts.

Being an Adult

Throughout time, many cultures have developed rites of passage that clearly demonstrate when a child has passed into adulthood. In the United States, it is a much more difficult issue. A person can vote at age 18 but cannot legally purchase alcoholic beverages until age 21. Most states allow a person to get a driver's license at or near age 16 but have a higher age limit for a person to get a commercial driver's license.

The age at which an alleged criminal can be tried as an adult is less than 18 in most states; however, most states prevent a youth under 18 from being a party to a contract or other legal instrument. Does marriage or having children define adulthood? Does holding a full-time job? Are 18-year-old college students adults because they live in apartments away from their parents? Are 30-year-olds living at home with their parents not adults?

All this background is provided to illustrate how the word *adult* is defined by our culture. An ***adult*** is an individual who has reached the stage in life to be personally responsible for themselves and who has assumed a productive role in society. This definition complies with the legal and cultural definitions we often use for being an adult in the United States. In short, an adult is expected to gain income for support and to support themselves and possibly other individuals. Adults typically assume productive roles in society.

Educating Adults

Adult education is much like *adult*—it is hard to define. People think of many things when they hear the term *adult education*. ***Adult education*** is the process of providing learning

FIGURE 18.2 Field trips for firsthand observations are often used in adult agricultural education.

opportunities to improve adults in a wide range of areas of their lives. Many people associate adult education with adult basic education, such as instruction in English and mathematics in order to receive a certificate of high school equivalency, commonly called the letters GED.

Adult basic education (ABE) is instruction for adults in basic and specific skills needed to function in society. It is for individuals who did not achieve or did not participate in education as a child. Most often, the instruction focuses on reading, mathematics, and other basic areas. However, adult education is much broader than ABE and includes a wide range of workshops, seminars, classes, tours, self-study, and other means of improving the education of adults.

Adult agricultural education is education for individuals who are engaged in the ordinary business of life. They typically have jobs in an area of the agricultural industry. They may have families and assume many responsibilities with their families and the communities in which they live.

The Cooperative Extension Service, through nonformal education, offers workshops, seminars, and classes on a variety of topics. Most religious organizations have adult education programs in the tenets of the faith as well as life topics such as finance, relationships, and personal growth. Businesses and agencies conduct training for employees to develop new skills related to their work. Colleges sponsor continuing education classes and activities. Apprenticeship programs exist for many of the trades. Local high schools and career centers offer adults instruction in languages, computers, construction, horticulture, welding, and much more.

According to the National Center for Education Statistics (NCES, 2007), about 46% of American adults participate in one or more adult education activities. The NCES placed the activities into the following six categories:

1. English as a Second Language, which includes language instruction for learners whose first language is not English, as well as instruction in basic American culture
2. Adult basic education, which includes instruction designed to assist adults in obtaining a high school diploma or its equivalent
3. Vocational or technical diploma programs, which are formal postsecondary programs leading to a degree or diploma in an occupational field
4. Apprenticeship and formal training for trades occupations
5. Courses related to work and advancing in one's career
6. Personal development courses, with hobbies, religion, health, and sports being examples

In agricultural education, adult education has experienced periods when it has thrived and others when it has faded. In some states, agriculture teachers are employed to work exclusively or primarily with adults engaged in production agriculture. These teachers may work from local public secondary school systems or community/junior college systems. In other instances, high school agriculture teachers offer night classes in areas related to their expertise and facilities, such as welding, construction, horticulture, landscape management, or financial management. Young Farmers, FFA Alumni and Supporters, or collaborations with the Cooperative Extension Service or Farm Bureau are additional adult education avenues for agriculture teachers.

Postsecondary Education

Postsecondary education is typically defined as education offered in grades 13 and 14 at technical schools, community colleges, or junior colleges. In some cases, it includes college- and/or university-level education.

Education beyond high school may or may not have an "adult focus." Some postsecondary education is a continuation of high school education. It may be designed to prepare students for entering a skilled occupation or for gaining additional education. Other postsecondary education focuses on the needs of adults. The needs of the individuals served often depend on the philosophy of those in charge of planning the educational programs at the postsecondary school.

Continuing Education

Continuing education is part-time adult education. It is education beyond typical levels of formal educational attainment, such as high school or college. Continuing education also refers to education that enhances skills and keeps people's knowledge and skills current. For example, schoolteachers take continuing education often referred to as professional development. Veterinarians, agronomists, floral designers, and many others do also.

Continuing education units may be offered for participation in programs of continuing education. These units are not college credits but, in some cases, are used in certification or recertification much as college credits are used.

CHARACTERISTICS OF ADULTS IN EDUCATIONAL SETTINGS

Some adult educators feel it is wise to discuss characteristics of adults in educational settings rather than dwell on what makes an individual an adult. Malcolm Knowles, known as the father of adult education, stated individuals should be treated as adults educationally if they behave as adults by performing adult roles and their self-concept is that of an adult.

FIGURE 18.3 Continuing education usually has a specific, hands-on focus and provides training in skills that can be immediately used.

The following four broad categories of adult learner characteristics are used:

Educational—how adults learn
Motivational—why adults use adult education
Cultural, physical, medical—what limits or enhances adults' learning
Personal—what makes an adult learner different from a child

Educational Characteristics

Adults bring a wealth of experiences into the educational setting. The adult educator can draw on these experiences to make the learning more relevant.

Because of their careers and life responsibilities, adults tend to want the knowledge and skills they are learning to be immediately applicable and useful. They may prefer visual and hands-on learning as opposed to auditory and note-taking—in other words, learning by doing. An old saying in adult education circles is adult learners "vote with their feet." They want value and appropriate instruction, and if they do not perceive this, they do not return for the next class.

Motivational Characteristics

Adults are motivated to participate in education for both external and internal reasons. They are often motivated by a real, crisis-driven need, such as a lost job, career change, retirement, or financial problems. The motivation can also come from requirements such as certifications, job entry, or court orders. Some adults may avoid education because of fear, such as anxiety over technology (too far behind or overwhelmed); discomfort (age, intellect, efficacy, knowledge); or peer support (presence or lack of).

Adults may be motivated by something they want (e.g., doing a training program to get higher pay) or something they do not want (e.g., taking a defensive driving course to reduce the impact of a traffic violation). Internally, adults may be motivated by a desire to enrich themselves or to learn a new skill. They may be pursuing a dream or may be motivated to change careers. Perhaps they now have the time to learn a new hobby. They may demonstrate anxiety, fear, or curiosity for things such as using new technologies.

Cultural/Physical/Medical Characteristics

Adults tend to be independent and make their own choices, which can greatly enhance the educational environment. Adults, as well as children, bring many cultural aspects into the educational setting. These include language, socioeconomic status, religious and cultural norms, and gender roles.

Older adults may have special needs related to the loss of stamina, hearing, eyesight, memory, dexterity, and/or mobility that require adaptation of the learning environment and elimination of learning barriers. Adults may have medical needs or limitations that require retraining or rehabilitation.

Personal Characteristics

With adults, timing and convenience are important. Adults, as opposed to children, are more likely to have full-time jobs, their own children or other dependents, community commitments, social obligations, and time restraints. With certain age groups or socioeconomic groups, transportation may be an issue.

As opposed to children who have compulsory education, adults engage in education for various reasons. Some do so for the social aspects—a chance to meet other people. Others, maybe through retirement, have extra time on their hands or want to learn a hobby. For others, technological changes and the ever-expanding knowledge base may necessitate their seeking further education and training.

ORGANIZATION OF ADULT EDUCATION

The delivery of adult education has evolved over many years. Several important approaches have shaped how adult education is carried out in the United States, particularly as related to agriculture.

True (1929) provided a detailed history of agricultural education in the United States. Adult agricultural education in the United States can be traced back to the colonies. Fairs, primarily for the sale of agricultural products, were held as early as 1644.

In 1785 societies were formed to promote and improve agricultural production. The Philadelphia Society for Promoting Agriculture had as its purpose to "print memoirs, offer prizes for experiments, improvements, and agricultural essays, and encourage the establishment of other societies through the country" (True, 1929, p. 7). The South Carolina Society for Promoting and Improving Agriculture and Other Rural Concerns was also organized in 1785. This movement grew so that in 1852 the United States Agricultural Society was organized.

Lyceums and Chautauquas

In 1826, Josiah Holbrook founded the first American lyceum in Millbury, Massachusetts. *Lyceums* were to be societies where the arts, sciences, agriculture, and public issues were presented and discussed. "In 1831, about 900 towns had lyceums, and for the next 20 years most public lectures were delivered before such organizations" (True, 1929, p. 32). After the American Civil War, the lyceum movement faded.

FIGURE 18.4 The USDA complex in Washington, D.C., is the headquarters for a number of national efforts in adult education in agriculture.

About the time the lyceum movement diminished in importance, the ***Chautauqua movement*** began. A Methodist Episcopal camp meeting in Chautauqua, New York, had held a summer Sunday-school institute for several years. In 1874, secular as well as religious instruction was included. The institute was such a success that local Chautauquas sprang up over the country, providing lectures and entertainment. These continued until the middle 1920s with about a dozen sites offering Chautauquas today. The original Chautauqua site still hosts programs and draws thousands of summer visitors (Chautauqua Institution, 2021).

Cooperative Extension Service

Federal legislation designed to promote adult agricultural education was enacted in the early twentieth century. Extension work had been conducted in the state of New York as early as 1894. In the early 1900s, Seaman Knapp, known as the founder of farm demonstration work, was traveling the South, teaching better practices to farmers through demonstrations (Bailey, 1945). These and other events led to the Smith-Lever Act of 1914, which established the Cooperative Extension Service. The act set the purpose of agricultural cooperative extension work as "the giving of instruction and practical demonstrations in agriculture and home economics . . . through field demonstrations, publications, and otherwise" (True, 1929, p. 288).

Public School Adult Agricultural Education

Public school–based adult agricultural education was recognized in the Smith-Hughes Act of 1917. The Vocational Education Act of 1963 broadened the scope of adult agricultural education to include nonfarm occupations in areas such as horticulture; agricultural mechanics; agricultural business, processing, sales, and services; and natural resources. The Carl D. Perkins Vocational Education Act of 1984 placed a greater emphasis on the education of adults. It focused federal funding on training and retraining adults, individuals who are single parents, those with limited English proficiency, those with disadvantages or disabilities, and men or women entering nontraditional occupations. The Perkins Act was amended in 1990, 1998, and 2006. The Carl D. Perkins Vocational and Technical Education Act of 1998 emphasized developing not only vocational and technical skills of secondary and postsecondary students but their academic skills as well. In 2018, the Strengthening Career and Technical Education for the 21st Century Act, also known as Perkins V, passed. Perkins V has a greater emphasis on addressing the needs of underserved audiences.

POSTSECONDARY PROGRAMS OF STUDY

Adult agricultural education on the postsecondary (after high school) level can be categorized into three areas: programs leading to an associate's degree, programs leading to transfer to a four-year baccalaureate degree, and short-term programs leading to certification. These programs tend to be offered at community or junior colleges; however, some four-year colleges and universities also have postsecondary programs. Many community and junior colleges tend to be for commuting students. Residence facilities may or may not be provided. The campuses easily accommodate both part-time and full-time students.

Two-year associate's degree programs concentrate instruction on specialized courses designed to prepare graduates for specific careers. Instruction tends to be career-focused and hands-on. Classes may be smaller than in typical university first-year and second-year courses. Many of these programs have internship requirements students complete within their areas of interest. Admission requirements tend to be more flexible than those for baccalaureate degree admission. Associate's degree students may also take some courses along with baccalaureate degree–seeking students. Individual institutions and states have guidelines for how many, if any, of the courses taken in these programs can transfer into a four-year program.

FIGURE 18.5
Adults like to be recognized for their educational accomplishments.

Career preparation programs may have agreements with major employers that provide facilities, up-to-date equipment, and internship opportunities.

Junior and community colleges also provide courses for students to transfer into baccalaureate degree programs at state and land-grant universities. These courses may include general education courses taken by all first- and second-year college or university students, such as English, mathematics, humanities, and basic sciences. They can also include basic lower-level courses in animal science, agronomy, or agricultural economics that are applicable to several majors. The junior or community college may have an articulation agreement with four-year colleges and universities to assure students courses taken will transfer and count toward degree requirements.

Junior and community colleges typically also offer courses for workplace training or retraining, personal growth, and lifelong learning. Some of these may be intensive, lasting less than one year, and lead to certifications. An employer may work directly with a junior or community college to tailor a course or series of courses specifically for that employer.

ADULT EDUCATION IN SECONDARY PROGRAMS

Secondary public schools may offer adult agricultural education in both career centers and comprehensive high schools. The instruction may be career-focused or interest-based.

Funding Programs

Funding for adult education on this level may take several forms. Secondary teachers may teach part-time adult agricultural education in addition to their in-school classes. In most cases, such agriculture teachers will receive a specific salary supplement for teaching adults in addition to their secondary student teaching load. This supplement may come from state funding, local funding, or user fees. Some states, such as Georgia, employ agriculture teachers specifically for adult instruction. Their salaries may be funded through the school systems or be supported by user fees. Other states, such as Connecticut, include an adult education component in the program assessment for state funding.

Nature of Instruction

The majority of adult instruction on this level is short term and topic specific. The adult educator must balance the time required to meet the instructional objectives adequately with the ability and interest of adults to devote time to the class.

Many agriculture teachers have found the winter months are a time of year when adults are more willing to participate in adult educational activities. Also, if classes can be structured to occur just before application is required, adults are more likely to participate. For example, classes on gardening, flowers, and landscaping are best conducted just before the spring gardening season begins.

Role of Secondary Agriculture Teachers

Should secondary agriculture teachers teach adults? This question is more relevant now than at any time in the past 20 years. With increasing class size in secondary classes, greater emphasis being placed on middle school agricultural education, and increased accountability pressures, do agriculture teachers have the time and resources to conduct quality adult education programs?

Many would argue that by engaging adults in agricultural education, greater support will be built for the secondary program. This could lead to increased resources or even the hiring of additional agriculture teachers. Others, however, espouse the view that agriculture teachers are employed to teach secondary students first and foremost. They believe anything that takes away from this focus is a detriment to the secondary students. Others point out the time commitments of agriculture teachers and teaching adults on nights and weekends takes away from personal/family time.

This debate must be resolved at the local level. The optimum adult education program is organized to meet the needs of the community, using the resources available, with benefit to both the adult learners and the agriculture teacher.

LAWS OF LEARNING FOR ADULT EDUCATION

Kahler et al. (1985) summarized the following eight laws of learning for adult education. These laws are based on psychological and educational research and have been confirmed by recent brain-based research. The laws provide a foundation for supporting adults' learning and meeting their interests and needs.

The desire to learn   This law states the learner must be motivated for effective learning to take place. The interest may stem from the learner wanting to delve into the subject or from the need to gain new knowledge. The desire to learn can be influenced by the teacher and can spread from one learner to others.

Understanding of the task   Learners need to know what is expected of them. This includes knowing the goals and objectives of the instruction, the degree to which performance is expected, and the activities required to master the task. The instructor needs to structure the learning situation so the task and content are at the educational level of the learner.

The law of exercise   Practice, practice, practice! Adults need the opportunity to learn by doing. This can occur through experiments, problem solving, or even home or workplace implementation of what was learned in the classroom. Breaking up the repetition cycle with periods of reflection or analysis tends to increase learning effectiveness. This law is closely associated with the principle that tasks not performed on a regular basis or information not used is forgotten.

The law of effect   This law states learning is affected by whether something provides satisfaction or annoyance. Results that are satisfying tend to strengthen learning, whereas results that are annoying tend to weaken learning. This law can be applied in the adult education classroom by making instruction interesting, having a comfortable environment, and structuring learning so learners see progress in their work.

The law of association   Learning is enhanced when new information can be associated with previous experiences or grouped with similar information. The law of effect is important here, as an adult educator should draw on learners' life and work experiences that resulted in satisfaction to form connections with new content.

Interest, vividness, and intensity   The greater the vividness and intensity of something, the more likely the learner is to show interest in and retain the learning. Vividness can be enhanced through the use of color, smells, emotions, movement, and clarity of words.

Intensity can be obtained through unexpected results, shock, emotions, tasks requiring concentration, and the finding of learners' passions.

Mindset   Learners bring with them a mindset that can either block learning or ease learning. Many adults have memories of past educational failures, fears of less-than-acceptable performance, prejudices, and biases. All these can hinder current learning. The adult educational environment needs to be one in which the learner has the freedom to explore new ideas and ways of thinking. The educational setting needs to be inviting and open.

Knowledge of success and failure   Learners need feedback to grow and develop. The adult educator provides feedback, but the learner also needs to be taught to conduct self-evaluations. Adult educators must be careful to provide any negative feedback in such a way that it promotes growth rather than discouragement.

METHODS OF TEACHING ADULTS

With some adaptation, many of the same teaching approaches and methods used with youth are also used with adults. However, certain teaching approaches work specifically with adults. Malcolm Knowles (1989) used the term *andragogy* to describe characteristics of adult education.

Andragogy is the study of processes and practices in adult education. It is based on the needs and readiness of the adult learners. As such, the instruction is more learner-focused, collaborative, and problem-centered than traditional pedagogy. The instruction takes advantage of learner experiences and recognizes adults need to know the purpose of the learning and have responsibility for the learning.

How does the adult educator decide what teaching methods to use? Through knowledge of the characteristics of adults and the laws of learning, the adult educator can best organize instruction to meet student needs. However, before the teaching method is selected, the first determination is deciding what should be taught.

Getting Organized

As detailed in Chapter 6, advisory committees and citizen groups are one means of determining subject matter, recruiting adult students, and otherwise assisting with an adult program. Advisory committees are to provide guidance regarding the total agricultural education program, which includes the adult education component. Figure 18.6 shows a community survey, another means that can be used to gather information on adult educational needs and interests. The survey can be placed online through a website, social media site, or emailed (if the school internet use policy allows electronic surveys), mailed to randomly selected community members, mailed to selected persons who have previously expressed interest in the program, placed in the community newspaper, or placed in selected local businesses. Depending on the outlet chosen, the survey will need to be modified appropriately.

Once the class topics have been mutually determined, appropriate teaching methodologies can be selected. Chapters 7 and 12 will be useful when selecting teaching methodologies. If the objective of the topic is on a knowledge/cognitive level, then some form of lecture supplemented with instructional materials, visuals, demonstrations, and group discussion is appropriate. You, as the adult educator, do not have to be the expert on every topic. Utilize resource persons, such as extension educators and local persons experienced in the topic, as

---

Adult Education Interest Survey

The Slate Valley Horticulture Department would like to offer adult education classes to the community. We want to plan programs that meet your interests and needs. Please take a few moments to complete this survey and return it by November 15. Thank you!

Rank your interest in learning about these topics from 1 to 5, with 1 being the topic you would most want to learn about and 5 being the least.

_____ Arranging cut flowers
_____ Exploring commercial greenhouse production
_____ Growing indoor plants
_____ Growing roses, shrubs, and ornamentals
_____ Using a small, backyard greenhouse

If the classes you ranked number 1 and number 2 were offered, what is the likelihood you would take the classes this January and February? (Check one only, please.)

_____ Definitely
_____ Probably
_____ Maybe
_____ No

Each class is designed for 8 to 10 sessions of 2 hours each. Please indicate the class structure that best fits your schedule. (Check one only, please.)

_____ 2 nights/week for 4–5 weeks
_____ 1 night/week for 8–10 weeks
_____ Saturdays 8 a.m. to noon for 4–5 weeks

FIGURE 18.6 Adult education interest survey form.

guest presenters. Give the adult learners frequent opportunities to apply the knowledge to real-life situations through scenarios, case studies, or discussions of problems experienced by class members.

Teaching Strategies

Adults respond best if active teaching strategies are used that show the usefulness of the information being taught. They want teachers to be well prepared and use formal methods of teaching, though often in informal ways.

Presentations are often used to provide new information. The presentations may include realia to show the relationships, step-by-step procedures, or comparisons. Resource persons are used to assure information is current, valid, and reliable.

Discussions are used to help assimilate the information and solve problems. Discussions should be carefully guided so students stay on the topic. A presentation may be followed with discussion time to clarify points and help class members assimilate the information.

Field trips and tours are used to introduce new technologies and practices. Planning to assure maximum learning is essential. Such events should be scheduled at times when adults are available.

When topics on the psychomotor level are involved, adults want to put them into action as soon as possible. Demonstrations followed by guided practice, experiments, or exercises are appropriate. Frequent feedback, including instructor evaluations and self-evaluations, is critical for the learners to develop their skills and achieve progress. Laboratory activities are used to develop specific skills. These are useful with some adults, especially those who want to learn the skills and who will find the new skills useful.

Newer technologies available through computers and the internet are appropriate with some adults. Some background instruction in computers and the internet may be needed.

STUDENT ORGANIZATIONS

Adult students in agricultural education are served by student organizations much as students at the secondary level are served by FFA.

NYFEA

The *National Young Farmer Education Association* (NYFEA) is the official adult student organization for agricultural education as recognized by the U.S. Department of Education. Though the initial focus was on younger individuals or those just beginning agricultural careers, the Young Farmers serves all ages at the local, state, and national levels. NYFEA is the organization that provides for the coordination of programs, the delivery of educational opportunities, and the recognition of members on a national level. NYFEA members are engaged in leadership development, business skill development, and community service. As stated in its constitution, NYFEA has as its mission to "educate agricultural leaders through leadership training, agricultural education and community service opportunities. The association seeks to provide support for the next generation by helping them understand the opportunities found in agriculture" (NYFEA, 2013).

Young Farmers clubs in Ohio began as early as 1927 (Phipps & Osborne, 1988). Over the years more clubs started and state associations were organized. These state associations collaborated on Winter Institutes, where host states invited Young Farmers from around the country to participate in educational seminars, take part in field trips to agricultural and cultural components of the host states, and socialize. Young Farmers clubs, activities, and state collaborations grew such that in 1982 NYFEA was incorporated.

About two thirds of the states have Young Farmers associations and affiliation with NYFEA. NYFEA sponsors two major national meetings per year. The NYFEA Institute continues the tradition of the Winter Institutes that have been held across the United States. Institutes feature tours, educational seminars/speakers, competitions, and entertainment.

A local Young Farmers chapter should be organized to meet the educational, social, and community service needs of its members. The agriculture teacher is the advisor and facilitator for the Young Farmers chapter. The Young Farmers chapter should not take an inordinate amount of the agriculture teacher's time. As adults, Young Farmers are able to organize and conduct most of their activities on their own initiative.

PAS

The *National Professional Agricultural Student Organization* (PAS) is one of the 11 student organizations approved by the U.S. Department of Education as integral parts of career and technical education. PAS (for students beyond high school) is an organization associated with agriculture, agribusiness, and natural resources in approved postsecondary institutions offering associate or baccalaureate degrees or vocational diplomas or certificates.

Draft bylaws for a national Post-Secondary Agriculture Students organization were formulated in 1979. PAS was officially founded in March 1980 in Kansas City, Missouri. The name changed in 2019. PAS has membership in chapters located in almost 20 states. It is available to students in postsecondary programs of agriculture/agribusiness/natural resources in approximately 550 institutions in all 50 states.

The motto of PAS is "Uniting Education and Industry in Agriculture" (PAS, n.d.). PAS provides opportunities for its members in individual growth, leadership, and career preparation. Activities are organized around these three areas. At the organization's national meetings, PAS members compete in the various career and leadership areas.

FFA Alumni and Supporters

National FFA Alumni and Supporters is affiliated with the National FFA Organization (National FFA Organization, 2021). It serves as one of the four types of membership in FFA. FFA Alumni and Supporters is organized at local, state, and national levels. It involves adults in educational ways that support FFA. (More about the National FFA Alumni Association is presented in Chapter 6 of this book.)

REVIEWING SUMMARY

Adult education in agriculture is an important component of a total agricultural education program. It takes many forms. It ranges from formal degree programs at colleges and universities to topic-specific presentations by the local agriculture teacher.

A clear definition of when a person becomes an adult is difficult to obtain. However, there are educational characteristics that determine when a learner should be treated as an adult. Although some of these characteristics can also be true of youth, they most appropriately apply to adults.

The eight laws of learning for adult education should be used to guide educators in structuring learning environments for adults. Selecting subject matter to teach adults should be a collaborative decision involving the teacher, the adult students, and the community. The agricultural education advisory committee, along with data obtained from community surveys, should provide input into the decision. Teaching methods should match the subject matter and objectives of the instruction.

Organizations such as NYFEA, PAS, and FFA Alumni and Supporters are able to assist agriculture teachers in organizing and delivering instruction to adults. NYFEA is primarily designed for out-of-school adults, whereas PAS is designed for students of postsecondary agricultural programs. Both organizations are recognized and supported by the U.S. Department of Education.

QUESTIONS FOR REVIEW AND DISCUSSION

1. How do adults learn, and how does this differ from how youth learn?
2. Who is an adult? Defend your answer.
3. What today serves the role of lyceums and Chautauquas?
4. Educationally, are college students more like adults or youth?
5. How did federal legislation influence adult agricultural education in the early 1900s?
6. How have the various Perkins Acts made changes for adult agricultural education?
7. What is the definition of postsecondary education?
8. How is adult agricultural education funded on the secondary level?
9. What does *andragogy* mean?
10. How are the eight laws of learning for adult education used in agricultural education?
11. What are the differences between the two student organizations for adult agricultural education?
12. What is continuing education?

ACTIVITIES

1. Investigate adult agricultural education in your state. Topics to cover include Young Farmers, PAS, and postsecondary agricultural colleges. Are secondary agriculture teachers paid for adult education, and if so, how?

2. Investigate current federal legislation regarding adult agricultural education. In addition to the latest Perkins Act, investigate acts for secondary and higher education.

3. Using the *Journal of Agricultural Education*, *The Agricultural Education Magazine*, and the proceedings of the National Conference of the American Association for Agricultural Education, report on research about adult agricultural education. Current and past issues of these journals can be found in the university library. Selected past issues can be found online at www.aaaeonline.org or www.naae.org.

4. Review the book entitled *Effective Adult Learning* (Birkenholz et al., 1999). Note the attributes of adults as learners and how educational programs are planned and delivered to maximize adult learning. Prepare a written report on your observations. Make a presentation in class.

REFERENCES

Bailey, J. C. (1945). *Seaman A. Knapp, schoolmaster of American agriculture.* Columbia University Press.

Birkenholz, R. J., Garton, B. L., Harbstreit, S. R., & Miller, W. W. (1999). *Effective adult learning.* Interstate Publishers.

Chautauqua Institution. (2021). Website. https://chq.org/

Kahler, A. A., Morgan, B., Holmes, G. E., & Bundy, C. E. (1985). *Methods in adult education* (4th ed.). Interstate Printers & Publishers.

Knowles, M. S. (1989). *The making of an adult educator: An autobiographical journey.* Jossey-Bass.

National Center for Education Statistics. (2007). *The condition of education 2007 (NCES 2007-064).* https://nces.ed.gov/programs/coe/

National Council for Agricultural Education. (2016). *National quality program standards for agriculture, food, and natural resources education: A tool for secondary (grades 9–12) programs.* https://thecouncil.ffa.org/

National FFA Organization. (2021). *Official FFA manual.* https://www.ffa.org

National PAS. (n.d.). *PAS.* https://www.nationalpas.org

National Young Farmer Educational Association (NYFEA). (2013). *NYFEA.* http://nyfea.org

Phipps, L. J., & Osborne, E. W. (1988). *Handbook on agricultural education in public schools* (5th ed.). Interstate Printers & Publishers.

True, A. C. (1929). *A history of agricultural education in the United States, 1785–1925.* U.S. Department of Agriculture, Miscellaneous Publication No. 36.

19

Evaluating Learning

Mrs. Thomas, agriculture teacher at Rose Hill High School, was at her desk grading a recent unit test on plant science. She was disappointed in the results as more than one-third of the class did not pass the test. Most students did well on the plant physiology questions and the plant nutrients section, but failed miserably on the plant diseases section of the test.

She thought back to the lessons on plant pests and their control. She seemed to recall students struggled with identifying the symptoms of plant diseases. Mrs. Thomas sat back in her chair and gazed out the window at the Agriculture Department's plant nursery adjacent to the classroom.

"I am going to have to reteach the plant disease lesson," she thought to herself. "And that is going to put me behind on the curriculum for this upcoming grading period. I could kick myself for not conducting more checks for understanding with students to see if they grasped the course material on plant diseases. I should have included more formative assessments in the lessons to gauge student progress at learning the concepts." Mrs. Thomas got up from her desk, exited the classroom, and walked out into the plant nursery. She noted some of the plants had developed symptoms of insect infestation and made a mental note to have the advanced horticulture class handle the problem in the morning. A sudden thought hit her.

"That's it!" she thought, "As part of the formative assessments, I'll provide students with the real-life experience of diagnosing and treating plant insects and diseases and measure their progress as they work through the diagnoses and treatments. A practicum would be a great form of authentic assessment."

With that, Mrs. Thomas returned to her desk in the classroom. As she began planning to reteach the plant diseases lessons, she thought, "Next time, I will use more formative assessments to save time, improve the quality of my teaching, and increase the chance of student success."

TERMS

achievement evaluation
authentic assessment
bell-shaped curve
criterion-referenced test
evaluation
forced-answer question
formative evaluation
grade
mastery learning

norm-referenced test
open-ended answer
performance-based evaluation
portfolio
pretest
product scale
scoring rubric
summative evaluation
table of specifications

OBJECTIVES

This chapter addresses the National Quality Program Standards for Agriculture, Food, and Natural Resources Education (National Council for Agricultural Education, 2016), specifically Standard 1D: Program Design and Instruction—Assessment. It has the following objectives:

1. Explain the meaning of evaluation, including formative and summative evaluation.
2. Define achievement evaluation.
3. Describe how to use performance-based evaluation in the classroom.
4. Prepare appropriate assessment instruments and test items.
5. Explain the use of alternative assessment and identify approaches to use.
6. Describe the importance of grading and identify proper grading practices.

THE MEANING OF EVALUATION

Evaluation is a process to analyze educational effectiveness (student achievement) by using measurement tools. These tools include quizzes/tests, observations, performances, and portfolios.

Evaluation serves multiple purposes. It can (1) evaluate academic achievement of individual students, (2) diagnose academic problems or deficiencies of individual students or classes, (3) evaluate the effectiveness of a curriculum or other educational product, and (4) evaluate the educational progress of large groups to inform public policy makers. Although evaluation is frequently teacher-driven, an overarching goal of the evaluation should be to hold students accountable for their learning. In this respect, evaluation is a feedback tool to assist students in advancing their education (Supovitz, 2012).

Evaluation occurring before or during instruction is called *formative evaluation*. The purpose of formative evaluation is to determine the starting point for teaching and give students and teachers feedback during instruction. Formative feedback helps teachers modify lessons to address deficiencies or problems.

Evaluation occurring after instruction is called *summative evaluation*. The purpose of summative evaluation is to determine how well students achieved and, in many cases, to provide a grade. Summative evaluation can give feedback to determine whether or not a student mastered the learning objectives.

As shown in Figure 19.1, evaluation is just one of many educational interventions for facilitating student learning. In this respect, evaluation becomes one of several tools the agriculture teacher uses to assist students in their education.

FIGURE 19.1 Education interventions for student learning.

Using Bloom's Taxonomy to Evaluate Learning

Achievement evaluation is the practice of measuring how much students know about a subject or how well they can use that knowledge. Tests to measure student achievement can be standardized, such as state competency tests or teacher-made. These can be paper-and-pencil tests or electronically delivered tests.

A *pretest* is a test given before instruction begins. It determines a student's readiness to learn in the subject area. Knowledge or skills in which students already have competence can be either lightly reviewed or omitted altogether. Achievement evaluations are norm-referenced or criterion-referenced.

Norm-Referenced Testing

A *norm-referenced test* is a test that compares the achievement of an individual student with an established group norm. Test preparers create norm-referenced tests so most individuals score in the average range and a few in the high and low ranges. Norm-referenced tests yield a *bell-shaped curve* in which student scores are distributed along a line in a bell shape. It is a graphic depiction of how a large number of scores would be allocated. The bell-shaped curve (see Figure 19.2) is known as a normal distribution curve. Norm-referenced tests compare individual test scores with those of the group and with other individuals.

Very few standardized tests have existed in agricultural education at the national level. Considering the diversity in agricultural regions and agricultural practices, it is difficult to determine what all agricultural education students should know. Some states have developed standardized tests for agricultural courses taught in those states.

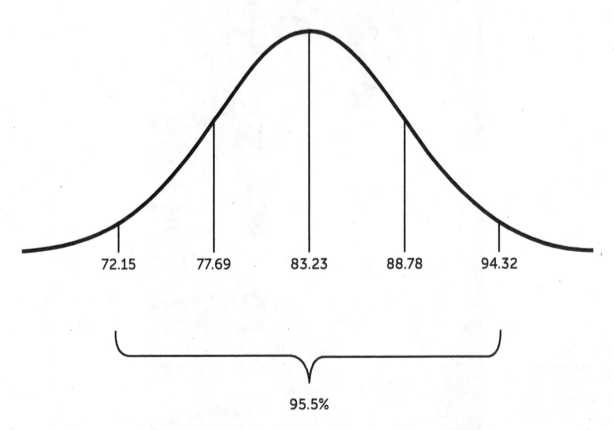

FIGURE 19.2 Bell-shaped curve.

Criterion-Referenced Testing

A *criterion-referenced test* is a test that compares the score of an individual student with established criteria or standards. This type of test determines how students' scores compare with established criteria rather than how they compare with the scores of other students.

Mastery learning is often associated with criterion-referenced tests. In *mastery learning*, students are given frequent opportunities to practice and multiple opportunities to improve their scores. An agriculture teacher who requires 100% correct on a safety test is using a criterion-referenced test with a mastery approach. An agriculture teacher who uses well-written learning objectives, as described in Chapter 12, to teach the content and develop the evaluation uses a criterion-referenced system. Agriculture teachers are typically concerned with knowing what students have learned rather than how students compare.

Mastery Learning

The *performance-based evaluation* uses approaches that measure how well students can do something under given conditions, such as weld a flat bead on a piece of mild steel. Typically, students demonstrate a skill or process, construct a product, or otherwise provide tangible evidence of their learning. Teachers observe and judge the evidence using tools such as rubrics, checklists, rating scales, or product scales.

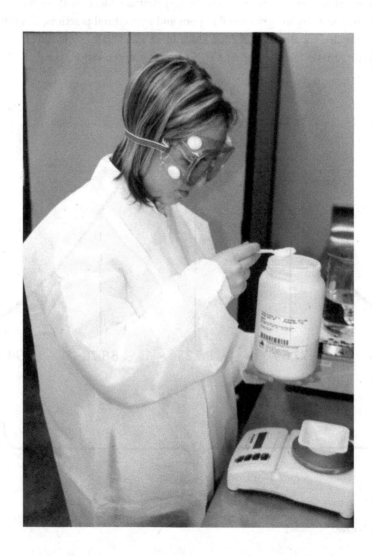

FIGURE 19.3 The ability of a student to follow the steps of the scientific method is an indication of performance. (COURTESY OF EDUCATION IMAGES.)

USING PERFORMANCE-BASED EVALUATION

Performance-based evaluation is perceived to have several advantages over achievement evaluation. These include clarity, authenticity, subjectivity, allowance for learner differences, and learners' engagement (Tanner, 2001).

Under performance-based evaluation, students have a more excellent knowledge of the level of performance required for success. This type of evaluation also has more meaning for the students in the real world. Although subjectivity can be a detriment, performance-based evaluation allows teachers to judge student performance in areas of effort, improvement, timeliness, and care. Performance-based evaluation can be tailored for an individual student, accounting for learner differences and ideally engaging the student in the assessment, making the evaluation a part of the overall learning process.

Agriculture teachers have numerous opportunities to use performance-based evaluation. Most laboratory situations require a teacher to evaluate how well students can apply and demonstrate what they know. Career-directed instruction lends itself to performance-based assessment. Supervised agricultural experience (SAE) and FFA Career Development Events are other examples of performance-based assessments.

Authentic assessment is a performance-based assessment that places the student in a situation or simulation that closely mimics the workplace or other life setting. In many career and technical education disciplines, the agricultural classroom and lab match the workplace setting as closely as possible. The tools, equipment, and tasks completed all model those of the workplace. Most supervised experience carries this a step further and places the student into the actual workplace.

Reproducing actual workplaces may not be practical. Schools may not have the budgets to purchase or build replicas of workplaces. Besides, innovative systems become obsolete quickly. Regardless, authentic assessment is still practical. Simulations, such as computer programs, models, and learning stations, can provide enough realism to evaluate students adequately.

FIGURE 19.4 The quality of plants produced by students in the school greenhouse is an indication of proficiency. (COURTESY OF EDUCATION IMAGES.)

DEVELOPING ASSESSMENT INSTRUMENTS AND TEST ITEMS

The purpose of assessment instruments is to measure the knowledge, skills, and dispositions of students accurately. A well-designed assessment instrument provides the agriculture teacher vital information regarding student learning in a particular subject area. However, assessment instruments can also measure other student variables. For example, sleepiness, hunger, bad mood, or unpleasant home life can negatively affect students' performance. And yet, there are deeper issues that may cause problems with student achievement

Tests may be prepared "on paper" or administered with a computer. Computer tests may be teacher-made or obtained from a provider. Some state departments of education prepare tests for all students enrolled in particular classes. Commercially prepared CD-based and online tests are available.

Students with lower than grade-level reading, writing, math, or thinking skills tend to have difficulty with assessments regardless of their knowledge of the subject area. Students may also be unmotivated toward the subject matter or negatively toward the class, the school, the teacher, or themselves. Educational problems such as poor study habits or failure to be attentive during instruction, practice, and review can also negatively affect students' assessment performance (Walvoord & Anderson, 2011). Phipps and Osborne (1988) describe a combination of student factors and instructional factors that may cause problems in student achievement. (See Table 19.1.)

When developing the assessment instrument, the agriculture teacher should consider the level of objectives and instruction. Particular evaluation techniques are best for particular groups of learning objectives. For example, essay questions are better than true/false questions for synthesis-level objectives. Teachers consider the nature of the subject matter when selecting evaluation techniques. For example, the proper welding of a joint is best evaluated through the performance of the task, not through a paper-and-pencil test.

FIGURE 19.5 Reviewing test performance with a student is one method for diagnosing problems with academic performance.

TABLE 19.1

Possible Causes of Problems in Student Achievement

Level One Causes	Level Two Causes
Inappropriate evaluation techniques	The type of evaluation is not parallel to objectives and instruction. The type of evaluation is not suited to the subject matter.
Poorly designed evaluation	Evaluation is graded unfairly. Questions are unclear or inappropriate. Evaluation questions are not parallel to objectives and instruction.
Lack of student understanding and skill	Objectives are not clear to students. Student motivation is low. Students do not have the background knowledge or skills to perform as expected. Students are not mentally or physically capable of performing at expected levels. Teaching is ineffective.

Adapted from Phipps & Osborne (1988).

Teachers must grade evaluation instruments without bias. Teachers must take care to construct questions for clarity and appropriateness. A lesson taught using knowledge- and comprehension-level objectives should have an assessment instrument written to test students on those levels.

The teacher must explain learning expectations to students. It is sometimes difficult to determine whether poor student performance results from causes stemming from the teacher's performance or the student's performance.

Table of Specifications

Teachers develop evaluation instruments weekly, if not daily. Examples of written instruments include quizzes, tests, and exams. Teachers need to ensure the evaluation instruments adequately and accurately address the amount and level of the subject matter reflected in the objectives and the instruction. One of the challenges teachers face is designing a test that adequately covers the scope of the material the students have learned. A *table of specifications* is a valuable tool for helping teachers accomplish this task. This table, or chart, is designed to determine how well test items align with lesson or unit objectives. Table 19.2 shows an example of a table of specifications, with an added section on how the teacher

TABLE 19.2

Table of Specifications for a 40-Item Test on Basic Animal Science

Instructional Objectives	Percentage (%) of Class Time on the Topic	Number of Test Items	Number of Questions at Levels of Bloom's Taxonomy—Cognitive Domain*					
			1	2	3	4	5	6
Nutrient needs of animals	20	8	2	2	3	1		
Animal health	20	8		3	4	1		
Breeds of livestock	15	6	4	2				
Basic genetics	20	8			3	2	1	2
Animal facilities and housing	25	10	2		3	2	3	
Total	*100*	*40*	*8*	*7*	*13*	*6*	*4*	*2*

*1 = knowledge, 2 = comprehension, 3 = application, 4 = analysis, 5 = synthesis, 6 = evaluation.

has distributed questions among the cognitive domain of Bloom's taxonomy. By analyzing the percentage of time spent in class for each objective, a teacher can determine how many questions should be devoted to each objective (Fives & Barnes, 2021).

Forced-Answer Question

A common type of test question is the *forced-answer question*, which requires the student to select the correct answer from given choices. Forced-answer questions include multiple choice, true/false, and matching.

Multiple Choice

Multiple-choice test items are the most frequently used question. Good multiple-choice questions can measure low-level to high-level objectives, be as effective as essay questions, and measure a specific content focus (Tanner, 2001).

The test designer constructs a multiple-choice question with a stem followed by four or five choices. Only one option is the correct or best answer; the rest are called distracters. The test designer writes the stem as a simple, straightforward sentence containing one idea. Teachers should avoid negative wording in the stem. The stem should not have vague terms like *sometimes, usually,* and *may,* or absolute terms, such as *always, never, none,* and *only.*

Writing good distracters is the key to an excellent multiple-choice question. Teachers should write distracters so students who have a general sense of the topic might think the distractors are plausible answers. Distracters should be logical, credible, and even attractive to those students. Distracters should be similar to the correct response in terms of length, grammar, and complexity. They should not give clues to correct answers to other questions. Occasionally, it is all right to make a distracter humorous, especially at the beginning of a test, to relax students.

Using the phrases "all of the above" and "none of the above" as choices should usually be avoided. Using these phrases as distracters is not effective if one of them is never the correct choice. On the other hand, if students can easily guess when one of these phrases is the right answer, the question loses its ability to discriminate between students who know the answer to the question and those who do not. Teachers also need to avoid patterns, such as the correct answer being choice "c" more than any other choice (Davis, 1993).

Multiple-Choice Example

1. Two taps of the chairperson's gavel
   a. signal members to be seated.
   b. signal members to stand up.
   c. announce a vote decision.
   d. call a meeting to order.

True/False

True/false questions have many advantages. They are easy to construct and grade. They are objective as a given question is either correct or incorrect. True/false questions require less reading than multiple-choice questions, so the test appears shorter. However, true/false questions also have disadvantages. Students are prone to guess if they do not know an answer since they have a 50% chance of being correct. Poorly constructed questions that are ambiguous or contain more than one idea can lead to student confusion and to arguments about whether a question is graded correctly. Also, poorly constructed questions may give too many clues, allowing students to answer the questions correctly without really knowing the content. The teacher may prepare true/false questions based on the recall of facts, as these questions are the easiest to write.

Every true/false item should have a statement sentence about one important concept. True and false statements should be the same length, and patterns should be avoided, such as three true followed by three false items. In general, the test should have an equal number of true and false or a greater number of false items. To indicate the choice of answer, a student should either circle a preprinted "T" or "F" or write "true" or "false" in a blank.

There are two standard methods to use for reducing the effect of guessing on true/false items. The first method, a formula, computes the student score by subtracting incorrect responses from correct responses (S = C – I). Items left blank do not count against the student, so it is better not to guess if they are not sure of an answer. The second method rewards the student for explaining why an item is false. For each item the student determines is incorrect, they write a statement telling why it is wrong or what would make the statement true. This second method requires more grading time for the teacher but encourages the student to think more deeply about the item.

True/False Example

T   F   1. To amend the main motion, a member rises, gains recognition from the chair, and says, "I wish to amend the motion by . . ."

Matching

Matching questions are popular on tests for evaluating students on lists of related information. They are easy for the teacher to write and quick for the students to answer. Matching questions do not take up much space on a page, so the test appears shorter than a multiple-choice test.

Matching does have some disadvantages. Matching is best suited for knowledge-level objectives, such as recall, define, and label. Instructions are more critical for matching than for other types of questions. Will answers be used more than once? Will there be more than one correct response to an item? Will some solutions not be used?

The following guidelines should help construct matching items: Place the words on the left and the list with the definitions or descriptions on the right. Each set of matching items should deal with a single concept. One list should have choices that will not be used so students will not be able to guess the last answer by the process of elimination. If possible, put one of the lists in alphabetical or chronological order or some other logical pattern. Place the entire matching set on the same page. The matching set should have between 5 and 12 items.

Matching Example

Match the FFA office to the emblem at the officer's station in the FFA Opening Ceremony.

_____ 1. emblem of Washington	a.	Parliamentarian
_____ 2. ear of corn	b.	President
_____ 3. American flag	c.	Reporter
_____ 4. handclasp of friendship	d.	Secretary
_____ 5. plow	e.	Sentinel
_____ 6. rising sun	f.	Treasurer
	g.	Vice President

Open-Ended Answer

Another common type of test question is the open-ended-answer question. The *open-ended-answer question* is the kind of item that requires the student to develop their answer to the question. Open-ended-answer questions include fill in the blank, short answer, and essay.

Fill in the Blank

Fill-in-the-blank questions are similar to the forced-answer questions described in the previous sections, except the student supplies the word or words to make each statement correct. Because of this, guessing is reduced or eliminated. Questions of this type are easy to write but rarely measure above the recall level. They are best suited for definitions, concepts, and critical terms. Fill-in-the-blank questions tend to be the most problematic of all the types. Although the teacher may believe a question is worded correctly, students will invariably give multiple technically correct answers, yet not what the teacher expected. The teacher also must decide whether to accept misspellings and synonyms. Teachers can eliminate some of these problems by providing the students a list of words or terms to be used. However, this adds to the difficulties of student guessing, extra test length, and the supplying of clues for the answers.

These guidelines should help construct fill-in-the-blank items: Construct each statement so the blank is at the end of the sentence. If too many words are left out, the statement loses its meaning, and students are confused. There is no need to make the length of the blank fit the size of the answer or put the same number of blanks as words in the response. It is necessary only to provide enough space for the student to write in the longest correct answer. Avoid using statements directly from textbooks or other curriculum materials. Doing this results in the statement being out of context and ambiguous.

Fill-in-the-Blank Example

1. The type of motion used to provide proper and fair treatment to all members is called _____.

Short Answer

Short-answer questions are similar to fill-in-the-blank questions but typically require slightly more writing and thought for the student. Teachers should design these questions to test a student's knowledge of definitions, series, names, and relationships. Short-answer questions allow the student more freedom in type and length of response.

Just as short-answer questions have advantages similar to those of fill-in-the-blank questions, they also have similar disadvantages. The teacher must decide how to grade misspellings, partially correct answers, and technically correct solutions. The teacher must also give clear directions regarding how much to write. Does the size of the white space indicate the expected response? Are one-word answers sufficient, or are explanations necessary? Box 19.1 lists guidelines a teacher should use to develop written tests.

Short-Answer Example

1. List the three ways to amend a motion and give an example of each.

Essay

The strength of essay questions exists in their capability to measure levels of objectives from simple to complex. This characteristic can also be a weakness. A teacher may write an essay question expecting to measure students' higher-order thinking abilities and receive answers

---

BOX 19.1 *Valuable Guidelines in Preparing a Teacher-Made Written Test*

- Cover in the test only what has been taught based on educational objectives.
- Provide instructions on how to complete the test.
- Use wording appropriate for the grade level.
- Spell all words correctly.
- Place a title and a date at the top of the first page.
- Provide a line on which the student can write their name.
- Use appropriate punctuation and grammar.
- Straightforwardly write test items—no tricky questions.
- Number all items.
- Provide sufficient space for the student to give answers.
- Proof the test before making copies.

---

on the recall level instead. Also, essays take a substantial amount of teacher time to grade. Other types of questions may be sufficient for testing lower-order objectives.

An essay question needs clear instructions. Is the expected answer to be one paragraph or multiple paragraphs? Are some topic areas required for full credit? Are students graded on spelling, grammar, and sentence structure?

These guidelines should help when constructing essay items: The teacher should decide before writing a question what objective the question is covering and what cognitive level the answer should reflect. The teacher should also develop a scoring rubric before administering the test. The rubric should include relevant content the teacher will look for in grading and the relative importance of each element of the answer. The teacher should grade all students on the same essay question before grading them on the following essay question. The pile of student tests should be reshuffled between questions, as some research has shown teachers tend to grade papers at the beginning and end of the stack of papers differently.

Essay Example

1. Explain how parliamentary procedure maintains order in an FFA meeting. Make sure to address the minority, the majority, and outside guests.

Alternative Assessment

Quizzes, tests, and exams are valid measures of what a student knows at a particular point in time. However, they are not the most effective tools for measuring progress over time or demonstrating creativity. Written tests are also not good at measuring affective and psychomotor objectives. Other assessment tools, such as portfolios and projects, are better suited to these purposes.

Portfolios and projects are examples of performance-based evaluation, discussed in an earlier section of this chapter. Depending on how they are structured, portfolios and projects are also examples of authentic assessments.

Portfolios

A *portfolio* is a collection of materials that represent a student's work. It may be contained in a notebook, a folder, a removable computer storage device, or on a cloud-based storage system. The portfolio usually includes a variety of samples of the student's work. These samples can be assignments (both graded and ungraded), tests, photos of projects or activities, and drawings; video and audio clips, reports, and computer projects; certificates and letters of commendation, and other items that document the student's progress and accomplishments.

Portfolios are organized according to their purpose. Portfolios typically serve three primary objectives: documentation of growth, documentation of best efforts, and compilation for assessment. A portfolio provides information on the whole student. Portfolios offer students the opportunity to use creativity in demonstrating all aspects of their learning (cognitive, affective, and psychomotor) together rather than in separate parts. Growth portfolios allow students to reflect on their efforts, leading to deeper understanding and more remarkable development of students' critical-thinking abilities.

Teachers using student assessment portfolios for grading purposes must develop clear, detailed criteria before students begin their portfolios. It is also critical teachers develop scoring rubrics early in the lesson planning process. Teachers need to take care the rubrics are fair, consistent, and detailed. Criteria and rubrics should not be too prescriptive, as this will stifle student creativity.

Agricultural education teachers have long used student portfolios in the classroom, in SAE, and in FFA. SAE record books, proficiency award applications, and degree applications are examples of portfolios. Although portfolio examples are less numerous in the classroom, some agriculture teachers have students develop portfolios for drawings, experiments, and projects.

Projects

A project is a task or problem completed by a student during or outside of class time. A project supplements instruction and allows a student to apply learned concepts. Examples of projects are numerous in agricultural education. Almost every area of agricultural mechanics uses student projects. The areas of natural resources, animal sciences, plant and soil sciences, horticulture, and agricultural business are replete with project examples.

FIGURE 19.6 This instructor utilizes an electronic portfolio to track student progress in FFA awards and recognition and supervised agricultural experience.

Planning by both the teacher and the student yields a successful student project. Students develop project drawings, schematics, flowcharts, and other planning tools and combine these into a project management plan that lists essential deadlines and the materials needed for projects. A key material that takes weeks to be delivered means the student must plan far in advance, alter the plan, work on other aspects of the project, or arrange for alternative class activities in the interim. Project procedures should be detailed, step-by-step, and in logical order. Many students, especially younger ones, will need help in this area. Depending on the purpose and type of project, the evaluation criteria may be student or teacher developed. It will also vary as to whether these are individualized or the same for all students. Figure 19.7 shows a student project planning sheet.

Figure 19.8 shows a project journal to be used when students are working on projects. It is designed for students to be self-accountable for the use of class time, progress toward goals, and quality of work.

Student name _____ Date _____

Title of project _____

Project description:

Materials needed:

Item description	Source	Quantity	Unit cost	Total cost

Project procedures:

Evaluation criteria:

FIGURE 19.7 A project planning sheet adaptable to a variety of agricultural subjects.

Student name	
Project title	
Date	What I accomplished today

FIGURE 19.8 Project journal used to encourage students to be accountable for their time and effort.

GRADING

Virtually all American public school systems require teachers to assign grades to student work regularly. A *grade* is a mark or descriptive statement that rates the extent to which a student has achieved a standard. The most common arrangement requires, at a minimum, assigning grades for the grading period, semester, and course. More frequent grading is often needed. Students are typically given either letter (A–F) or numerical (0–100) grades, with numerical grades usually later converted to letter grades. School systems typically specify the format and structure of grading.

The agriculture teacher has several grading decisions to make. How will individual assignments be graded? What weight will teachers give to each task? How will grades be reported to students and parents? What about qualitative assessments that do not easily convert to a letter or numerical grades? Should SAE and FFA activities be graded?

Scoring Rubrics

A *scoring rubric* is a set of guidelines for judging student work. The term *rating scale* is analogous. For a rubric to be effective, the agriculture teacher needs to make several decisions before and during the construction of the rubric. How much precision is required to discriminate between products of differing quality? Sometimes a 3-point scale (high, middle, low) is sufficient, while other times, a 10-point scale may be most appropriate.

The teacher should use language that describes each level on the scale selected. Consideration must be given to the highest point on the scale and to the point where the above is acceptable and below is unacceptable. Defining these two points is critical and will make constructing the remainder of the scale easier. The difference between any two points on the scale should be equivalent to the difference between any other two points. For example, the difference in quality between level 4 and level 5 performances should be the same as the difference in quality between level 2 and level 3 performances.

Scoring rubrics must give the grading criteria or categories, levels of performance (the scale), and descriptions of what constitutes performance under each group. The more specific and detailed the descriptions, the better students understand what is expected of them. Detailed descriptions also reduce or eliminate arguments over the assignment of grades and provide feedback for improvement. Figure 19.9 is an example of a rubric used to grade agricultural science fair projects.

A particular type of rubric is a product scale. A *product scale* is a series of actual objects representing each level of quality along the scale. Each object has a numerical or letter grade indicating its position on the scale. A student's product is then compared with the product scale and receives the number or letter of the scale item most closely matching the student's product. Teachers use product scales in agricultural mechanics for evaluating welds. The agriculture teacher can develop a board containing welds of varying qualities, from exceptional to inferior. In a mastery learning approach, the student evaluates their weld and continues working until they produce a weld of the desired quality.

Qualitative Assessments

The mission of agricultural education is to prepare the student for a successful career and a lifetime of informed choices in the plants, animals, and natural resources systems. Agriculture teachers must provide their students with qualitative assessments on attitudes, behaviors, and efforts to accomplish this mission. Qualitative assessments include rating scales, checklists, and essays.

Category	Proficient	Average	Improvement Needed	Much Improvement Needed
Scientific concepts	The report illustrates an accurate and thorough understanding of the scientific concepts underlying the lab.	The report illustrates an accurate understanding of most scientific concepts underlying the lab.	The report illustrates a limited understanding of scientific concepts underlying the lab.	The report illustrates an inaccurate understanding of scientific concepts underlying the lab.
Materials	All materials and setup used in the experiment are clearly and accurately described.	Almost all materials and the setup used in the experiment are clearly and accurately described.	Accurate descriptions exist for most of the materials and the setup used in the experiment.	Many materials are described inaccurately OR lack description.
Experimental design	Experimental design is a well-constructed test of the stated hypothesis.	Experimental design is adequate to test the hypothesis but leaves some unanswered questions.	Experimental design is relevant to the hypothesis but is not a complete test.	Experimental design is not relevant to the hypothesis.
Safety	The lab is carried out with full attention to relevant safety procedures. The lab setup, the experiment, and the lab teardown posed no safety threat to any individual.	The lab is generally carried out with attention to relevant safety procedures. The setup, experiment, and teardown posed no safety threat to any individual, but one safety procedure must be reviewed.	The lab is carried out with some attention to relevant safety procedures. The setup, experiment, and teardown posed no safety threat to any individual, but several safety procedures need to be reviewed.	The student ignored safety procedures, or some aspect of the experiment posed a threat to the student's safety or safety of others.

FIGURE 19.9 Sample rubric used to evaluate agriscience fair projects.

Rating scales and checklists help assess attitudes and behaviors. As the name implies, this is a checklist of characteristics, qualities, attributes, or behaviors. The teacher places a checkmark or note beside observed items. Checklist items should be worded in a parallel manner—for example, all items are stated positively to avoid confusion about whether a checkmark is desirable. If the checklist determines a grade, directions must include the number of checkmarks required for a specific grade. Essays allow students to provide background information and demonstrate higher-order thinking skills. Students can give examples and explanations more easily with essays. Finally, essays give students practice in writing skills that apply to numerous occupations.

Calculating Grades

Course grades are indicators of student accomplishment, growth, and level of understanding of course content. Fair, consistent, and accurate calculating of student grades is essential.

The variety of grading periods at schools is astounding. There are seven-period days, trimesters, full-year blocks, semester blocks, and variations on these. Grading periods may occur at four-week, six-week, nine-week, or other intervals. The teacher calculates and distributes grades at the end of the grading period and often at the midpoint. School systems

TABLE 19.3

Sample Weightings for Assigning Grading-Period Grades

Category	Weighting (%)*
Quizzes	10–15
Tests	25–35
Homework	10–15
Daily grades	10–15
Laboratory exercises	25–35
Projects	25–35
Grading-period exam	10–15
Notebook	5–15
SAE	5–15
FFA	5–15

*Actual weightings used will add to 100%.

may require teachers to provide grades to students and parents at the middle and the end of the grading period. School systems using a learning management system (LMS) and/or a student information system (SIS) that parents can access may require weekly graded assignments and posting of these on the LMS/SIS.

Many teachers call or email a different group of parents each week to update them about their students' performance and progress. Some schools require FFA members to verify passing grades in their classes before they are allowed to participate in out-of-school activities.

How does an agriculture teacher determine a student's grading-period grade? Electronic grade books simplify calculating grades; however, teachers must still assign values and weights to each assignment. Table 19.3 presents sample weightings.

Grading Supervised Agricultural Experience and FFA

Many questions arise related to grading SAE and FFA participation. Should agriculture students receive grades for SAE? Many factors enter into answering this question. Do students receive academic credit on their high school transcripts for supervised experience? Are all agricultural education students required to conduct foundational and/or immersion SAE? What is the philosophy of the local program? Are there state requirements concerning SAE? Do school administrators, parents, and advisory boards support the grading of students on activities conducted outside of regular school hours?

The supervised experience must be graded as pass/fail at a minimum because it is a part of the academic program. Otherwise, the agriculture teacher should discuss the preceding questions with local administrators, the agricultural education advisory board, the state agriculture teachers association, the agricultural education teacher education program, and the state department of education agricultural education staff.

Should agriculture students receive grades for participation in FFA activities? This question also requires consideration of many of the same factors as supervised experience. The agriculture teacher should consult with the same entities described in the preceding paragraph to make this decision. Additional questions to be considered regarding grading participation in FFA activities include the following: Are all students FFA members, even those in semester courses? Do students have equal opportunities to participate in activities? Will some students be excluded from activities because of access or financial constraints? Will students participating in sports or other organizations be able to participate in enough FFA activities to earn the highest possible grade? Can a grading system be developed that is not biased in favor of FFA officers to the detriment of students who are not officers?

REVIEWING SUMMARY

Evaluation is an integral part of the education process. It involves more than tests and grading and ideally leads to students taking responsibility for their learning. Evaluation should occur before, during, and at the end of instruction.

Achievement evaluations are typical in today's educational environment. Both standardized and teacher-made achievement tests are used throughout the curriculum. Norm-referenced tests help determine what individual students know in comparison to the group or other individuals. Criterion-referenced tests help determine how much students learn in comparison to standards and established criteria.

Teachers have used performance-based evaluations for a long time. The use of authentic assessments, portfolios, projects, and qualitative assessments provides a nearly complete picture of the total student. Although these may be more difficult to score, they yield rich data about the student.

Teacher-made assessment instruments are used on an almost daily basis in schools. Teachers must take the time and effort required to construct valid instruments. The type of question, the wording of question items, and the scoring rubric must all be carefully selected.

QUESTIONS FOR REVIEW AND DISCUSSION

1. What are examples of formative evaluations used in agricultural education?
2. Why do teachers rarely use norm-referenced standardized tests in agricultural education?
3. Does a mastery learning approach encourage students to put forth less than their best effort?
4. What are examples of performance-based evaluations used in agricultural education?
5. What is the relationship between authentic assessment and supervised experience?
6. What are the possible causes of students performing poorly on a teacher-made test?
7. What are examples of appropriate uses of true/false questions in agricultural education assessment instruments?
8. What are the advantages and disadvantages of open-ended questions?
9. What are examples of uses of student portfolios in agricultural education?
10. When is it appropriate for students to receive grades for their SAE or FFA participation?

ACTIVITIES

1. Investigate the debate over the use of standardized testing in U.S. public schools. Prepare a written report on your findings. Be sure to include the pros and cons of standardized testing in your report.
2. Develop a table of specifications for a unit of instruction you might teach. From the table, develop the assessment instrument.
3. Explore an online assessment system. Prepare a report on your findings.

REFERENCES

Davis, B. G. (1993). *Tools for teaching*. Jossey-Bass.

Fives, H., & Barnes, N. (2021). *The SAGE encyclopedia of educational research, measurement, and evaluation* (pp. 1655–1657; vol. 1–4). Sage. https://doi.org/10.4135/9781506326139

National Council for Agricultural Education. (2016). *National quality program standards for agriculture, food, and natural resources education*. National Council for Agricultural Education. https://thecouncil.ffa.org/

Phipps, L. J., & Osborne, E. W. (1988). *Handbook on agricultural education in public schools* (5th ed.). Interstate Printers and Publishers.

Supovitz, J. (2012). Getting at student understanding—The key to teachers' use of test data. *Teachers College Record*, 29.

Tanner, D. E. (2001). *Assessing academic achievement*. Allyn and Bacon.

Walvoord, B. E., & Anderson, V. J. (2011). *Effective grading: A tool for learning and assessment in college*. John Wiley & Sons.

20

Meeting the Needs of Diverse Students

Welcome to Ms. Kim's fifth-period Agricultural Biotechnology class. She has 24 students who are 10th through 12th graders. The students are rather diverse in many ways. She has students identify as nonbinary, female, and male. Dexter is going to be the valedictorian of their class. Alisha is the star hitter on the volleyball team and 11th-grade class president. Jose is the first member of his family born in the United States, and Spanish is the language spoken at home. Ten of Ms. Kim's students have individualized education plans (IEPs). She has one autistic student, one who is visually impaired, and one emotionally disturbed student who is prone to violence.

Today's lesson is on gene sequencing, and Ms. Kim is excited and ready to get every student engaged in the topic. How does her knowledge of students' backgrounds help her prepare? How would you engage a diverse group of students?

OBJECTIVES

This chapter addresses the National Quality Program Standards for Agriculture, Food, and Natural Resources Education (National Council for Agricultural Education, 2016), specifically Standard 1B: Program Design and Instruction—Instruction. It has the following objectives:

1. Discuss the meaning and importance of diversity.
2. Explain multicultural education and describe useful approaches.
3. Relate the meaning of prejudice and discrimination in education.
4. Summarize the meaning and practice of the least restrictive environment in the education of students with special needs.
5. Identify approaches that are appropriate with academically gifted students.
6. Explain considerations with students who have physical disabilities.
7. Describe teaching variations with students who have mental or emotional disabilities.

TERMS

cultural pluralism
disability
discrimination
diversity
handicap

learning disability
least restrictive environment
multicultural education
prejudice
stereotype

FIGURE 20.1 Public schools serve students from all backgrounds.

DIVERSITY

Diversity is the variety of differences within a category or classification. For example, a school described as ethnically diverse may have students who identify as African American, Latinx, Hmong, Vietnamese, and people of other ethnic backgrounds.

Although ethnicity, gender, and socioeconomic status are the most discussed forms of diversity, others are critical for the agriculture teacher to bear in mind. Geography, family situation, religious and value beliefs, mental and physical abilities, age, and language also need to be considered.

Many people stereotype agricultural education students as White male farmers; however, this image is no longer true. Students whose families make 100% of their income from farming are a small percentage of agricultural education students. Today, agricultural education students are also from urban and suburban backgrounds, rural backgrounds not connected to production agriculture, and farms where most family income is nonfarm. Nationally, females make up about one half of agricultural education students. More than 60% of agricultural education students are White.

Just as students may be stereotyped, certain attributes may also be ascribed to teachers. An example is the notion that all agriculture teachers are White males. All genders teach agriculture. These individuals represent a wide range of ethnic, social, and cultural backgrounds. Diversity among teachers should be respected and appreciated just as it is among students.

Diversity brings unique talents, experiences, and learning styles to the agricultural education classroom. A teacher should use these differences as a basis for enhancing instruction. Some differences do present definite challenges to a teacher. These differences, however, provide the opportunity for the teacher to gain personal satisfaction from seeing students achieve and be successful in their schoolwork.

FIGURE 20.2 Students learn best in a resource-rich learning environment. This shows a teacher giving instructions on the dissection of a fetal pig in an agricultural biology class.

MULTICULTURAL EDUCATION AND APPROACHES

"Multicultural education incorporates the idea all students—regardless of their gender and social class, and their ethnic, racial or cultural characteristics—should have an equal opportunity to learn in school" (Banks, 1989, p. 2). Concurrent with this is the teacher's belief that all students can learn. Banks explained that *multicultural education* is an idea, an educational reform movement, and a process.

The ideology of multicultural education is that the purpose of schools is for all students to have the opportunity to learn. Unfortunately, some schools have been configured so students with certain characteristics have a better opportunity to learn than others. This disparity requires a reform movement to change the educational system. Finally, multicultural education is a process that takes time, effort, and a willingness to change.

A major goal of multicultural education is the improvement of academic achievement for all students. Multicultural education strives for all students to acquire the reading, writing, and math skills necessary to compete in the global economy (Banks, 2008). A second goal is "to foster unity with diversity" (Banks, 2009, p. 10). All students should appreciate the common American identity while also being respected and able to express their unique cultural identity. Bullying and discrimination in all forms are not tolerated.

Sleeter and Grant (1994) described the following five approaches to multicultural education:

1. Teaching the exceptional and the culturally different
2. Human relations
3. Single-group studies
4. Multicultural education
5. Education that is multicultural and social reconstructionist

Single-group studies, such as women's studies, Jewish studies, or African American studies, are typically on the collegiate level and will not be discussed here. All the following approaches have valid teaching advice as well as potential problems. In all cases, the agriculture

FIGURE 20.3 Students work together in studying the skeleton of a rabbit.

teacher is encouraged to treat all students as individuals who desire to learn and are capable of learning.

Teaching the Exceptional and the Culturally Different

Teaching the exceptional and the culturally different is an approach that views the dominant American society as basically good. The purpose of education is to ensure all students are capable of becoming productive citizens who fit into the culture. Differences should be viewed simply as differences, not as deficiencies. Teachers who follow this approach will take students where they are and work to build them up to where they need to be. For students who are English language learners, teachers may use learning materials in the students' first language until the students are able to learn in English. These teachers will try to structure their instruction to accommodate students of varying learning styles and cognitive abilities. They will use examples and visuals students relate to and understand. A criticism of this approach is that it values the mainstream culture to the exclusion of all others.

Agriculture teachers following this approach may structure their classes in this way: They believe every student should be welcome and feel safe in agriculture classes and the FFA. They work with special education teachers to give students with special needs the extra help they need to succeed. Step-by-step instructions for projects, assignments, laboratories, and such are given in both oral and written form. Students who speak limited English may be paired with students who are bilingual. Not only do agriculture teachers screen visuals and materials to eliminate their being stereotypical, racist, or sexist, but they also make an effort to show females and people of color in agricultural careers.

Human Relations

The human relations approach aims to reduce stereotypes, facilitate positive relationships, and promote tolerance and acceptance. Whereas the first approach targeted those students needing assistance, the human relations approach targets all students. Teachers using this approach will combine students into varying work groups. These teachers will make sure all students are involved in classroom activities and everyone gets along. They will not tolerate name-calling or hateful speech. They will work to break down barriers between groups and to refute stereotypes.

Teachers using the human relations approach structure the classroom so every student can succeed; that is, they make sure the success of one student is not at the expense of another student. The human relations approach has received several criticisms. Some criticize it for emphasizing relationships at the expense of academics. Others see it as a simplistic answer to deep social and societal problems. They see the human relations approach as merely concentrating on "Can't we all just like each other?" without addressing the issues of poverty, discrimination, and hate.

An agriculture teacher following this approach may structure their classes in this way: Respect as an individual is shown to all regardless of identities or characteristics. The first week of school, do activities that demonstrate how much in common the students have with each other. Posters around the room are motivational and value the uniqueness of individuals and groups. The FFA chapter has activities that are attractive to a variety of students and does not hold activities that exclude students. The FFA chapter emphasizes community service and personal development of all members over competition or development of positional leaders only.

Multicultural Education

The multicultural education approach aims to change education and society such that equality of all and cultural pluralism are promoted. *Cultural pluralism* is the social philosophy that the culture of countries such as the United States is a product of the cultures of the various immigrant groups to the country. This approach targets not only all students but also the school and the community.

In a pluralistic society, all groups are free to maintain and develop their own cultural systems. Teachers using this approach are proactive in eliminating equality concerns. They will advocate making all facilities and activities accessible to students and parents with physical disabilities. They will work to change disciplinary procedures that punish one group unfairly.

Teachers following this approach would support students wearing items of religious significance. In addition, they would make accommodations for students observing religious holidays that are not school break days. They would voice their concerns in favor of faculty and administrator demographics reflecting the gender, ethnic, and disability diversity of the community.

The multicultural education approach also has its criticisms. Some criticize it for enhancing divisiveness by encouraging differences. By emphasizing educational content from a multitude of cultures, students may miss needed content from the mainstream culture. Others argue the multicultural education approach changes cognitive knowledge but not affective behaviors. If this is true, students would know more about other cultures yet still have the same attitudes, prejudices, and biases. Teachers can structure their classrooms to address these criticisms.

Agriculture teachers following this approach may structure their classes using fundamentals of multicultural education. They would evaluate their agriculture program to identify and eliminate any inequality and discrimination. They would actively recruit students from minority groups and advocate that new faculty hires reflect diversity. Teachers would also encourage students to celebrate their differences and to accept differences in others. Students would be taught to analyze content critically for biases and work to eliminate any found. Instruction would be student-centered, addressing all learning styles. Students would be taught about the agricultural accomplishments of people from minority groups. As an example, not only would students learn the history of New Farmers of America, the African American student organization until 1965, but they would also explore the aspects of the

FIGURE 20.4 FFA participation should promote diversity. (COURTESY OF NATIONAL FFA ORGANIZATION.)

FFA absorbing the NFA that were detrimental to African American teachers and students. (More about the NFA is in Chapter 23.)

Multiculturalism and Social Reconstructionist Approach

The multicultural and social reconstructionist approach is similar to the multicultural education approach. It also aims to change education and society such that equality and pluralism are promoted. However, it goes a step further and advocates teaching students how to work toward changing society. Teachers using this approach organize their instruction around current social issues, such as racism, ageism, sexism, and classism. They explore oppression and power and guide students to work toward alleviating oppression and equalizing power. Students are a part of the democratic decision-making process of the classroom. Diversity is not only accepted but expected.

An agriculture teacher following this approach may structure their class in this way: A unit on agricultural lending may be taught from the perspective of African American farmers who were discriminated against. After exploring the issues of privilege, discrimination, power, social class, and poverty, the agriculture teacher would lead the class to analyze lending practices in the community. The students would then actively work to reverse any inequities. Another unit might look at fruit and vegetable production from the perspective of the migrant worker. The same procedure would be used as described above. A third example would be exploring the importation of cut flowers. Students would look at issues such as environmental impact and worker exploitation in the exporting country.

AGRICULTURAL EDUCATION DIVERSITY, EQUITY, AND INCLUSION EFFORTS

The decades of the 1960s through the 1980s saw much societal change in the areas of equity and inclusion. School-based agricultural education also experienced change. Integration and busing brought students from different races/ethnicities together who previously had attended separate schools. Vocational, now career and technical education, programs actively recruited students from genders not traditionally associated with occupations in that area. For example, more females enrolled in agricultural education.

The 1990s to 2010s were a time of increased research in diversity and task forces to increase diversity in school-based agricultural education. For example, the American Association for Agricultural Education (AAAE) in the early 1990s formed a Population Diversity Work Group that produced the monograph *Enhancing Diversity in Agricultural Education* (Bowen, 1993). This led to research studies on barriers to enrollment, models of recruitment, identification of biases, and mentoring of students from underrepresented groups (Bowen, 1993; Esters & Bowen, 2005; Foster, 2001; Foster, 2003; Jones & Bowen, 1998; Talbert & Larke, 1995a; Talbert & Larke, 1995b; Warren & Alston, 2007; Wiley et al., 1997). Concurrently, the National FFA Organization created several task forces to address the lack of diversity in membership. The 2010s saw a new focus on inclusion (LaVergne et al., 2011; LaVergne et al., 2012; Vincent et al., 2012) and exploring perceptions and professional development needs of agriculture teachers (Vincent et al., 2014; Vincent & Kirby, 2015; Vincent & Torres, 2015).

Agricultural education in the 2020s is working to address inequalities in access, representation, and involvement. These diversity, equity, and inclusion (DEI) efforts include curriculum, instruction, SAE, FFA, and professional organizations. The most recent effort began in 2018 with a task force and steering committee. The task force included representatives from all components of agricultural education representing various aspects of human diversity. A major outcome of the DEI effort was public statements committing the profession to action. Concurrent with these DEI efforts were the release of SAE for All and Agricultural Education for All. SAE for All is further discussed in Chapter 22 and Agricultural Education for All is further discussed in Chapter 23. Both open agricultural education opportunities to all enrolled students regardless of background and previous agricultural experience. Both have as goals for students to feel welcomed, safe, and accepted in agricultural education. The professional associations for teachers and researchers each have commitment to DEI statements. The National Association of Agricultural Educators (NAAE) states its DEI focus and effort are an intentional choice that requires commitment. The American Association for Agricultural Education (AAAE) Strategic Plan Goal 1 is to build a more inclusive culture within the organization.

PREJUDICE AND DISCRIMINATION

Prejudice is a preference or idea formed without having complete information. It is closely related to a stereotype. A *stereotype* is a conception resulting from the assignment of oversimplified characteristics to a whole group.

Everyone has prejudices. A person who does not eat vegetables because of a dislike for the taste of broccoli is prejudiced against vegetables. They have taken a stereotype of broccoli and applied it to all vegetables. Others may be prejudiced against a particular smartphone brand because they do not like a certain feature that was on an older phone. These prejudices may be harmless; however, prejudices against people tend to harm human relationships and serve to drive groups apart. Prejudices against people or groups tend to be irrational and hurtful. Everyone deserves to be treated as an individual worthy of respect and dignity. This is not possible if prejudices are maintained. Stereotypes keep people from learning about others as individuals and can lead to irrational judgments and actions based on generalizations.

Discrimination is the act of treating one group differently from another group. Prejudices many times lead to discriminatory practices. For example, a male superintendent may be prejudiced, believing all women are weak disciplinarians and indecisive in their decision making. This could lead him to discriminate against qualified women applicants for administrative roles in the school system.

FIGURE 20.5 FFA provides opportunities for students to work together on common goals. (COURTESY OF EDUCATION IMAGES.)

Most forms of discrimination are illegal in the school system. Title IX of the 1972 Education Amendments prohibits discrimination in educational programs or activities on the basis of gender. The Civil Rights Act of 1964 prohibits discrimination in voting, public accommodations, schools, and employment on the basis of race, color, religion, gender, or national origin. The No Child Left Behind Act of 2001 upheld the right to a public education for children of migrant workers. The purpose of the Every Student Succeeds Act of 2015 is for all children to receive a fair, equitable, and high-quality education. Public Law 94-142, the Education for All Handicapped Children Act of 1975, requires a free, appropriate public education for all handicapped children. This law and the use of the word *handicapped* are discussed in detail later in this chapter.

Over the years, agricultural education has addressed discriminatory policies and practices. Several of these efforts were influenced by legal and/or societal changes. In 1965, during the time of the Civil Rights Act, New Farmers of America (NFA) and Future Farmers of America (now FFA) were combined. This brought the African American student organization, NFA, and the predominately White student organization, FFA, together under the same umbrella. However, African American students and teachers were not afforded the same opportunities, especially in leadership positions. The NFA and its absorption is covered in more detail in Chapter 23. Over the years, females had enrolled in small percentages in agricultural education. The opening of national FFA membership to females in 1969 greatly expanded the number of females enrolling in agricultural education. Beginning in earnest in the 1970s and continuing today, students with disabilities and students with educational handicaps have been a part of agricultural education classrooms. In 2011 National FFA published the first version of its official manual in Spanish. It was not until 2017 that National FFA removed gender-specific titles from its official dress guidelines and 2019 that a National FFA officer stated they were a part of the LGBTQ (lesbian, gay, bisexual, transgender, queer/questioning) community while in office.

STUDENTS WITH SPECIAL NEEDS

Students with certain characteristics have been identified as needing special instruction within the school system. In 1975, PL 94-142, the Education for All Handicapped Children Act, was passed. PL 94-142 is also known as the Individuals with Disabilities Education Act, or IDEA. This act had far-reaching implications for education. It required students with handicaps be educated in the *least restrictive environment* or the closest-to-typical educational setting, meaning students with handicaps and students without would be in the same classrooms. Students had to be evaluated before they could be placed in special education programs. The law also required special education students to have individualized education plans (IEPs). PL 94-142 greatly opened the doors of education to students with handicaps and the channels of communication with their parents.

The 1997 IDEA was designed to enhance and expand the original 1975 act. It raised academic expectations for students with disabilities, increased parental involvement in the educational process, involved regular classroom teachers more in the planning process, included all students in assessment and public reports, and supported greater professional development (USDE, n.d.). The Individuals with Disabilities Education Improvement Act (IDEIA) of 2004 more closely aligned the requirements of IDEA with the No Child Left Behind Act of 2001. The Every Student Succeeds Act of 2015 updated and amended the 2004 IDEIA.

Agriculture teachers are expected to participate fully in the IEP development process. They may have more students with disabilities in their classrooms because current practice is to place these students in regular classrooms whenever possible. However, agriculture teachers should have greater access to support services to assist them in teaching students with disabilities. They also need specific professional development to address the needs of students with disabilities and to fully involve those students in the agricultural education model (Giffing et al., 2010). Finally, agriculture teachers should have regular interaction with parents of students with disabilities.

The 1997 IDEA also changed the terminology from "handicapped children" to "children with disabilities." This is a subtle change but an important one. In changing the order of the words, the focus changes to the child rather than the condition. According to Woolfolk (1998), a *disability* is the inability to do something specific, such as hear or see. A *handicap* is a disadvantage under certain situations. For example, being paraplegic is a disability for movement of the legs, but is a handicap only if the situation requires walking. However, it is not a handicap for tasks that require speaking or writing. The modification of using a wheelchair can eliminate the handicap from many situations.

The IDEIA of 2004 listed 13 categories of disability for which children who have one or more need special education and related services. The categories are preschool disabled, intellectual disability, hearing impairment, speech or language impairment, visual impairment, emotional disturbance, orthopedic impairment, autism, traumatic brain injury, other health impairment, specific learning disability, deaf-blindness, or multiple disabilities.

STUDENTS WHO ARE HIGH ABILITY

Students are classified by schools as high ability if they exhibit exceptional gifts, talents, motivation, or interests and show the potential for performing at outstanding levels of accomplishment in intellectual, creative, academic, arts, or interpersonal areas. They may demonstrate above-average intellectual ability, creativity, and/or motivation to achieve (Woolfolk, 1998). Although completing class assignments may come easily to them, that is not a defining characteristic of students who are high ability. Instead, students who are high ability tend to master an area earlier than their peers, learn in ways different from their peers, and have a high internal motivation to excel in the area for which they are gifted (Santrock, 2008). Educational interventions for these students include special classes or programs, accelerated classes, alternative assignments, and outside-of-school opportunities.

FIGURE 20.6 Students are challenged by the endless opportunities in agricultural education. (COURTESY OF LEDYARD AGRI-SCIENCE & TECHNOLOGY PROGRAM.)

Agriculture teachers can provide challenges and opportunities to students who are high ability. With any agricultural education course that is project-based, these students can choose a more demanding project—one that requires greater abstract thinking and creativity. The possibilities for independent study also exist. Many universities now offer college-level agriculture courses via distance education. Also, with the close working relationship agriculture teachers have with Cooperative Extension educators and their state's agricultural university(ies), students who are gifted could be involved in field-based agricultural research projects.

FFA and SAE also provide opportunities to students who are high ability. The nearly 50 proficiency award areas allow students to specialize in their areas of interest and challenge themselves to excel. The Agriscience Student Recognition and Scholarship Program provides opportunity for students to conduct research projects. The entrepreneurship aspects of agricultural education can be attractive to students who are high ability, as entrepreneurship allows students to exercise fully their creativity, abilities, and motivation to achieve.

STUDENTS WITH PHYSICAL DISABILITIES

The mainstreaming requirements of PL 94-142, IDEA, IDEIA, and the Carl Perkins Acts have led to more students with physical disabilities being in agricultural education courses. The tasks of the agricultural education classroom necessitate modifications for students with physical disabilities. Agriculture teachers also need to be aware that socialization may also be a need of students with physical disabilities.

Students who use orthopedic devices, such as wheelchairs or crutches, need access to buildings and facilities. Ramps may need to be added for classrooms or greenhouses. Outdoor laboratories and natural resources trails may need hardened paths, bridges over wet areas, and signage at a readable height. Agricultural mechanics equipment may need to be height-adjusted to allow students in wheelchairs to fit under worktables and to use tools.

Communications disabilities include hearing, vision, and speech impairments. Students with communications disabilities require classroom modifications also. Many of these modifications are good teaching methods for all students. Box 20.1 outlines recommended modifications.

FIGURE 20.7 All students can be active participants in agricultural education. (COURTESY OF NATIONAL FFA ORGANIZATION.)

BOX 20.1 *Teaching Modifications for Students With Communications Disabilities*

- Speak clearly.
- Speak to the class, not to the chalkboard/whiteboard. When writing on the board or facing away from the students, do not speak. Keep your mouth uncovered so lip readers can see your mouth movements.
- Write plainly and in large letters when writing on the board or projecting your handwriting.
- If necessary, provide printed copies of notes for students with visual impairments. The school may provide a special magnification reader for the student.
- Teach other students to be understanding, accepting, and helpful. Do not tolerate belittling or hurtful comments.
- If a task involves multiple steps, give instructions in both written and oral form.

STUDENTS WITH MENTAL OR EMOTIONAL DISABILITIES

Students with mental disabilities have received their education by various means over the years. Before PL 94-142, the majority of these students, especially those with significant mental disabilities, were placed in special public or private facilities. Often these facilities provided inadequate educational and social training.

Beginning in the 1970s and accelerating during the 1980s, students with mental disabilities were educated in public schools but in self-contained special education classrooms. This still did not provide adequate socialization and "real-world" experiences. Gradually these students were mainstreamed into more and more classes. This process was accelerated by IDEA 1997. Each successive rewrite of the Carl Perkins Act encouraged greater mainstreaming of students with disabilities into career and technical education programs, including agricultural education.

Intellectual Disability

Intellectual disability is defined as considerable limitations in both intellectual functioning and adaptive behavior (American Association on Intellectual and Developmental Disabilities, 2021). Limitations in adaptive behavior may include the inability to communicate, take care of self, gain employment, socialize with others, or provide for personal health and safety. Intellectual disability presents itself before the age of 18. Many school systems divide stages into mild, moderate, severe, and profound.

Students with intellectual disability need varying levels of support (Santrock, 2008). The lowest level is intermittent, in which support is needed only at intervals. The next level is limited, in which support is needed on a more consistent basis over time. The third level is extensive, in which regular or day-to-day support is needed. The highest level is pervasive, in which high-intensive support is needed in all life situations, including those that are life-sustaining.

An agriculture teacher is likely to have students requiring intermittent and limited support. These students will probably be below grade level in reading, mathematics, and general knowledge. They may also have difficulty in reasoning and understanding abstract concepts. Box 20.2 provides guidelines for teaching students with intellectual disability.

BOX 20.2 *Teaching Modifications for Students With Intellectual Disability*

- Assess readiness. However little a student may know, they are ready to learn the next step. State and present objectives simply. This helps students know what they are to do and learn.
- Base specific learning objectives on an analysis of a student's learning strengths and weaknesses. Present material in small, logical steps. Practice extensively before going on to the next step. Work on practical skills and concepts based on the demands of adult life.
- Include all steps. Students with average intelligence can form conceptual bridges from one step to the next, but students with intellectual disability need every step and bridge made explicit. Make connections for the students. Do not expect them to "see" the connections.
- Be prepared to present the same idea in many different forms.
- Go back to a simpler level if you see a student is not following.
- Be especially careful to motivate students and maintain attention.
- Use materials that are appropriate for the students. A middle school student may need the low vocabulary of a beginning reader picture story book but will be insulted by the age of the characters and the content of the story.
- Teach for success. Focus on a few target behaviors or skills so you and the students have a chance to experience success. Everyone needs positive reinforcement.
- Be aware that these students must overlearn, repeat, and practice more than students of average intelligence. They must be taught how to study, and they must frequently review and practice their newly acquired skills in different settings.

Adapted from Woolfolk (1998).

Learning Disabilities

One third of the students with disabilities served under IDEIA are classified as having learning disabilities (National Center for Education Statistics, 2021). Of public-school kindergarten through 12th-grade students, 5% were identified as having a specific learning disability. Although difficult to define, a *learning disability* is typically considered to be the failure or incapacity to function at the age-appropriate level in language, reading, spelling, mathematics, and/or other areas of learning. Most educators agree learning disabilities include a heterogeneous group of disorders affecting students of average intelligence who have significant academic problems and perform well below what is expected.

Students with learning disabilities typically have their academic problems in a specific area, such as reading, writing, listening, mathematics, or reasoning, and perform at average or above-average levels in other areas. Students with learning disabilities need help overcoming frustrations, fear of failure, and other educational difficulties associated with their disabilities. A simple and first approach is to observe attendance patterns and study habits. Students who miss class frequently or have poor study habits tend to fall behind their peers and become discouraged. Changing these patterns and habits may be the only modification some students need. Other students will need additional support from special needs teachers in their area of learning disability—mathematics or reading comprehension, for example.

How does an agriculture teacher help students with learning disabilities? First, they should work closely with the special needs teachers. These experts will have suggestions and modifications for the agriculture teacher to use. Next, the agriculture teacher should observe the students for signs, such as frequent absences, frustration, or giving up. When these signs are present, the agriculture teacher needs to work with the student and the special needs teacher to remedy the situation. Finally, the agriculture teacher must use good, individualized teaching methods. What helps one student to be successful will be different from what helps another student.

Emotional or Behavioral Disabilities

Students with emotional or behavioral disabilities present problems in the classroom for both themselves and their peers. For a student to be classified with a behavioral disability, the behavior must be serious, persistent, and age-inappropriate (Eggen & Kauchak, 1997). Students with behavioral disabilities have trouble with self-control of behaviors and self-management of emotions.

Hallahan and Kauffman (1994) described two categories of behavioral disabilities. The first category is those behaviors that are external. These include aggression, uncooperativeness, and cruelty. Students with behavioral disabilities of this type may also have academic problems and are at high risk to drop out of school. The second category is those behaviors that are internal. Students with behavioral disabilities of this type are withdrawn or may suffer from anxiety or depression. They tend not to have social friends and may be quiet or shy in class.

Students with behavioral disabilities need positive reinforcement of their successes and accomplishments. Students with internal characteristics also need a teacher who will engage them as individuals and respect them without judging them. Students with external characteristics need firm and consistent rules with consequences that are spelled out. Students with behavioral disabilities may need a part of the classroom they can go to as a "gain control" spot. This allows them to remove themselves from the situation and defuse any escalating emotions or conflicts. The wise saying of "praise in public, discipline in private" is extra important for these students.

REVIEWING SUMMARY

The agricultural education classroom serves a diverse body of students. All students can benefit from the agricultural education program. It is the role of the teacher to structure the program to best meet student needs.

Gender, ethnic, language, cultural, socioeconomic, and geographic diversity exist within the agricultural education classroom. The agriculture teacher is encouraged to treat all students as individuals who desire to learn and are capable of learning. There are five approaches to multicultural education; however, they are not mutually exclusive. Awareness of the need for multicultural education and a desire to meet the needs of all students as individuals are key to educating students in today's diverse classroom.

Prejudices and stereotypes are harmful to teacher–student relationships as well as to learning. Teachers should work to identify and eliminate prejudices and stereotypes they hold. Discrimination is illegal and has no place in the agricultural education program. Policies and procedures should be put in place to ensure that all students have equal access to the agricultural education program.

More and more, students with special needs are included in the agricultural education classroom. Agriculture teachers should work with special needs teachers to develop appropriate modifications for these students. Agriculture teachers should be continually updating their knowledge of students with special needs and of recommended practices for facilitating their learning.

Students with learning disabilities are the largest group of students with disabilities. These students typically have academic problems in a specific area but perform at average or above-average levels in other areas. Agriculture teachers are encouraged to work individually with their students who have learning disabilities. What works for one student may not for another.

QUESTIONS FOR REVIEW AND DISCUSSION

1. What is multicultural education?
2. What are examples of prejudice and discrimination within agricultural education from the past?
3. Which of the five approaches to multicultural education best describes agricultural education within your community or state?
4. What is the ethnic, gender, and geographic (farm, rural, suburban, urban) diversity of agricultural education within your state?
5. What are examples of program changes and actions agriculture teachers could make to ensure gender, ethnic, and socioeconomic diversity and respect within their classroom?
6. What impacts have the 1975, 1997, and 2004 Individuals with Disabilities Education Acts had on agricultural education?
7. What is the difference between a disability and a handicap?
8. What are examples of modifications agriculture teachers could make for students who are gifted in their classes?
9. What are examples of modifications agriculture teachers could make for students with physical, mental, or emotional disabilities in their classes?
10. What are examples of modifications agriculture teachers could make for students with learning disabilities in their classes?

ACTIVITIES

1. Investigate students with special needs and agricultural education. Topics to cover include modifications in FFA and SAE. Are there national resource materials to assist students with special needs in agricultural education? Possible references include the National FFA Organization at www.ffa.org, your state's teacher education institution(s), your state's agricultural education staff, and your state's agriculture teachers' professional association.

2. Investigate current federal legislation regarding students with special needs. You will want to include the latest Perkins Act, Individuals with Disabilities Act, Elementary and Secondary Education Act, and Higher Education Act, among others.

3. Using the *Journal of Agricultural Education*, *The Agricultural Education Magazine*, and the proceedings of the National Conference of the American Association for Agricultural Education, report on research about diversity, equity, and inclusion in agricultural education. Current and past issues of these journals can be found in the university library. Selected past issues can be found online at www.aaaeonline.org or www.naae.org.

REFERENCES

American Association on Intellectual and Developmental Disabilities. (2021). *Definition of intellectual disability*. http://www.aamr.org

Banks, J. A. (1989). Multicultural education: Characteristics and goals. In J. A. Banks and C. A. McGee Banks (Eds.), *Multicultural education: Issues and perspectives* (pp. 1–26). Allyn & Bacon.

Banks, J. A. (2008). *An introduction to multicultural education* (4th ed.). Pearson Education.

Banks, J. A. (2009). *Teaching strategies for ethnic studies* (8th ed.). Pearson Education.

Bowen, B. (Ed.) (1993). *Enhancing diversity in agricultural education*. American Association for Agricultural Education.

Eggen, P., & Kauchak, D. (1997). *Educational psychology: Windows on classrooms*. Prentice Hall.

Esters, L. T., & Bowen, B. E. (2005). Factors influencing career choices of urban agricultural education students. *Journal of Agricultural Education, 46*(2), 24–35. https://doi.org/10.5032/jae.2005.02024

Foster, B. (2003). Profiling female teachers of agricultural education at the secondary level. *Journal of Career and Technical Education, 19*(2), 15–27.

Foster, B. B. (2001). Choices: A dilemma of women agricultural education teachers. *Journal of Agricultural Education, 42*(3), 1–10. https://doi.org/10.5032/jae.2001.03001

Giffing, M. D., Warnick, B. K., Tarpley, R. S., & Williams, N. A. (2010). Perceptions of agriculture teachers toward including students with disabilities. *Journal of Agricultural Education, 51*(2), 102–114. https://doi.org/10.5032/jae.2010.02102

Hallahan, D., & Kauffman, J. (1994). *Exceptional children* (6th ed.). Allyn & Bacon.

Jones, K. R., & Bowen, B. E. (1998). A qualitative assessment of teacher and school influences on African American enrollments in secondary agricultural science courses. *Journal of Agricultural Education, 39*(2), 19–29. https://doi.org/10.5032/jae.1998.02019

LaVergne, D. D., Jones, W. A., & Larke, A., Jr. (2012). The effect of teacher demographic and personal characteristics on perceptions of diversity inclusion in agricultural education programs. *Journal of Agricultural Education, 53*(3), 84–97. https://doi.org/10.5032/jae.2012.03084

LaVergne, D. D., Larke, A., Jr., Elbert, C. D., & Jones, W. A. (2011). The benefits and barriers toward diversity inclusion regarding agricultural science teachers in Texas secondary agricultural education programs. *Journal of Agricultural Education, 52*(2), 140–150. https://doi.org/10.5032/jae.2011.02140

National Center for Education Statistics. (2021). *Fast facts: Students with disabilities.* https://www.nces.ed.gov/fastfacts

National Council for Agricultural Education. (2016). *National quality program standards for agriculture, food, and natural resources education: A tool for secondary (grades 9–12) programs.* https://thecouncil.ffa.org/

Santrock, J. W. (2008). *Educational psychology* (3rd ed.). McGraw-Hill.

Sleeter, C. E., & Grant, C. A. (1994). *Making choices for multicultural education: Five approaches to race, class, and gender* (2nd ed.). Macmillan.

Talbert, B. A., & Larke, A., Jr. (1995a). Factors influencing minority and non-minority students to enroll in an introductory agriscience course in Texas. *Journal of Agricultural Education, 36*(1), 38–45. https://doi.org/10.5032/jae.1995.01038

Talbert, B. A., & Larke, A., Jr. (1995b). Minority students' attitudes toward agricultural careers. *NACTA Journal, 39*(1), 14–17.

USDE. (n.d.). *IDEA: Individuals with disabilities education act.* https://sites.ed.gov/idea/

Vincent, S. K., Henry, A. L., & Anderson, J. C. II. (2012). College major choice for students of color: Toward a model of recruitment for the agricultural education profession. *Journal of Agricultural Education, 53*(4), 187–200. https://doi.org/10.5032/jae.2012.04187

Vincent, S. K., & Kirby, A. T. (2015). Words speak louder than action? A mixed-methods case study. *Journal of Agricultural Education, 56*(1), 32–42. https://doi.org/10.5032/jae.2015.01032

Vincent, S. K., Kirby, A. T., Deeds, J. P., & Faulkner, P. E. (2014). The evaluation of multicultural teaching concerns among pre-service teachers in the south. *Journal of Agricultural Education, 55*(1), 152–166. https://doi.org/10.5032/jae.2014.01152

Vincent, S. K., & Torres, R. M. (2015). Multicultural competence: A case study of teachers and their student perceptions. *Journal of Agricultural Education, 56*(2), 64–75. https://doi.org/10.5032/jae.2015.02064

Warren, C. K., & Alston, A. J. (2007). An analysis of diversity inclusion in North Carolina secondary agricultural education programs. *Journal of Agricultural Education, 48*(2), 66–78. https://doi.org/10.5032/jae.2007.02066

Wiley, Z. Z., Bowen, B. E., Bowen, C. F., & Heinsohn, A. L. (1997). Attitude formulation of ethnic minority students toward the food and agricultural sciences. *Journal of Agricultural Education, 38*(2), 21–29. https://doi.org/10.5032/jae.1997.02021

Woolfolk, A. E. (1998). *Educational psychology* (7th ed.). Allyn & Bacon.

21

Using Laboratories

The Chestnut Grove High School agriculture department has a variety of laboratory facilities. An agriscience laboratory has state-of-the-art equipment. A greenhouse is used for propagating and growing selected ornamental plants. A land laboratory is used by the agriculture and science departments and by the nearby elementary school. An animal sciences laboratory has Japanese quail, aquaculture tanks, and small animal cages.

Agricultural educators often talk about hands-on learning. We know that laboratory facilities should support the curriculum and promote mastery learning. Efficiently using laboratory facilities requires planning and instructional skill. How do agriculture teachers make the best use of laboratories?

OBJECTIVES

This chapter addresses three of the National Quality Program Standards for Agriculture, Food, and Natural Resources Education (National Council for Agricultural Education, 2016):
Standard 1B: Program Design and Instruction—Instruction
Standard 1C: Program Design and Instruction—Facilities and Equipment
Standard 1D: Program Design and Instruction—Assessment
It has the following objectives:

1. Identify the kinds and uses of laboratories in agricultural education.
2. Explain the goals of laboratory instruction in agricultural and science education.
3. Discuss practices in laboratory management.
4. Identify various sources of funding to support purchases of laboratory equipment and supplies.
5. Discuss how to develop laboratory activities.
6. Describe the use of experiential learning theory in planning and facilitating laboratory activities.
7. List and discuss safety considerations in laboratory teaching.
8. Describe forms of assessment in the laboratory setting.

TERMS

authentic assessment
consumables
corrosive material
employability skills
expenditures
experiential learning theory
flammable material
grouping
laboratory
laboratory activity
laboratory instruction
laboratory superintendent

learning center
limiting factor
OSHA
oxidizer
performance-based assessment (PBA)
personal protective equipment (PPE)
purchase order
reactive
readiness level
sharps
toxic material

FIGURE 21.1 A Connecticut teacher instructs students in the school's hydroponic greenhouse how to test and adjust for the pH in the nutrient solution of the NFT table.

LABORATORIES AND INSTRUCTION

Agricultural education programs may have several kinds of laboratory facilities. A *laboratory* is a space for individual or group student experiments, projects, or practice. In agricultural education, the activities may be in the areas of agriscience, agricultural mechanics, animal science, food science, horticulture, plant and soil science, and natural resources.

Laboratories in agricultural education vary according to community, school, and state. They range from multipurpose large rooms, to single-purpose stations within a room, to outdoor land, pond, or greenhouse locations. Chapter 9 covered the space, equipment, and storage requirements for laboratory facilities. This chapter covers how best to use these facilities.

Laboratory instruction is organized instruction occurring in laboratories that, compared with classroom instruction, provides for greater freedom of movement for students, focuses on psychomotor skill development, uses special tools and equipment, and uses student self-directed learning (Phipps & Osborne, 1988). Laboratory instruction allows students to inquire into the content. The principles and concepts learned through classroom instruction are tested through application and hands-on learning. Students may conduct experiments, practice skills, and simulate real-world experiences.

A *laboratory activity* is an application of concepts and principles. The definition of a laboratory activity includes experiments, hypothesis testing, demonstrations, learning exercises, application exercises, and skills practice. For example, a lab in an agricultural mechanics laboratory may be wiring a circuit board with a single-pole light switch. A lab in an agriscience laboratory may be testing the effects of temperature on plant growth rates.

A lab in a natural resources laboratory may be using a Biltmore stick to determine the volume of standing trees.

Purpose of Laboratory Instruction in Agricultural Education

As discussed in Chapter 3, school-based agricultural education has its roots in vocational education. While agricultural education initially sought to primarily prepare White male students in the workforce as farmers, today's agricultural education seeks to elevate learning opportunities in agriculture for all students, regardless of race, gender, socioeconomic status, sexual orientation, or ability status. Today's career and technical education in agriculture focuses on preparing individuals to be college- and career-ready through the development of technical, job-specific skills, core academic skills, and employability skills (ACTE, 2019).

Agricultural education has been described as a vehicle for teaching the content of agriculture (technical knowledge and skills), as well as teaching core academic and employability skills within the context of agriculture to develop agriculturally literate citizens (Roberts & Ball, 2009). Some agriculture teachers view agricultural education's purpose as multifaceted and overlapping, including ideas of career and college preparation, agricultural literacy, life skills development, and as a way to engage students in individualized learning (Rice & Kitchel, 2017). These shifting purposes of agricultural education ultimately guide curriculum and instruction in agriculture. Embedded within instructional decisions are the teacher's personal educational philosophies, as well as the aims of the local school district and community. Ultimately, secondary education should seek to create learners who can think critically in whatever context of life they find themselves after graduating from high school (Littky & Grabelle, 2004). Laboratory instruction in agriculture provides infinite opportunities for students to develop these skills and more.

Laboratory instruction engages students in hands-on practice and application of skills in agriculture. Without the opportunity to engage in this psychomotor skill development, the agriculture class would likely be considered more of an agricultural awareness class that increases agricultural literacy. While increasing agricultural literacy is exceptionally important and may be more of a focus for some programs, without clearly developed laboratory instruction, an agriculture course would not be considered CTE. Thus, laboratory instruction in agriculture is a hallmark feature of any school-based agricultural education program.

Laboratory instruction illustrates the "why" and "how" of concepts and principles in agriculture. Laboratory learning activities can provide students a space to practice or observe agricultural phenomena, or to design and conduct experiments. Not only do labs provide a space to deepen learning, but the opportunity to see and do things that are not in a textbook can also serve to motivate students to more richly engage in the class and program. Ultimately, laboratory instruction in agriculture serves three main goals (Phipps et al., 2008):

1. Development of conceptual skills
2. Development of procedural skills
3. Development of personal skills

Figure 21.2 lists specific skills developed within each of these three categories.

As agricultural education increasingly integrates STEM and agriscience applications, it is useful to examine how science education labs promote learning in labs. The National Research Council (2006) identified the following goals of science laboratory instruction:

1. Strengthening science subject matter mastery
2. Developing scientific reasoning skills

3. Increasing understanding of how to navigate the ambiguity of conducting science
4. Developing practical skills
5. Strengthening understanding of the nature of science
6. Stimulating an interest in science and learning about science
7. Strengthening teamwork skills

To accomplish these goals, laboratory instruction should be planned with clear learning outcomes that are intentionally sequenced within the broader curriculum to enhance the flow of instruction. Integration of science content and knowledge of the process of doing science should also be explicitly integrated into instruction. Finally, instructors should plan to incorporate ongoing reflection and discussion to solidify student learning.

Clearly, agricultural education labs and science labs have similar goals and instructional processes. Instructional decisions related to the use of laboratory instruction should begin with these end goals in mind to ensure learning outcomes of these labs are aligned with the purposes of laboratory instruction in agricultural education. Further details on planning laboratory instruction utilizing experiential learning theory will be discussed in a later section of this chapter.

Determining When to Use Laboratories in Teaching Agricultural Education

The factors used in determining when and how to use agricultural education laboratories include curriculum, student, facilities, and teacher. Each of these categories will be discussed in more detail in the following sections.

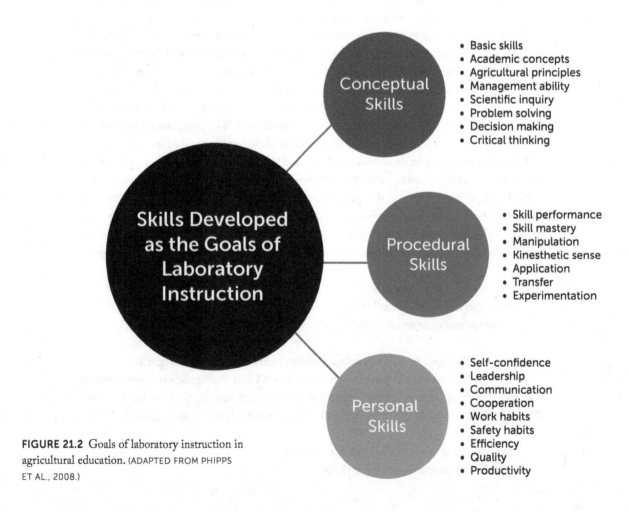

Conceptual Skills
- Basic skills
- Academic concepts
- Agricultural principles
- Management ability
- Scientific inquiry
- Problem solving
- Decision making
- Critical thinking

Skills Developed as the Goals of Laboratory Instruction

Procedural Skills
- Skill performance
- Skill mastery
- Manipulation
- Kinesthetic sense
- Application
- Transfer
- Experimentation

Personal Skills
- Self-confidence
- Leadership
- Communication
- Cooperation
- Work habits
- Safety habits
- Efficiency
- Quality
- Productivity

FIGURE 21.2 Goals of laboratory instruction in agricultural education. (ADAPTED FROM PHIPPS ET AL., 2008.)

Curriculum

The curriculum is a major determining factor in using agricultural education laboratories. As an example, a curriculum with a strong emphasis on agriscience will include instruction on plants. Instructional areas may include cells, plant functions, propagation, and various biotechnology processes. Following brain-based learning theory, discussed in Chapter 11, students might conduct a learning exercise to see the wholeness of a concept, such as how plants take up and transport water. They would then receive classroom instruction to learn the content and principles of plant water transportation. The instruction would end by using the laboratory for experiments on plant water transportation.

Another curricular factor is the degree of learning and the purpose of learning. In the example above, the exercises and experiments could be conducted in a classroom with tables and borrowed equipment. However, if the curricular focus is career-related, aimed at gaining the knowledge, skills, and dispositions required for employment in biotechnology fields, then a fully equipped agriscience laboratory is needed. A career-related focus would also require more time be spent in the laboratory than an awareness-related focus. Students expected to gain competence in an area need additional practice in the skills connected with that area.

Student

The maturity and developmental levels of the students must be taken into account when deciding how to use laboratories, as was covered in Chapters 16 and 17.

Middle school students may use laboratories frequently; however, their uses will probably be short in time and varied in focus. Laboratories provide this age group with concrete examples of abstract concepts. Physical activity is also involved, which can enhance learning.

High school students in advanced classes, on the other hand, may use laboratories a significant amount of time and focus on specific skills. Laboratories can be used to extend instruction into real-world settings using equipment found in the workplace. Some instructors utilize a project-based approach in which students are learning content through exploration of a "messy, ill-structured" problem. Project-based learning often involves extensive use of laboratory facilities in conjunction with student-driven exploration of knowledge and information needed to carry out the project.

While hands-on learning is a hallmark feature of agricultural education, it is important for instructors to take a "hands-on, minds-on" approach to ensure there is purpose and focus to the laboratory learning activities. It is necessary to strike a balance between classroom and laboratory-based learning activities to allow students to learn the "what" and the "why" of the content, as well as the "how." Without emphasizing the "what" and "why" behind the applied learning activities, student motivation and engagement will likely decrease.

When planning laboratory instruction, it is also important to remember some of your students, particularly those of middle school and early high school age, may be physically quite small. Operating tractors, power tools, and other large equipment may be very intimidating to them, and in some cases, not possible. For example, many new garden tractors are equipped with a safety feature that does not allow the engine to start if the person sitting in the operator's seat does not weigh at least 100 pounds. Some of your students may not be able to start the tractor, not because they are afraid, but because of the safety shut-off.

The loud sounds of power tools and heavy equipment may not only be intimidating to some students, but also trigger panicked reactions from students who are sensitive to sudden loud noises, such as students with autism. In addition to demonstration and walking students through the safety steps in operating a tool, providing them the opportunity to practice the physical motions of operating a tool prior to turning it on can help them be more comfortable with what to expect when the tool is in operation.

FIGURE 21.3 Operating large equipment and machinery, such as a tractor, may be intimidating to some students. It is important for the teacher to consider students' physical and emotional maturity when designing hands-on learning experiences to ensure a positive and safe learning experience.

The agriculture teacher must determine the ***readiness level*** of students for laboratory instruction. Do the students possess the requisite background knowledge in English, mathematics, or science? Has the requisite background knowledge in agriculture been learned to sufficient depth? Are the students physically capable of performing the required tasks? Are the students emotionally, mentally, and behaviorally ready? These are important questions for which the answers may be different among class periods during the same school year and between school years. Teachers can utilize data from standardized assessments such as state testing or district literacy tests to contribute to their decision making, in addition to their observations, classroom assessments, and conversations with their students and their families.

Facilities

Facilities are another factor in determining laboratory use. The first consideration is inventorying what facilities are available for the agriculture teacher to use. Obviously, in a single-teacher agriculture program, all laboratory facilities are available for use. However, within a multiple-teacher program, a schedule must be developed for using the laboratories. There may be other laboratories within the school available for the agriculture teacher to use. Common shared facilities with science programs include small greenhouses, science laboratories, and outdoor nature laboratories. Common shared facilities with other career and technical education programs include mechanics laboratories, food science laboratories, and computer facilities. Computer and other mediated-instruction facilities are many times shared by the entire school and coordinated by the school's media specialist.

The condition of laboratory facilities will affect their usefulness. Facilities containing out-of-date or broken equipment lead to less than desirable laboratory instruction. In many cases, laboratories that were built for class sizes of 20 students now must accommodate class sizes of 30 or more. This strains both the space and the equipment, and can lead to safety concerns. Students may not get the practice opportunities required for skill mastery.

In other instances, laboratory facilities built for one purpose can be converted to serve other purposes. Many aquaculture and hydroponics laboratories are housed in facilities once built for agricultural construction. Before converting facilities, the agriculture teacher should go through the program planning process described in Chapter 5. This will help the agriculture teacher determine whether the new use is the best for the program.

Teacher

The teacher's comfort level with agricultural laboratories influences how these laboratories are used. It is important during teacher preparation that preservice students are exposed to and get experience with a variety of agricultural education laboratories. Some of this experience will occur in college classes, but much must be obtained by students in field experiences, in student teaching, and on their own initiative. Prospective agriculture teachers should reflect on their areas of need regarding laboratory instruction and develop plans for obtaining necessary knowledge and experiences.

The care and maintenance of agricultural laboratories also affect the teacher. Does a laboratory require night, weekend, holiday, or summer care? If so, is this the responsibility of the teacher? The answer is yes, even if students do most of the work as part of SAE or other arrangements. The teacher has the ultimate responsibility for the health and well-being of the plants and animals in agricultural education laboratories. Teachers may need to plan for how animals or plants may be cared for during the summer months, particularly if they do not have an extended contract. Some teachers receive compensation or compensatory days for the outside-of-regular-hours work they do in connection with agricultural education laboratories.

MANAGING LABORATORIES

Managing agricultural laboratories requires the teacher to plan, organize, and be efficient. The teacher is in charge of assuring proper use and care of laboratories. The skills needed are in addition to the instructional skills of an agriculture teacher.

The agriculture teacher must secure and maintain supplies, tools, and equipment for laboratory use. They must organize both the laboratory facilities and the instruction. Student behavior and safety are important laboratory considerations. Supplies, tools, equipment, organization, and student behavior are discussed in the following sections. Safety is discussed later in this chapter.

Supplies, Tools, and Equipment

Supplies, tools, and equipment of the proper type and amounts are critical for successful laboratory instruction. Among the duties of the teacher is to see these are available when needed, correctly used, and properly stored when not in use.

Many agricultural education programs keep supplies, tools, and small equipment within an enclosed tool room. The advantages of this system include the agriculture teacher having greater control over item use, less chance for unauthorized use, and less laboratory space taken for item storage. Other programs keep these items in wall or floor cabinets within the laboratory space. The advantages of this system include having the items readily available in the areas where they are used, the ability to see quickly if items are missing, and the elimination of students grouping in a small storage room at the beginning and ending of the class period.

Regardless of which system is used, organization is a priority. Supplies must be ordered so they will be received by the time they are needed. Otherwise, instructional time is wasted

FIGURE 21.4 Grooming supplies are kept organized on this shelf to ensure that students have easy access to the tools they need when performing dog grooming labs at Lyman Hall High School in Wallingford, Connecticut.

waiting for materials. Supplies, such as chemicals, must be checked for viability, as their expiration dates may have passed since the last time the items were used. Instructions and Safety Data Sheets (SDS) should be reviewed and filed for future reference. With the storage of all supplies, fire and health codes must be adhered to strictly.

Securing and Budgeting for Supplies

Managing supplies also includes deciding what funds will be used to purchase them. The types of funds available to agricultural education programs varies significantly from school to school and state to state. The majority of school-based agricultural education programs in the United States are found in public schools, so public funding supports the cost of running and maintaining the program. Ideally, all programs should be provided the appropriate funds to purchase and maintain equipment and to purchase consumable supplies each year. *Consumables* are teaching supplies that are used up each year during instruction. Examples

FIGURE 21.5 It is important for teachers to regularly update and organize their supplies and ensure that Safety Data Sheets (SDS) are available for all chemicals.

of consumables include nitrile gloves, welding rods, pH buffer solution, animal feed, and many more. Schools often require teachers to complete a *purchase order* to request purchases of supplies. A purchase order is a form that requests expenditure of school funds and that typically includes all information needed to place an order with a company including item numbers, item description, cost per item, number of items, and estimated shipping costs. Capital *expenditures* are equipment purchases that can be used for more than one year, such as a welder, livestock scale, or lab benches. The type of funding or account used to purchase either of these kinds of supplies depends on the type of supplies or equipment you would like to purchase.

Capital Accounts

Capital accounts are typically used for equipment purchases for long-term use. Teachers may not regularly have the chance to purchase equipment using capital funds, but administrators may ask for a list of equipment you would like to buy in the next three to five years. There is no guarantee you will be able to make these purchases, but submitting a well-researched "wish list" of supplies as well as a rationale for how these items will impact student learning can situate you to be prepared when the opportunity arises.

Instructional Supplies

Instructional supplies are typically anything that is consumable within one year of instruction. These supplies can include both laboratory consumables and day-to-day educational supplies such as flip charts, markers, and curriculum subscription services. An agricultural education program usually has a line item for supplies in its annual budget. Your school might provide you with pens, pencils, and paper, or you may be expected to utilize your yearly instructional supplies budget to make these purchases. Care must be taken to distribute the funds wisely across all courses and throughout the school year. To accurately budget for your yearly instructional supplies expenditures, it will be necessary to review your curriculum for laboratory activities you anticipate conducting throughout the year. Then, estimate the number of students who will be using the supplies and calculate the cost using price quotes from various vendors. Some companies specialize in the sale of instructional supplies specifically for agricultural education or science laboratories and put together laboratory kits that reduce the amount of work a teacher would have to do to assemble laboratory equipment each year. In other cases, it will be necessary to secure supplies for specific agricultural equipment from companies that supply the agricultural industry, and not necessarily education. Most equipment and supplies can be purchased through online ordering systems, but your school may prefer to work with local brick-and-mortar vendors as well. In some cases, you may be able to get a better price on supplies working with vendors in your community, or it is more practical to order locally.

Federal Funding—Perkins V

The Carl D. Perkins Act of 2006 was most recently reauthorized for federal funding in 2018. This legislation (now called Perkins V or the Strengthening Career and Technical Education for the 21st Century Act) provides almost $1.3 billion annually to support CTE funding across the United States. School districts may designate an individual as a director of CTE funds to distribute Perkins dollars to CTE programs. Teachers should coordinate with this CTE director to identify how they can best utilize and access the funds. CTE funds can be used for a variety of purposes, but must always be used to strengthen academic and technical skills of students enrolled in a state-approved CTE course. Perkins funds are often used to purchase equipment that might otherwise be purchased using capital accounts. Since this funding is contingent on current legislation that is subject to be updated, be sure to stay abreast of happenings in CTE funding by remaining an active member of your state affiliate of the National Association of Agricultural Educators and the Association for Career and Technical Education.

State Funding

Some states have separate budgetary funding available to CTE programs. This varies significantly from state to state, so be sure to be in communication with your state CTE or agricultural education supervisor to identify what state funds may be available to your program, and how to secure those funds.

Lab Fee Systems

Sometimes a program operates on a fee system in which each agriculture course has a lab fee associated with it. Funds from this fee are used to purchase supplies only for that course. Lab fee systems should be avoided whenever possible as they may serve as a barrier to students participating in agricultural education classes.

Fundraiser Accounts

In some cases, laboratories generate income that is used to purchase supplies. Examples include greenhouses, aquaculture ponds and tanks, and outdoor laboratories such as farm plots. Depending on your school district's policies, these funds may go back into your instructional supplies account, or might be deposited into a separate FFA account for the FFA chapter's use.

Grants

A myriad of educational grants are available to teachers at all levels. Grants usually have a list of specific requirements or intended target audiences for which they aim to supply grant funds. Some grants specifically target increasing environmental or food education, while others seek to uplift historically underserved populations or communities. Grants may be as small as a few hundred dollars, or upward of several thousand dollars. Grant applications usually take a significant investment of time and coordination of stakeholders, but they can be an outstanding source of funds for purchasing equipment or expanding educational facilities.

Grants can be found at the federal, state, and local level. They are often supported by nonprofit or community foundations. The National FFA Organization funds a variety of grant programs each year, including Living to Serve grants that can be used to conduct service-learning projects that address challenges within the local community. Usually the greater the value of the grant, the more application paperwork is involved. Partnering with another teacher or your business office manager can be a great way to help you navigate the process of applying for a grant.

FFA Alumni and Supporters

Membership in the National FFA Alumni and Supporters is open to any adult who is interested in supporting agricultural education, whether or not they graduated from an agricultural education program. Maintaining an FFA Alumni and Supporters chapter at your program can help set up an additional source of support for both the teachers and students, as well as graduates. Each FFA Alumni and Supporters chapter may have different goals, but they may be a source of support for fundraisers that could support the purchase of equipment or supplies for your program. The national organization facilitates a variety of grant programs to support agricultural education, including some that address funding purchases of equipment and supplies.

Community Donations

Occasionally, you may be contacted by area organizations or businesses with an offer of supplies or resources for your program. Smaller donations such as leftover seeds from an area feed store can be easily collected by the teacher, whereas larger donations, such as hydroponic growing equipment, would likely need to involve school business office staff to coordinate tracking of equipment. Sometimes, it may not be necessary to purchase equipment, especially those pieces that may be used only once or twice per year. In cases such as this, borrowing equipment, such as a mulch layer, can be coordinated with area farmers. Building positive relationships in your community will help add to the support and relevance of the programming offered.

Students

Finally, some programs require students to bring in or purchase their own supplies. This is especially true of agricultural mechanics, particularly if students are doing individual projects. In situations such as this, it will be important for the instructor to consider alternative resources for students who do not have access to project materials to bring into school.

Organization and Cleanup

Proper organization of laboratories and daily care are essential for efficient and safe use. Keeping a laboratory clean and neatly organized also reflects well in areas of public relations. Intentionally planning for time at the end of class to allow for cleanup and organization of supplies used that day aids the teacher in keeping the facilities well kept, but also helps students develop personal responsibility and other employability skills.

Classroom or Lab Arrangement

The arrangement of the classroom or laboratory environment can impact the flow of instruction, student collaboration, demonstration of skills, and much more. If you have the ability

FIGURE 21.6 In this food science lab in Connecticut, the agriculture teacher clearly posts student responsibilities for organization and lab cleanup. She places student names on a Velcro-backed tag and assigns names to each of the four main responsibilities to ensure all students have an opportunity to serve in each role. You can also see the grading rubric posted to the right of the bulletin board, which she uses to assess lab participation.

to move lab tables, you might consider rearranging desks for particular lab activities to encourage student communication within groups, or perhaps space them out to set up specimen identification stations.

When determining the arrangement of your teaching space, it will be important to consider the following factors.

Safety   Regardless of the layout of your teaching space, how will you ensure students can exit the room in the case of an emergency? Or, how will you be able to facilitate an appropriate response in a school lockdown or active shooter drill? If you are going outside for your lab activity, be sure to bring a handheld two-way radio to communicate with your central office in the event of an emergency. It may also be advisable to bring a first aid kit outdoors with you. Consider which workstations might make students be most prone to injury, and be sure to situate those stations in an area easily accessible and observable to you.

Quantity of students   Do you have enough space for students to safely engage in laboratory activities? Unfortunately, many agricultural education classes find themselves in the position of being assigned more students than seats in their classrooms. This can create safety hazards in both space and the number of students a teacher can safely monitor at any given time. You may need to plan for labs in which half the class is engaging in a lab activity, while the other half is working at their seats on a related assignment, being sure to flip the groups to allow all students to experience both learning activities. Long-term planning may require developing relationships with your administration and guidance departments to emphasize the need to limit class sizes for student safety.

Mode of instruction   Will students be watching the teacher demonstrate and model a skill, and then practicing it in groups or independently? If so, arranging the room so all students can clearly see the instructor will be important. If you are facilitating a lab activity in which students work in small groups and retrieve equipment from a central location, you will want to make sure the lab benches are arranged to allow students to work closely together, without necessarily focusing on the teacher in a specific location in the room. Will you be simulating a veterinary office or board room meeting? If so, then rearranging the layout of the room can help recreate the particular setting you seek to emulate.

Location of supplies   Where will supplies be located for students to access during the lesson? Will you have a central location and designate one student per group to retrieve their group's necessary materials? Or, will you have supplies set up at each station ready for student use? The decision on where to locate your supplies likely will depend on the number of students in your class and the number of supplies available. For example, there may only be one EC meter available, and you have six groups needing access to the sensor. Limited quantities of supplies usually leads to lost instructional time due to students needing to share equipment.

Access to power, water, and gas sources   Will your lab activity require students have access to electrical outlets? For example, if you are using microscopes, then it is likely you will need both a power source and a level surface upon which to position the microscope. Will you need to have running water or sinks available to conduct your activity or to assist in cleanup? Consider the location and availability of these resources when setting up the arrangement of your teaching space.

Wi-Fi access Will you need access to the internet to conduct particular lab activities? Do you have access to Wi-Fi in all parts of your school, or will you need to prioritize a certain location or setup to ensure you can still conduct your lab?

Ventilation Will lab activities necessitate the use of a hood or fan? Will certain supplies create strong odors or vapors? Do you have access to outdoor areas where your activity might better be facilitated?

Accessibility How will you ensure whatever lab activity you are planning is accessible to students who have physical disabilities that may affect their movement or fine motor control? If you have students who have hearing or vision loss, how will you make sure the arrangement and location of the lab is accessible to them?

Storage of long-term or ongoing projects Some labs require experiments or supplies be left out for an extended time frame, such as when needing to dry out soil to conduct a soil test. Do you have a space set aside for these supplies so they remain undisturbed?

Assessment If you are conducting a lab assessment, will you be setting up stations? If so, consider the spacing of the stations to allow enough space to help prevent inadvertent or intentional cheating.

Learning Centers

Many agricultural education laboratories do not have enough room or pieces of equipment for all students to do the same lab at the same time. For example, an agricultural mechanics laboratory may have only 10 arc welding stations for a class of 25 students. Or a greenhouse may be too small for all students to be at the potting station at the same time. In these and other instances, a rotation schedule using learning centers is required.

A *learning center* is an area designed for one activity or lab and allows small groups of students to complete the activity or lab at the same time. A learning center is typically self-contained with all the equipment, tools, supplies, and instructions available in one place.

Learning centers allow for efficiency in both time and space. All students are engaged in active learning for the entire class period rather than having to wait for equipment to become available. Learning centers also require less space, as a smaller quantity of each item is needed. By working in groups at learning centers, students can observe and learn from peers.

Grouping, or placing students together to accomplish a common task or goal, is an important consideration in using learning centers. Students must receive prior instruction in the background content, concepts, and principles for all learning centers so the rotations will move efficiently. Because of this, the number of centers should be limited or designed around related content.

At certain points within the laboratory instruction, students may need additional classroom instruction or reteaching. At all times, the agriculture teacher should be alert to the need for instruction at a particular learning station. An example of learning centers used in an agriscience laboratory is given in Figure 21.7.

How should students be grouped? The first task is to design learning centers with activities that take approximately the same amount of time to complete. This reduces the chance for some students to get done early and become bored and for other students to feel rushed. The second task is to place students into the groups. If possible, groups should be of equal size so all students have equal opportunity to learn and participate. In most cases, heterogeneous

Group 1	Group 2	Group 3	Group 4
Dicot seed dissection	Monocot seed dissection	Flower dissection	Stem dissection
Monocot seed dissection	Flower dissection	Stem dissection	Dicot seed dissection
Flower dissection	Stem dissection	Dicot seed dissection	Monocot seed dissection
Stem dissection	Dicot seed dissection	Monocot seed dissection	Flower dissection

FIGURE 21.7 An example of a student rotation schedule using four learning centers in an agriscience laboratory.

rather than homogeneous grouping should be used according to student ability levels. This grouping strategy contributes to student achievement gains, reduces frustration and boredom, and provides the opportunity for peer teaching (Phipps et al., 2008).

Cleanup

Laboratories must be cleaned, with tools and supplies properly stored, every class period. The responsibility for this belongs to the students. Cleanup provides the opportunity to check the condition of tools and equipment as well as to restock supplies if needed. Students need instruction on why cleanliness and order are important. Several systems exist for assigning cleanup duties to students.

With laboratories that involve few tools and little debris, such as agriscience labs, students should be responsible for their own areas. Other laboratories, such as those for agricultural mechanics, have more tools used by multiple students and produce more debris. These require the cleanup tasks be scheduled among the students. An example is shown in Figure 21.8. The number of students assigned to each task can be increased or decreased based on the number of students in the class and the particular requirements of the laboratory. The purpose of a cleanup schedule is to ensure all tasks get accomplished and all items placed back where they belong.

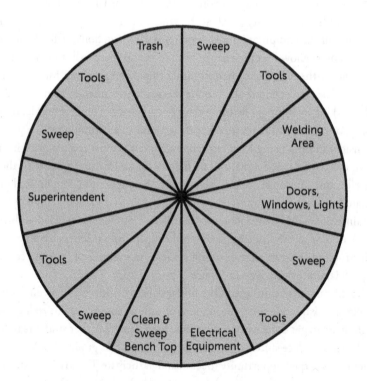

FIGURE 21.8 An example of a laboratory cleanup schedule wheel. Student names are written around the outside for each activity, or the students can be orally assigned to a group.

One student can be designated *laboratory superintendent* to assist the teacher in making sure students do their tasks and to help out on tasks assigned to students who are absent. The laboratory superintendent can also check to make sure areas are clean, with tools and supplies properly put away. The agriculture teacher may assign one student the responsibility of tool manager. This person checks tools out and back into the tool storage room and checks the condition of tools when returned.

Restocking Supplies

In some lab spaces, supplies will need to be replaced regularly. For example, in an animal science facility, it will be important to develop a system to track feeding and watering of animals, as well as a way to determine when to order more feed. Teachers can collaboratively develop a system with their students, or have a predetermined system for students. Regardless of the approach, it will be important for the students to communicate to the instructor when feed gets low so the instructor has adequate time to pick up or order more feed. If feed is not ordered within an appropriate time frame, it may be necessary to make emergency alternative plans for securing feed.

Student Behavior Issues

Laboratories involve special considerations in terms of student behavior. Inappropriate student behavior in a laboratory can cause accidents resulting in injury and harm. For this reason, great emphasis needs to be placed on students acting in a businesslike, professional manner. Horseplay cannot be tolerated. Safety considerations are discussed in another section of this chapter. (Classroom management was covered in detail in Chapter 14.)

In laboratories, students must take greater responsibility for their own actions. Laboratory instruction tends to be more individualized, so most of the time the agriculture teacher is focusing attention on the instruction of other students. Students must be self-directed and capable of performing multiple steps without prompting or teacher guidance.

Laboratories often require students to work with expensive, sensitive equipment and tools. Students must use these items properly and carefully to avoid damaging or breaking them. Students who do not follow safety protocols should not be allowed to participate in labs, as this poses a serious threat to themselves, their peers, and the instructor. The instructor should be sure to keep records of any safety violations and communicate with the student's parents/guardians as well as the school nurse and administration to keep them apprised of student behavior. Safety violations that lead to injury can easily escalate into litigious activities, so it is wise to communicate clear expectations up front, require safety examinations prior to laboratory use, keep evidence of safety training on file, and record any violations of safety.

DEVELOPING LABORATORY ACTIVITIES

Laboratory activities must be selected and planned for efficiency in the teaching and learning environment. Such selection and planning are the teacher's responsibility. It is important to note that simply because laboratory activities often take place in agricultural laboratory settings or are hands-on in nature, learning does not automatically occur. It is essential for the agriculture teacher to intentionally structure laboratory learning activities to promote quality learning experiences. Through quality learning experiences, students can more readily develop critical thinking, skill mastery, work habits, and scientific inquiry.

Experiential Learning in the Laboratory

Experiential learning is often described as a foundational approach to pedagogical design in school-based agricultural education. It is frequently associated with "learning by doing" and supervised agricultural experiences. However, experiential learning can occur across all three components of an agricultural education program, not just in SAE instruction (Baker et al., 2012). Agriculture teachers should intentionally design or select learning experiences to more effectively develop designated skills. There are various theories explaining how experiential learning should best work, rooted in the early work of famed education philosopher John Dewey. Perhaps the most widely cited version of experiential learning theory is Kolb's (1984) experiential learning theory (ELT).

Experiential learning theory describes learning as resulting from the transformation of experience (Kolb, 1984). The following six key premises serve as the foundation of experiential learning (Kolb & Kolb, 2005):

1. Learning is a *process* rather than a *product*.
2. All learning is relearning.
3. Conflict, dissonance, and disagreement drive learning.
4. Learning is a whole body and mind experience.
5. Learning results from interactions between a learner and their experiences.
6. Learning is the process of creating knowledge.

The model in Figure 21.9 illustrates how learning occurs through experiences, using the above principles as a foundation.

Each of these four learning modes of CE, RO, AC, and AE exist in tension with one another, and learners can move back and forth among each of the stages as they are going through learning experiences. The process begins with a concrete experience, which then leads to learners engaging in reflective observation of their experiences, leading to abstract

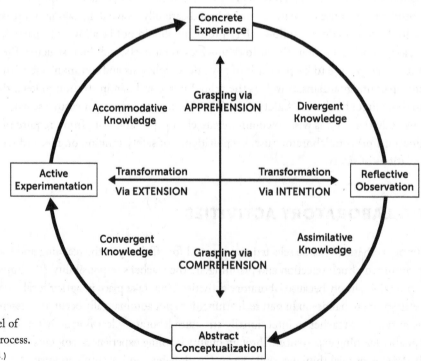

FIGURE 21.9 Kolb's model of the experiential learning process.
(ADAPTED FROM KOLB, 1984.)

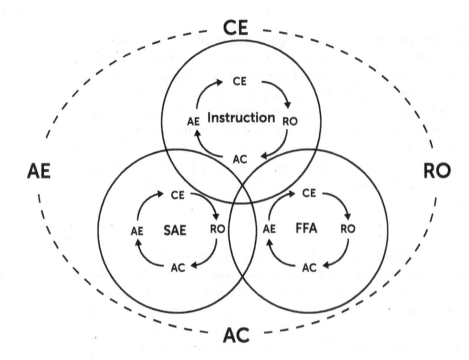

FIGURE 21.10 A comprehensive model for agricultural education in which experiential learning experiences are included within and across each component of a school-based agricultural education program. (REPRODUCED WITH PERMISSION FROM BAKER, M. A., ROBINSON, J. S., & KOLB, D. A. [2012]. ALIGNING KOLB'S EXPERIENTIAL LEARNING THEORY WITH A COMPREHENSIVE AGRICULTURAL EDUCATION MODEL. *JOURNAL OF AGRICULTURAL EDUCATION,* *53*[4], 1–16.)

conceptualization of the phenomena they may be observing or experiencing. Finally, drawing from their observations and experiences, learners test new ideas through active experimentation, which then leads to new concrete experiences. Reflection upon experiences is central to moving through these stages.

Following are recommendations on how to best apply Kolb's theory of experiential learning to agricultural education (Baker et al., 2012; see Figure 21.10):

- Experiential learning should be present in all three components of the agricultural education model, including classroom and laboratory instruction, SAE, and FFA.
- Instructors should purposefully provide support to help facilitate student learning through the experiential learning process.
- To best help students learn content and how to learn content, instructors can guide students through the experiential learning process to develop metacognitive skills.
- Curriculum planning and instructional assessment should be purposeful in incorporating experiential learning to maximize learning outcomes.

To ensure that meaningful learning persists, educators must help students purposefully process their experiences. Going into the greenhouse to water plants can easily become a mindless task. But if the instructor provides students with an experience where plants have been over- or underwatered, and students work to identify the cause through experimentation and reflective discussion, the importance of watering properly becomes more solidified. It is essential teachers do not merely provide for psychomotor hands-on experiences, but also account for opportunities for reflection, abstraction, and experimentation.

Factors in Developing Laboratory Activities

Five factors should be considered when developing lab activities. These are purpose, student readiness, methodology, resources, and time. Each of these is covered in the following sections.

Purpose of Activity

Every laboratory activity must contribute toward student attainment or growth toward established learning outcomes, or exploring unit essential questions. A teacher may use several questions to assess whether an activity should be included. How does it meet learning objectives? How does it teach or expand content, concepts, and principles? Beginning with your end goals in mind at the onset of planning ensures your lab activities align with your intended outcomes and assessment. All lab activities should be structured to meet learning objectives and to accomplish educational purposes. Although students may view labs as fun, different from classroom instruction, or exciting, this should not be the primary reason for conducting a lab. A lab should teach a specific concept in itself or expand students' knowledge of the content.

Student Readiness

Efficient use of student and teacher time and school resources requires students to be at a stage of readiness. Do the students have the background knowledge, skills, and dispositions to conduct the lab? Many lab activities are conducted after students have received classroom instruction, although this is not always the case (see inquiry-based instruction discussed in Chapter 12). Once students have learned the basic concepts and principles, they are better prepared to apply them in the laboratory. Another consideration is student maturity. Are the students physically capable of performing the gross and fine motor skills required by the lab? If not, continuing with the lab as planned may lead to student frustration and accidents.

Methodology

Another factor to consider when developing a lab activity is lab methodology. What are the steps involved in the lab? Can the lab be finished in one class period, or will it involve several class periods? Does the teacher need to demonstrate, or can students successfully complete the activity without seeing it first? Does the lab involve a mastery of skills requiring repeated practice? Content in Chapter 12 will be helpful in determining how best to structure the lab instruction.

Resources

The proper resources must be available to carry out activities. These can be limiting factors. A *limiting factor* is a resource whose unavailability or time requirements may make conducting a lab impossible or require its modification. Each of the following should be considered carefully. If the agricultural education program does not have the resources, can resources be borrowed from other programs within the school? Would a field trip to a site with appropriate resources facilitate the laboratory instruction? Are there alternative labs that will yield equivalent student learning? Can some steps be done ahead of time to allow better utilization of class time during the lab activity?

Time

As just mentioned, time is often a limiting factor in relation to resources. The time available for students to be engaged in a learning activity is important. School years and class periods are limited. Deciding whether an activity should be used requires consideration of what other learning may need to be left out of the teaching calendar. What is the best use of teaching and learning time? Is sufficient time available to carry an activity to a reasonable level of completion?

Planning Essentials

Laboratory instruction should follow a lesson plan just as should classroom instruction. The class period should be structured for maximum student learning and appropriate sequencing of activities. Because of the special characteristics of laboratory instruction, the lesson plan may include information that is slightly different from one for classroom instruction. For example, a lab lesson plan should have notes reminding the teacher how far in advance to order and receive supplies, along with any special care instructions. A lab lesson plan may also contain group rotation and laboratory cleanup notes. Evaluations of student performance on laboratory activities are essential and must be included in the lesson plan. Finally, it is always recommended the agriculture teacher actually do each lab before the students do it. This allows the teacher to see areas where questions will arise, problem areas, and areas where results may differ.

SAFETY CONSIDERATIONS IN LABORATORY TEACHING

Laboratory teaching requires attention to safety. All safety considerations must follow fire, health, and safety codes applicable to the school environment. Many of the safety standards are set by the federal Occupational Safety and Health Act and managed through the Occupational Safety and Health Administration, also known as *OSHA*. (Chapter 9 discussed safety in instructional environments.)

Safety rules, standards, and procedures are for the protection of both the student and the teacher. Not only must hearing and eye protection be provided, but its use must be enforced. Students who do not use hearing and eye protection when it is required should not be allowed to continue in the laboratory activities. Eye protection provided must be of the appropriate type. The various types of welding, for example, require differing levels of tinting.

Agricultural education laboratories may require *personal protective equipment* or PPE. Personal protective equipment is worn to reduce exposure to hazards that may cause injury. Examples include gloves, headgear, chaps, and aprons. Personal protective equipment supplied by the school such as safety glasses, welding helmets, or chainsaw chaps should be regularly inspected for damage and should be kept well maintained. Teachers should be sure to teach students the rationale for PPE and when to use it; proper use, care, and adjustment of the equipment; and limitations of the equipment. As with all aspects of safety, the agriculture teacher must be a positive role model in the use of safety equipment. Always use appropriate PPE when demonstrating or supervising laboratory activities.

Students may need to bring in a set of work clothes or coveralls so their regular clothes do not get dirty or damaged. The agriculture teacher may also consider keeping a cotton lab/shop coat available to put on over their clothing to model appropriate use of PPE. In other cases, it is appropriate for the teacher to wear coveralls to protect their clothing. Footwear is also a concern, as many agriculture labs include use of heavy equipment, working with livestock, or use of chemicals. Sandals, athletic shoes, and ballet flats do not provide the required protection for many agricultural education lab activities. Students and the teacher should plan to wear sturdy, closed toe and closed heel shoes that have a nonslip tread. Any clothing or footwear the students are expected to supply should be clearly communicated to students and their parents/guardians at the beginning of the course (preferably in the course syllabus) so all parties are aware of the safety expectations. If a student does not have appropriate attire to participate in the lab, the teacher will need to find an alternative assignment or location for the student, or have supplies readily available for student use. In some cases, it might be helpful for the school to purchase a set of lab coats or coveralls and boots for students to use during class time to avoid putting students from lower-income families in the position to need to supply additional clothing for class.

Students must also protect their respiratory functions. The appropriate dust mask should be worn when sanding, grinding, or spray painting. If students are involved in applying pesticides or other chemicals, special care should be taken to ensure the appropriate respiratory protection is used. The teacher should be sure all activities involving a possible respirator hazard are conducted in locations with appropriate ventilation/filtration systems or outdoors.

HAZARDS IN LABORATORY SETTINGS

Several kinds of hazards may be present in an agricultural education laboratory. These vary with the nature of the instruction and the supplies or materials used. School boards may have regulations about the presence of some materials on the school grounds. Always know and abide by the regulations of the school board.

Corrosive Materials

A *corrosive material* is one that can cause chemical reactions that damage metal and can injure people. Corrosives are of three kinds: acids, bases, and other miscellaneous materials.

Some of the most corrosive acids in school laboratories are sulfuric acid, hydrochloric acid, acetic acid, nitric acid, and phosphoric acid. Common bases include sodium hydroxide (lye), potassium hydroxide, and aqueous ammonia. Miscellaneous materials include iodine, mercuric chloride, bromine, phenol, and ferric chloride. Besides being corrosive, materials containing mercury are toxic.

Fertilizers containing nitrogen, sulfur, and other materials can pose corrosion problems. Such fertilizers should be properly stored. Equipment for fertilizer application should be thoroughly cleaned after use.

Flammable Materials

A *flammable material* is a material that easily catches on fire and burns. Many factors, such as flash point and ignition temperature, should be taken into account with a flammable material. Flash point is the lowest temperature at which a chemical will give off vapors to form a flammable mixture in the air. Ignition temperature is the lowest temperature at which the vapor will ignite and continue to burn without heat from another source.

Gasoline, kerosene, diesel fuel, acetic acid, propane, and diethyl ether are examples of highly flammable materials in the agriscience and technology laboratory.

Organic solvents give off vapors that are flammable. Common examples of such solvents in the laboratory are acetone, benzene, ethanol, and toluene. Paints, varnishes, paint removers and thinners, and similar materials are often found in agricultural mechanics laboratories.

Toxic Materials

A *toxic material* is a substance that is poisonous to humans, other animals, and plants. Toxic materials are usually kept in containers marked with a skull and crossbones. Most pesticides are toxic. Not only should they be used properly, but they should be stored properly.

It is a good idea to know how various substances can enter the body. Toxins can enter the body through inhalation. These irritate the nose and lungs and find their way into the bloodstream. Once in the bloodstream, they can cause injury in other parts of the body. Chemicals used as fumigants in greenhouses and other places can be particularly dangerous.

Some toxins are ingested. Most people would not think of purposefully eating a chemical; however, not washing their hands can lead to chemicals entering their bodies on food or in other ways.

Absorption occurs when chemicals enter the body through the skin. Rubber gloves, aprons, boots, and other protective devices should be worn to keep chemicals off the skin.

The effects of chemicals may occur immediately after exposure, or they may take a while to develop. They may range from mild nausea to cancer.

Teachers should be well aware of the school procedures regarding student exposure to toxic materials. The number of the nearest poison control center should be posted by the telephone.

Oxidizers and Reactives

Oxidizers and reactives are chemicals that can explode or react violently in other ways. An *oxidizer* is a material that can initiate and promote combustion. A *reactive* is a material that reacts with water or other materials. Many pyrophoric materials are reactives.

Some of the common fertilizers and chemicals used in agriculture are explosive. Ammonium nitrate fertilizer is an example. Other explosives include isopropyl ether, potassium nitrate, nitric acid, potassium chlorate, potassium, and sodium.

Some of these chemicals are very dangerous and become increasingly unstable with age. The agriculture teacher should obtain only those chemicals that will be needed and only in the amounts needed. Specialists in chemical disposal should be consulted when potential problems exist.

Electric Shock

Electric shock hazards are present in many agricultural education laboratories. Students and teachers may be exposed to these when using electricity to heat, view, and prepare items in a laboratory. Only approved equipment should be used. Electrical service should be installed according to the electrical codes. Ground fault circuit interrupters (GFCIs) should be used if there is a possibility of water being in the work area.

Aquaculture and hydroponics laboratories often have a dangerous mix of water and electricity. All electrical work should be done by licensed electricians. All electrical devices should have GFCIs and be kept in good operating condition.

Learning activities in applied physical science may involve the study of electricity. Such activities should be carefully planned and monitored to ensure student and teacher safety. Whenever possible, low-voltage power sources should be used.

Sharps

Depending on the nature of the laboratory, agricultural education activities may involve the use of *sharps*, or devices that have sharp points or edges that can cut skin. Sharps are typically found in medical settings and include items like needles, syringes, and scalpels. Used sharps should be disposed of using an appropriate sharps disposal container.

Moving Blades and Parts

Working in the agricultural mechanics laboratory space presents several opportunities for students to work with hand tools and power tools. With this opportunity also comes increased opportunity for injury. Agriculture teachers should take great care to ensure blades and equipment are in good working order, and they are appropriately sharpened to ensure safety. Students should be trained and closely supervised in the safe use of this equipment. Well-maintained tools present less of a risk for injury than poorly maintained equipment.

Cutting meat and processing vegetables also may involve the use of hand tools such as knives, or power tools such as band saws. In addition to blade safety, students should also be mindful to ensure they are following good agricultural practices (GAPs) to assure food safety.

Finally, working in the land laboratory may expose students to moving blades and parts. Cutting brush with a machete, cutting down branches with a chainsaw, using a hedge trimmer to shape bushes, attaching a mower to the PTO of a tractor, all actively involve students using hand and power tools in these settings. The agriculture teacher should be sure to provide appropriate instruction in the safe use of these tools, provide appropriate PPE, and ensure students put away tools at the end of use to avoid rusting.

SAFETY WITH SPECIMENS

Lab activities may involve animals, plants, and various microorganisms that pose potential hazards. Students and teachers should exercise safety. They should also assure the well-being of animals used in the activities.

Animals

Agricultural education courses may use live animals, animal tissues, and animal specimens for instructional purposes. Livestock may be kept in pens, barns, or pastures. Small animals may be kept in cages or pens. Fish may be kept in tanks or ponds. Preserved animal tissues and specimens may be purchased from supply houses. Fresh animal tissues may be secured from processing facilities and kept frozen in a chest freezer designated for specimen storage.

Live Animals

With any use of animals, care needs to be taken for the safety and health of students. Live animals may injure students through kicks, bites, scratches, and other bodily harm. Students may also get diseases, parasites, or other contaminations from animals. Some students may be allergic to animal hairs or skin, and animal odors can affect students as well.

All live animals must be properly fed, watered, housed, and maintained. Students should receive handling instructions before working with live animals and should wear appropriate PPE. If students are injured, they should report to the school nurse immediately. The school policies regarding reporting and caring for injuries should be followed.

It is important to remember that while the agriculture teacher is utilizing live animals to engage students in instruction, those animals will become an attraction for other students, faculty, and staff in your school to come visit the agriculture program facilities. This can be a great way to build relationships and increase public relations efforts, but it can also be a challenge to manage from a biosecurity standpoint. Given the likelihood that many people visiting your animal facility will develop emotional attachments to the animals, it is even more important to ensure the public understands how the welfare of the animals is being assured. Clearly posted signage regarding safety and wellness procedures for the animal facilities can be one way to help communicate this to both students and the public.

Animal Specimens

Agriculture teachers might plan activities for students to process carcasses in a meat science unit in which students are working with exsanguinated animal tissues. It is important for students to have appropriate PPE when using knives that can cut through sinew and muscle, including chain mail gloves, cut-resistant gloves, and latex or nitrile gloves, as well as a laboratory overcoat.

FIGURE 21.11 Students tending animals in school labs should be instructed in proper safety.

In other situations, students might be dissecting preserved or fresh animal tissues such as a heart or reproductive tract. Dissection involves the use of sharp tools students should be trained how to use safely prior to use in the lab setting. Specimens should also be stored and disposed of in accordance with appropriate handling instructions. If fresh specimens are used, they should be stored in a labeled freezer designated only for dissection specimens. Students should wear appropriate PPE including latex or nitrile gloves and safety glasses. It is important to have both latex and nitrile gloves available for students, as some students may have an allergy to one or the other.

Plants

Although plants cause fewer complications than animals, care should still be taken when working with them in the laboratory. Safety problems associated with plants include poisonous plants and allergic reactions. Teachers working with vegetables such as hot peppers should be mindful of the effects of capsaicin when processing peppers and provide appropriate PPE and ventilation. Poison ivy is the most common example of a plant that causes an allergic reaction. Table 21.1 gives examples of plants that are poisonous to humans.

TABLE 21.1

Examples of Toxic Plant Injuries

Symptom	Plant*
Damage to internal organs (stomach, heart, liver, and kidneys)	Azalea, bird-of-paradise, castor bean, daffodil, foxglove, holly, hyacinth bulbs, mistletoe, nightshade, poinsettia, rhododendron, sweet pea, and wisteria
Skin rashes and blisters	Amaryllis; carnation; chrysanthemum; daffodil; geranium; iris; pencil cactus; poinsettia; poison ivy, oak, and sumac; weeping fig; and tulip bulbs
Swelling of mouth, upset stomach, and breathing difficulties	Calla lily, dieffenbachia, and philodendron

Data source: Lee (2000).

*Plant species listed here have been known to cause hazards to some individuals, companion animals, and other animals. Plants may be safe to some individuals and cause reactions in others. If an individual ingests any of these plant materials and appears ill, do not induce vomiting until instructed to do so by a poison control expert. If harmful plants come into contact with the skin, wash the area immediately with soap and water.

Microorganisms

Bacteria, fungi, protozoa, and other microorganisms are often used in, or are present in, the laboratory. Some are kept or produced for useful purposes, such as in the biofilter of an aquaculture facility. Others are pests transported into the facility in some way, such as with the purchase of new plants for the school arboretum. Students should always follow appropriate procedures when working with microorganisms. These usually include wearing protective gloves and masks, washing hands thoroughly, and not having food or drink in the laboratory.

ASSESSING LEARNING IN THE LABORATORY SETTING

Chapter 19 described how teachers evaluate learning in agricultural education. It is important to plan how you intend to assess student learning at the beginning of instruction so subsequent learning activities are aligned with your assessment.

There are two major types of assessment: formative and summative. Formative assessment checks for student understanding throughout the learning process, whereas summative assessment evaluates student learning at the end of an instructional unit. Considering much of laboratory-based instruction engages students in hands-on learning experiences, assessment in the lab setting is often performance-based or authentic assessment.

Performance-Based Assessment

In *performance-based assessment*, the instructor evaluates how well a student can do something under given conditions. For example, students may be asked to demonstrate proper tomato trellising techniques in the hydroponic greenhouse, or asked to perform various welds. Since several students are being evaluated in this setting, the instructor usually has to be efficient in their evaluation. Consequently, many teachers use rubrics or checklists.

Teachers might also choose to use a longer term, more in-depth performance-based assessment to assess learning. These are often structured in the form of a group project in which students are provided a scenario, and working collaboratively they develop recommended solutions to the problem. For example, students in a veterinary science course might be tasked with identifying potential alternative therapies to add to a growing veterinary practice. They then work together to select a therapy and justify the introduction to the practice by putting together a sales pitch presentation to the lead veterinarian(s) of the practice. Teachers could serve as the veterinarian in this situation, or might invite veterinary professionals in to serve as judges. Team problem-solving presentations such as these are frequently used as the team event portion of several National FFA Career Development Events.

Performance-based assessment pairs well with project-based learning approaches. In this case, a whole unit or course uses a project-based approach.

Authentic Assessment

Authentic assessment is a type of performance-based assessment that places the student in a situation or simulation that closely mimics the workplace or other life setting. Since agricultural laboratories often are designed to simulate agricultural facilities in industry, teachers can use these spaces to replicate situations in which students demonstrate career readiness. When time, space, or management challenges prevent a teacher from using laboratory facilities to integrate authentic assessment, an alternative approach would be to use computer simulations or provide students with scenario-based questions on tests or quizzes.

Product Versus Process

When designing your assessments, it is important to consider what you are actually assessing. In product-based assessment, the teacher is evaluating the quality of the desired product. For example, in a floriculture class, students might be assessed on the quality of their final arrangement, comparing the product to the expectations outlined in a rubric. In contrast, process-based assessment evaluates the learning process that led up to creation of a product. In this approach, the floral arrangement itself might not be the focus of assessment, but rather the process students took to get to the final product: Did they use the appropriate tools without prompting? Select appropriate flowers and greenery? Set about their work without prompting?

Process-based assessment is increasingly used to assess development of *employability skills*, or those "personal qualities, habits, and attitudes that influence how you interact with others" (Minnesota State CAREERwise, 2021). Laboratory instruction is ideally situated to integrate opportunities for students to develop and apply these skills. To learn more about employability skills, visit https://cte.ed.gov/initiatives/employability-skills-framework. Integrating various FFA resources such as LifeKnowledge into your instruction can easily address these skills in your laboratory instruction.

Daily Participation or Engagement

Considering hands-on learning experiences are a key piece of agricultural education, many teachers want to assign grades to assess student engagement in these activities. It is important to establish evaluation criteria that are clearly communicated to students and their parents/guardians in advance. Arbitrary assignment of daily participation should be avoided. To more clearly address development of career-readiness in their program, the Rockville High School Agricultural Education Program in Rockville, Connecticut, developed a rubric along with input from members of their advisory committee. Figure 21.12 (p. 408) is an example of how to integrate assessment of employability skill development into assessment of student performance.

REVIEWING SUMMARY

Laboratory instruction in agricultural education is important for teaching applications of concepts and principles. In the laboratory, students gain needed skills and expanded knowledge regarding the content. In most cases, laboratory instruction involves the use of psychomotor skills, which are not used as often in classroom instruction.

Laboratory instruction in agricultural education provides for multiple opportunities to structure experiential learning opportunities for students. Teachers should be sure to address all four stages of Kolb's (1984) experiential learning model to help best ensure lasting learning.

Deciding when and how to use agricultural education laboratories involves looking at several factors. A major consideration is how a laboratory fits into the curriculum. The agriculture teacher must also decide to what extent students are ready mentally and physically for the laboratory instruction. Facilities also play a role in the decision process. A final factor is the teacher and their ability to manage the laboratory.

The management of laboratory space and supplies is an important responsibility. The agriculture teacher must decide arrangement of the teaching space, management procedures, purchasing plans, organization, and care. As much as possible, the agriculture teacher should make students responsible for the maintenance of laboratory organization and cleanliness. Students should rotate among tasks and should be taught the reasons behind the tasks they are doing.

Class _____

Employability Skills Rubric

Directions: Each student will complete the following rubric for each 4-week unit of instruction. The student self-grade will be taken into consideration when the teacher formulates the student's participation grade. Employability skills (or class participation) will represent 20% of the unit grade.

Criteria	Grade			Student Self Grade	Teacher Grade
	3 Excellent	2 (Room for Improvement)	1–0 (Needs Improvement)		
Attendance	Student has no *unexcused* absences and is always on time	Student has no *unexcused* absences and has been tardy 1 or 2 times	Student has 1 or more *unexcused* absences and/or is frequently tardy to class		
Attitude	Student is *consistently* positive, motivated, and enthusiastic	Student is positive, motivated, and enthusiastic *most of the time*	Student is *rarely* positive, motivated, and enthusiastic		
Cooperation	Student *always* cooperates with students and teacher and *consistently* demonstrates self-control	Student *almost always* cooperates with students and teacher and *almost always* demonstrates self-control	Student *rarely* cooperates with students and teacher and *frequently* lacks self-control		
Follows Directions	Student *always* follows directions	Student *almost always* follows directions	Student *frequently* does not follow directions		
Communication Skills	Student *always* uses active listening skills and social cues, asks for help when needed in an appropriate manner	Student *almost always* uses active listening skills and social cues, asks for help when needed in an appropriate manner	Student *rarely* uses active listening skills and social cues, asks for help when needed in an appropriate manner		
Safety	Student *always* follows safety rules	Student *almost always* follows safety rules	Student *rarely* follows safety rules		
Responsibility	Student is *always* dependable, honest, and on task	Student is *almost always* dependable, honest, and on task	Student is *rarely* dependable, honest, and on task		
Self-Starter	Student *always* demonstrates appropriate initiative	Student *almost always* demonstrates appropriate initiative	Student *rarely* demonstrates appropriate initiative		
Organization	Student *always* prepared for class: materials are organized and available	Student *almost always* prepared for class: materials are organized and available	Student *rarely* prepared for class: materials are *rarely* organized and available		
Dress and Grooming	Student *always* follows RHS Student Handbook concerning dress and personal hygiene; industry-accepted attire/dress at all times	Student *almost always* follows RHS Student Handbook concerning dress and personal hygiene; industry-accepted attire/dress at all times	Student *rarely* follows RHS Student Handbook concerning dress and personal hygiene; attire/dress *rarely* industry accepted		

FIGURE 21.12 This employability skills rubric is an example of how to assess student engagement in laboratory activities across a unit of instruction. (ADAPTED FROM ROCKVILLE HIGH SCHOOL AG ED PROGRAM, ROCKVILLE, CONN., 2009.)

It is important to plan ahead as much as possible when securing and budgeting for supplies and equipment necessary to utilize agricultural education laboratories. Teachers should work with their CTE director, administration, and school district business office/treasurer to coordinate budgets and paying for supplies.

Safety is critical in agricultural laboratories. Agriculture teachers are responsible for following fire, health, and OSHA codes and regulations. They are also responsible for modeling proper safety habits and practices.

Assessment in the laboratory setting is often performance-based. Instructors should consider if they are assessing the process of learning or the product of learning when designing assessments.

QUESTIONS FOR REVIEW AND DISCUSSION

1. Why are laboratories important in agricultural education instruction?
2. What are the factors for determining when and how to use agricultural education laboratories?
3. What are some differences in using laboratories with middle school and high school students?
4. How does the teacher influence laboratory activities?
5. What are different systems for storing supplies, tools, and small pieces of equipment?
6. What are the advantages and disadvantages of heterogeneous grouping of students according to ability levels?
7. Why are laboratory cleanliness and organization important?
8. Why is student behavior of special concern in laboratory situations?
9. What protective equipment and clothing might be necessary in an agriscience laboratory?
10. How are laboratory activities developed for effective instruction?

ACTIVITIES

1. Investigate the fire and health codes and requirements for agricultural education laboratories in your state. Begin by interviewing the safety coordinator or person in charge of safety in a local school district.
2. Choose an agricultural education laboratory facility, such as one for agriscience or agricultural mechanics. Design a learning centers rotation schedule. In addition, put together a list of supplies, tools, and equipment needed for the learning centers. Include quantity, cost, and potential suppliers.
3. Using the *Journal of Agricultural Education*, *The Agricultural Education Magazine*, and the proceedings of the National Conference of the American Association for Agricultural Education, write a report on research about laboratory instruction within agricultural education. Current and past issues of these journals can be found in the university library. Selected past issues can be found online at www.aaaeonline.org or www.naae.org.
4. Explore the tenets of Kolb's (1984) experiential learning theory more closely. Intentionally plan a lesson or unit that leverages each of the four stages of the experiential learning cycle and identify ways the teacher can support student metacognition of their own learning.
5. Assemble a folder or binder of Safety Data Sheets (SDS) for common chemicals found in a laboratory setting of your choice. Familiarize yourself with the handling and storage, and emergency procedures for each item.

REFERENCES

Association for Career and Technical Education. (2019). What is career and technical education? *ACTE Online.* https://www.acteonline.org/wp-content/uploads/2019/03/What_is_CTE_infographic _2019.pdf

Baker, M. A., Robinson, J. S., & Kolb, D. A. (2012). Aligning Kolb's experiential learning theory with a comprehensive agricultural education model. *Journal of Agricultural Education, 53*(4), 1–16. https://doi.org/10.5032/jae.2012.04001

Kolb, A., & Kolb, D. A. (2005). Learning styles and learning spaces: Enhancing experiential learning in higher education. *Academy of Management Learning & Education, 4*(2), 193–212.

Kolb, D. A. (1984). *Experiential learning: Experience as the source of learning and development.* Prentice Hall.

Lee, J. S. (2000). *Program planning guide for agriscience and technology education* (2nd ed.). Pearson Prentice Hall Interstate.

Littky, D., & Grabelle, S. (2004). *The big picture: Education is everyone's business.* Association for Supervision and Curriculum Development (ASCD).

Minnesota State CAREERwise. (2021, June 29). *Employability skills.* https://careerwise.minnstate .edu/careers/employability-skills.html

National Council for Agricultural Education. (2016). *National quality program standards for agriculture, food, and natural resources education: A tool for secondary (grades 9–12) programs.* https:// thecouncil.ffa.org/

National Research Council. (2006). *America's lab report: Investigations in high school science.* Washington, DC: National Academies Press. https://doi.org/10.17226/11311

Phipps, L. J., & Osborne, E. W. (1988). *Handbook on agricultural education in public schools* (5th ed.). Interstate Printers & Publishers.

Phipps, L. J., Osborne, E. W., Dyer, J. E., & Ball, A. (2008). *Handbook on agricultural education in public schools* (6th ed.). Thomson Delmar Learning.

Rice, A. H., & Kitchel, T. (2017). Teachers' beliefs about the purpose of agricultural education and its influence on their pedagogical content knowledge. *Journal of Agricultural Education, 58*(2), 198–213. https://doi.org/10.5032/jae.2017.02198

Roberts, T. G., & Ball, A. L. (2009). Secondary agricultural science as content and context for teaching. *Journal of Agricultural Education, 50*(1), 81–91. https://doi.org/10.5032/jae.2009.01081

ADDITIONAL SOURCES

Davis, O. L. (1998). From "hands-on" to "minds-on." *Journal of Curriculum and Supervision, 13*(2), 119–122. http://www1.ascd.org/publications/jcs/winter1998/Beyond_Beginnings@_From_%E2%80% 9CHands-On%E2%80%9D_to_%E2%80%9CMinds-On%E2%80%9D.aspx

Froschauer, L., & Bigelow, M. L. (2012). *Rise and shine: A practical guide for the beginning science teacher.* National Science Teachers Association Press.

International Institute for Education Planning Learning Portal. (2021, June 29). *Glossary.* United Nations Educational, Scientific, and Cultural Organization. https://learningportal.iiep.unesco .org/en/glossary/c

Shoulders, C. W., & Myers, B. E. (2013). Teachers' use of experiential learning stages in agricultural laboratories. *Journal of Agricultural Education, 54*(3), 100–115. https://doi.org/10.5032/jae.2013.03100

U.S. Department of Labor. (2021, June 29). *Personal protective equipment.* https://www.osha.gov /personal-protective-equipment

U.S. Food and Drug Administration. (2018, August 30). *Safely using sharps (needles and syringes) at home, at work and on travel.* https://www.fda.gov/medical-devices/consumer-products/safely -using-sharps-needles-and-syringes-home-work-and-travel

Part 4

Supervised Agricultural Experience, FFA, and Community Resources

22

Supervised Agricultural Experience

Jose, a freshman agriculture student, is starting his agricultural career exploration and planning. He is starting the foundation for a career pathway in the field of agriculture.

Cheehlu, a sophomore agriculture student, is spending a Saturday shadowing Ms. Matsumoto, the owner of Matsumoto Floral. Cheehlu would like to start her own floral shop when she finishes high school.

Asia, a junior agriculture student, is spending the evening working on completing the skills gained summary in his records. Asia has a small flock of purebred Suffolk ewes and one ram. He plans to expand his sheep flock over the next few years and sell project lambs to 4-H and FFA members in his community.

Kelsie, a senior agriculture student, is devoting her weekend to completing the written report for her Agriscience Fair project, which is a part of her ongoing research on plants. She is planning to attend college next year and will major in plant science. She plans to become a plant breeder and would eventually like to work for the USDA or a land-grant university.

What do these four students have in common? They are all participating in supervised agricultural experience activities that may eventually lead them into an agricultural career.

Experience is important to all of us. Employers want employees who have had experience. Learners like the notion of firsthand experience. Agricultural educators place emphasis on experience for students.

OBJECTIVES

This chapter addresses the National Quality Program Standards for Agriculture, Food, and Natural Resources Education (National Council for Agricultural Education, 2016), specifically Standard 2: Experiential, Project, and Work-Based Learning Through SAE. It has the following objectives:

1. Explain the meaning and importance of supervised agricultural experience.
2. List and distinguish between the types of SAE.
3. Describe how to develop SAE, including training plans and agreements.
4. Discuss the role of supervision.
5. Describe record keeping and the kinds of records kept.
6. Explain how to evaluate SAE.
7. Relate SAE to FFA achievement.

TERMS

experiential learning
FFA Agriscience Fair
FFA degree program
foundational SAE
immersion SAE
National FFA Scholarship Program
ownership/entrepreneurship
supervised agricultural experience
placement/internship supervised
agricultural experience
proficiency award

research supervised agricultural experience
school-based enterprise supervised agricultural experience
service-learning supervised agricultural experience
Star Award
supervised agricultural experience training agreement
training plan
training station

FIGURE 22.1 On-site supervision and instruction of a student supervised experience promotes proficiency development.

To experience something, an individual must be actively involved in sensing it or making it happen. Many students learn best by actually doing a particular activity or job. Learning by doing is often called *experiential learning*. Learning through actual experience is usually fun; additionally, students also learn from their mistakes.

Knowledge gained through experience is often of greater depth than that gained through memorization. For example, students can read about how to drive a car, but until they actually get behind the wheel and try to drive, they do not fully understand what is required to be a successful motor vehicle operator. The "behind the wheel" experience follows general instruction on how to operate a vehicle safely.

THE MEANING AND IMPORTANCE OF SUPERVISED AGRICULTURAL EXPERIENCE

In agricultural education, supervised agricultural experience (SAE) has traditionally been thought of as the world of work learning component of the program. It is a "student-led, instructor supervised, work-based learning experience that results in measurable outcomes within a predefined, agreed upon set of Agriculture, Food and Natural Resources (AFNR) Technical Standards and Career Ready Practices aligned to a career plan of study" (National Council for Agricultural Education, 2017a, p. 2; 2017b, p. 2). Supervised agricultural experiences should improve agricultural literacy and/or the skills and abilities required for a student's career.

Supervised agricultural experience is the part of agricultural education that provides students with the opportunity to actually apply and build on what they learn in the classroom. SAE activities range from growing plants and animals to using technology in conducting an in-depth study of a particular topic of interest to an individual student. More advanced supervised agricultural experiences include actual employment in a business or the creating

FIGURE 22.2 Supervised agricultural experience provides the opportunity for students to develop hands-on skills.

and running of students' own business operations. Supervised agricultural experience is sometimes referred to as the work-based learning component of agricultural education.

The objective of supervised agricultural experience is to provide planned, goal-oriented, and practical activities that help students develop the skills needed to be successful in the workplace or in life. Supervised agricultural experiences should be planned around the goals and career plans of the individual student.

Teachers supervise the experiences to make them more meaningful and relate them to the classroom instruction. Teachers should involve and cooperate with parents and other adults to provide the most meaningful experiences for students. Older students may work in businesses and are supervised by their employers in cooperation with the teacher.

Students should be actively involved in deciding and planning the supervised agricultural experiences that are most interesting and important to them. Each student should list career goals before beginning to plan supervised agricultural experiences.

In 2010 the National Council for Agricultural Education began the process of revising SAE, culminating in 2017 with the complete updating of SAE titled *SAE for All* (Gossen, 2019). In addition to a revised definition, SAE was reimagined with *foundational SAE* the starting point for all agricultural education students. A foundational SAE consists of five components: career exploration and planning, employability skills for college and career readiness, personal financial management and planning, workplace safety, and agricultural literacy. Levels are awareness grades 6–9, intermediate grades 9–11, and advanced grades 11–12. All students enrolled in an agricultural education course have a foundational SAE at a minimum level of awareness. As students take additional agricultural education courses, they continue learning and gaining skills in the five components on intermediate and advanced levels.

Agricultural education students can further enrich their experiences through *immersion SAE*. The five types of immersion SAE are placement/internship, ownership/entrepreneurship, research, school-based enterprise, and service-learning. Immersion SAE has three levels of motivation, which are graded, recognition, and career-ready. Whereas all agricultural education students have a foundational SAE with all five components, immersion SAE is chosen by students as an enrichment of the agricultural literacy component. The types and levels of motivation for immersion SAE are discussed later in this chapter.

Planned Programs

The *supervised agricultural experience* (SAE) program is a series of individualized practical learning activities planned by the student, with the guidance and support of the teacher, the student's parents/guardians, and, if applicable, the employer. The program meets established minimum criteria, is supervised by a qualified agriculture teacher, and develops competencies related to the interests and career goals of the individual student. The things learned in the classroom are applied in real-life situations through supervised agricultural experiences. Carefully planned supervised agricultural experiences make in-class instruction more meaningful. Supervised agricultural experience allows the agriculture teacher to expand the boundaries of the school classroom to include the whole community as an instructional facility. All the community's agriculturalists become potential resource persons. A number of incentives for immersion SAE achievement are available to students. Many of these incentives provide FFA award recognition through contributions of National FFA Foundation sponsors.

Supervised agricultural experience, similar to FFA, is designed to accomplish identified objectives of the agricultural education program and is conducted under the supervision of the agriculture teacher. It, too, is an integral part of agricultural education instruction and considered to be intracurricular. Supervised agricultural experience adds to the instruction received during class time. It allows each student to enrich their education by focusing on the area or areas of their greatest interest.

Making decisions about the kind of supervised agricultural experience to focus on and then planning the activities of the supervised agricultural experience helps students learn about addressing real-life situations. Many students find it easier to understand classroom lessons when they apply the lessons though supervised agricultural experiences.

Supervised agricultural experiences are important for increasing students' understanding of agriculture and for developing skills and abilities they need for careers. Through supervised agricultural experience programs, students learn to deal with problems and events that occur in everyday life. They often find they need to learn more about a particular topic and are directed to specific areas where they need additional study. Some students are motivated by the opportunity to earn money, some find SAE an enjoyable way to learn, and others gain personal satisfaction from their individual accomplishments.

Benefits of Supervised Agricultural Experience

Supervised agricultural experience benefits students, teachers, employers, agricultural education programs, the local community, and the agricultural industry. Benefits for students are the most important. They include the following:

- Development of decision-making skills, including career and personal choices
- Improved self-confidence and human relations skills
- Application of knowledge learned in the classroom
- Knowledge of a variety of occupations and careers
- Development of time management and record-keeping skills
- Document of experience needed on job applications
- Discovery of areas of personal interest
- Practice of responsibility and development of independence
- Development of pride through personal accomplishment

How Supervised Agricultural Experience Came Into Being

Since its inception in public secondary schools in 1917, agricultural education has included some form of experiential education as a teaching strategy. In the early days, when "vocational

FIGURE 22.3 Teachers often provide individual instruction and guidance with supervised agricultural experience.

agriculture" students came from farms or ranches and were preparing to return there after completing their high school education, experiential learning usually consisted of production enterprises in livestock, poultry, crops, and so on, conducted on their home farms or ranches. The purpose of these "projects" was threefold: (1) to provide opportunity to develop through experience, supervised by the teacher, knowledge and skills required to conduct profitable production agriculture enterprises, (2) to demonstrate to the community modern agriculture practices, and (3) to help students (future farmers, as they were then called) to begin their actual establishment in farming.

Beginning with the 1950s, and especially after 1963, it became generally recognized that agriculture included farming and more. The U.S. Office of Education proposed classification of agriculture by occupational clusters. These were production agriculture (farming and ranching), agricultural supplies and services, agricultural mechanics, agricultural products and processing, ornamental horticulture, agricultural resources, and forestry.

With this broadened concept of agriculture, agricultural education's mission was similarly expanded. It now had the task of preparing individuals for occupations in all seven occupational clusters, and the original threefold purpose of "agricultural education projects" was no longer valid.

Today's agricultural education programs offer a wide variety of supervised agricultural experiences from which students plan their individual supervised agricultural experience programs. The important concepts for modern supervised agricultural experience programs include (1) supervision by a qualified agriculture teacher, (2) quality practical experience that provides hands-on learning opportunities for the student, and (3) activities and experiences planned and conducted around the student's personal interests and career goals.

Characteristics of Supervised Agricultural Experience

A supervised agricultural experience has the following characteristics:

1. The activity is identified with or seeks to identify a specific agricultural enterprise, occupation, career, or problem and involves the student in hands-on experiences directly associated with an enterprise, occupation, career, or problem.
2. The student's involvement in this experience includes time outside of the school's usual agricultural education class hours.
3. The student's experience may be located and/or conducted on the school premises.

4. The student plans the supervised agricultural experience with the assistance of an agriculture teacher and conducts it under the supervision of that teacher.

5. The agriculture teacher allocates a part of their time to the supervision of students' supervised agricultural experiences.

6. The student keeps records pertaining to their experience as prescribed by the teacher, and these records are periodically reviewed by the teacher.

7. The student may be engaged individually in supervised agricultural experience or be engaged cooperatively with other students.

8. The student's plan for the supervised agricultural experience includes goals and provisions for growth in scope and complexity.

Purposes of Supervised Agricultural Experience

In SAE, the agriculture teacher's main function is to serve as a manager, coordinator, or consultant of learning for guiding students as they seek careers in agriculture. Supervised agricultural experience can certainly lead to establishment in farming, but that is no longer the only goal. Purposes of today's supervised agricultural experience programs include the following:

- Providing opportunities for hands-on experience in skills and practices that lead to successful personal growth and employment in an agricultural career
- Providing opportunities to gain documented experience that can provide references for future education and/or agricultural employment
- Providing opportunities for students to identify, develop, and demonstrate personal characteristics required for successful personal growth and employment (e.g., initiative, responsibility, dependability, and self-reliance)
- Providing opportunities for students to observe and participate in the "world of work"
- Capturing, retaining, and focusing student interest in agriculture
- Providing opportunities for students to discover and deal with financial realities

TYPES OF SUPERVISED AGRICULTURAL EXPERIENCE

A foundational SAE is a part of every student's agricultural education instruction whether they enroll in only one semester-long course or multiple courses during their high school years. Every student deserves instruction, experiences, and guidance in the five components of foundational SAE.

Through career exploration and planning, students research and explore agricultural careers. Students become aware of career opportunities in the Agriculture, Food, and Natural Resources career cluster. They explore education and training requirements, potential salary and benefits, and job responsibilities. They develop a career plan, which provides a road map to obtaining their desired career.

All students are expected to be college and career ready upon high school graduation. This component of a foundational SAE helps the student develop skills in communication, critical thinking, and other related areas necessary for success in college and careers.

Personal financial management and planning is a life skills component. The goal of this component is to help students develop knowledge, skills, and practices that will lead to their financial independence.

Workplace safety is the fourth component of a foundational SAE. Many agricultural jobs fall under the classification of hazardous occupations. Workplace safety is an essential habit for all workers to develop.

FIGURE 22.4 Foundational supervised agricultural experience allows students to have firsthand contact with a wide range of areas.

As discussed in Chapter 15, agricultural literacy is important for consumers and decision makers, but especially for those employed in agricultural occupations. A foundational SAE provides the understanding about agriculture necessary for informed choices and decision making. As stated earlier in this chapter, immersion SAE is an enhancement of the agricultural literacy component of a foundational SAE. Immersion SAE is discussed next.

There are five types of immersion SAE a student may incorporate into their overall supervised agricultural experience program. A student may choose to focus on a single type of immersion SAE, or might decide to do activities and experiences in two or more types:

1. Placement/internship
2. Ownership/entrepreneurship
3. Research: experimental, analysis, or invention
4. School-based enterprise
5. Service-learning

Placement/internship is the type in which the student is employed for compensation or the work can be unpaid. A planned program of experiences is used. Placement may be on a farm or ranch, at an agricultural business, or at another community facility. The purpose is to provide practical experience and develop skills needed to enter and advance in a particular occupation. The location where a student is placed is called a ***training station***. A student with this type of SAE may, for example, work as a floral designer, fish hatchery assistant, or farm supplies store clerk. Unpaid means the student works in a job for experience only and is not compensated in any other manner for hours of labor. Placement may be in the same or similar training stations as those used for paid placement. Younger students are often in unpaid placements and then later advance to paid placements as they gain experience in particular occupations. An example of this type of SAE is working without pay for a landscaper, farmer, or local agribusiness.

FIGURE 22.5 A goat is part of this student's owner-ship SAE as well as an FFA showing activity. (COURTESY OF LEDYARD AGRI-SCIENCE & TECHNOLOGY PROGRAM.)

Ownership/entrepreneurship is the type in which students develop skills needed to own and manage enterprises. The students own materials and other resources for their enterprises. These enterprises may be individually owned, or they may be partnerships, cooperatives, or enterprises involving other forms of group ownership. They may be conducted on school property or off. The key feature is that the students engaged in this type of supervised agricultural experience have financial investment or risk in their enterprises. Examples of entrepreneurship/ownership SAE include owning a lawn care service, raising fish, growing and selling flowers, and operating a custom harvesting business.

Research can be one or more experiments, analyses, or inventions. It is the type in which the student carries out an investigation into a problem using scientific approaches. The experiences usually involve identifying a particular problem, searching for information, and then conducting a scientific experiment or using other research procedures to arrive at conclusions. The student then makes recommendations about how to solve the particular problem. Students conducting a research SAE often exhibit at FFA agriscience fairs and/or general science fairs. Research supervised agricultural experiences may involve cooperating with a science teacher or a scientist in the local community. An example of this type of SAE is studying water pollution, feed nutrients, or tissue culture.

School-based enterprise is the type in which a group of students work cooperatively at the school. The enterprise may be a product or a service. The school's resources, including facilities and equipment, are available for students to use in their SAE. A school-based enterprise SAE is especially for students who are unable to engage in other types of immersion SAE. The SAE may be funded by the agriculture program, FFA chapter, or school but it may also be funded through the students' own funds. Students may or may not share in

FIGURE 22.6 Research and experimentation supervised agricultural experience may be carried out using the school's facilities.

profits returned from these activities. An example is growing plants in a school greenhouse or fish in a school aquaculture facility. Some schools have land and livestock laboratories for group projects with field crops and livestock.

Service-learning is the type in which one or more students design one or more projects to benefit the school, community, or public organization. Service-learning SAE is not designed to benefit the FFA chapter. The SAE only works when local stakeholders provide input and participate in decision making from the beginning of the project. Service-learning SAE is designed for student learning and must include reflection and evaluation.

PLANNING AND CONDUCTING SUPERVISED AGRICULTURAL EXPERIENCE PROGRAMS

Supervised agricultural experience is the most challenging component of the agricultural education program model to implement. Large classroom enrollments, student diversity, teacher confusion about middle school SAEs, and availability of resources are often cited as barriers to student participation in SAE. Most agriculture teachers value SAE as an important part of the program but have difficulty with its implementation. To overcome these barriers, SAE for All was developed. By clearly separating FFA recognition from foundational SAE, all agricultural education students can engage in an appropriate progression of SAE activities and experiences for their own growth and development. For those students seeking advanced FFA degrees and SAE award recognition, agriculture teachers need to ensure these students know about the requirements for the advanced degrees and FFA awards based on SAEs.

The agriculture teacher's primary responsibility for supervised agricultural experience is to assure it is an essential, effective component of the overall agricultural education program. The teacher should ensure all agricultural education students are aware of its values, purposes, characteristics, and opportunities and they participate in it.

The first step in assisting students is to teach an introductory unit on supervised agricultural experience. This unit should cover the components and levels of supervised agricultural experience, emphasizing every student enrolled in the class benefits from having an SAE. The teacher should cover how supervised agricultural experience works and what students gain from it. The relationship of supervised agricultural experience and FFA should also be discussed. A list of possible supervised agricultural experiences available at the school and in the community should be developed and provided to each student. Many teachers have older students visit the class to discuss their supervised agricultural experience programs.

A training agreement and a training plan should be completed for each immersion SAE project. Templates for both are in the *SAE for All Teacher Guide* as well as part of most online and paper-based record-keeping systems. Copies of the training agreement and training plan should be made and provided to all parties involved.

A *training agreement* is a written statement of the exact expectations, understandings, and arrangements of all parties involved in the supervised agricultural experience program. It is signed by the student, student's parent(s) or guardian(s), training employer, and teacher (on behalf of the school). The training agreement also describes the resources and materials used in the SAE along with the working conditions. It is essential the youth is working within the allowed time and only with allowed equipment for their age as specified by child labor laws. The U.S. Department of Labor and your state department of labor have detailed youth work requirements and restrictions provided on their websites. Additionally, all parties need to complete an SAE risk assessment. An online assessment is available from the Safety in Agriculture for Youth website.

A *training plan* is a written statement that documents the specific experiences and competencies in which the student is expected to participate during supervised agricultural experience. The experiences and competencies are often based on the occupational competencies needed for the student to enter their tentative occupational objective. The training plan provides space for recording when the experiences/competencies have been accomplished and any school-related instruction. It may be signed by the student, student's parent(s) or guardian(s), and employer.

Each type of immersion SAE needs additional information as part of the training plan. A placement/internship SAE requires planning the method of supervision and performance evaluation. Agreeing on these prior to starting the SAE helps to eliminate confusion. Ownership, entrepreneurship, and school-based enterprise SAEs need marketing, operating, and financial plans. These are planning tools used by successful businesses. Research SAEs must have hypotheses, literature review, and research methods developed prior to beginning the SAE. Determining these prior to beginning a research project helps to strengthen the research. Service-learning SAE planning includes identifying the need, setting goals and outcomes, determining needed resources, and setting a plan of action. Planning these ahead of time ensures the service-learning benefits both the student(s) and local stakeholders.

THE ROLE OF SUPERVISION

Teacher supervision of SAE programs should be a high priority in high-quality agricultural education programs. According to the Local Program Success Initiative, teacher involvement is key to bridging the gap between the classroom and the workplace and has a direct correlation to SAE program quality and student success (National FFA Organization, 2002). However, with today's increasing student enrollments, lack of teacher release time, and decreasing funds for travel expenses, supervision of SAE has multiple different formats (Stahley, 2019). Some states and schools still provide financial incentives to encourage teacher supervisory visits. For example, many schools provide the agriculture program with a vehicle (often a pickup truck) for use by the agriculture teacher(s). Other schools reimburse teachers for their mileage to make supervisory visits using personal vehicles. If the school provides neither, then school-based supervision, virtual supervision, and group supervision should be considered.

Teachers should keep records of supervisory visits for income tax purposes, to provide documentation for maintaining financial support for visits, and to document the evaluation of student performance in supervised agricultural experiences. The *LPS Guide* contains forms to assist teachers in documenting SAE visits. Some states and local departments have their own forms. Regardless, always know and follow school policy.

After teaching the class about the types of supervised agricultural experiences available, the teacher should make individual visits to each student's home or placement station to help the student develop their own supervised agricultural experience plan. During these visits the teacher should be prepared to offer supervised agricultural experience suggestions based on the student's particular interests, goals, and available opportunities. The teacher should provide information about how to find materials, facilities, and other items needed for the supervised agricultural experience program.

It is important to involve parents/guardians in planning the supervised agricultural experience program because a younger student will likely need their assistance in conducting the supervised agricultural experiences. Involving parents/guardians early in the planning process allows them to assist in determining the best supervised agricultural experiences for the student. If parents/guardians are personally interested in their student's supervised agricultural experience, they are more likely to assist with ideas, transportation, financial support, and other items that will ensure a successful supervised agricultural experience program. A home visit to all new students entering the agricultural education program allows teachers to meet the parents, review available facilities, and discuss supervised agricultural experience requirements and opportunities.

After the student has established the supervised agricultural experience program, the teacher should schedule organized and purposeful visits to observe the student's activity and assure the student's experiences are of high quality. The frequency of supervisory visits by the teacher will vary among the students according to the complexity of their supervised agricultural experience. However, the teacher should make periodic visits throughout the duration of the activity to help ensure each student has a successful supervised agricultural experience.

For a student employed in an agricultural job as part of their supervised agricultural experience program, the teacher should consider the employer as a co-supervisor. The teacher and the employer should work together to ensure the experience helps the student prepare for a career.

A student may conduct their supervised agricultural experience at home. In this case, the teacher has an opportunity to include a parental visit with the task of observing the student's supervised agricultural experience activity. The teacher should take advantage of this opportunity to become better acquainted with the parents/guardians. For a student who conducts their supervised agricultural experience away from home, the teacher should incorporate into the visitation schedule at least one parental/home visitation each year. The purposes of home visitations are as follows:

- To demonstrate to parents/guardians that the teacher is interested in the development of their child
- To form an alliance with parents/guardians for the career and personal guidance of their child
- To become acquainted with home conditions that may have a bearing on the student's performance

- To inform the parents/guardians of program purposes, expectations and activities, and their child's performance

When students carry out supervised agricultural experience activities in school facilities, the teacher is responsible for maintaining a safe environment in the facilities and for assuring the students conduct themselves safely. Teachers often use classroom time on student sharing and discussion of experiences, since these experiences are an extension of the classroom instructional program.

RECORD KEEPING AND KINDS OF RECORDS KEPT

The teacher is responsible for ensuring all agricultural education students incorporate record keeping as an important segment of their SAEs. Such records document the experiences of students and serve as the basis for applying for various advancements and awards in FFA.

The teacher must teach students how to keep appropriate records related to their experiences and ensure the records are kept current. Most states have adopted specific record systems for use by students. This is helpful when evaluating student records for both grading purposes and competitive awards at the local and state levels. Teachers should become familiar with their state's record-keeping system and develop lessons and activities to ensure their students keep accurate and up-to-date records. One of the most difficult tasks a teacher can undertake is trying to help students apply for advanced degrees when records are not accurate and up-to-date.

The records kept may vary from one state or school to another. Some states or schools require comprehensive online management systems; others require specially prepared record books. Most state record systems include templates for training agreements and training plans.

Students need to keep records on time, finances, and accomplishments. Time includes hours worked in the SAE, especially placement/internship, research, and service-learning. Students

FIGURE 22.7 An online supervised agricultural experience record-keeping system allows the student to make entries anywhere at any time.

should also record hours in skills learned, community service, and FFA. Finances include budget(s), income and expenses, capital items, inventory, and financial statements. Animal-related SAEs may require students to keep breeding and management records. Accomplishments in the SAE, FFA, and other youth organizations or school activities need to be recorded.

Besides the records kept by each student, the teacher should maintain records in the department files that include the following:

- Individual student supervised agricultural experience agreements and training plans
- Individual student records of kind, size, growth, and performance
- Information on visitations, including dates, contacts, mileage, and major observations
- School-wide summarization of student supervised agricultural experiences by type and scope

The purposes for keeping records in the department files are (1) to provide documentation of the teacher's time and expenses incurred in making SAE visits; (2) to assist the teacher in making preparations for visits by reviewing agreements, training plans, and past visitation recommendations; (3) to allow the teacher to summarize SAE activities for reporting to the administration, local community, and state leaders; and (4) to provide documentation for grading student SAE performance and identifying student potential for award recognition and advanced degrees.

EVALUATING SUPERVISED AGRICULTURAL EXPERIENCE PROGRAMS

The agricultural education activities conducted through supervised agricultural experience are intracurricular. Therefore, the evaluation of student performance in agricultural education should include consideration of the student's level of involvement and performance in those activities. With its outside-of-class activities and purpose of skill development, SAE is a good area to implement mastery learning (Fuelling & LaRose, 2021).

Grading Supervised Agricultural Experience

A grading system for supervised agricultural experience should be based on the premise of every student enrolled in the agricultural education class being able to attain the highest grade possible. In multiple-teacher departments, the grading system should be agreed upon by the agricultural education teachers and uniformly applied. It should be possible for students to be informed at any time concerning their particular grade status. Visible records, such as grading charts, point award systems, or rubrics, may be used for this purpose. The grading system should be explained to every student enrolled in agricultural education so it is thoroughly understood. The system should be a matter of record and be incorporated into the department's management plan. Because of the interrelationship of SAE and FFA activity to the instructional program, these areas may constitute a percentage of the total grade. (Chapter 19 discusses evaluation procedures in more detail.)

Out-of-class-time participation can be reasonably viewed as agricultural education homework. As such, full credit for the agricultural education course(s) in which the student is enrolled plus the grade earned in the related supervised agricultural experience activity should be dependent upon satisfactory, measured participation.

Student participation in classroom/laboratory, FFA, and supervised agricultural experience is essential for a student to have access to the full curriculum of the agricultural education program. This makes documentation of student progress in supervised agricultural experience and FFA just as important as documentation of student progress in the classroom.

Enhancing the Quality of Supervised Agricultural Experience

In their 100+ years of including supervised agricultural experience as an instructional strategy, agricultural educators have developed many proven practices, some of which have already been mentioned in this chapter. Additional proven practices are listed here for teachers desiring to increase the quality of their supervised agricultural experience programs:

- Prepare and distribute to students a handbook describing the school's requirements for supervised agricultural experience, listing the kinds of projects that can be included in the supervised agricultural experience, explaining how supervised agricultural experiences are evaluated, and giving examples of good-quality supervised agricultural experiences that show progress from year to year. Additionally, all students should have access to the *SAE for All Student Guide*.
- Be aware that the term "supervised agricultural experience" may not be understood by some students. A simpler term, such as "work-based learning," can be used, although it has limited meaning compared to all that foundational and immersion SAE provides.
- Require a written training plan for every student's immersion SAE. That plan should be reviewed annually by the student, the teacher, the student's parents/guardians, and, if applicable, the employer.
- Utilize National FFA agricultural proficiency and achievement award systems.
- Incorporate accomplishments into FFA chapter point award systems.
- Emphasize the honor of FFA State and American Degrees; recognize chapter members who attain these degrees.
- Encourage participation in "project competition" programs—local level and above.
- Solicit local organizations to provide support, such as livestock "chains."
- Develop local sources for project funding—for example, banks and credit institutions, booster club loan funds, community foundations, and other groups.
- Provide school facilities for supervised agricultural experiences.
- Encourage cooperative projects for students with limited resources.
- Maintain regular written and oral communications with parents/guardians.
- Provide project tours for parents/guardians, prospective students, and other interested persons.
- Adjust home visitation hours to coincide with times when parents/guardians are home.
- Involve parents/guardians in school laboratory work days and improvement projects.
- Maintain a visible record of teacher supervision visits as a means of keeping supervised agricultural experiences in the minds of students and visitors to the agriculture department.
- Plan a visitation schedule to assure equitable supervision of all students' supervised agricultural experiences.
- Take beginning students on tours of successful projects.
- Utilize summer months to contact all first-year students and their parents/guardians to discuss supervised agricultural experience plans.
- Take steps to help assure the success of each student's first project.
- Use third- and fourth-year students as peer advisors to beginning students.

- Utilize the assistance and experience of other teachers whose students have successful supervised agricultural experiences.
- Provide the school board and administration with special presentations.
- Invite school board members and administrators to serve as local judges for project competition.

RELATIONSHIP OF SUPERVISED AGRICULTURAL EXPERIENCE AND FFA

Recognizing students for their supervised agricultural experience achievements is important. A major part of the recognition program is acknowledgment by the teacher, parents/guardians, employers, and other students. FFA provides additional opportunities to recognize students conducting quality supervised agricultural experience programs. For many students, FFA is their best opportunity for receiving recognition.

FFA Degrees and Proficiency Awards

The *FFA degree program* is designed to encourage students to establish and work toward career goals in the agricultural industry. Degrees are also designed to promote and recognize student achievement in leadership, personal growth, scholarship, and career development. A student's achievement in these areas should be documented in the supervised agricultural experience records. One of the criteria for advancement in the FFA degrees at the high school level and beyond is based on planning and developing a quality supervised agricultural experience to gain the skills needed to be successful in an agricultural career. The five FFA degrees are listed and described in Chapter 23.

A *proficiency award* is an award provided through FFA to a member who excels in the development of specialized skills in a specific agricultural area. Proficiency awards provide

FIGURE 22.8 A student whose supervised experience is beekeeping competes in the FFA proficiency award area of Specialty Animal Production.

opportunities for students to be recognized at the chapter, state, and national levels in areas related to supervised agricultural experience programs. There are three major categories for proficiency awards. One is "placement"; students who are employed in agricultural businesses (including farms) are eligible to apply for the placement awards. Another category is "entrepreneurship"; students who own production enterprises or agribusinesses may apply for these awards. Some award areas are "combined," which means students can include both placement and entrepreneurship records in the application.

Examples of proficiency award areas are agricultural communications, agricultural processing, agricultural services, agriscience research and experimentation, beef production, diversified crop production, fruit production, food science and technology, goat production, home and/or community development, landscape management, and wildlife management.

FFA Star Awards

A *Star Award* is considered the top award provided by the National FFA Organization to an individual student in FFA. The award is based on a student's overall achievements in FFA, including outstanding accomplishments in supervised agricultural experience.

Star Awards recognize outstanding FFA members at each degree level, beginning at the seventh or eighth grade with the Star Discovery Award. The recipient is the local member who has been most active and shown leadership.

The first Star Award at the high school level is the Star Greenhand Award. This award goes to a chapter's most active Greenhand member. The award is based on the member's supervised agricultural experience program and demonstrated leadership abilities.

The Chapter Star Farmer Award goes to a member with outstanding supervised agricultural experience in production agriculture who has demonstrated the most involvement in all chapter activities. The Chapter Star in Agribusiness Award is the same as the Chapter Star Farmer Award except the recipient's supervised agricultural experience must be in agribusiness rather than production agriculture.

The Chapter Star in Agricultural Placement Award goes to the top member with outstanding supervised agricultural experience involving placement in production agriculture, agribusiness, or directed laboratory that is not agriscience-based. The placement may be for either paid or unpaid experience.

The Chapter Star in Agriscience Award goes to the top chapter member who has outstanding supervised agricultural experience that involves natural resources and/or who has a research/experimentation-based or science-based laboratory experience.

At the state level, Star Awards are provided for State Star Farmer, State Star in Agribusiness, State Star in Agricultural Placement, and State Star in Agriscience. The state awards are also based on outstanding supervised agricultural experience and on participation in various FFA activities at the chapter level and above.

The National FFA Organization recognizes the American Star Farmer, American Star in Agribusiness, American Star in Agricultural Placement, and American Star in Agriscience. The national awards are based on a member's supervised agricultural experience and participation in activities from the chapter level through the national level.

The National FFA Foundation provides funds for plaques and medals for chapter Star Award winners. The foundation also provides funds for medals, plaques, and cash awards for all the Star Award winners at the state and national levels.

Scholarship Awards

The *National FFA Scholarship Program* provides opportunities for financial assistance for students continuing their education at the postsecondary level. Each scholarship has specific criteria defined by the sponsor when the scholarship is established. Some scholarships are based on student career goals. Outstanding SAE accomplishments may be included on the scholarship application form. Additional information about the National FFA Scholarship Program can be found at the FFA website: www.ffa.org.

Other Awards

The *FFA Agriscience Fair* recognizes middle school and high school students who are studying the application of science and technology in agricultural enterprises. Participation begins at the chapter level and continues to state and national levels. The FFA Agriscience Fair provides recognition opportunities for students conducting research supervised agricultural experience.

Grants are available through the National FFA and provide small amounts of start-up cash to students with limited resources. A grant is awarded based on sponsor guidelines and must be used to establish or reinforce a student's supervised agricultural experience. Application information and other details about SAE grants can be found on the FFA website at www.ffa.org.

Recognition Is a Motivational Tool

While awards and incentives should not be the primary reason for conducting supervised agricultural experience, they do provide valuable opportunities for teachers to gain publicity for students and the agriculture program.

Recognition is an important motivational tool, and for many students, their supervised agricultural experience provides avenues for them to achieve recognition through FFA. It may be from exhibiting winning livestock or other agricultural products at a local, county, or state fair, or it may be from winning a local, state, or national proficiency or other FFA award. Good teachers excel at motivating students, and FFA provides many motivational tools for teachers to use.

REVIEWING SUMMARY

Supervised agricultural experience is one of the three major interrelated components of the agricultural education program. It involves students, teachers, parents/guardians, and employers and extends the agricultural education program outside the classroom walls into the local community. When properly conducted, supervised agricultural experience is based on student interests, needs, and career goals, is supervised by a qualified agriculture teacher, is based on the curriculum, is cooperatively planned, is documented, and includes recognition for student achievement.

Several types of supervised agricultural experience programs are available, and students may choose the type best suited to their personal goals and career aspirations. Teachers are responsible for teaching students, parents/guardians, employers, and others about supervised agricultural experiences and their value. They are also responsible for providing overall supervision of student experiences, documenting and evaluating the experiences, and ensuring that students are recognized for their achievements.

QUESTIONS FOR REVIEW AND DISCUSSION

1. What is supervised agricultural experience?
2. What are the benefits of supervised agricultural experience programs?
3. What are the types of supervised agricultural experience and the associated components and categories?

4. Who are the key people involved in planning and conducting supervised agricultural experiences?
5. What are the purposes of supervised agricultural experience programs?
6. Why are supervised agricultural experience training agreements important?
7. What supervised agricultural experience records should be maintained in the department files?
8. What are some important considerations in evaluating supervised agricultural experience programs?
9. What are some ways to recognize students for supervised agricultural experience achievements?
10. How are FFA and supervised agricultural experience related?

ACTIVITIES

1. Investigate supervised agricultural experience. The National FFA Organization, National Association of Agricultural Educators, and *Journal of Agricultural Education* websites are good resources.
2. Prepare a presentation on supervised agricultural experience in your state. Topics could include types of supervised agricultural experience available and student participation by type, comparison of electronic and paper record systems, and the history of supervised agricultural experience in your state.

REFERENCES

Fuelling, K., & LaRose, S. E. (2021). How can mastery learning elevate your classroom? *The Agricultural Education Magazine*, *93*(6), 23–24.

Gossen, L. (2019). Something had to change. *The Agricultural Education Magazine*, *92*(3), 5–6.

National Council for Agricultural Education. (2016). *National quality program standards for agriculture, food, and natural resources education: A tool for secondary (grades 9–12) programs*. https://thecouncil.ffa.org/

National Council for Agricultural Education. (2017a). *SAE for all: Student guide*. https://thecouncil.ffa.org/sae-resources/

National Council for Agricultural Education. (2017b). *SAE for all: Teacher guide*. https://thecouncil.ffa.org/sae-resources/

National FFA Organization. (2002). *A guide to local program success* (2nd ed.). https://ffa.app.box.com/v/LPSguide

Stahley, J. (2019). Meaningful career conversation—More than an SAE visit. *The Agricultural Education Magazine*, *92*(3), 13–14.

23

FFA

Imagine these scenes. What do they all have in common? They each represent an aspect of FFA.

- Twenty-five seventh-graders are busily drawing the FFA emblem on poster boards to place on their lockers proudly announcing their involvement in their middle school FFA chapter.
- A dozen agricultural education students are planting annuals in a street roundabout on a Saturday morning for an FFA chapter community service project.
- Six FFA members go through one more practice after school before area competition in the Parliamentary Procedure Leadership Development Event (LDE).
- Two hundred FFA members, parents/guardians, and invited guests attend the local annual chapter awards banquet.

FFA's mission is to make a positive difference in the lives of students by developing their potential for premier leadership, personal growth, and career success through agricultural education (National FFA Organization, 2021d). This mission is manifested in the leadership programs, Career Development Events, scholarship programs, and community development programs offered to FFA members at the local, state, and national levels. Let's learn more about this organization that from its official beginning in 1928 to today has benefited millions of young people in their social, emotional, and professional development through their participation in the FFA experience.

OBJECTIVES

This chapter addresses the National Quality Program Standards for Agriculture, Food, and Natural Resources Education (National Council for Agricultural Education, 2016), specifically Standard 3: Leadership and Personal Development Through FFA. It has the following objectives:

1. Discuss the history and purpose of FFA.
2. Explain how FFA is structured as an organization.
3. Discuss FFA basics that are important in a local chapter.
4. Explain the FFA degree system.
5. Explain how to develop an FFA program of activities.
6. Identify best practices in managing an FFA chapter.

TERMS

American FFA Degree	growing leaders
building communities	New Farmers of America
Chapter FFA Degree	program of activities
Discovery FFA Degree	Public Law 81-740
FFA Creed	Public Law 105-225
FFA emblem	Public Law 116-7
FFA Motto	State FFA Degree
Greenhand FFA Degree	strengthening agriculture
Henry C. Groseclose	

FIGURE 23.1 FFA represents opportunity for students to travel, learn, and develop important personal skills. (COURTESY OF NATIONAL FFA ORGANIZATION.)

WHY JOIN FFA?

At some point in an agriculture teacher's career, a student will ask them, "Why should I join the FFA?" The FFA is, after all, an organization based on voluntary membership. What would cause a student to want to be a part of an organization that often takes extra time and effort beyond the school day? The agriculture teacher should be ready to answer this question in terms the student can understand, but the actual reason why FFA is important to the agricultural education program is rooted in human motivation theory. One of the most prominent theorists on human motivation was Abraham Maslow, and his hierarchy of needs theory of human motivation explains the value of the FFA experience for students.

The human being is an integrated organism. It is impossible to separate the various components of a person's self. When an individual experiences hunger, it is their whole self that is hungry and not just selected physiological components (Maslow, 1970). Until the basic physiological needs are met, the human is motivated to satisfy these needs. Once basic physiological needs are met, a different set of needs become evident. Maslow referred to these as the safety needs and characterized them as stability, security, and protection from harmful external conditions. If both the physiological and safety needs are met, then love and affection needs emerge. Maslow (1970) categorized these needs as the need for contact and intimacy and a sense of belonging.

The next level of need is identified by Maslow as the esteem needs. These needs are characterized by a person's desire for status, fame, dominance, and importance. The satisfaction of self-esteem needs will lead the individual to feel self-confident, worthy, and useful in their environment. Resting upon esteem needs are cognitive needs characterized by a person's search for knowledge and understanding. If these needs are met, the individual progresses to aesthetic needs that are identified as a person's desire for order and beauty. At the top of the hierarchy is self-actualization. Even if all of the other needs are met, the individual will develop a sense of restlessness and discontentment unless they are accomplishing goals true to themself (Maslow, 1970).

By the time a young person reaches high school age, they have matured to the point in Maslow's hierarchy where there is a need for a sense of belonging and contact with others.

A student may have progressed beyond these needs and reached a point whereby they have a strong need for status, recognition, and importance. Rose et al. (2016) found FFA membership and involvement provided students a place and opportunities to meet the needs identified by Maslow. FFA members reported a sense of love and belonging through FFA activities, teams, and travel. FFA recognitions and honors helped students meet their self-esteem needs. Finally, respondents in the Rose et al. study reported higher aspirations for themselves as a result of FFA, which could contribute to self-actualization.

HISTORY AND PURPOSE OF FFA

Boys and Girls Clubs for Rural Youth

The movement to develop organizations for agricultural youth began at the turn of the 20th century. There is some question as to when boys and girls agricultural clubs were established in the United States. A. B. Graham organized boys and girls clubs in January 1902 in the Springfield Township school community in Clark County, Ohio. Club meetings were held once per month in an assembly room of the county building. These were corn clubs. Later the clubs were broadened to include vegetable projects (McCormick & McCormick, 1984).

Another possibility is that W. B. Otwell created the first boys and girls clubs in agriculture in Macoupin County, Illinois. Otwell created a corn yield contest for local boys as a way to encourage attendance at farmer's institutes. Farmers were normally reluctant to attend these training institutes, but the corn yield competitions sparked their interest. The first year's contest involved 500 boys. There is also evidence the first boys club may have been organized in Holmes County, Mississippi, by W. H. Smith, the local school superintendent. Agricultural clubs for girls sprang up in South Carolina in 1910 (True, 1969).

Agricultural clubs were incorporated into public schools as a means to provide social interaction among youth and to encourage their interest in academic subjects. These early school clubs were highly organized, with monthly meetings focused on agricultural subjects. J. B. Berry (1924) reported elementary children were participating in junior project clubs. In his handbook on agricultural education, Berry (1924) discussed the important role of the schoolteacher in advising and the activities of the clubs. Berry said, "The wise teacher utilizes pupil activities to as great extent as possible, thereby developing leadership qualities in pupils" (Berry, 1924, p. 196).

The Beginning

The movement to create the FFA organization started long before the passage of the Smith-Hughes Act in 1917 and the incorporation of FFA in 1928. Many states already had some type of system for providing training in the agricultural and mechanical arts to students below the college level. Clubs and organizations designed to encourage and support farm youth grew out of the agricultural education movement, but there was no national effort to coordinate the activities of the individual clubs.

A Student Organization Emerges

With the passage of the Smith-Hughes Act, the national coordination of agricultural education made it convenient for a student organization to emerge with the goal of encouraging farm boys and their families to adopt the best practices of agricultural production. Between 1917 and 1928, the agricultural education profession began the work to create a national organization.

One of the most notable efforts to begin an organization was in Virginia. Walter S. Newman, state supervisor for agricultural education in Virginia, was a pioneer in the development of FFA. He believed rural farm boys were not getting the same opportunities for personal growth and advancement as their city-bred counterparts. The isolation inherent to farm life made it difficult for farm youth to experience the cultural and social life so readily available to urban youth. This isolation bred discontent and stunted the development of a positive self-image in farm youth. Newman believed some type of organization was necessary to provide opportunities for personal advancement and the growth of self-confidence in rural farm youth. In 1925, he joined with colleagues Henry C. Groseclose, Edmund Magill, and H. W. Sanders in developing the Future Farmers of Virginia.

Before the creation of FFA, regional and national livestock and dairy judging activities provided opportunities for competition among agriculture students. In 1926, federal agricultural education officials reached an agreement with the American Royal Livestock Show to host a livestock judging contest in Kansas City, Missouri. Later that same year, the National Congress of Vocational Agriculture Students was also established in Kansas City, Missouri. This three-day convention included tours of local agricultural businesses, banquets, and meetings designed to encourage students to learn more about the agricultural industry.

Future Farmers of America Formed

By 1928, the movement for a national agricultural youth organization had grown so strong the Federal Board for Vocational Education was asked to assist in the development of the youth organization, its governance structure, and its bylaws. The National Congress of Vocational Agriculture Students and the American Royal Rodeo and Livestock Show also included a special convention held for the purposes of establishing the Future Farmers of America.

On November 20, 1928, 33 delegates from 18 states officially adopted the constitution and bylaws of the new organization and elected Leslie Applegate of New Jersey as the national president. C. H. Lane was chosen to serve as the National FFA advisor, and *Henry C. Groseclose* was elected to serve as the executive secretary/treasurer. The old Baltimore Hotel where the organization was founded is no longer standing, but a monument erected by the citizens of Kansas City, Missouri, to honor the establishment of FFA in that city exists on the site of the hotel. Until 1999, the National FFA Convention was held in Kansas City, Missouri.

Expansion of FFA

As the National FFA Organization grew, it became necessary to develop more effective business practices to keep pace with the needs of a growing number of members. Delegates to the 1938 National FFA Convention set aside funding to establish an FFA camp. In the process of doing this, they also created a board of trustees to oversee the construction and development of this camp and subsequent properties. In 1939, these funds were used to purchase 25.5 acres of land in Alexandria, Virginia, for the National FFA camp. This property, once owned by George Washington, was to serve as the headquarters of FFA for 59 years. The National FFA Foundation, Inc., was established in 1944 to support the activities, programs, and services of the National FFA Organization. In 1948, the National FFA Supply Service opened its doors on the Alexandria property, providing official FFA jackets and other paraphernalia for FFA members and chapters.

In 1952, *National Future Farmer Magazine* began publication and all FFA members received it as part of their membership benefits. In 1953, the U.S. Postal Service honored the FFA with a special commemorative postage stamp. In 1959, a new headquarters building was constructed on the site of the old National FFA camp.

By the 1960s, FFA was maturing as an organization. Many new programs and services were being offered to its members, but the social climate of America had progressed to the point where it was impossible for FFA to continue to limit its membership to farm boys. In 1965, FFA absorbed the New Farmers of America (NFA) to become one organization. In 1969, girls were officially admitted to FFA membership, even though they had been unofficial members for years. One method by which FFA advisors were able to secure FFA membership for girls before 1969 was to list only their first initials and last names on the official FFA roster. The National FFA Organization was none the wiser that a certain "G. Bradley" of a local FFA chapter in North Carolina was actually a female agriculture student, Genie Bradley. Numerous such cases exist in the history of the FFA organization. The last set of NFA national officers was in 1964–1965; however, it was not until 1973 that the first African American was elected to a national FFA office. Although girls were officially admitted to FFA membership in 1969, it was not until 1976 that the first female was elected to a national FFA office.

In 1971, the National FFA Alumni Association was established to encourage friends of FFA to continue their support of FFA through their time and resources.

The 1980s and 1990s were a period of rapid change for the National FFA Organization. In 1988, delegates at the National FFA Convention, in an effort to reflect the diverse nature of FFA members, changed the name of the Future Farmers of America to the National FFA Organization. In 1989, *National Future Farmer* became *FFA New Horizons*. The National FFA Convention was held in Louisville, Kentucky, from 1999 to 2005 and 2013 to 2015. Indianapolis, Indiana, hosted the convention from 2006 to 2012, beginning again in 2016 and scheduled through 2031. The National FFA Center moved to Indianapolis, Indiana, in 1998.

The 2000s and 2010s broadened FFA participation. In 2000, the Discovery FFA Degree was approved, opening the award structure to members in grades 7 and 8. In 2003, the first person with a native language other than English and first Puerto Rican was elected as National FFA president. The first female National FFA executive secretary was named in 2013. During this time was the first live webcast of the National FFA Convention, and in 2020 the first virtual national convention was delivered via multiple media.

THE NEW FARMERS OF AMERICA

The *New Farmers of America* (NFA) was the companion organization to FFA that served Black youth from 1926 to 1965. NFA was established during the winter of 1926 in Virginia at the urging of Dr. Walter S. Newman, state supervisor of agricultural education, and Dr. H. O. Sargent of the Federal Board for Vocational Education. Sargent's primary responsibility for the federal board was to coordinate agricultural education programs for Black students. Both Newman and Sargent recognized the need for an organization for Black students, and in May 1927, the New Farmers of Virginia held its first state convention in Petersburg, Virginia. In 1935, after forming a national organization from the various state associations of New Farmers, the New Farmers of America held its first national convention at the Tuskegee Institute in Alabama.

The New Farmers of America was in most regards very similar to the Future Farmers of America. Both organizations had similar contest and award programs, degree programs, emblems, leadership development programs, business operations, and student leadership structures. The adult leadership for both NFA and FFA was provided through the Federal Board for Vocational Education.

FIGURE 23.2 The emblem of the New Farmers of America.

For many Black Americans, the New Farmers of America added dignity to the pursuit of agricultural occupations. While the United States struggled through years of segregation and the American education establishment operated under the concept of "separate but equal" schools, NFA offered young Black students the incentive to pursue agricultural careers while building their self-confidence and technical skills (Wakefield & Talbert, 2003). Both NFA and FFA served their respective clientele for many years.

The passage of the Civil Rights Act of 1965 also marked the end of the coexistence of NFA and FFA. That year the NFA ceased to exist and its members were absorbed into the FFA. As schools across America were desegregating, the ensuing years were not easy. Members of NFA were proud of their heritage and felt they had the most to lose in the combined organization. Many NFA members were afraid they would not receive adequate recognition and representation in the National FFA Organization or within their state FFA associations (Wakefield & Talbert, 2000). In many respects the NFA members were justified in their concerns. Although state and federal leaders promised to retain African American adult leaders and representation of NFA aspects in the FFA, little was accomplished. Jones et al. (2021) found many African American agriculture teachers lost their jobs, were put in middle schools, or were demoted to a lower teacher category. Former NFA members were neither selected by their states as national officer candidates nor received recognition for awards on the national level. Today, fewer Black students are involved in agricultural education and FFA nationwide than there were prior to the combining of the two organizations (National FFA Organization, 2021b; Wakefield & Talbert, 2000).

In 2020, National FFA unveiled the Agricultural Education for All initiative. The initiative included a road map to support inclusion, diversity, and equity (IDE). Action items supported include diversity positions, updated language in materials, IDE curriculum, and IDE professional development. National FFA committed itself to promoting anti-bullying efforts, celebrating diversity, and valuing tolerance and respect of differences.

HOW FFA IS STRUCTURED

The foundation of FFA activity is the individual FFA member. Because FFA is an organization founded upon the premise of preparing young people for careers in agriculture, certain minimum qualifications must be met to be an FFA member. An FFA member must be enrolled in at least one agricultural education course per academic year in a school that has an FFA chapter in good standing and have a supervised experience program that prepares the member for a career in agriculture.

Membership in the National FFA Organization can be through one of two methods. The first is through individual students paying dues themselves. This places the membership decision on the individual person. The other method is called affiliate membership. In this method, all enrolled agricultural education students are dues-paying members by virtue of the chapter paying a set membership fee. The advantage of this method is that all students have the opportunities, recognitions, and other benefits of FFA membership open to them. Affiliate membership also addresses any financial impediments to individuals joining FFA. Sheehan and Moore (2019) studied the impact of affiliation on FFA membership and asked whether affiliation should be the only membership option. It is expected this will be a much debated topic in the future.

Intracurricular Organization

FFA is an intracurricular organization. An intracurricular organization is a component that is a part of the curriculum—not an organization outside the curriculum or an extracurricular activity.

In the early years of FFA, the SAE requirement was met without much ado. Students who enrolled in agricultural education courses also received instruction in FFA and were required to develop and maintain quality SAE programs. Thus, this minimum requirement was automatically met in most cases. Students in agricultural education completed a series of courses in the high school agricultural education curriculum and then pursued careers upon graduation from high school. In recent years, however, more and more students are pursuing higher education upon graduation from high school. Agriculture has evolved into a highly technical and specialized industry, and many agricultural education students need college educations for successful entry into agricultural occupations. To be competitive in the college admissions process, students must often complete additional college preparatory work while in high school. This additional coursework could reduce the number of elective courses students can complete in high school, therefore making it difficult for students to enroll in agriculture courses and participate in FFA.

When a student is denied admission into an agriculture class because of conflicts with college preparatory classes or other valid reasons, the student has the option of completing a planned individualized course of study in an agriculturally related field. To be an FFA member, a student who cannot enroll in a regular agriculture class must complete a planned course of study. This could be an independent study course that includes instruction in an agricultural subject. It should also include the development of an SAE program for the student. There is a great deal of flexibility in developing this plan of work, but the best plans include the following:

- Work goals and objectives
- Clear assignments with due dates
- Schedule of visits by the agriculture teacher for individualized instruction

- List of responsibilities for all parties involved (i.e., what the student, parents, and teacher agree to do with regard to this plan)
- Grading scale
- Calendar of topics or competencies to be mastered by the student
- List of resources needed to complete lessons

Because a fair degree of flexibility exists within the planned course of study, the possibility also exists that the course of study will not be managed effectively. Often official course credit is not granted to a student who completes the planned course of study, so the experience may not be a quality one for the student. Teachers may also see this as a way to involve students in FFA competitive events without requiring them to take an agricultural education course. To protect the integrity of the teacher, the student, and the FFA organization, the planned course of study should be thoroughly prepared in written form and be agreed to by the student, the student's parents, and the agriculture teacher. Furthermore, the student and the teacher should keep detailed records of the student's progress through the planned course of study. It is important to note some state FFA associations may have specific rules regarding the administration of the planned course of study.

In addition to enrollment in an agricultural education class and the SAE requirement, students must also participate in the activities of the local FFA chapter and conduct themselves in a manner congruent with the aims and purposes of FFA. FFA members are also usually required to pay annual dues. Annual membership dues fall into three categories: national, state, and local. Students usually pay unified dues to all three levels of the organization at one time.

Agricultural education students begin their journey through FFA by joining the local FFA chapter at their school. The local FFA chapter is where students develop interpersonal and leadership skills for agricultural careers. FFA members participate in local chapter activities, such as Career Development Events, proficiency awards, and community service projects. FFA members also participate in chapter recreational activities and attend chapter business meetings. The local FFA chapter is advised by the agricultural education teachers at the school and led by a team of chapter officers elected by the chapter membership.

Charters and State Advisement

The state FFA association charters local FFA chapters, and the National FFA Organization charters state associations. The state FFA association has the responsibility for administering all the state-level FFA activities and providing assistance to local FFA chapters. There are state FFA associations in all 50 states. Puerto Rico and the Virgin Islands also have FFA. A state FFA advisor and a state FFA executive secretary (may be the same person) administer each state association, and technical assistance is provided by the agency responsible for administering the agricultural education program in the state.

The state FFA advisor has traditionally been the chief administrator of agricultural education in the state. This person generally serves as the chief executive officer of the FFA association. The state FFA executive secretary serves as the chief operating officer and handles most of the daily administrative matters of the association.

A team of state FFA officers provides leadership in the youth activities of the state FFA association. The major duties of state FFA officers are member leadership development, recruitment, and retention. Boards of directors provide the leadership for some state FFA associations. A state board of directors usually includes FFA advisors, who represent the agricultural education profession, and state FFA officers, who represent the members. This board provides the overall leadership and policy for the state association according to the guidelines set forth in the state FFA constitution. Some states also have active FFA foundations that raise funds to support state and regional FFA activities.

A state FFA association plays an important role in determining the quality of the FFA experience for students in the state. State FFA leaders determine the standards of performance for Career Development Events, provide funding for the training and development of chapter FFA advisors and student leaders, and serve as a clearinghouse for national FFA membership dues. While state FFA associations cannot have lesser standards than those set forth in the National FFA Constitution, they can be more rigid and exacting.

The National FFA Organization requires FFA members earn the State FFA Degree before they can receive the American FFA Degree. State associations can set very rigid standards for the state degree, thus effectively influencing the quality of supervised experience programs by which students qualify for the state degree and controlling the number of FFA members eligible for the award.

Another important role of state associations is to certify teams and individuals for participation in National FFA competitive events. Before any student can participate in a National FFA Career Development Event, the state FFA advisor or state FFA executive secretary must approve the student. The National FFA Organization also gives the state association the authority to grant and revoke the charters of local FFA chapters.

National FFA

The National FFA Organization provides the overall leadership and direction for FFA. As FFA grew in the years before World War II, questions began to arise as to how the organization should be administered. Planning and administering education is primarily the responsibility of the states, but the federal government has traditionally exerted some influence. The Smith-Hughes Act made the federal government a partner in the administration of agricultural education programs in the states, but the act did not specifically define FFA's role in agricultural education. The Smith-Hughes Act was signed into law 11 years before the National FFA Organization was founded, and local and state boards of education were faced with the problem of trying to determine how FFA should be administered. FFA members were participating in field trips, judging contests involving livestock, and other activities that created liability issues for boards of education.

FIGURE 23.3 The headquarters of the National FFA Organization is located in Indianapolis, Indiana. (COURTESY OF EDUCATION IMAGES.)

It had been maintained since the inception of FFA that the organization was an integral part of instruction in vocational agriculture. However, state and federal employees were administering FFA even though it was a private organization. Before FFA, local agriculture clubs were not coordinated through a national organization. Once these clubs became FFA chapters, however, there was considerable concern by schools and school officials as to their degree of responsibility and liability for FFA activities. As pressure mounted from school boards and school officials, state and national leaders recommended that the Federal Board for Vocational Education assume administrative leadership for FFA and the board appoint a special committee of state leaders in agricultural education to advise the National FFA officers and convention delegates (Tenney, 1977).

The Federal Board for Vocational Education adopted these suggestions. The comptroller general of the United States supported the action of the federal board by determining FFA was an integral component of vocational agriculture, and it was a necessary, practical, and legitimate use of state and federal funds to allow state and federal officials to administer the FFA organization.

A. W. Tenney, in his history of FFA, wrote, "Education historically had been a primary responsibility of the states, but education which increases the productive capability of the nation makes it necessary for the federal government to define its role and determine how it will participate" (Tenney, 1977, p. 78). In 1946, the George-Barden Act (Public Law 79-586) further expressed the integral nature of FFA and agricultural education by authorizing the expenditure of funds for attending meetings of the FFA organization.

From its beginning in 1928, FFA had been a private organization incorporated in the state of Virginia. As the organization expanded across the United States, it became necessary to strengthen the relationship between FFA and the federal government. In 1950, Congress passed **Public Law 81-740**, which granted FFA a federal charter. This charter provided for incorporation of the National FFA Organization and recognized the integral nature of FFA in agricultural education. The National FFA Organization is not the only youth-serving organization to have a federal charter. For example, Girl Scouts, Little League Baseball, and Scouts BSA are granted federal charters. A unique aspect of FFA's charter was the requirement for the national FFA advisor to be a U.S. Department of Education, formerly Office of Education, employee.

Specifically, the 1950 incorporation did the following:

- Created the Board of National Officers. These six student officers advise the FFA Board of Directors on all matters but have no voting privileges.
- Set policies governing the powers of the Board of Directors.
- Protected the name "FFA" and the emblem from copyright infringement.
- Established the headquarters of the FFA corporation in Washington, D.C. Public Law 116-7 allowed the National FFA Organization to establish its headquarters and principal office location through its constitution and by-laws.

Public Law 105-225 in 1998 provided technical amendments to update the charter. Over the next 20 years it became evident the charter needed a new revision. One of the most pressing issues was removing the specificity for U.S. Department of Education involvement. The resulting legislation, **Public Law 116-7**, passed in 2019, addressed this and other issues. The national director for agricultural education had traditionally served as the National FFA advisor and chairperson of the National FFA Board of Directors. The charter revisions of 2019 charged the National Council for Agricultural Education with naming the National FFA advisor and chairperson of the National FFA Board of Directors.

The revised charter further specified the integral nature of FFA to the instruction of agricultural education in the United States. The law reads, "to be an integral component of instruction in agricultural education, including instruction relating to agriculture, food, and natural resources" (National FFA Organization, 2021c). However, the federal charter does not explicitly state that every agricultural education program must have an FFA chapter. Despite the intentions of the federal charter, some state departments of education and school boards do not interpret Public Law 116-7 as a compelling case for the inclusion of FFA in the agricultural education program (National FFA Organization, 2021c). Of the approximately 1 million students enrolled in agricultural education in 2020, only 760,113 were members of FFA (National FFA Organization, 2021a).

The FFA Board of Directors consists of the national FFA advisor, four state supervisors of agricultural education, and four representatives of the U.S. Department of Education with expertise and job duties in agricultural education. The National FFA advisor serves as the board chair. The four U.S. Department of Education seats typically are filled by agriculture teachers, teacher educators, and other appropriate appointees. This board works with the National FFA officers as the administrative board of FFA. A National FFA secretary and National FFA treasurer serve as ex-officio nonvoting members of the board.

FFA BASICS

One of the first things new students do upon entering an agricultural education class is learn the basics of the FFA organization. The assimilation process requires students to learn the history, ceremonies, and member responsibilities if they are to gain maximum benefits from membership.

Emblem

One of the best-known icons of the organization is the official emblem. The *FFA emblem* is a graphic symbol that represents or identifies the FFA organization. It has undergone very few changes over the years. The first emblem was derived from a design created for the Future Farmers of Virginia in 1927, but the origin of the design likely comes from a Danish agriculture organization in the 1920s. This emblem featured an owl perched on a spade with the rising sun in the background.

The FFA emblem today is composed of the image of an owl sitting on a moldboard plow in a field with the rising sun in the background (see Figure 23.4). The owl represents knowledge and wisdom in the agricultural sciences, the plow represents labor, and the rising sun represents a new day in agriculture. This image, which rests on a background resembling the cross section of an ear of corn, is framed by the words "Agricultural Education" and displays the letters "FFA." Corn is a field crop common to all states, and the cross section of corn on the emblem symbolizes the common bond created by agriculture for all students enrolled in agricultural education. The words "Agricultural Education" and the letters "FFA" are included together to represent the integral nature of FFA to agricultural education. An eagle at the top of the emblem symbolizes the national scope of the organization. In 1950, Public Law 81-740 ensured the sole rights of FFA to use the emblem, and registered trademarks continue to protect its use. These rights extend to state associations and local chapters, bound by the National FFA board policy on FFA trademarks.

Colors

National blue and corn gold were adopted in 1929 as the official colors of the National FFA Organization. National blue is derived from the blue field on the U.S. flag and represents national unity. Corn gold comes from the color of grain fields at harvest time and is symbolic of national pride in the highly productive agricultural industry. These colors appear in

FIGURE 23.4 The FFA emblem. (COURTESY OF
NATIONAL FFA ORGANIZATION.)

the official FFA jacket, official publications, and other uses deemed appropriate by the organization.

Dress

The Fredericktown FFA Chapter of Fredericktown, Ohio, was one of the first to produce and wear an FFA jacket. Shortly afterward, in 1933, the blue corduroy jacket was adopted as official dress. The FFA jacket has the student's name on the front, FFA emblems on the front and back, and the name of the student's FFA chapter and state association. The blue corduroy jacket is the most recognized symbol of FFA (National FFA Organization, 2021a). However, the jacket may also be seen as exclusionary and a barrier. For example, corduroy as a fabric is not a popular clothing choice. The emblems and symbolism on the jacket can be confusing to newer members and those considering membership. Additionally, FFA members with chapter-owned jackets may be embarrassed if their jacket has no name or a previous student's name.

Official dress for members is the FFA jacket zipped to the top, black slacks and black socks/nylons or black skirt and black nylons, a white-collared blouse or white-collared shirt, official FFA tie or official FFA scarf, and black dress shoes with closed heel and toe. FFA members may wear official garb of recognized religions with official dress. They may also wear appropriate personal protective equipment with official dress.

FFA advisors do not have an official dress code to follow when attending FFA events. It is strongly recommended teachers set a positive example for students by dressing professionally at all times. It does nothing for the professional image of a teacher to be underdressed in comparison with their students.

Creed, Motto, and Salute

The creed, motto, and salute are important components of routine FFA activities.

Creed

The *FFA Creed* is a statement of beliefs held by every FFA member with regard to their place in the industry of agriculture. In the first two years of the organization's existence, FFA had at least two creeds for members to use in local chapter activities. The *Official FFA Manual*

The FFA Creed

I believe in the future of agriculture, with a faith born not of words but of deeds—achievements won by the present and past generations of agriculturists; in the promise of better days through better ways, even as the better things we now enjoy have come to us from the struggles of former years.

I believe that to live and work on a good farm, or to be engaged in other agricultural pursuits, is pleasant as well as challenging; for I know the joys and discomforts of agricultural life and hold an inborn fondness for those associations which, even in hours of discouragement, I cannot deny.

I believe in leadership from ourselves and respect from others. I believe in my own ability to work efficiently and think clearly, with such knowledge and skill as I can secure, and in the ability of progressive agriculturists to serve our own and the public interest in producing and marketing the product of our toil.

I believe in less dependence on begging and more power in bargaining; in the life abundant and enough honest wealth to help make it so—for others as well as myself; in less need for charity and more of it when needed; in being happy myself and playing square with those whose happiness depends upon me.

I believe that American agriculture can and will hold true to the best traditions of our national life and that I can exert an influence in my home and community which will stand solid for my part in that inspiring task.

FIGURE 23.5 The FFA Creed.

included one version of the creed that was dissimilar to the one used most frequently by members in chapter activities. A college professor from Wisconsin, E. M. Tiffany, wrote the version of the FFA Creed adopted by the National FFA Convention delegates in 1930. The FFA Creed has experienced little change over the years and is still recited by Greenhand candidates as they work toward attainment of the degree.

Reciting the FFA Creed is sometimes thought to be an excellent introduction to public speaking for young people. More important, reciting the FFA Creed gives some students their first introduction to character and values education. Agriculture teachers should take the time not only to have every student recite the FFA Creed but also to have a class discussion or lesson that dissects the creed and explains the importance of values and character. The FFA Creed is shown in Figure 23.5.

Motto

The *FFA Motto* is a short expression of the character of FFA and represents a quick and easy way to remember the mission of FFA. It reflects the pragmatic philosophy of work prevalent when C. H. Lane, the first National FFA advisor, recommended it to the National FFA Convention delegates. The FFA Motto is shown in Figure 23.6.

The FFA Motto
Learning to Do,
Doing to Learn,
Earning to Live,
Living to Serve.

FIGURE 23.6 The FFA Motto.

The FFA Code of Ethics

FFA members conduct themselves at all times to be a credit to their organization, chapter, school, community and family.

As an FFA member, I pledge to:

1. Develop my potential for premier leadership, personal growth and career success.
2. Make a positive difference in the lives of others.
3. Dress neatly and appropriately for the occasion. (See page 23 & 24 of the Official FFA Manual for proper use of Official Dress)
4. Respect the rights of others and their property.
5. Be courteous, honest and fair with others.
6. Communicate in an appropriate, purposeful and positive manner.
7. Demonstrate good sportsmanship by being modest in winning and generous in defeat.
8. Make myself aware of FFA programs and activities and be an active participant.
9. Conduct and value a supervised agricultural experience program.
10. Strive to establish and enhance my skills through agricultural education in order to enter a successful career.
11. Appreciate and promote diversity in our organization.

FIGURE 23.7 FFA Code of Ethics. (COURTESY OF NATIONAL FFA ORGANIZATION.)

Salute

The FFA salute is the Pledge of Allegiance to the flag of the United States: "I pledge allegiance to the Flag of the United States of America, and to the Republic for which it stands, one Nation under God, indivisible, with liberty and justice for all." The FFA salute is usually part of the ceremony used to close FFA meetings.

Code of Ethics

The FFA Code of Ethics helps guide FFA members as they go about activities and assures ethical conduct and appropriate representation to other FFA members and the general public. It helps members move toward adulthood. Adopted in 1952, the code sets the standard by which all FFA members should conduct themselves. The code was revised in 1995 to reflect an appreciation for diversity in the FFA organization. The FFA Code of Ethics is shown in Figure 23.7.

FFA PROGRAMS

FFA programs fall into three basic categories: career development, leadership and personal development, and FFA chapter development. Under the career development umbrella are Career Development Events, proficiency awards, and the FFA degree program.

Career Development Events are competitive activities designed to reinforce the skills students learn from instruction in the agricultural sciences and through independent study in supervised experience. All the Career Development Events are competitive activities, and most are team activities. For instance, the Forestry Career Development Event requires a team of four FFA members to complete a series of activities relevant to the forestry industry. Team members are evaluated on their ability to identify important timber species, demonstrate forest inventory practices, complete a written test, and reveal through an interview their knowledge of current forestry issues. Participants with qualifying scores generally proceed through a local event to a regional or state event before reaching the national level. The National FFA Foundation provides the funding and support for Career Development Events at the national level and, in some cases, funding at the state and local levels. In addition to the national Career Development Events, state associations have the option to create Career Development Events and Leadership Development Events specific to their respective needs. These special state-created CDE/LDEs do not advance to national competition. Information on CDE/LDEs is provided each year by the National FFA Organization.

Students who have outstanding supervised experiences have the opportunity to compete in the FFA proficiency awards program. This program involves transferring data from SAE records to a proficiency award application and submitting that application for judging. Based upon the nature of their SAE, students choose to participate in the placement or entrepreneurship category of a proficiency award. The placement category is designed for those students who primarily earn wages through their supervised experience, while the entrepreneurship category is for students who own enterprises.

The applications in a particular area are judged, and the top-ranked ones proceed to the next level of judging. The process for judging proficiency applications varies from state to state.

The National FFA Organization judges those applications certified as state winners and recognizes national finalists and winners at the National FFA Convention each fall.

FFA DEGREES

The FFA degree program is designed to encourage students to establish and work toward career goals in the agricultural industry. The degrees of membership provide students with a measure of their progress toward agricultural careers and reward them for their efforts. There are five degrees of membership in FFA.

Discovery Degree

The *Discovery FFA Degree* is designed for students enrolled in a middle grades (7 and 8) agricultural education program. It is the first degree a student can earn in FFA. The Discovery FFA Degree is optional and not required for advancement to the next level. Besides submitting a written application for the degree, the student must meet the following requirements in order to wear the bronze and blue Discovery Degree pin:

- Be enrolled in at least one agricultural education class for at least a portion of the school year while in grade 7 or 8.
- Pay the required membership dues.
- Participate in at least one extracurricular local FFA chapter activity.
- Have knowledge of careers available through the agriculture industry.
- Have some basic knowledge of the FFA chapter program of activities.
(National FFA Organization, 2021c)

The Star Discovery Award is presented to a student who has excelled in leadership and participation in FFA at the middle grades level.

Greenhand Degree

Until the creation of the Discovery Degree, the *Greenhand FFA Degree* served for many years as the first degree a member could attain. The Greenhand Degree name is derived from an early American colloquialism that describes a novice on the farm. The term was

FIGURE 23.8 Greenhands in the Smyrna, Delaware, FFA chapter developed a bulletin board with autographed handprints. (COURTESY OF EDUCATION IMAGES.)

applied to the newest FFA members as well because they are novices in the FFA organization. Greenhand members are usually first- and second-year students enrolled in a high school agricultural education program.

The Greenhand Degree is awarded by the local FFA chapter. Besides submitting a written application for the degree, the student must meet the following requirements in order to wear the bronze Greenhand Degree pin:

- Be enrolled in agricultural education and have a planned supervised agricultural experience.
- Learn and explain the basics of FFA: the creed, motto, salute, mission statement, emblem, colors, and code of ethics, as well as the proper method of wearing the FFA jacket.
- Demonstrate knowledge of key events in FFA's history, knowledge of the chapter constitution and bylaws, and knowledge of the chapter program of activities.
- Have access to the *Official FFA Manual* and the *FFA Student Handbook*.

(National FFA Organization, 2021c)

A local chapter or state association may set more stringent requirements for the degree but may not set less rigorous guidelines than those of the National FFA Organization. The Star Greenhand is awarded to Greenhands who demonstrate leadership, participation, and significant progress in SAE during the membership year. Students who receive the Greenhand Degree are encouraged to attain the Chapter FFA Degree.

Chapter FFA Degree

The ***Chapter FFA Degree*** is the highest degree awarded by a local FFA chapter. It is earned by those students who have well-established SAE and have distinguished themselves in FFA activities during the academic year. Recipients of the Chapter FFA Degree must have demonstrated leadership abilities and significant academic progress in all coursework. In addition to submitting a written application for the degree, a candidate must meet the following qualifications:

- Have earned and received the Greenhand FFA Degree.
- Have satisfactorily completed the equivalent of at least 180 hours of systematic school instruction in agricultural education at or above the ninth-grade level and be enrolled in an agricultural education course.
- Operate an approved supervised experience program and develop plans for continued growth and improvement in the supervised experience program.
- Participate in the planning and conducting of at least three official FFA chapter functions.
- Earn and productively invest at least $150 or work at least 45 hours in excess of scheduled class time, or a combination thereof.
- Lead a group discussion for 15 minutes and demonstrate five procedures of parliamentary law.
- Show progress toward individual achievement in both academic work and the FFA award programs.
- Complete at least 10 hours of documented community service, above and beyond any hours completed as part of a supervised agricultural experience program.

(National FFA Organization, 2021c)

For those students who have demonstrated exceptional leadership ability, outstanding progress in their SAE, and growth in personal development through participation in FFA activities at the chapter level and above, the Star Chapter Awards await them. There are four areas in which a student can earn additional recognition through the Star Awards program:

Chapter Star Farmer—Awarded to a student who excels in production agriculture.
Chapter Star in Agribusiness—Awarded to a student who excels in agribusiness.
Chapter Star in Agricultural Placement—Awarded to a student who excels in a wage-earning SAE.
Chapter Star in Agriscience—Awarded to a student who excels in an agricultural science SAE.

State FFA Degree

State FFA associations award the *State FFA Degree* to those students who have excelled in their FFA experience. Each state association sets the requirements for the State FFA Degree, provided those requirements are not inconsistent with the National FFA Constitution. Some state associations award the degree to students based upon their achievement of a standard set of requirements. In these states, every student who meets the minimum requirements will earn the degree. Other states choose to award the degree based upon a quota system that allows only a certain percentage of members to receive the degree each year. In these states, students may meet the minimum requirements but might not earn the degree if they are not in the top percentage of members submitting the application in a particular year. A candidate for this degree must meet the following qualifications:

- Have earned and received the Chapter FFA Degree.
- Have been an active FFA member for at least two years (24 months) at the time of receiving the State FFA Degree.
- Have completed the equivalent of at least two years (360 hours) of systematic school instruction in agricultural education at or above the ninth-grade level.

- Earn and productively invest at least $1,000 or work at least 300 hours in excess of scheduled class time, or some combination of hours and dollars that adds up to the minimum earnings requirement, in a supervised experience program.
- Perform 10 procedures of parliamentary law and present a six-minute speech on an agricultural or FFA topic.
- Serve as an officer, as a committee chair, or as an active participant on a chapter committee.
- Have satisfactory academic grades as certified by the local agricultural education instructor and a school administrator.
- Participate in the chapter program of activities and in at least five different official FFA activities above the chapter level.
- Complete at least 25 hours of documented community service, above and beyond any hours completed as part of a supervised agricultural experience program. These 25 hours must be earned through participation in at least two community service activities.

(National FFA Organization, 2021c)

State FFA Degree candidates with the best SAE program and stellar records of leadership and participation in FFA are awarded the State Star Awards. Like the Chapter Star Awards, the State Star Awards recognize excellence in four areas: production agriculture, agribusiness, placement, and agriscience.

American FFA Degree

The *American FFA Degree* is the highest degree awarded to an active FFA member. For many students, earning this degree is the crowning achievement of their FFA experience. The requirements for the degree are rigorous and challenging. A candidate for this degree must show significant growth and development in supervised experience and have a distinguished record of leadership and personal development that continues beyond the years in high school. The earliest a student could be awarded this degree is one year after graduation from high school. To receive this degree, a candidate must meet the following requirements:

- Have earned and received the State FFA Degree.
- Have held active membership for the three years (36 months) immediately preceding the application for the degree.
- Have actively participated in FFA activities at the chapter level and above.
- Have completed a minimum of three years (540 hours) of systematic secondary school instruction in an agricultural education program or, if less than three years of instruction were available at the school last attended, have completed the program in agricultural education offered by that school.
- Have graduated from high school. There is a waiting period of one academic year after high school graduation before the degree can be granted. This allows the student to become established in an agricultural career or in postsecondary education. A student may work toward the degree during the waiting period.
- Have a current supervised experience program through which the member has exhibited excellence in its management and operation by earning at least $10,000 and investing at least $7,500. A student may also work some combination of paid and unpaid hours to earn the degree, provided at least $2,000 is earned and 2,250 hours are worked outside of scheduled class time. The value of unpaid hours is established at

a set factor, which may increase in subsequent years as the degree requirements are revised.

- Have a record of high academic achievement and of outstanding leadership and community involvement.
- Complete at least 50 hours of documented community service, above and beyond any hours completed as part of a supervised agricultural experience program. These 50 hours must be earned through participation in at least two community service activities.

(National FFA Organization, 2021c)

The top American FFA Degree candidates can receive American Star Awards in four areas: production agriculture, agribusiness, placement, and agriscience. Upon receipt of the American FFA Degree, a student's active membership in FFA is over. By receiving the American FFA Degree, the student has, in effect, "graduated" from FFA by achieving the highest honor available to any FFA member. Membership in the FFA Alumni and Supporters now awaits the American Degree holder who wishes to continue to be a part of the organization.

PROGRAM OF ACTIVITIES FOR AN FFA CHAPTER

A *program of activities* (POA) is the map by which a local FFA chapter charts its course for the school year. This program serves as the curriculum guide for FFA by providing student activities that develop leadership, citizenship, and career skills. The program of activities is designed to provide activities for the individual member, the FFA chapter, and the school community. A common fallacy is that the program of activities is only a written document that lists all the things a chapter will do during the school year. This limited view of the program of activities tends to cause teachers and students alike to see the process as tedious. By its very nature, a program of activities is a living thing. It encompasses every activity that an FFA chapter does during the year and includes activities that are fun, educational, and of benefit to the community.

The National FFA Organization has established criteria for appropriate chapter activities. Activities must provide the following:

- Educational experiences both inside and outside the classroom
- Opportunities for learning in emerging areas reflecting current and future workforce needs
- Opportunities to learn and practice leadership skills and participate in personal development activities
- Accessibility to all students with diverse backgrounds, regardless of their present skills, abilities, and experiences
- Opportunities for students to develop self-confidence, responsibility, citizenship, and cooperation skills
- Experiences that connect students to agriculture and natural resources

(National FFA Organization, 2017)

A high-quality program of activities provides an FFA chapter with a basis of operation from year to year. The POA helps the chapter develop a sensible budget and calendar along with a list of human and fiscal resources needed to carry out a year of valuable activities and services for members. The National FFA Organization (2017) divides the program of activities into three major areas: growing leaders, building communities, and strengthening agriculture.

Growing Leaders

The *growing leaders* part of a program of activities should provide members with opportunities to learn new leadership skills and develop teamwork skills through participation in leadership training experiences. Activities to promote a healthy lifestyle and lifelong learning are also a part of this area. Finally, activities to promote member personal development and career readiness are a part of growing leaders. Through careful execution of a POA, the FFA chapter provides opportunities for students to attend local, state, and national leadership conferences such as the following:

- 360 Leadership Conference, a chapter leader training conference.
- 212 Leadership Conference, a personal leadership development conference.
- The Washington Leadership Conference (WLC), which is sometimes said to be the flagship of leadership training and development in FFA. This is a weeklong leadership training conference held annually in Washington, D.C. Students attending this conference learn essential leadership skills and have the opportunity to see leadership in action by visiting the U.S. Congress and other institutions in the nation's capital.
- State Officer Leadership Continuum, which is a variety of trainings and experiences designed to prepare and develop state officers for their leadership roles.

State FFA associations and local FFA chapters also design and deliver leadership training according to the goals outlined in their respective programs of activities. The growing leaders part of a program of activities also provides experiences for students that encourage a healthy lifestyle, scholarship, and career skill development. Following are examples of activities that an FFA chapter may deliver for students:

- Attending an FFA camp or leadership conference
- Holding a local agriscience fair to allow students with agriscience SAEs to exhibit their work
- Establishing a scholarship program that pays the fee for students to attend leadership conferences
- Holding an FFA awards banquet

Building Communities

The *building communities* area of a program of activities provides the framework for members to perform service-learning and engagement activities that benefit the community. Areas of emphasis are environmental, human resources, citizenship, stakeholder engagement, and economic development. Collaboration and involvement need to be key characteristics of each activity.

Strengthening Agriculture

The *strengthening agriculture* part of a program of activities focuses on FFA chapter development, safety, agricultural advocacy, and agricultural literacy. A healthy FFA chapter has positive relationships with stakeholder groups and is growing in membership and member participation. A focus on safety benefits the chapter members and the community. The local FFA chapter should conduct activities to inform and educate the public about agriculture and promote agriculture. Agricultural literacy is further explained in Chapter 15.

BEST PRACTICES IN MANAGING AN FFA CHAPTER

Success with the FFA component of agricultural education requires attention to selected areas of its programs as related to the local curriculum. Researchers have attempted to identify the "best practices" in managing an FFA chapter. Several best practices are covered here.

Local Program Success Emphasis

State and national FFA program success is dependent upon local program success. Several efforts over the years have been developed to assist local FFA advisors in this goal. The National FFA Organization (2002) published the *Local Program Success Guide* and a *Local Program Resource Guide*. As discussed in Chapter 5, National Quality Program Standards can be used in conjunction with the local advisory committee to assess the current status of the local program and determine priority action items.

The FFA Advisor

The single greatest resource an FFA chapter can have is a highly motivated and well-qualified agriculture teacher as the advisor. The teacher is the essential key to the organization because they provide all the initial training and information about FFA to members. Furthermore, the role of the FFA advisor is specific to the agriculture teacher. No one else—parent, alumnus, sponsor, or administrator—can serve in that role. Because FFA is a component of instruction, the FFA advisor is the gatekeeper to all the things FFA members can do in the organization. Therefore, the best interest of students is served by having an advisor with a good understanding of FFA and the leadership skills necessary to help students grow personally and professionally through their FFA experience.

FFA advisors should develop the habit of setting aside time for professional development by attending workshops, reading the most current literature about FFA and agricultural education, and participating in a variety of FFA activities. FFA advisors should seek leadership positions in their professional organizations and on FFA boards and committees to broaden their knowledge. An FFA chapter will be only as good as the FFA advisor. So, what do FFA advisors do?

FIGURE 23.9 An FFA advisor adjusts the official dress of a member.

FFA Advisors Supervise the Activities of the FFA Chapter

The FFA advisor is also the agricultural education instructor. Since the primary role of the teacher is to supervise all instructional activities, it naturally falls to the teacher to supervise all FFA activities as well. This is especially true when one considers FFA activities are, in reality, instructional activities. FFA advisors may enlist the help of parents, colleagues, and alumni members to assist in supervising activities, but the primary responsibility for safe FFA programming falls on the shoulders of the agriculture teacher as FFA advisor.

FFA Advisors Are the Primary Resource for Current FFA Programs and Services

Many students walk into their first agricultural education classroom with no idea of the depth and breadth of FFA programs available to them. It is the FFA advisor's responsibility to make the students aware of leadership, personal development, and career-related programs offered by the FFA. The FFA advisor trains a new FFA officer team every new school year. Furthermore, the FFA advisor serves as the "institutional memory" of the FFA chapter by explaining to new students how the FFA chapter operated in years past. This collected knowledge is useful to students as they plan future FFA activities. The experienced FFA advisor will have a network of resource persons who can be called upon to assist in FFA programs. The state and national FFA administrative staffs often communicate pertinent program information to students through the FFA advisor. The FFA advisor also provides information to parents, colleagues, school administrators, and program supporters about FFA programs. It is critical that the agriculture teacher/FFA advisor keep up-to-date on FFA programs, so the local FFA chapter remains a contemporary and relevant organization for its members (National FFA Organization, 1998).

FFA Advisors Encourage Youth Development Through the FFA

One of the most important roles for the FFA advisors involves youth development. The FFA advisor provides instructional resources that develop the leadership potential for students. Students often need encouragement to develop their leadership potential. The FFA advisor takes the first step by providing leadership training for all students in the agricultural education program. The FFA advisor also provides opportunities for students to attend regional, state, and national leadership programs.

FFA Advisors Build Support for the Agricultural Education Program

The local agricultural education program needs the support of the community to provide a relevant instructional program for students. The FFA advisor builds support in the community by informing citizens about FFA program activities. It is also important to market the agricultural education program within the school as well, so colleagues understand how the agricultural education program supports the overall school mission. The FFA has traditionally held an excellent reputation for leadership, professional development, and service. This reputation is a powerful tool in explaining the importance of the agricultural education program.

FFA Advisors Include All Students in FFA Programs

The FFA advisor encourages all students to be involved in FFA chapter activities. This is accomplished by making sure a variety and sufficient number of FFA activities are provided for all students. No student can participate in every FFA activity, but every student should have the opportunity to participate in most FFA activities of interest to them. The FFA advisor also includes students in FFA through Career Development Events (CDE) and Leadership Development Events (LDE) and awards programs. These competitive activities provide valuable leadership and career-based training for students. By offering a variety of CDEs and LDEs, the FFA advisor can involve a significant number of students in instructional activities beyond the regular school day.

The FFA advisor needs to ensure all students, regardless of background or characteristics, are welcomed in FFA. This includes eliminating bias and discriminatory speech, and having zero tolerance for bullying. However, including all students requires further steps. FFA chapter activities need to be critically reviewed. The FFA advisor and members need to ask whether an activity excludes groups of students and if so, how it can be modified to be more inclusive. Additionally, all students must be a part of the program of activities development including having their ideas included.

The Agricultural Education Program

Another important component of a successful FFA chapter is a "customized" agricultural education program. The agriculture teacher is bombarded almost daily with new programs, services, rules, or regulations. It does not take long for a teacher to feel buried in the avalanche of information that appears on their doorstep. To combat this and create an effective and relevant agricultural education program, the teacher, with the support of an advisory group, should choose the FFA activities, programs, and services to incorporate into the instructional program. If the closest dairy operation is 50 miles away from the school, no student is currently interested in a career in dairy science, and dairy production is not part of the curriculum, then it is perfectly acceptable to eliminate the dairy production CDEs from the list of activities for an FFA chapter. Agriculture teachers are human, and there are limited hours in which they can and should supervise FFA activities. The best agriculture teachers carefully choose the most important FFA services and offer them to the members.

Some teachers choose to delegate responsibilities for some FFA services to others who are more knowledgeable and have more time to supervise the members. An example of this is having a local beef producer train the livestock judging CDE team. Delegating tasks to others is acceptable because it leverages the FFA chapter's capacity to serve the broader interests of students. It is important to note delegation does not take the place of supervision. Agriculture teachers should continue to provide supervision over all FFA activities to assure program quality and student safety. There is a finite amount that can be delegated in an FFA chapter, so the FFA advisor should pay careful attention to make sure their workload does not become unmanageable. Having an appropriate life–work balance is further discussed in Chapter 25.

A Carefully Organized FFA Chapter

An FFA chapter that operates effectively and efficiently is a joy to advise, but a chapter does not become this way on its own. The first step is to have the chapter adopt a constitution and a set of bylaws that explain the mission, purpose, and direction of FFA at the school. These documents should clearly define the roles of leaders and the expectations of members and provide general guidelines for operation. Every FFA member should have a personal copy of the chapter constitution or have ready access to it. There is a sample chapter constitution online at the National FFA Organization website, www.ffa.org. From this starting point, the FFA chapter can develop a high-quality program of activities. Box 23.1 provides additional advice for having a successful FFA experience for local members.

Training Student Leaders

Part of the mission of FFA is to develop premier leadership through agricultural education. The agriculture teacher, in their role as FFA advisor, teaches students leadership skills and assists the FFA chapter in structuring and organizing activities to promote leadership

---

BOX 23.1 *Advice for Providing a High-Quality FFA Experience*

- Link FFA activities to high-quality agricultural education curriculum.
- Recruit and retain a diverse membership.
- Inform every student about the diverse opportunities in FFA.
- Elect capable officers and train them well.
- Ensure all members share responsibilities and have access to leadership and other opportunities.
- Formulate a workable constitution and bylaws.
- Develop a challenging program of activities.
- Secure adequate financing.
- Build school and community support.
- Conduct fun, well-planned, regularly scheduled chapter meetings.
- Maintain proper equipment and records.

---

development of all students. This training occurs inside and outside the classroom. The following practices promote student leadership development:

- Be a facilitator of student learning through FFA. Provide multiple opportunities for students to practice leadership skills. Think of FFA as independent practice of learned concepts.
- Be engaged, but not controlling. Ask questions, gently nudge, but do not do the work for the students. As students develop new leadership skills through FFA involvement, allow them to make mistakes. The FFA advisor has a role in helping students reflect on and grow from their mistakes.
- Be the adult advisor. Guide positional FFA leaders, such as the officer team and committee chairs, to be inclusive of all students. Push them to take learning risks.
- Be focused on learning goals and outcomes. The best FFA chapters provide a high-quality learning experience for every student. This learning experience supplements and complements the classroom and laboratory instruction and the SAE.

REVIEWING SUMMARY

FFA is an integral part of an agricultural education program in a secondary school. The agriculture teacher is the FFA advisor and is responsible for having an active FFA chapter. The overall mission of FFA is to make a positive difference in the lives of young people.

The history of FFA is parallel with the emergence and development of agricultural education. Many efforts to form a student organization followed passage of the Smith-Hughes Act in 1917. Walter S. Newman, H. W. Sanders, Edmund McGill, and Henry C. Groseclose, of Virginia, are credited with much of the early planning and leadership for a student organization, specifically the Future Farmers of Virginia.

The primary organizational structure of FFA is at three levels: local, state, and national. The local level is in the local school system as a component of the agricultural education program. The state level is primarily operated through state education agencies. The national level is through the U.S. Department of Education, with the National FFA Organization providing important program leadership and support. The federal charter for FFA is said to be a major enactment that provides viability for FFA.

All teachers and students should be well aware of FFA basics, such as the emblem, colors, official dress, creed, motto, salute, and code of ethics. Some of these reflect the history of FFA, but the code of ethics was created more recently to meet needs of students and teachers participating in FFA events.

FFA has Career Development Events (CDEs), leadership and personal development events (LDEs), and chapter development activities. Students with outstanding SAE programs advance with FFA proficiency awards.

A system of FFA degrees is used to motivate participation and recognize the accomplishments of members. The degrees of membership are Discovery FFA Degree, Greenhand FFA Degree, Chapter FFA Degree, State FFA Degree, and American FFA Degree.

Each local chapter should prepare a program of activities. This is the plan that a chapter establishes to guide its work for the year. Students use a committee structure to prepare the program of activities. Most programs of activities have three major areas: student development, chapter development, and community development. Initiatives of the national and state FFA levels are often part of a local program of activities.

Local chapters tend to show the most promise if teachers use the best practices identified by leaders in agricultural education. The Local Program Success initiative is much a part of these best practices.

QUESTIONS FOR REVIEW AND DISCUSSION

1. How did FFA come into existence?
2. What is the mission of FFA? Is it relevant to current student needs?
3. How is FFA organized at the national and state levels?
4. What is the role of the National FFA Organization?
5. What role did NFA serve and how was it absorbed into the FFA?
6. What is the importance of the federal charter for FFA (Public Law 105-225)?
7. Many agricultural education students are not members of FFA. Why?
8. What are the major parts of the FFA emblem?
9. What are the FFA colors?
10. Why is official dress used at FFA events? How relevant is the FFA jacket?
11. Why should members learn and recite the FFA Creed?
12. What is the FFA Code of Ethics? Why is it important?
13. What are the three major areas of FFA programs? Briefly describe each.
14. What are the FFA degrees? Identify and briefly explain each, including how requirements increase as the degrees become more advanced.
15. What is a program of activities? What are the three major areas of a program of activities?
16. What are some of the "best practices" in managing an FFA chapter?

ACTIVITIES

1. Research what PL 81-740 and its successors, PL 105-225 and PL 105-225, are designed to accomplish. In particular, explain the connection between classroom/laboratory instruction and FFA.
2. List on one page the top twenty moments in FFA history. Include the year and a one-sentence description of each.
3. Create a diagram describing the organizational structure of FFA from the national level to the local level.

4. Prepare a report describing the history and development of New Farmers of America. Include information about its absorption into FFA.
5. Prepare a one-page summary describing Career Development Events and a one-page summary describing Leadership Development Events to new students in your agriculture program. Include the purpose for their existence.
6. Explore the National FFA Organization's website, www.ffa.org. Investigate the contents of the site as related to student members and to teachers. Also, investigate the links to other agricultural education organizations.

REFERENCES

Berry, J. B. (1924). *Teaching agriculture.* World Book.

Jones, W. A., LaVergne, D. D., White, C. D., & Larke, A., Jr. (2021). The voices of African American agriculture teachers in one southern state regarding the NFA/FFA merger. *Journal of Agricultural Education, 62*(2), 39–52. https://doi.org/10.5032/jae.2021.02039

Maslow, A. (1970). *Motivation and personality* (2nd ed.). Harper & Row.

McCormick, V., & McCormick. R. (1984). *A. B. Graham: County schoolmaster and extension pioneer.* Cottonwood Publications.

National Council for Agricultural Education. (2016). *National quality program standards for agriculture, food, and natural resources education: A tool for secondary (grades 9–12) programs.* https://thecouncil.ffa.org/

National FFA Organization. (1998). *Agriculture teacher's handbook.* https://www.ffa.org/resource_tag/agriculture-teachers-manual/

National FFA Organization. (2002). *A guide to local program success* (2nd ed.). https://ffa.app.box.com/v/LPSguide

National FFA Organization. (2017). *National chapter award program: 2017–2021.* https://www.ffa.org

National FFA Organization. (2021a). *FFA.* https://www.ffa.org/

National FFA Organization. (2021b). *Growing leaders: 2021–22 fact sheet.* FFA. Retrieved February 7, 2022, from https://ffa.app.box.com/s/lgrbxeltnzsnmaw4agsz08mnuqrvqto1

National FFA Organization. (2021c). *Official FFA manual.* https://www.ffa.org

National FFA Organization. (2021d). *Vision, mission, motto.* https://www.ffa.org/about/who-we-are/mission-motto/

Rose, C., Stephens, C. A., Stripling, C., Cross, T., Sanok, D. E., & Brawner, S. (2016). The benefits of FFA membership as part of agricultural education. *Journal of Agricultural Education, 57*(2), 33–45. https://doi.org/10.5032/jae.2016.02033

Sheehan, C. Z., & Moore, L. L. (2019). Trends and impact of FFA affiliation on National FFA Organization student membership: A secondary analysis of existing data. *Journal of Agricultural Education, 60*(2), 210–221. https://doi.org/10.5032/jae.2019.02210

Tenney, A. W. (1977). *The FFA at 50, 1928–1978.* National FFA Organization.

True, A. C. (1969). *A history of agricultural extension work in the United States: 1785–1923.* Arno Press.

Wakefield, D. B., & Talbert, B. A. (2000). *Exploring the past of the New Farmers of America (NFA): The merger with the FFA.* Paper presented at the National Agricultural Education Research Conference, San Diego, CA.

Wakefield, D. B., & Talbert, B. A. (2003). A historical narrative on the impact of the New Farmers of America (NFA) on selected past members. *Journal of Agricultural Education, 44*(1), 95–104. https://doi.org/10.5032/jae.2003.01095

24

Community Resources

Octavius Miller is reflecting on the past school year. "I need to thank all the local businesses and organizations that have supported our program this year," he thought. As an experienced agriculture teacher, Octavius knew the value of involving the community in the agriculture program and wanted to make sure this collaborative relationship continued in the future.

Octavius had assessed the agriculture in the area. He made sure to involve the new orchards, small animal businesses, flower shops, animal shelters, and the many other emerging agricultural businesses and resources within a short distance of the school.

As you become a teacher, how will you determine the resources in a community? Further, how will you use these resources to improve your agricultural education program?

OBJECTIVES

This chapter addresses the National Quality Program Standards for Agriculture, Food, and Natural Resources Education (National Council for Agricultural Education, 2016), specifically Standard 4: School and Community Partnerships. It has the following objectives:

1. Discuss the importance of community resources in classroom/laboratory instruction, supervised agricultural experience, and FFA.
2. Explain how to identify community resources.
3. Describe resources provided by the school for conducting supervised agricultural experiences.
4. Describe ways of relating community resources to an agricultural education program.
5. Describe how to promote and market an agricultural education program.

TERMS

agricultural community
business community
community

community resource
resource
school community

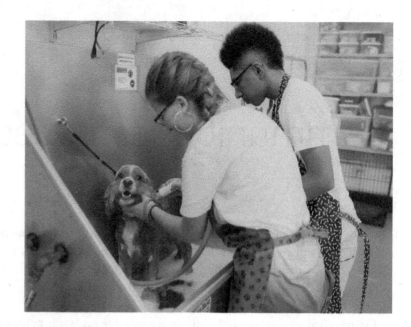

FIGURE 24.1 Visiting local agricultural businesses, such as companion animal care, helps determine community resources that might be available. (COURTESY OF LEDYARD AGRI-SCIENCE & TECHNOLOGY PROGRAM.)

THE IMPORTANCE OF COMMUNITY RESOURCES IN AGRICULTURAL EDUCATION

A *resource* is something that can be turned to for support or help. A *community resource* is a resource found in the area served by the school that may be capable of supporting agricultural education in some way. Because of the hands-on "learning by doing" methodology of agricultural education and the diversity of the agricultural industry, agriculture teachers need multiple resources to be successful.

Future teachers should understand they are not expected to know everything and do everything on their own. Using resources is not an indication of teacher weakness. The identification and wise use of resources is an indication that the teacher is informed and connected to local community expertise. Using available resources makes the teacher's job easier and more enjoyable.

Agriculture is a highly scientific and complex industry. As professional educators, agriculture teachers need the assistance of laypersons to provide high-quality educational programs. The increasing need for rapid change in an age of advanced technology complicates the teacher's role. Employment opportunities in agriculture are constantly changing, and new technologies are continually being developed and incorporated into the nation's agricultural industry and educational institutions. Effective teachers reach out into the surrounding community for assistance in keeping abreast of all these changes.

Teacher Support

In today's world, teachers need the support and cooperation of others to make a major difference in the lives of their students. Teamwork, partnerships, and cooperation are the keys to modern success. Working well with parents, building and maintaining good community relations, and working well with support groups are important in being a successful agriculture teacher (Roberts & Dyer, 2004). However, agriculture teachers are cautioned to maintain a balanced and healthy life, which includes refraining from becoming overinvolved in community responsibilities (Eck et al., 2019).

Successful agriculture teachers are good managers of the resources available to them and do not try to do everything themselves. There are a number of key groups that can help "make or break" a successful agriculture program. Therefore, agricultural education teachers should find and meet key people in the community and strive to involve them in the agricultural education program.

Training Opportunities

Students must be trained for today's jobs as well as new opportunities that become available. The need is constantly increasing for individuals trained in specialized technical occupations in agricultural fields. The *agricultural community* includes farmers, agricultural businesses, and others connected to the agricultural industry. This group of individuals can help teachers stay abreast of changing employment trends and opportunities. Developing and maintaining programs that include internships, work-study, and other types of on-the-job training requires close coordination with industry representatives.

Parents are important partners. Their involvement and support is essential in planning and conducting classroom and laboratory instruction, supervised agricultural experience programs, and for supporting FFA events and activities. The parents of first-year students are especially important, as they are often the source of the financial, transportation, and motivational support required to start supervised agricultural experience programs and to get their children involved in FFA functions.

Employers are also key partners for providing agriculture students with training stations and other opportunities for real-world experiences. Employers are often major financial contributors to the local FFA chapter for scholarships, achievement awards, and other FFA activities. They are also excellent resource people to speak to students in classroom and laboratory settings.

Local government officials and civic organization leaders can provide support for the agriculture program and FFA chapter through community improvement opportunities. They may also be sources of supervised agricultural experience opportunities. The mayor and other key community members have an interest in community development. They are often eager to involve young people in learning about local government and in making the community a better place to live.

FIGURE 24.2 The grounds superintendent of a golf course demonstrates to a student the use of a Stimpmeter in assessing turf quality. (COURTESY OF EDUCATION IMAGES.)

Alumni and Supporters are another source of expertise. They can help train CDE/LDE teams, speak to classes about the importance of FFA and supervised agricultural experience, serve as chaperones for trips, provide employment opportunities, and collectively provide financial and human resources for the local program. FFA Alumni and Supporters is discussed in more detail in Chapter 6.

Partnerships

Creating partnerships within the school is also important. When mainstreamed into agricultural education courses, students who are disadvantaged or who otherwise have special needs often require teachers to reach out for assistance in providing them with quality educational programming. The special education teachers in the school system are key resource people for assistance in this area. Creating partnerships with school counselors often leads to increased enrollments for the agricultural education program. Collaboration with science teachers and other faculty members can increase the quality of the instructional program and increase student achievement.

Another high priority of the agriculture teacher should be to keep the school administration informed about the agricultural education program and its value to the school and the community. Most administrators want to be connected with successful programs in their school and are more generous in supporting the programs they value.

By developing community partnerships, the agricultural education teacher increases school and community support for the program, eases their individual workload, and improves the overall quality and impact of the agricultural education program.

IDENTIFYING COMMUNITY RESOURCES

The *community* is a group of people living in the same locality and under the same government. The community consists of parents, employers, employees, government officials, and others living in the area served by the school. Teachers need to become familiar with the community and meet key resource people living there.

FIGURE 24.3 A sign demonstrates the strong community support for an agricultural education program with a school farm. (COURTESY OF EDUCATION IMAGES.)

School Community

When thinking about the communities that can be most helpful to agricultural education, a teacher should first consider the resources available through the school. The *school community* includes the students and all employees working in a particular school and the students and personnel of all the schools in the entire school system.

The beginning teacher should explore the school community to gain knowledge of the people and places that can provide assistance. Information is needed regarding a school's personnel, departments, and buildings. Identifying individuals important to the success of the agriculture department is essential. These people may be at the teacher's school or at other schools in the community. The teacher should become familiar with the educational system from the elementary through the postsecondary levels. Of importance to the agriculture teacher are the district and school administrators, business office personnel, guidance and counseling personnel, clinic personnel, media specialists, front office staff, cafeteria personnel, transportation staff, custodial and maintenance staff, other faculty members, and school support staff.

Key personnel can be identified from personnel directories published by the school system. Organizational charts show the position structure and lines of authority and are helpful in understanding how the school system works. A school map shows the school's departments, offices, and room numbers. If a map is not available, the agriculture teacher and/or agriculture students can develop one. A school district map shows the locations of the schools that make up the entire system.

When seeking resources outside the school community, the teacher should look first to those most closely connected to the students. Parents can be one of the most important allies of the agriculture teacher. Parental involvement in students' education has benefits for all involved: the student, the teacher, the parent, the school, and the community (Hornby & Lafaele, 2011). The agriculture teacher can become acquainted with parents through home visits, parent–teacher meetings, and school activities. Students' and parents' names and contact information should be collected by the teacher and filed for future reference.

After becoming familiar with the school community and the parents of the students enrolled in the program, the teacher should survey the surrounding community to determine the resources available in close proximity to the school. Several strategies can be used to become familiar with the community around the school. The members of the agriculture advisory committee and/or the FFA chapter officers can assist the teacher in learning about the local community.

Other strategies for identifying community resources include conducting a windshield survey by driving around the community to determine the names and locations of agricultural businesses. The teacher can check the yellow pages of the local phone directory, review the chamber of commerce website, read the local newspapers, attend community meetings, visit the mayor's office, and learn where and when the "movers and shakers" in the community meet in both formal and informal settings. Another good strategy is to meet with the county agricultural agent and ask about the agricultural organizations and key farmers in the community. The Farm Bureau may have a local presence and can be a community resource. As the teacher networks in the community, their list of contacts will increase until they are familiar with the entire community.

Business Community

The *business community* is the group of individuals who manage businesses in the area, including those closely affiliated with agriculture. This group is especially important because these individuals are a prime source of training stations for student supervised agricultural experience programs. They are also potential employers of the program graduates.

FIGURE 24.4 Agriculture varies in each area. This corn sweetener facility can be a resource for plant science, food science, and agribusiness courses.

The agriculture teacher should meet as many of the community's employers as possible, with a priority on those related to agriculture. If the community has a chamber of commerce, the teacher should become acquainted with the executive director of the chamber. Becoming involved with the chamber of commerce provides opportunities to meet many employers.

Government and civic organization leaders are another important resource group. Some of these individuals may already be identified, as some will be parents, school employees, and/or businesspeople.

Once the available resources have been identified, they should be entered into a database that can be continually updated. This will ensure easy access to the names and correct contact information for the agricultural education program partners.

RESOURCES SCHOOLS PROVIDE FOR SUPERVISED AGRICULTURAL EXPERIENCE PROGRAMS

Supervised agricultural experience programs provide a link between the agricultural education program and the local community. The entire school should value this linkage, as it contributes to positive school and community relationships. Therefore, it is important for the school to provide resources for conducting supervised agricultural experience programs.

The school's major responsibility for the supervised agricultural experience segment of the agricultural education program is to provide the services of the teacher for supervising students in their supervised agricultural experiences. The teacher should receive compensation for on-site supervised agricultural experience visitations. The individualized supervised agricultural experience instruction the teacher provides during the summer months

FIGURE 24.5 Specialized school facilities provide opportunities for learning and gaining experiences.

is a major justification for extending the employment of the agriculture teacher beyond the regular school year.

Because supervised agricultural experience programs are located throughout the community, the agriculture teacher is required to travel to provide on-site supervision. The school should provide the transportation, either by furnishing a district-owned vehicle and fuel or by paying mileage for supervisory travel when the teacher furnishes the vehicle and fuel. Often a school provides a pickup truck for the agriculture teacher, because part of the teacher's responsibility is to assist students in obtaining livestock, feed, seed, fertilizer, and equipment used in their supervised agricultural experience programs.

The school should also provide specialized equipment and facilities required for the successful operation of supervised agricultural experience programs that may not be available to students from other sources in the community. Examples include portable scales, greenhouses, land, and livestock facilities. Many schools supplement the funds they have available for these items with additional resources from booster clubs, service clubs, and private donations.

RELATING COMMUNITY RESOURCES IN AGRICULTURAL EDUCATION

Community resources provide support for agricultural education. Following are several examples of such support.

Support of Parents or Guardians

Support of parents or guardians is important to students, the FFA chapter, and the entire agricultural education program. To involve parents, the teacher should first obtain their names and addresses. Then contact each student's parents and ask them to assist in educating their child. All involved in parenting a student should be contacted, whether they live together or not.

The teacher should personally invite parents to visit during school open houses and parent–teacher conferences. Good teachers also offer to meet parents at other times to accommodate those unable to attend specific school functions. Good teachers also make accommodations for parents new to the community or those who speak limited English.

Parents should be asked to help with field trips and other events held in the community. They should also be invited to the annual FFA chapter banquet and other official events. Parent volunteers should be welcomed to the classroom and to FFA activities. Other strategies for involving parents include creating a parent booster club or inviting parents to become involved with the FFA Alumni and Supporters affiliate.

Educating parents about the agricultural education program should be a high priority of the teacher. This is necessary for gaining parental support for the teacher's guidance and direction of supervised agricultural experiences and FFA activities.

Support of Businesses

Working with the business community is extremely important. This group provides training opportunities for students, while the agriculture program provides the business community with current and future workers, along with future leaders of the agricultural industry and the local community. The business community is a source of guest speakers on a variety of topics; advisory committee members; competitive event judges; experiential learning opportunities, including supervised agricultural experience programs; field trip sites; student employment; graduate placement; borrowed resources (equipment, materials, books, etc.); and financial support. The teacher and the students should develop a plan for informing employers about FFA and supervised agricultural experiences to build awareness and gain commitment for the agricultural education program.

Teachers can create and maintain relationships with the agricultural community by meeting business, industry, and agricultural organization leaders, staying in contact on a regular basis, and keeping them informed about the agriculture program. Teachers and students should invite them to program events and functions, ask for assistance and offer assistance, and recognize and thank them as often as possible.

Support of Former Students and FFA Alumni and Supporters Members

Former FFA members and others in the community can also provide assistance and support for the agricultural education program. Their contributions can often be maximized through an active FFA Alumni and Supporters affiliate or Young Farmers chapter. The FFA Alumni and Supporters' role should be to support the local FFA chapter in meeting the needs of its members and to assist the agricultural education program in achieving its vision, mission, and goals.

Alumni and Supporters members and FFA members should be encouraged to work together. This interaction provides valuable learning experiences for students. Examples of support include raising funds for various events, awards, and activities; providing transportation, chaperons, judges, guest speakers, coaches, mentors, and other human resources for agricultural improvement and FFA chapter programs; and providing advocacy for agricultural education at school board and community meetings.

Local National Young Farmer Education Associations (NYFEA) may also be a resource for local FFA chapters. Joint meetings of FFA, FFA Alumni and Supporters, and local NYFEA provide opportunities for students to gain additional "real-world" information and contacts. NYFEA members can provide practice facilities and coaching for FFA teams, serve as resource persons for various events and activities, offer opportunities for farm and agribusiness tours, and provide leadership for service projects that promote agriculture.

Support of Educational and Research Institutions

All high schools have access to the resources of colleges and universities. These resources may be related to teaching, research, or outreach. Branch campuses or experiment stations often have well-trained scientists, economists, or others who can assist as valued resource persons.

Scientists can be invited to the school to give demonstrations and lead discussions. Proper preparation of students is needed. This includes instructing students in how to relate to resource persons.

School groups can take field trips to laboratories, test plots, and other facilities to learn from the scientists and technicians who are carrying out the research. It is not always necessary to travel, as video conferencing can be used. Students may establish dialogue with scientists on particular projects, such as one being done for FFA Agriscience Fair. Websites may be used to acquire information recommended by the cooperating scientists.

PROMOTING AND MARKETING AN AGRICULTURAL EDUCATION PROGRAM

Promoting or marketing the agricultural education program is important for increasing access to the available community resources. Strategies used by successful agriculture teachers for marketing a local program include the following:

- Having a visible and updated social media presence
- Holding an agricultural education program open house
- Meeting and working with the local media
- Preparing and distributing agriculture program print and electronic newsletters
- Working with the FFA reporter to prepare timely news releases
- Promoting the program with signs, posters, calendars, websites, and other media

A key strategy for maintaining ongoing involvement of the community in the local agricultural education program is to recognize and thank the individuals, groups, and organizations that contribute time, money, and other resources to the program. This can be accomplished through news releases, thank-you letters from the teacher and the students, appreciation certificates and plaques, honorary membership in the FFA chapter, and other special recognition. Providing recognition and appreciation to those contributing to the program should be an ongoing priority of the agricultural education department, and care should be taken to ensure all who contribute to the program receive appropriate recognition and thanks.

REVIEWING SUMMARY

Identifying and utilizing community resources connects the agricultural education program with the community. Community resources are important for providing students with state-of-the-art learning opportunities that prepare them for employment or for postsecondary education. Community partners may be school-based or from the community surrounding the school.

The agricultural education teacher and students should place a priority on identifying resources available in the community and on educating key partners about the agricultural education program. This will generate support and cooperation in providing a variety of educational opportunities and experiences.

Creating and maintaining long-term partnerships increases the value and impact of the agricultural education program in the local community.

QUESTIONS FOR REVIEW AND DISCUSSION

1. Why are community resources important to the agricultural education program?
2. What are some ways for teachers to identify community resources?
3. What resources should the school provide for supervised agricultural experience programs?
4. How can teachers connect community resources with the agricultural education program?

ACTIVITIES

1. Use the internet to investigate the agricultural businesses in a local community.
2. Prepare a report describing potential supervised agricultural experiences available in that community.
3. Prepare a report that describes how to develop links between an agricultural education department and the local community.
4. Develop a marketing plan for building rapport between an agricultural education department and the local community.
5. Use the local telephone directory or online directories to develop a list of potential resources in the school district. Cluster the resources to meet the needs of your report. Examples of clusters include horticultural businesses, farms and ranches, equipment dealers and repair services, forestry, environmental and natural resources, animal care, agricultural supplies, agricultural services, and agricultural products processing and marketing. Arrange to visit a small sample of the businesses you have identified.

REFERENCES

Eck, C. J., Robinson, J. S., Ramsey, J. W., & Cole, K. L. (2019). Identifying the characteristics of an effective agricultural education teacher: A national study. *Journal of Agricultural Education*, 60(4), 1–18. https://doi.org/10.5032/jae.2019.04001

Hornby, G., & Lafaele, R. (2011). Barriers to parental involvement in education: An explanatory model. *Educational Review*, 63(1), 37–52. https://doi.org/10.1080/00131911.2010.488049

National Council for Agricultural Education. (2016). *National quality program standards for agriculture, food, and natural resources education: A tool for secondary (grades 9–12) programs.* https://thecouncil.ffa.org/

Roberts, T. G., & Dyer, J. E. (2004). Characteristics of effective agriculture teachers. *Journal of Agricultural Education*, 45(4), 82–95. https://doi.org/10.5032/jae.2004.04082

Part 5

Career Stages in Agricultural Education

25

Progressing Through the Profession

The Rockville High School FFA banquet was over, with everyone saying it had been a great banquet. The FFA banquet committee had cleaned up and put everything away. The students, parents, and other guests had already left for the night. Ms. Leslie Rayburn, the agriculture teacher, was the only one left in the building. When she got to her office, she paused for a moment. She took stock of the teaching awards hanging on the walls and the pictures of successful FFA members on her desk. She glanced into the classroom and took pride in the state and national banners won by FFA members in CDE/LDE competitions and proficiency awards. She could point to each of these as confirmation of her status as a successful agriculture teacher.

Ms. Rayburn was in her seventh year of teaching, all of which had been at Rockville High School. "It has taken a lot of time and effort to get to this part of my career, and certainly, I have been successful," Ms. Rayburn thought to herself. This led her to remember a journal article (Traini et al., 2019) she had read for one of her graduate courses. As she drove home, she contemplated the journal article questions: What does it mean to be a successful agriculture teacher? What does a healthy work–life balance look like? Can agriculture teachers be both successful and have a healthy balance between work and life?

TERMS

alternatively certified teachers	induction
content knowledge	National Association of
cooperating school	Agricultural Educators
cooperating teacher	pedagogical content knowledge
curricular knowledge	preservice teacher
early field experience	education program
life–work balance	student teacher

FIGURE 25.1 For novice teachers, in-service training is a crucial element for career advancement. (COURTESY OF EDUCATION IMAGES.)

THEORETICAL ASPECTS OF TEACHER GROWTH AND DEVELOPMENT

Chapter 1 explains the teacher certification process by which a person becomes an agriculture teacher. This chapter describes the concepts that characterize a person's progression through the teaching career.

Some teachers see the beginning of their teaching career as an indication that the formal education process is over. They completed their student teaching internship experience and met the requirements for teacher certification. They now want to focus their attention on what comes next—preparing lessons plans, learning to use the shop and lab equipment at their school, and learning to manage the greenhouse and outdoor lab spaces. Agriculture teachers quickly realize the learning process is not over. The process is just beginning as it transitions to professional development and self-directed learning.

Theory of Self-Directed Learning

Adults are motivated to learn when they perceive a need to do so (Lindeman, 1926). The theory of self-directed learning (SDL) has been influenced greatly by Malcolm Knowles, introduced in Chapter 18 as a key thinker in adult education, and Allen Tough, who viewed SDL as a process. Adults seek knowledge when they have an interest or a question that can be satisfied by learning, with adults able to implement their own learning strategies. They organize learning differently from the way traditional PK–12 education is structured. Adults structure learning based upon life experiences and not necessarily by subject matter. For example, an agriculture teacher who needs additional knowledge to construct a carpentry project is not likely to see the need to attend classes in physics (structural concepts), mathematics (measurement and calculation of the bill of materials), or materials science (selecting the suitable construction materials). Instead, the teacher may seek a refresher workshop on carpentry skills. For adults, the most valuable resource is accumulated experience. The question then becomes, how does one reflect on that experience and analyze it for its best use? Additionally, as adults mature, their learning needs may change, along with their preferred learning methods. What may have been an effective learning strategy for a 23-year-old teacher may be ineffective for that same teacher at age 45 (Knowles et al., 2020).

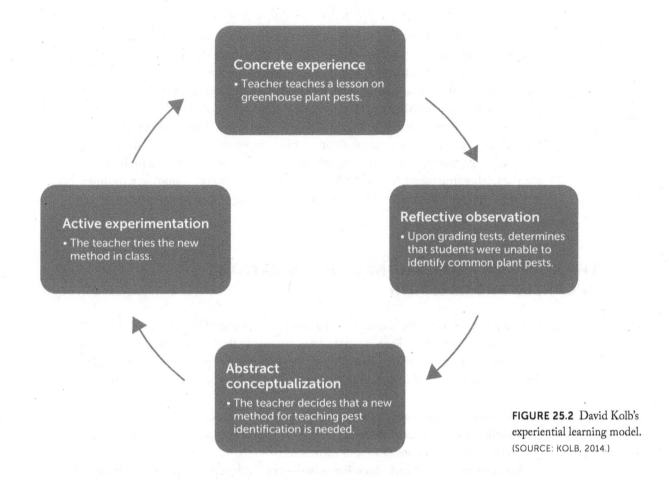

FIGURE 25.2 David Kolb's experiential learning model. (SOURCE: KOLB, 2014.)

Theory of Experiential Learning

In Chapter 21, Kolb's experiential learning model was discussed as it applies to school-based agricultural education students. Teachers often use the same learning model for their professional development. Kolb's model (Kolb, 2014) is based on our interpretation of our experiences and how we might modify similar future experiences after careful reflection and planning.

Figure 25.2 is an example of an experiential learning model for an agriculture teacher. The teacher has a concrete experience of teaching a lesson. After the lesson, the teacher has a reflective observation while grading assessments on the lesson. During abstract conceptualization, the teacher decides what is needed to address the observed problem. During active experimentation, the teacher reteaches the lesson using a different teaching strategy.

Transformative Learning Theory

Transformative learning theory states adults have a problem, challenge, or crisis that requires them to seek to make meaning of that experience (Mezirow, 1981). This learning process progresses as the problem forces the adult to challenge their long-held beliefs and seek to understand the dilemma of their understanding of a concept not matching reality. For example, consider this situation. A beginning agriculture teacher was successful as an FFA member in high school with a proficiency in forest management and products, competed in the Forestry CDE, and competed in the Agriscience Fair in Environmental Services/ Natural Resources Management. This teacher's first position is in a school with no forested land nearby and students who have no interest in forestry. This is a disorienting dilemma for the teacher.

Once the agriculture teacher recognizes the disconnect between their prior experiences and beliefs and this new perspective, they can critically assess the situation and explore options. This teacher decides to expand their knowledge base through several means. First, they participate in professional development organized by the state agriculture teachers association on SAE for All. Next, they participate in professional development organized by the state agricultural education program leader on Agricultural Education for All. Finally, the teacher seeks advice from their student teaching university supervisor. As the teacher's knowledge base builds, strategies emerge to mitigate the problem or concern. In this example, the teacher structures SAE and FFA for the resources of the community and the interests of the students. The process of a transformative learning process repeats itself throughout a teacher's career.

THE ROLES OF PROFESSIONAL ORGANIZATIONS IN CAREER PROGRESSION

Learning is very much a social endeavor. Teachers learn in the social environment offered by professional organizations. For most agriculture teachers in the United States, the ***National Association of Agricultural Educators*** (NAAE) is their primary professional organization. Teachers join the NAAE and other professional organizations for three main reasons:

Professional development—to learn and improve the craft of teaching
Building communities of shared practice—to share best practices, new ideas, and
    classroom innovations in a social environment
Advocacy for agricultural education policy—to provide input for policy makers on
    a policy that affects agricultural education and the work of agriculture teachers in
    the classroom; collectively through an organization

Professional organizations hold professional development conferences and special events designed to bring teachers together to share ideas and enjoy the camaraderie among teachers. Professional organizations also provide resources to improve teaching and provide awards and recognition for teachers who excel. Professional organizations offer scholarships to preservice teachers and grants for teachers to try new and innovative methods in the learning environment. These organizations may also provide benefits such as professional liability insurance and life insurance. Finally, through participation in a professional organization, teachers can develop leadership skills to advance in their careers.

Professional organizations provide a valuable structure to the social framework needed for teachers to survive and thrive in the classroom. From beginning teacher induction programs to leadership programs for seasoned teachers, professional organizations provide a valuable service to agriculture teachers, regardless of their length of service in the profession. Professional organizations help teachers as they progress through the stage of their careers.

CAREER STAGES OF AGRICULTURAL EDUCATION TEACHERS

Stage 1: Preservice Education

Most agriculture teachers begin their teaching career in a ***preservice teacher education program***. That is, completing a baccalaureate program or graduate program in a college or university leads to certification to teach agriculture. Approximately 86 higher education–based

FIGURE 25.3 The student-teaching internship is an opportunity to develop the knowledge, skill, and disposition to teach. (COURTESY OF EDUCATION IMAGES.)

agriculture teacher education programs exist in the United States (Swortzel, 2016). Two-thirds of new teachers arrive in the classroom due to a teacher preparation program in higher education. One-third of teachers are certified through *alternative* means, which are described in Chapter 1. Teacher education programs differ widely across the United States regarding entrance and exit requirements and progression through the program. For example, one university's teacher education program places student teachers in their cooperating school site three weeks before starting the official student teaching semester. Then, the student teachers come to the university campus at the end of the third week to learn approved practices for developing high-quality lessons and learning experiences for youth. This three-week preliminary session allows student teachers to build questions and concerns that can be addressed in their teaching methods courses. Other teacher education programs have students report to the university campus for instruction before starting their student teaching experience. *Cooperating schools* are those where student teachers are placed to conduct student teaching practicums and internships under the supervision of an experienced educator or *cooperating teacher*. A *student teacher* is typically a student in their last year of college preparing for or already placed in a cooperating school site for their internship. While there are many differences between agriculture teacher education programs, there are some similar traits. Teacher education programs usually require students to complete an *early field experience* that provides students with a shadowing experience in the first few years of their academic program. The student teaching internship occurs in the final year of their educational program.

Stage 2: Induction

Induction is the process of guiding and supporting beginning teachers in their first years in the classroom. Induction programs vary widely but may include a new teacher orientation, a mentoring program, support teams, and workshops for new teachers. Throughout the induction process, evaluation measures identify areas of strength and areas where professional development is needed.

FIGURE 25.4 Continuous improvement is the hallmark of the professional educator. (COURTESY OF EDUCATION IMAGES.)

Between 20% and 40% of new teachers leave the teaching profession during their first five years of teaching (Gray & Taie, 2015). Crutchfield et al. (2013) stated agriculture teachers leaving the profession left due to family circumstances, low efficacy, low motivation, or burnout. Induction programs support teachers in those fragile first years so they grow into the teaching role successfully.

Stage 3: Building Competency

By the fifth year of teaching, a teacher has mastered the basics of teaching and the particular characteristics of the bureaucratic school system. They know the procedures for requesting permissions for field trips, have learned the policies associated with handling funds from program fundraisers, and have developed classroom management skills congruent with those of the school. In short, by year five, a teacher has learned how things work in a school. They become more focused on becoming better teachers and more successful teachers.

Agriculture teachers in this career stage are concerned with contributors and stressors to job satisfaction (Sorensen & McKim, 2014; Sorensen et al., 2016). Contributors include compensation, working conditions, family/personal factors, and employment factors. Stressors include high weekly work hours and an unhealthy work–life balance. Hainline et al. (2015) recommended agriculture teachers prioritize family matters/obligations, explicitly structure family time, eliminate nonobligatory duties, delegate, and use a schedule and time management strategies.

Beyond Teaching: Career Paths for Professional Advancement

In stage three, teachers begin to settle into the teaching career, but some also start exploring or working toward other career options in education. Some teachers complete additional formal education that allows them to begin pursuit of a specialist role in education.

Educational specialist careers include school-based career and technical education administration or subject matter specialists within the state education agency's workforce development or CTE division. Still others begin to develop leadership skills that will lead them to become department heads within their school. Agriculture teachers have advanced in their careers to become principals and school superintendents, and some have become teacher educators at a college or university.

National Board Certification

Some teachers pursue certification by the National Board of Professional Teaching Standards (NBPTS). Teachers who complete the certification process develop improved teaching skills. Certification can also lead to higher pay and more leadership responsibility in the school. Board certification improves the quality of teaching, and that leads to better academic performance for students. An indirect result of the certification process is that schools report heightened morale among teachers and a better retention rate among experienced teachers (NBPTS, 2021).

Stage 4: Mastering Teaching

By the 10th year of teaching, professional development and classroom experience have begun to pay dividends. Master teachers have developed strong skills in the craft of teaching and have established themselves as teacher leaders in the school and profession. Master teachers are often called upon to mentor novice teachers and provide leadership for school-based improvement initiatives. They have mastered all three areas of "teacher knowledge":

Content knowledge—The conceptual and theoretical facts and principles of the agricultural sciences.

Pedagogical content knowledge—The ability to represent and communicate content in a manner that provides the most significant gain in student learning.

Curricular knowledge—The conceptual and theoretical principles of how content is structured and sequenced in the course of instruction.

FIGURE 25.5 Teachers who possess a thorough knowledge of agricultural subject matter know more options for teaching the material. (COURTESY OF EDUCATION IMAGES.)

Eck et al. (2019) recommended that for agriculture teachers to be effective they need to understand student needs, have a balanced agricultural education program, have a healthy work–life balance, refrain from working extra hours, support diversity and inclusion, have professionalism, develop positive dispositions, and limit involvement as community leaders. In the area of work–life balance they recommended committing to a balanced life, learning to say no, and asking for help.

Stage 5: Frustration and Stagnation Versus Career Satisfaction

Teaching is a "flat" profession. That is, there are few levels of advancement within teaching without leaving day-to-day classroom teaching. Regardless of how many years a teacher teaches, they have the same job title and significant responsibilities as a first-year agriculture teacher. Toward the end of a teacher's career, they may enter a period of stagnation and frustration with the job. They may begin to see the work as routine, unexciting, and unappealing. The position may not be as challenging as it once was.

External factors also play into a teacher's frustration. School policies are constantly changing, and new procedures may be too complicated for an experienced teacher to adopt. Teachers who have won accolades for their teaching performance may feel a sense of confusion about what the next great challenge will be. Not every teacher experiences this stage. For those who do, this stage often propels the teacher rapidly toward exiting the teaching career.

Clark et al. (2014) studied agriculture teachers who remained in the profession past becoming eligible for retirement. They found these agriculture teachers had a transformational experience leading to career sustainability. They had an abundance of support from students, parents, their administration, and the community. They also developed and maintained a balance between work and life in which they were satisfied. These factors led them to be successful throughout their career.

Much has been discussed and written lately about achieving an appropriate work–life balance. Many have proposed that to truly achieve a balance the focus needs to be reversed and stated as *life–work balance*. The past decade has seen an emphasis in helping agriculture teachers determine their preferred balance and providing suggestions for achieving this balance.

Throughout a teacher's career, there is a struggle to balance personal and family needs with the needs of the job. Novice teachers discover upon entering the profession that there is more to do than can be done in the agricultural education program. Numerous research studies report teachers are working well beyond the amount of time for which they are compensated. The average workweek for agricultural education teachers is 57 hours. This is the equivalent of working two full workdays beyond a standard 40-hour week. As teachers become more involved in teaching career demands, the struggle to meet family and personal obligations intensifies (Murray et al., 2011; Sorensen & McKim, 2014). If unattended to, the buildup of stress and frustration leads to burnout and job dissatisfaction. Teachers facing burnout often have no option but to depart the profession.

The Agricultural Education Magazine devoted entire issues to life–work balance in 2011 and 2013. Authors in the 2011 issue recommended that agriculture teachers use a system to control their time management, implement organizational systems, and use coping strategies to manage stress. Suggestions included determining priorities and devoting time to these, which means other activities receive less time or no time. Open and frequent communication with those significant in your life was emphasized throughout the issue. One article recommended maintaining a healthy lifestyle, which contributes to positive stress management.

The 2013 issue focused on the family aspect in a life–work balance. Authors recommended building your "village" to support you and your agricultural education program as this promotes less work stress and less excess time at work. Multiple authors emphasized making

> ### *Strategies for Balancing Life and Work Roles and Responsibilities*
>
> **Set boundaries**—Decide how much time is necessary to do the job well and limit work-related tasks to those hours.
>
> **Learn to say no**—There is never enough time to do everything. Focus time and energy on the most important things and do those first. Then you can do the less important things if time and resources permit.
>
> **Live with imperfection**—For teachers striving to improve their teaching competency, it may be tough to adopt a "less than perfect" attitude. Perfection often causes teachers to expend great effort for small improvements.
>
> **Be patient with professional development**—A teaching career is more like a marathon than a sprint. Develop a professional learning plan that allows for incremental and manageable improvement.

family and your personal life a priority and making sure you care for those important to you. Recommendations to support a healthy life–work balance included involving your family in agricultural education and FFA activities when possible and appropriate, informing family members well ahead of time of weekend and overnight commitments, and taking advantage of work time-savers such as premade teaching materials and delegation to others when able and appropriate.

The myth of the superhuman agriculture teacher is just that, a myth. Plan time to rest and recuperate from periods of intense work. By alternating periods of work and rest, productivity actually increases.

Stage 6: Exiting the Career of Teaching

In this stage, teachers prepare to transition out of the profession and into retirement or another career. This stage can be a pleasant period for many teachers when they reflect on the many positive memories and achievements during their careers.

For most teachers, the choice to depart the classroom belongs solely to them. Retirement is an opportunity to slow down the pace of life and catch up on leisure activities, reconnect with friends and family, and perhaps become more fully involved in civic endeavors. The choice to retire is influenced by any number of factors, including these:

- Age, health, and family factors
- Dissatisfaction with school administration policies
- Increasing workload, or a workload incompatible with a teacher's preferences
- Societal and community changes incompatible with a teacher's preferences

Regardless of the reasons, a voluntary departure from the teaching profession occurs for most teachers. Schools are bureaucracies with many layers of administration, rules, policies, and regulations. Exiting the classroom requires notifying school administrators and resigning based upon the procedures set forth by the school system. This is a formal process designed to ensure benefits and pension funds begin to flow to the retiree promptly. Retiring teachers will start to analyze pension plans and benefits to make sure they can maintain their preferred standard of living.

Departing the classroom is often an emotional experience for teachers who have devoted much time to developing the agricultural education program and the education of youth. A period of recognition for exemplary service will follow once a teacher indicates the desire to retire or depart the classroom.

In retirement, teachers may remain involved in FFA Alumni and Supporters activities at the school and serve as judges for FFA Career Development Events. Over time, these roles will diminish as the retired teacher focuses on different interests beyond the field of education.

REVIEWING SUMMARY

Teachers progress through specific stages in their career, from novice teacher to master teacher. Each stage has its unique characteristics and challenges. Teachers go through the novice stage to build competency and become master teachers before exiting the classroom. Some teachers move from the classroom environment and into administrative or higher education roles. A teacher's success depends upon how well they embrace their lifelong learning opportunities. Teachers gain benefits from the social and collegial interaction afforded by professional organizations.

QUESTIONS FOR REVIEW AND DISCUSSION

1. Which stage of a teacher's career seems to be the most challenging to you?
2. What kinds of emotional responses would you expect from the teacher exiting their career after three decades in the classroom?
3. How does induction differ from preservice education?
4. What strategies for balancing life and work seem to be the easiest to accomplish? Which one would be the most difficult to achieve, in your opinion?
5. How might a teacher combat career stagnation and frustration?
6. Describe Kolb's theory of experiential learning.
7. Describe the theory of transformational learning.
8. What are the benefits of being a member of a professional organization?
9. What are some careers in education that teachers may choose to pursue beyond the role of teaching?
10. What are the unique characteristics of the master agriculture teacher?

ACTIVITIES

1. Shadow a novice teacher and describe the particular challenges they face and the rewards they receive at this stage of their career.
2. Interview a master agriculture teacher. How do they perceive their role in the agricultural education profession?
3. Read research articles that focus on one or more stages of the professional growth and development of teachers. What do you find most interesting in the readings?

REFERENCES

Clark, M. S., Kelsey, K. D., & Brown, N. R. (2014). The thornless rose: A phenomenological look at decisions career teachers make to remain in the profession. *Journal of Agricultural Education*, 55(3), 43–56. https://doi.org/10.5032/jae.2014.03043

Crutchfield, N., Ritz, R., & Burris, S. (2013). Why agricultural educators remain in the classroom. *Journal of Agricultural Education, 54*(2), 1–14. https://doi.org/10.5032/jae.2013.02001

Eck, C. J., Robinson, J. S., Ramsey, J. W., & Cole, K. L. (2019). Identifying the characteristics of an effective agricultural education teacher: A national study. *Journal of Agricultural Education, 60*(4), 1–18. https://doi.org/10.5032/jae.2019.04001

Gray, L., & Taie, S. (2015). *Public school teacher mobility and attrition in the first five years: Results from the first through the fifth waves of the 2007–08 beginning teacher longitudinal study* (NCES 2015-337). U.S. Department of Education, National Center for Education Statistics. https://nces.ed.gov/

Hainline, M. S., Ulmer, J. D., Ritz, R. R., Burris, S., & Gibson, C. D. (2015). Career and family balance of Texas agricultural science teachers by gender. *Journal of Agricultural Education, 56*(4), 31–46. https://doi.org/10.5032/jae.2015.04031

Knowles, M. S., Holton III, E. F., Swanson, R. A., & Robinson, P. A. (2020). *The adult learner: The definitive classic in adult education and human resource development* (9th ed.). Routledge. https://doi.org/10.4324/9780429299612

Kolb, D. A. (2014). *Experiential learning: Experience as the source of learning and development.* FT Press.

Lindeman, E. (1926). *The meaning of adult education.* New Republic.

Mezirow, J. (1981). A critical theory of adult learning and education. *Adult Education, 32*(1), 3–24. https://doi.org/10.1177/074171368103200101

Murray, K., Flowers, J., Croom, B., & Wilson, B. (2011). The agricultural teacher's struggle for balance between career and family. *Journal of Agricultural Education, 52*(2), 107–117. https://doi.org/10.5032/jae.2011.02107

National Board for Professional Teaching Standards (NBPTS). (2021). *National Board certification.* https://www.nbpts.org/

National Council for Agricultural Education. (2016). *National quality program standards for agriculture, food, and natural resources education: A tool for secondary (grades 9–12) programs.* https://thecouncil.ffa.org/

Sorensen, T. J., & McKim, A. J. (2014). Perceived work-life balance ability, job satisfaction, and professional commitment among agriculture teachers. *Journal of Agricultural Education, 55*(4), 116–132. https://doi.org/10.5032/jae.2014.04116

Sorensen, T. J., McKim, A. J., & Velez, J. J. (2016). A national study of work-family balance and job satisfaction among agriculture teachers. *Journal of Agricultural Education, 57*(4), 146–159. https://doi.org/10.5032/jae.2016.04146

Swortzel, K. (2016). *American Association for Agricultural Education 2015–2016 faculty salary report.* http://aaaeonline.org/

Traini, H. Q., Claflin, K., Stewart, J., & Velez, J. J. (2019). Success, balance, but never both: Exploring reified forms of success in school-based agricultural education. *Journal of Agricultural Education, 60*(4), 240–254. https://doi.org/10.5032/jae.2019.04240

Glossary

A

academic software kinds of software that are useful to teachers and learners in the teaching and learning processes.

academic subjects courses or areas of study required of all students, including English/language arts, mathematics, science, and social studies.

accessibility process of ensuring activities, materials, and online resources are usable by all students.

accountability the state of being accountable or answerable; the holding of schools, programs, or teachers responsible for student learning and achievement; standardized tests in the core academic areas have become popular instruments for measuring accountability in education.

achievement evaluation the practice of measuring how much students know about a subject or how well they are able to use their knowledge; this measurement usually comes in the form of standardized or teacher-made tests.

activity manual an ancillary resource that has activities organized around a basal resource, such as a textbook.

administration the organization that manages or directs an institution, such as a school or a college, composed of administrators or individuals who manage the procedures used to implement policies in the institution.

adult an individual who has reached the stage in life to be personally responsible for himself or herself and who has assumed a productive role in society.

adult agricultural education education for men and women who are engaged in the ordinary business of a society and who typically have jobs in an area of the agricultural industry.

adult basic education (ABE) instruction intended to provide adults with basic and specific skills needed to function in society; instruction for individuals who did not achieve a certain level of, or participate in, education as children.

adult education the process of providing learning opportunities to improve adults in many areas of their lives.

advisory committee a group of laypersons empowered to give advice on educational programs but not actually to make decisions.

affective influenced or affected by the emotions.

agricultural awareness consciousness or knowledge of the agricultural industry; awareness enhanced by agricultural education programs that teach students about the nation's plants, animals, and natural resources systems.

agricultural community the community consisting of farmers, agricultural businesses, and others connected to the agricultural industry; the group of individuals with knowledge of changing agricultural trends and opportunities.

agricultural education a program of instruction in and about agriculture and related subjects commonly offered in secondary schools, though some elementary and middle schools and some postsecondary institutes/community colleges also offer such instruction.

agricultural education program the total offering of agricultural instruction in a school ranging from one to several subject areas or classes and including supervised agricultural experience and FFA.

agricultural literacy education about agriculture; above-average knowledge pertaining to the plants, animals, and natural resources systems.

agriculture the science, art, business, and technology of the plants, animals, and natural resources systems.

Agriculture in the Classroom (AITC) one of the best-known examples of an agricultural literacy initiative; provides instructional materials, human resources, and teacher training opportunities; exists in every state and has a federal presence within the U.S. Department of Agriculture.

agriscience laboratory a facility used in teaching science and math principles and concepts associated with agriculture; should adhere to standard recommendations of floor space, storage, and available utilities.

alternatively certified teachers teachers who are licensed without having completed traditional programs of teacher education preparation.

American Farm Bureau Foundation for Agriculture a national foundation focusing on agricultural literacy information.

American Federation of Teachers (AFT) a national education organization with more than 1 million members; unlike NEA, AFT is a recognized union and part of the AFL-CIO.

American FFA Degree the highest degree awarded to an active FFA member; the requirements for this degree are rigorous and challenging; the earliest a student can be awarded this degree is one year after graduation from high school.

ancillary instructional resource secondary or supplementary instructional material that accompanies primary instructional material, most often a textbook.

andragogy the study of processes and practices in adult education; it is based on the needs and readiness of the adult learners and thus is more learner-focused, collaborative, and problem-centered.

app/mobile application app is short for mobile application, which is software designed to operate on a mobile device.

aquaculture laboratory a facility used for education in fish farming and related areas; inside laboratories should be equipped with tanks or vats, while outside laboratories may use ponds or raceways.

area career and technical education center a specialized type of high school designed to prepare students for careers; this type of school tends to offer a great variety of career and technical education courses that are typically taught by teachers who have documented work experience in the areas in which they teach.

Aristotle a pupil of Plato's, introduced to the world around 335 BCE; sometimes called the father of scientists; his work yielded both the scientific method and the problem-solving method of instruction; he believed in realism rather than idealism.

articulation agreement a formal agreement between public high schools and postsecondary institutions to document the alignment of the high school curriculum with the postsecondary institution, resulting in equivalent postsecondary academic credit earned.

Association for Career and Technical Education (ACTE) the national professional organization for advancing the education of youth and adults for careers; consists of more than 30,000 members and has 12 divisions, one of which is agricultural education.

Association for Middle Level Education the primary professional organization for middle grades instructors and administrators.

asynchronous learning activities that occur on demand at any time that works best for the student's schedule.

authentic assessment an evaluation procedure that places the student in a situation that closely mimics the workplace or other life setting and then assesses the student's performance.

autonomy self-government or self-direction; local autonomy has always been a part of the agricultural education program philosophy to ensure agricultural education programs are structured around the needs and resources of their local communities.

B

backwards design an instructional design approach emphasizing beginning with identifying desired results, next establishing assessment evidence of successful attainment of the results, and ending with designing learning experiences to prepare students to master these results.

basal instructional resource a material that provides the base or foundation for the organization of a subject—for example, a textbook.

behavioral learning theory a theory of learning that focuses on observable changes in outward behavior and on the impact of external stimuli to effect change; behavioral learning objectives are formulated based on this theory.

bell-shaped curve a graphic depiction in which student scores are distributed along a line in the shape of a bell; shows how a large number of scores would be distributed; used to compare individuals with the group and with other individuals; also called a normal distribution curve.

blended learning refers to a variety of modes of face-to-face instruction that integrate a component of online learning.

block scheduling a class scheduling system in which students take fewer classes but in which each class period occupies a larger block of time, usually between 85 and 100 minutes; in this system, students usually take four classes each day and earn eight Carnegie units per school year.

Bloom's taxonomy a theory put forth by Benjamin S. Bloom in 1956 that classified educational objectives into three domains of learning: cognitive, affective, and psychomotor; the three domains are further broken down into more specifically definitive categories that may help educators pinpoint student learning behaviors.

board of education a group, provided for by the laws of the state, that has the legal authority to take action and use resources in the operation of a school district; members may be elected or appointed.

brain-based learning theory a theory developed in the 1990s by Renate Nummela Caine and Geoffrey Caine that united thought from multiple disciplines, including biology and psychology; led to the further development of the 12 principles of brain/mind learning.

building communities area of FFA program of activities that provides the framework for members to perform service-learning and engagement activities that benefit the community.

business community the group of individuals who manage businesses in an area, including those closely affiliated with agriculture; valuable in helping develop the students and employ the graduates of the agricultural program.

C

career cluster a broad grouping of occupations that have similar characteristics; all possible occupations are placed into one of 16 career clusters, such as the cluster of Agriculture, Food, and Natural Resources.

career exploration the use of investigative activities to help students identify potential career interests and gain the information relevant to those interests; offers students the opportunity to learn about agricultural careers and the skills needed in those careers.

career guidance the process designed to help a person choose a career and proceed through the development stages necessary for that career; involves providing students with career information and helping them make wise career decisions based on their strengths and interests.

career ladder the steps in an individual's general progression in a career from entry to retirement; additional education, expanded career responsibilities, experience, and other conditions are beneficial to an individual's movement upward.

Career Pathways System a structure for students enrolled in secondary and postsecondary career and technical education to develop skills and credentials leading eventually to viable jobs. The system requires industry, postsecondary education, and secondary education to coordinate their career and training efforts.

Carnegie unit the amount of credit a student is awarded for a course that meets one instructional hour per day for 180 school days; important for determining high school graduation and college entrance requirements; a course that meets for 50 minutes per day for the entire 180-day school year would be worth one Carnegie unit.

case study stories with an educational message.

chapter development the part of a program of activities that empowers individual members and the school to leverage their own resources in carrying out the FFA program.

Chapter FFA Degree the highest degree awarded by a local FFA chapter; earned by distinguished students who meet a number of requirements.

charter school a publicly supported school given freedom to operate outside of state and local bureaucratic control and thus able to experiment and try educational innovations; such a school is still required to meet local and state performance standards for its students.

Chautauqua movement an educational movement in the late 19th and early 20th centuries that combined secular and religious instruction; named for Chautauqua, New York, where the movement was born; died out in the 1920s.

Chromebook a version of a netbook that utilizes Google's Chrome operating system.

classroom a school facility designed for group instruction; ideally equipped with a number of items, including individual desks or tables and chairs, and at least one large display/writing surface.

cloud/cloud storage a network of online computer servers for online software access and usage and where data is stored in virtualized pools by third parties.

cognitive relating to "stored" information; having a basis in factual knowledge.

cognitive development the development of an individual's thought processes; includes adaptation to the environment and assimilation of information.

cognitive learning theory a theory of learning that focuses on the internal mental processes, how they change, and how they affect external behavior changes.

coherent refers to curriculum that is intentionally organized to ensure student learning builds upon prior learning and leads to attainment of established goals.

Comenius, Johann Amos (1592–1670); philosopher who wrote that teaching should promote a student's natural tendency to learn and learning should proceed from easy subjects to more difficult ones; one of the first to promote education for career preparation.

community a group of people living in the same area and under the same government; different types of communities include business communities and school communities.

community development the part of a program of activities that provides the framework for members to perform service-learning tasks that benefit the community.

community resource a resource found in the area served by the school that may be capable of supporting agricultural education in some way.

community survey a process to collect comprehensive and specific local information from various areas of the agricultural industry; the collected data is used to evaluate the local agricultural education program and help make a wide array of decisions concerning the needs of the program and its students.

comprehensive high school a high school that encompasses the full range of subjects and activities in academic, career and technical, citizenship, and personal development areas; also offers citizenship development activities, such as clubs and organizations, and personal development activities, such as sports and driver's education.

computer-based module instructional material that involves the use of units or modules guided by a computer program.

concrete operations the stage identified by Piaget in which the child develops organized and logical thinking skills.

connecting activity an activity that establishes relationships between school and life; FFA, for example, provides connecting activities for students of agricultural education.

constructivism a process of learning whereby the student constructs knowledge from his or her past experiences; a set of learning theories that emphasize how students actively make sense of the information they receive; teachers who utilize constructivism encourage their students to make sense of the outside world by actively engaging in their environment.

consumable teaching supplies that are used up each year during instruction.

content knowledge the conceptual and theoretical facts and principles of the agricultural sciences.

contextual learning instruction given within a perspective to which students can easily relate; effective agricultural education instruction often contextualizes content using plants, animals, and natural resources systems.

continuing education education that brings a participant up-to-date in a particular area of knowledge or skills; education received after an individual has completed a general level of formal education.

continuing renewal credits credits that fulfill or help fulfill requirements for the renewal of teaching licenses; awarded for attendance and participation at workshops, conferences, and seminars and for participation in other in-service activities.

cooperating school in teacher preparation the host secondary school for field experience, student teaching, and internship students.

cooperating teacher in teacher preparation the host agriculture teacher for field experience, student teaching, and internship students.

corrosive material a material with the potential to cause chemical reactions strong enough to damage metal and injure people; there are three kinds of corrosive materials: acids, bases, and miscellaneous materials such as iodine.

Council for the Accreditation of Educator Preparation an organization that accredits college and university teacher education programs. It uses a performance-based system and verifies that certain standards have been met.

counseling advice and guidance for persons with problems by appropriately trained individuals, and help in finding satisfactory answers; an important service and component in today's school systems.

craft committee an advisory group that is specific to a particular area, such as horticulture or forestry; often a subcommittee organized within a program advisory committee.

credentialing the process of determining and certifying a person can perform to required standards and meet competencies necessary to their profession.

creed a system of general beliefs, principles, or opinions that guide individuals in a profession or organization.

criterion-referenced test a test that compares the score of an individual student with established criteria or standards; designed to determine how students compare with established criteria rather than how they compare with others.

CTE concentrator refers to a student who has successfully completed at least three courses within a single CTE program of study.

CTE participant refers to a student who has successfully completed at least one course within a single CTE program of study.

CTE programs of study/program of study (POS) a sequence of CTE courses covering academic and technical content within a career area. Agribusiness, nursing, and construction are examples of CTE programs of study.

culturally responsive teaching a teaching approach in which teachers build upon the cultural knowledge, experiences, and frames of reference of culturally diverse students to create more relevant and meaningful learning experiences that are validating, empowering, and transformative.

cultural pluralism the social philosophy that the culture of a country, such as the United States, is a product of the cultures of that country's various immigrant groups; a teacher using a pluralistic approach would be an advocate for diversity in a school.

curricular knowledge the conceptual and theoretical principles of how content is structured and sequenced in the course of instruction.

curriculum the list of all courses offered in a school; also a group of related courses, such as the agricultural education curriculum; sometimes defined more specifically as the learning goals, transfer goals, and learning experiences a student has within classes and across the entire agricultural education program experiences.

curriculum alignment the process of ensuring the skills and learning outcomes within and across courses build upon each other, creating program coherence.

curriculum development the process of deciding what should be taught, how to teach it, how to assess learning, and an estimated time frame and sequence in which to teach it.

curriculum guide/curriculum map an outline of areas/units of instruction, lesson objectives, and scope and sequence that is aligned with appropriate educational standards.

curriculum integration the process of combining academic curriculum with career and technical education curriculum so learning is more relevant and meaningful to students.

D

demeanor the outward manner of an individual; the way in which an individual behaves; individual attributes that can be changed by the individual if necessary.

demonstration a teaching process that involves showing students how to do something before they do it for themselves or that involves showing them what would happen as the result of a particular action.

desktop computer a personal computer with limited portability, essentially for use in the same location all the time.

development the orderly, lasting change that occurs in humans as a result of maturation, learning, and life experiences; advancement from a simpler to a more complex stage.

device name for equipment especially related to computing such as laptops, smartphones, printers, screens, and such.

De Witt, Simeon (1756–1834); in 1819, proposed colleges for teaching agricultural subjects and conducting experimental research; introduced the concept of agricultural colleges.

Dewey, John (1859–1952); a pragmatist; believed in "learning by doing"; believed in inquiry as an orderly and scientific process that consisted of hypothesizing a problem, producing possible solutions to the problem, and attempting to solve the problem using those solutions; the father of progressivism.

differentiated instruction (DI) a systematic approach to how teachers design instruction to meet individual learning needs of academically diverse students.

digital citizenship a concept that describes the traits of an appropriate, responsible, and empowered user and consumer of technology.

digital learning connecting the internet and other digital technology with educational technologies to develop a wide range of approaches to learning to create individualized learning opportunities for students.

digital native person who has always known a world where internet access is omnipresent, social media plays a key role in how they experience the world, and technology is a necessary component of functioning in a global society.

directed laboratory supervised agricultural experience the type of supervised experience in which experiences in a practical activity are planned by the agriculture teacher; geared toward students who are unable to engage in other forms of supervised experience; often funded through some source other than the students' own funds.

disability a disadvantage or deficiency; the inability to do something specific, such as hear or see.

discipline teaching structure, routine, and behaviors to students to facilitate classroom learning.

Discovery FFA Degree the first degree a student can earn in FFA; designed for students enrolled in a middle grades agricultural education program; this degree is optional and not required for advancement to the next level.

discrimination the act of treating one group differently from another group; often the result of prejudice; most forms of discrimination are illegal in the school system, as stated in such laws as Title IX, the Civil Rights Act of 1964, and the No Child Left Behind Act of 2001.

discussion a teaching process less formal than a lecture that depends on student involvement to disperse information; typically, both teacher and student ask and answer questions, offer opinions, and debate.

display panel an electronic presentation device connected to a computer, camera, or other equipment for the presentation of visual information.

distance learning/virtual learning/online learning distance learning is technically learning that separates the learner from the instruction in time and/or location; however, has become synonymous with virtual learning or online learning, which is learning that occurs through internet-based instruction.

diversity the variety of differences within a category or classification; most often refers to differences of gender, ethnicity, and socioeconomic status, though other forms of diversity, including geography, religious beliefs, and language, need to be considered.

document camera electronic device at a teacher's station used to present visual information to the group.

dual credit course in which students earn college academic credit while enrolled in high school.

E

early field experience provides students with a shadowing experience in the first few years of their academic program.

e-book reader a mobile electronic device designed for reading of digital e-books and periodicals.

educational standard the expectation of what students should know and be able to do after completing a given area of instruction.

e-learning electronically based teaching and learning that usually involves some application of computer technology; also refers to a day of instruction that occurs either planned or unplanned, in which students learn from a distance.

elective course/program a course or program that is not mandated by state policies or standards and students are not required to take to graduate; courses or programs in agricultural education are electives, and whether they are offered is up to the local board of education.

electronic-based material instructional material accessed by the student or teacher through the use of a computer; typically includes CDs or DVDs.

elementary school typically a school with grades 1 through 6.

employability skills those "personal qualities, habits, and attitudes that influence how you interact with others" (Minnesota State CAREERwise, 2021, https://careerwise.minnstate.edu/careers/employability-skills.html).

engagement act of a student making a psychological investment in their own learning.

enrollment barrier an element in the school or community that prevents enrollment in agricultural education; some barriers include schedule conflicts, stricter enrollment requirements, and negative perceptions of the program.

entrepreneurship/ownership supervised agricultural experience the type of supervised experience in which students develop skills needed to own and manage enterprises; students engaged in this type of supervised experience have financial investment or risk in their enterprises.

epistemology the philosophical study of the nature and origin of knowledge.

equipment implements used to aid in the completion of a task; includes such instructional resources as welding machines, radial arm saws, trimmers, soil mixers, and components of irrigation systems.

Erikson, Erik (1902–1994); proposed the psychosocial theory of development, which is based on the idea that people have the same basic needs and that a purpose of society is to provide for those needs; he theorized that all people go through eight stages of psychosocial development, with the outcome of each stage having an effect on later stages.

evaluation in education, a process to analyze educational effectiveness and student achievement by using measurement tools, such as tests, observations, and performances.

experiential learning learning by doing; knowledge gained through experience. Experiential learning theory describes learning as resulting from the transformation of experience.

experiment a trial that uses a definite and planned procedure; the process of conducting a test for the purpose of demonstrating a truth or examining a hypothesis.

exploratory supervised agricultural experience an educational experience that provides the student with an opportunity to investigate a number of areas in agriculture; designed to help students gain information for making decisions about their future education and careers; job shadowing is an example of an exploratory supervised experience.

F

facility a building, area, or item that is not movable and that is built or established to serve a particular purpose.

feeder school a middle school or junior high school whose students will attend a specific high school for which the feeder school provides preparation.

FFA Agriscience Fair recognizes middle school and high school students who are studying the application of science and technology in agricultural enterprises. The FFA Agriscience Fair provides recognition opportunities for students conducting research supervised agricultural experience.

FFA Alumni/FFA Alumni and Supporters an adult group within the National FFA Organization that supports agricultural education and is open to anyone who has an interest in supporting FFA; it currently has approximately 52,500 members; its mission is very similar to that of FFA.

FFA Creed a statement of beliefs held by every FFA member with regard to his or her place in the industry of agriculture; in 1930, E. M. Tiffany, an educator from Wisconsin, wrote the version of the FFA Creed adopted by the Third National FFA Convention.

FFA degree program a program of the National FFA Organization to encourage students to establish and work toward career goals in the agricultural industry; promotes advancement in FFA.

FFA emblem a graphic symbol that represents the FFA organization; the first emblem was derived from a design created for the Future Farmers of Virginia in 1927; FFA has sole rights to use the heavily symbolic emblem.

FFA Motto a short expression that reflects the character and mission of FFA; it also reflects the pragmatic philosophy of work prevalent when C. H. Lane, the first National FFA advisor, recommended it to the National FFA Convention delegates.

field trip a learning experience that involves traveling away from the school site, for example to an agricultural experiment station to observe professional research projects.

flammable material a material that easily catches on fire and burns; gasoline and kerosene are two examples.

flipped classroom a teaching model in which the teacher creates various forms of content delivery such as a recorded lecture, or curated readings for students to review outside of class time. Instead of focusing on content delivery during class time, the teacher prioritizes application and extension activities for students to clarify and deepen their understanding of the content.

Food for America a program sponsored by FFA that provides elementary and middle school students with instruction on a variety of agricultural literacy topics, including food, fiber, and natural resources.

forced-answer question a common type of test question that requires the student to select the correct answer from given choices; includes multiple-choice, true/false, and matching questions; sometimes known as fixed-response question.

formal education education or training provided in an orderly, logical, planned, and systematic manner and often associated with school attendance.

formative evaluation evaluation occurring before or during instruction; used to determine the starting point for instruction and to provide feedback for students and teachers during the instruction.

foundational SAE consists of five components: career exploration and planning, employability skills for college and career readiness, personal financial management and planning, workplace safety, and agricultural literacy.

free elective an optional course that does not fulfill any educational requirements.

G

game-based learning use of electronic games in instruction with the priority on education, not entertainment.

gamification involves integrating elements of game play such as points, levels, and badges into learning activities, but in a manner that is not meant to entertain.

global competence the disposition and knowledge to understand and act on issues of global significance.

grade a mark or descriptive statement that rates the extent to which a student has achieved a standard; typically assigned by either letter (A–F) or number (0–100); the format and structure of grading are typically specified in school system policy manuals.

Greenhand FFA Degree until the introduction of the Discovery Degree, this was the first degree a member could attain; Greenhand members are usually first- and second-year students enrolled in a high school agricultural education program.

Groseclose, Henry C. (1892–1950); the first National FFA executive secretary/treasurer, elected on November 20, 1928; sometimes said to be the father of FFA because of his work in authoring important FFA documents.

grouping placing students together to accomplish a common task or goal.

group teaching method a teaching approach that instructs students together as a group; advantages include overcoming such teaching concerns as time and resource requirements; group instruction teaches all students the same content.

growing leaders part of FFA program of activities that provides members with opportunities to learn new leadership skills and develop teamwork skills through participation in leadership training experiences.

guidance a process consisting of different techniques designed to help students make wise decisions regarding choices or changes; counseling is a popular technique for providing guidance to students.

H

Hamlin, Herbert M. (1894–1968); professor of agricultural education at the University of Illinois; felt that decisions about local agricultural education programs should be made only with citizen input; author of *Public School Education in Agriculture: A Guide to Policy and Policy-Making* (1962).

handicap a physical or mental disability that creates a disadvantage in certain situations; for example, blindness is a handicap if a situation requires sight.

hardware physical equipment that allows users to interact with digital tools.

Hatch Act an act passed in 1887 that established agricultural experiment stations and educated the public about the implications of the research conducted at these experiment stations.

hidden curriculum values not explicitly articulated within curriculum, but communicated through what is considered acceptable or unacceptable behaviors.

hierarchy of human needs developed by Abraham Maslow in 1970, this hierarchy prioritized human needs into seven layers, starting with the most necessary: physiological needs, safety needs, belonging needs, esteem needs, and self-actualization.

high school a school that includes grades 9 or 10 through 12; most high school agriculture courses are designed to develop specific skills in agriscience, agribusiness, technology, leadership, and human relations.

home school education carried out by parents and others outside of traditional school buildings and typically in the home until completion of high school.

horizontal articulation articulation within the same grade level among the various courses being taken by the student; allows continuity from subject to subject, such as from science to language arts.

Hughes, Dudley (1848–1927); sponsor of the Smith-Hughes bill in the U.S. House of Representatives.

I

idealism the action or practice of envisioning things in an ideal form; the theory that something is in its purest and ideal form as an idea.

immersion SAE chosen by students as an enrichment of the agricultural literacy component of foundational SAE. Types are placement/internship, ownership/entrepreneurship, research, school-based enterprise, and service-learning. Levels of motivation are graded, recognition, and career-ready.

improvement project an experience or group of experiences, usually involving home or community work, carried out in conjunction with one of the supervised experience programs.

independent study a learning activity that, for any of a variety of reasons, is not part of an organized class learning experience; in almost all independent study activities, the student takes on more responsibility for his or her instruction.

individual teaching method an approach that recognizes that each student learns at a different pace and possibly at a different level; a primary advantage of individual instruction is the potential for greater student interest and deeper learning.

induction the process of guiding and supporting beginning teachers in their first years in the classroom.

infuse to introduce or instill one thing into another so as to affect it; in relation to agricultural literacy, to include agricultural lessons, examples, and topics in courses for other subjects.

inquiry-based instruction teaching method that meshes the natural curiosity of students along with the scientific method to cultivate critical thinking skills and scientific knowledge.

instructional material material that contains the information being taught, for example, a textbook, transparency, electronic presentation, or magazine.

instructional materials adoption process used by school districts and state education agencies in selecting materials and allocating resources.

instructional resource material a teacher uses to provide instruction and students use to learn; examples are lab instruments, supplies, and safety equipment.

instructional resource guide a collection of material, often containing such resources as detailed lesson plans and sample tests, that helps a teacher plan and deliver instruction.

instrument a device used for a particular purpose, usually to facilitate work, for example, a caliper used to make precise measurements.

integration the process of combining academic curriculum with career and technical education curriculum so that learning is more relevant and meaningful to students; designed to eliminate distinction between academic and career and technical education.

Internet of Things (IoT) a network of items that have digital sensors embedded within them to collect data that can be shared with other devices on a network to improve performance and efficiency.

interactive whiteboard a presentation medium that uses a computer connected to a projector to display materials on a white surface that acts as a touch screen; components may be connected wirelessly or via USB or other cables.

intracurricular literally, "within the curriculum"; an integral part of the program or curriculum, as opposed to an extracurricular program or club; FFA is an intracurricular component of agricultural education.

inventory a complete and itemized list of all equipment, tools, and supplies; should be taken at least yearly; helpful in monitoring the condition of items and their frequency of use.

J

James, William (1842–1910); asserted that pragmatism is about searching for truth, a departure from Charles Peirce's belief that pragmatism is about finding meaning in a concept; believed that the test of an idea is in its truth or workability.

job shadowing spending time with and observing an experienced person or persons performing work in a specific occupation.

junior high school a school that includes grades 7 and 8 and sometimes grade 9; specific agriculture instruction may be offered at this level.

L

laboratory an area equipped for student experimentation, research, or study; different educational courses may require different kinds of laboratories.

laboratory an application of concepts and principles that includes experiments, hypothesis testing, demonstrations, learning exercises, application exercises, and skills practice.

laboratory instruction instruction occurring in laboratories that provides freedom of movement, psychomotor skill development, the use of special tools and equipment, and student self-directed learning; educational emphasis placed on application and hands-on learning through the conducting of experiments and the simulation of real-world experiences.

laboratory superintendent a teacher's assistant (usually a student) who helps with various laboratory duties, such as making sure students do their tasks and making sure that areas are clean, with tools and supplies properly put away.

Lane, Charles Homer (1877–1944); the first National FFA advisor, chosen November 20, 1928.

laptop a personal computer that is easily transported for mobile use and has the capabilities of a typical desktop computer; may have a rechargeable battery.

layperson an individual who is not an agricultural educator or other professional or certificated person employed by the school.

LEAP (Licensure in Education for Agricultural Professionals) a certification program in agricultural education that is delivered online; administered by the Agricultural and Extension Education Department at North Carolina State University; accredited by the Council for the Accreditation of Educator Preparation (CAEP).

learning the experience of gaining knowledge or skill that results in a permanent change of behavior.

learning center an area designed for one activity or lab that allows small groups of students to complete the activity or lab at the same time; typically self-contained, with all the equipment, tools, supplies, and instructions available in one place.

learning disability any of a variety of disorders, such as dyslexia, that interfere with a student's ability to learn; often leads to the failure or incapacity of the student to function at the age-appropriate level in language, reading, spelling, mathematics, and/or other areas of learning.

learning environment the various conditions that promote or impair teaching and learning; the teacher's responsibility is to provide a quality learning environment for the students.

learning management system (LMS) an online portal that serves as an access point for students and teachers to interact with class materials.

learning module a self-contained activity that a student can complete independently; most often used for cognitive and psychomotor development; advantages of learning modules are that students can work at their own pace and instruction can be modified to meet the needs of individual students.

learning objective a carefully written statement of the intended outcomes of a learning activity; always written in terms of what the student will be able to do at the conclusion of the instruction.

learning stop a location in a lab where students stop and read information, do a task, or observe some phenomenon.

learning style a description of how a person prefers to learn and within what environment he or she prefers to learn; learning styles vary from student to student and can include a variety of factors, such as whether a student is intrinsically or extrinsically motivated and whether the student prefers to learn by auditory, visual, or kinesthetic means.

least restrictive environment the idea that students with disabilities be allowed the closest-to-normal educational setting possible; required under the Individuals with Disabilities Education Act of 1975, which aimed to end in schools the discrimination against students with disabilities.

lecture an oral presentation by the teacher or other individual that may include a wide range of techniques; an instructional method of delivering information to an audience or class.

lesson plan a road map used by teachers to ensure effective, efficient, and empowering instruction.

lesson plan library a collection of teaching plans focused on a broad area, such as landscaping or wildlife management, that aids the teacher in delivering instruction; lesson plans include summaries of content, suggested instructional strategies, sample tests, lab sheets, and transparencies or electronic-presentation images.

life–work balance comparison of the amount and quality of time spent in personal and family pursuits with that spent in work.

limiting factor a resource whose unavailability or time requirements may make conducting a lab impossible or require its modification.

literacy the condition of being able to read and write; the quality of having more than average knowledge in a particular field.

Local Program Success (LPS) a national initiative designed to enhance the quality and success of local agricultural education programs through the employment of four strategies: (1) program planning, (2) partnerships, (3) marketing, and (4) professional growth.

lyceum an institution introduced in 1826, when Josiah Holbrook founded the American lyceum in Millbury, Massachusetts; a society where the arts, sciences, agriculture, and public issues were presented and discussed; the lyceum movement faded after the Civil War.

M

magnet school a specialized high school designed to attract students for a specific purpose, such as intense studying of the arts, science, mathematics, or even agriculture; usually, one must apply to attend a magnet school, and competition for enrollment is high.

mastery learning an educational approach in which students are given frequent opportunities to practice and multiple opportunities to improve their scores; often associated with criterion-referenced tests.

middle school a school that commonly includes grades 6 through 8; agriculture instruction at this level is broad, focusing on helping students understand the plants, animals, and natural resources systems and helping them make informed choices about their future.

mindset what a person believes about the underlying nature of their abilities.

misbehavior a person's inappropriate, yet purposeful and motivated, attempt to cope with situations in the environment; behavior that violates social norms within the classroom.

mission statement a written statement detailing the purpose of an organization and how the organization will go about achieving that purpose.

Morrill, Justin (1810–1898); congressman from Vermont and proponent of the American farmer and the working-class American; worked to pass a landmark land-grant bill that would be known as the Morrill Act of 1862.

Morrill Act of 1862 act signed into law on July 2, 1862, by President Abraham Lincoln; led to the creation of colleges for the teaching of the agricultural, mechanical, and military arts; named after Congressman Justin Morrill, who worked the legislation through Congress.

Morrill Act of 1890 act that set aside funds for teacher education in agricultural and mechanical arts; restricted funds to only those colleges where no distinction was made on the basis of race or color in student admissions; appropriated funds from the sale of public lands.

motivation the energy and direction given to behavior; an inducement or incentive, whether intrinsic or extrinsic.

multicultural education education that provides all students with equal opportunities to learn, regardless of their race, class, or culture; a reform idea that promotes equality and the use of various perspectives in education; a major goal of multicultural education is the improvement of academic achievement for all students.

N

National Association of Agricultural Educators (NAAE) the national professional organization for agriculture teachers; provides advocacy for agricultural education and for all agricultural educators; consists of more than 7,600 members in 50 state associations.

National Board for Professional Teaching Standards (NBPTS) an independent, nonprofit organization that provides advanced certification to teachers who meet its standards; the standards are based on five core propositions.

National Council for Accreditation of Teacher Education (NCATE) an organization that accredits college and university teacher education programs; uses a performance-based system and verifies that certain standards have been met; successful completers receive teaching licenses from the state of North Carolina. They are then eligible to teach in almost all 50 states through a process called reciprocity.

National Education Association (NEA) the largest national organization for educators, with more than 2.7 million members; NEA is involved on local, state, and national levels doing such work as bargaining, government monitoring, and lobbying.

National FFA Scholarship Program an initiative of the National FFA Organization to recognize excellence by awarding scholarships for students to pursue education beyond high school.

National Middle School Association (NMSA) an organization founded in 1973 that is dedicated to improving the education of young adolescents; a leading advocate for middle grades education.

National Quality Program Standards an online assessment tool for local agricultural education programs to use in analyzing and developing goals/objectives for program growth.

National Young Farmer Education Association (NYFEA) the official adult student organization for agricultural education as recognized by the U.S. Department of Education; its stated mission is to "promote the personal and professional growth of all people involved in agriculture."

needs assessment a program planning process used to collect and analyze community data, including statistics from various sources on such things as local agricultural products, employment, and employers.

netbook a category of small and inexpensive laptop computer that has found some favor for educational purposes.

New Farmers of America (NFA) the companion organization to FFA that served Black Americans from 1926 to 1965; offered young Black students the incentive to pursue agricultural careers while building their self-confidence and technical skills; the Civil Rights Act of 1965 brought an end to NFA by requiring that it merge into a single organization with FFA.

Newman, Walter Stephenson (1895–1978); a state supervisor for agricultural education in Virginia and a pioneer in the development of FFA; believed that some type of organization was necessary to provide opportunities for personal advancement and the growth of self-confidence in farm youth; in 1925, developed the Future Farmers of Virginia with the help of colleagues Henry C. Groseclose, Edmund Magill, and H. W. Sanders; later urged the development of NFA.

nonformal education education or training offered outside of school curriculums and in settings other than those that involve the implementation of structured standards and fulfillment of diploma or degree requirements.

norm-referenced test a test that compares the achievement of an individual student with an established group norm; constructed to yield a bell-shaped curve, in which the majority of students have scores in the average range, with a few students having scores in the high and low ranges.

O

one-to-one device each student has a personal device to use for educational work.

open-ended-answer question a question that requires the student to develop his or her own answer; fill-in-the-blank, short-answer, and essay questions are constructed-response questions; sometimes known as constructed-response question.

operant conditioning behavior modification in which the desired behavior is encouraged through positive or negative reinforcement.

OSHA (Occupational Safety and Health Act) a federal act that requires certain fire, health, and safety standards be met in the school environment.

oxidizer a material that can initiate and promote combustion; chlorine is an oxidizer.

P

Page, Carroll (1843–1925); Vermont senator in the early 1900s with a passion for vocational education; believed it was the duty of the federal government to provide for job-specific training for young people; instrumental in establishing the Smith-Hughes Act.

paid placement supervised agricultural experience a supervised experience in which the student is employed for compensation while gaining practical experience and developing skills needed to enter and advance in a particular occupation.

paper-based material any material printed on paper, either in color or in black and white, and almost always bound in some way.

pedagogical content knowledge the ability to represent and communicate content in a manner that provides the most significant gain in student learning.

Peirce, Charles Sanders (1839–1914); the first well-known proponent of pragmatism; believed true meaning evolves by way of a three-stage process through which one moves from an unconscious understanding of a concept to awareness of the characteristics of a concept and, finally, to the application of what one knows about a concept to other situations involving that concept.

performance-based assessment (PBA) instructor evaluates how well a student can do something under given conditions.

performance-based evaluation the evaluation of how well students can do something under given conditions; students provide tangible evidence of their learning to teachers, who observe and judge the evidence using tools such as rubrics, checklists, rating scales, or product scales.

Perkins, Carl (1912–1984); senator from Kentucky and sponsor of the Vocational Education Act of 1963; in part responsible for the Perkins Acts (Federal Vocational Education Acts of 1984, 1990, 1996, 1998, 2006, and 2018), which attempted to modernize vocational education further and to expand its emphasis as career and technical education.

personal protective equipment (PPE) worn to reduce exposure to hazards that may cause injury. Examples include gloves, headgear, chaps, and aprons. Personal protective equipment supplied by the school such as safety glasses, welding helmets, or chainsaw chaps.

Pestalozzi, Johann Heinrich (1746–1827); developed the principle of informal education, such as hands-on learning in the sciences.

Phi Delta Kappa International a professional organization for educators on all levels, including undergraduates preparing to become teachers; an international association with more than 600 local chapters, most of which are based on college campuses.

philosophical foundations the basic beliefs that guide people's actions; the guide by which educational programs are designed and delivered; developed by individual thinkers or philosophers.

philosophy a discipline that attempts to provide general understanding of reality and interpret the meaning of what is observed; a system of values held by an individual or group.

Piaget, Jean (1896–1980); a Swiss psychologist who developed a theory that there are four stages of cognitive development: sensorimotor, preoperational, concrete operational, and formal operational.

Plato (circa 428–347 BCE); an early Greek philosopher and star student of Socrates who continued in the dialectic tradition through his writings and teaching; author of *Republic*, which attempted to define the nature of the ideal state; established the first known college of higher education, called the Academy, where he taught idealism.

policy a general guiding principle, usually based on a law; established by individuals or groups that have the authority to do so, for the purpose of guiding the management, procedures, and especially the decision making of government bodies, such as school boards.

portfolio a collection of materials that represent a student's progress and accomplishments, usually contained in a notebook or a folder or on a CD-ROM or a website.

postsecondary education typically defined as education offered at technical schools, community colleges, or junior colleges, though it sometimes includes college and/or university-level education.

postsecondary program a program that is sequenced after the high school level, such as a program at a community college or at a career and technical center; agricul-

tural education in one of these programs is more specialized and provides students with advanced technical competencies to enter and progress in specific agricultural careers.

pragmatism a philosophy centered on the concept of "knowing by doing"; a relatively young philosophy, it views knowledge as a practical activity and understands questions about meaning and purpose within this context.

prejudice a preconceived preference or idea formed without complete information or examination of the facts.

Premack principle the psychological principle that places the less preferred activity before the preferred activity and uses the preferred activity as positive reinforcement for the less preferred activity.

preoperational the stage identified by Piaget in which the child begins to understand the symbolic nature of things.

prepared curriculum instructional materials created by sources other than the teacher. They can include curriculum maps, lesson plans, and instructional materials needed to implement the lessons.

preservice teacher education program a baccalaureate program or graduate program in a college or university leads to certification to teach agriculture.

pretest a test given before instruction is begun to determine previous student learning in the subject area; helpful in deciding which material needs to be covered in detail and which can be either lightly reviewed or omitted altogether.

principal a person who has controlling authority or is in a position of presiding rank, especially the head of a local school site.

procedure the way a policy is to be implemented or carried out; a set of established forms or methods for conducting the affairs of an organized body, such as a school board.

product scale scoring rubric a scale on which the different levels of quality are represented by actual objects, with each object having an assigned numerical or letter grade, depending on its position on the scale; for example, in agricultural mechanics the teacher can develop a product scale containing welds of varying qualities, from exceptional to inferior, by which to grade the students' welds.

profession an occupation that requires training and advanced study in a specialized field and adheres to a code of ethics to guide the individuals in the occupation; the body of qualified persons of one specific occupation or field.

professional an individual who has acquired the necessary traits ascribed to the profession they practice; an individual who has an assured competence in a particular field or occupation.

Professional Agricultural Student Organization (PAS) an organization associated with agriculture, agribusiness, and natural resources in approved postsecondary institutions offering associate's degrees or vocational diplomas or certificates; one of the 10 student organizations approved by the U.S. Department of Education as integral parts of career and technical education.

professional code of ethics a statement of ideals, principles, and standards related to the conduct of individuals within a profession.

professional growth plan a teacher licensure renewal option in which the teacher sets goals for professional development with points awarded for attendance and participation at workshops, conferences, and seminars and for participation in other in-service activities.

proficiency award an award through the National FFA Organization to recognize a student who excels in skill development.

program advisory committee a committee that may serve to provide advice for the overall educational program; may consist of several subcommittees, sometimes called craft committees, that focus on specialty areas within agricultural education.

program components constituent elements of a program; the three main agricultural education components are classroom and laboratory instruction, supervised agricultural experience, and involvement with FFA.

program development the process of identifying and developing the necessary areas of formal instruction and other components in a local secondary school agricultural education program.

program evaluation an assessment of progress in achieving stated goals and outcomes of the program plan; effective evaluation includes the collection of necessary data, the comparison of that data against the goals and outcomes set forth in the program plan, and recommendations, complete with action steps and implementation strategies, for improvement.

program of activities (POA) the curriculum guide for FFA that works to develop leadership, citizenship, and career skills in students; provides activities for the individual member, the FFA chapter, and the school community.

program planning a process that documents all the activities needed to design and implement local school education.

progressive education focuses on students' extraordinary and unique experiences, interests, and abilities as the driving force behind instruction.

progressivism educational theory marked by emphasis on the individual child, informality of classroom procedure, and encouragement of self-expression; based on John Dewey's ideas of inquiry and the creative use of relevant theory in solving problems.

project an activity of educational value with one or more definite goals; advantages of projects include the opportunity to maximize student interest by allowing the student to design his or her own project and the ability to cater to the needs of the student.

project-based learning (PBL) an approach to learning and student engagement that provides opportunities for students to investigate an authentic and complex question, problem, or real-world challenge in a personally meaningful way.

promotional material a product or approach, such as a brochure or a newspaper article, used to elicit desired responses from people; an effective way to relay information to a target group.

Prosser, Charles (1871–1952); an elementary school teacher, principal, and school superintendent in New Albany, Indiana, who eventually became the secretary of the National Society for the Promotion of Industrial Education, a position from which he would contribute to the establishment of the Smith-Hughes Act; appointed director of the Federal Board for Vocational Education, which ensured the provisions of the Smith-Hughes Act were being carried out according to law.

protective clothing clothing that serves to reduce the risk of danger or personal injury; examples include gloves, headgear, chaps, and aprons.

proximity control the shaping of student behavior by being present in the learning environment; often the simple presence of the teacher will cause the students' behavior to improve.

psychomotor a domain of behavior marked by an individual's skill in manipulating something physically.

Public Law 116-7 passed in 2019 by the Congress of the United States to update the FFA federal charter and address issues in the previous charter. **Public Law 105-225** passed in 1998 by the Congress of the United States providing technical amendments to **Public Law 81-740**. PL 81-740, passed in 1950, recognized FFA as an integral part of agricultural education and granted a federal charter to the FFA.

punishment a penalty imposed to discourage socially unacceptable behavior; different acts of misbehavior call for different modes of punishment.

purchase order a form that requests expenditure of school funds that typically includes all information needed to place an order with a company including item numbers, item description, cost per item, number of items, and estimated shipping costs.

R

reactive a material that reacts with water or other materials; sodium is a reactive.

readiness level the status of a student's emotional, mental, behavioral, and physical state; knowing a student's readiness level allows a teacher to decide whether that student is able to complete a particular task or receive particular instruction.

realism the doctrine that the truth of an idea rests in its form; a philosophy that urges the thinker to seek the cause of a concept, on the belief that the true meaning of the concept lies in the cause.

reciprocity the recognition by one state of the validity of a teaching license issued by another state; does not relieve a teacher from state-specific requirements.

recruitment the process of seeking and soliciting students to enroll in agricultural education courses; this process consists of six key variables: the agriculture program, the recruitment program, student characteristics, parents, school support, and community support.

reference material that refers to something more advanced than students would normally use; may be found in such formats as books, manuals, DVDs, and online.

required courses courses needed to meet core requirements for graduation or admission to an institution of higher education; these courses are determined through local school board policy.

research and experimentation supervised agricultural experience the type of supervised experience in which a student carries out an investigation into a problem using scientific approaches and then makes recommendations about how to solve the particular problem; results are often exhibited at FFA agriscience fairs.

resource something that can be turned to for support or help.

resource people experts in a particular area who can provide knowledge and insights that the teacher may not have; are also valuable in lending credibility to information or instruction that students doubt or about which they have heard conflicting opinions.

restorative practice model of handling student misbehavior that involves the school community in the righting of wrongs committed. The school or classroom community is restored by performing service.

retention reenrollment of a student in agricultural education classes; leads to increased enrollment in the program and

to the student accumulating experience, knowledge, and skills relative to agricultural education.

reward a pleasant or satisfying consequence of an action.

role-playing the act of an individual assuming the real or imagined role of another individual; enables the individual to assume different perspectives and better comprehend different experiences.

S

safety a condition or state in which there is a reduced risk of danger or personal injury; safety must be a primary concern in agricultural education, as many items used for instruction in this program can be dangerous if misused.

Safety Data Sheets (SDS) contain information about hazardous materials in a consistent, easily understood format.

salary schedule a list or table that shows the salary at each experience level and educational level; increments are added for each year of teaching experience, though some schedules may not go beyond 15 or 20 years; the typical basis for teacher compensation.

SAMR model of technology integration model to help teachers make instructional decisions about integration of technology into their teaching.

scaffolding the help, prompts, and questions an adult gives a child to assist the child's cognitive development.

scenario short stories focused on a particular group of individuals that describe the nature of the individuals involved, the context, and the goals they want to achieve.

schema an abstract guide used to organize an experience or concept and assist in responding to it.

scheme an adaptable mental representation an individual forms of objects and events of the world.

school-based enterprise SAE the type of immersion SAE in which a group of students work cooperatively at the school. The enterprise may be a product or a service.

school community the students and employees in a particular school; on a larger scale, the students and personnel of all the schools in the entire school system.

school district a geographical area under the supervision of a given school board.

school management software a computer-based program used in schools to achieve a wide range of functions related to delivering education, recording data, and reporting information.

scope the extent, depth, or range of the curriculum; an important concept in preparing a course of study; for example, when determining the scope of a curriculum, one thing that will be decided is whether emphasis should be placed on a wide breadth of topics or whether fewer topics should be taught in greater depth.

scoring rubric a set of guidelines for judging student work; also called a rating scale.

secondary school typically a school with grades 7 through 12.

self-efficacy the belief that one can accomplish something and have a positive outcome.

sequence the order or progression of the material covered; sequence in agricultural education may be determined in any of several ways, such as from simple to complex or by production cycle; an important concept in preparing a course of study or a curriculum.

service-learning SAE the type of immersion SAE in which one or more students design one or more projects to benefit the school, community, or public organization. Service-learning SAE is not designed to benefit the FFA chapter.

Seven Cardinal Principles educational objectives put forth by the Cardinal Principles Report of 1918 that dramatically shaped secondary education; the principles are command of the fundamental processes, worthy home membership, health, vocation, citizenship, worthy use of leisure time, and ethical character.

sharps devices that have sharp points or edges that can cut skin.

simulation learning activities that are used to replace and amplify real experiences with guided ones in a controlled setting.

smartphone a mobile telephone with a computing platform that allows features such as internet and email capability far beyond telephone conversations.

Smith, Hoke (1855–1931); sponsor of the Smith-Hughes bill in the U.S. Senate.

Smith-Hughes Act act passed in 1917 that organized agricultural education so it would be available to every student who wished to study it; among other provisions, it provided funding to the states for the purpose of training and employing teachers in agricultural education, industrial arts education, and home economics education.

Smith-Lever Act act passed in 1914 that created the Cooperative Extension Service, a partnership between the federal government and the land-grant colleges established for the purpose of extending knowledge about the best practices in agriculture to rural communities.

Socrates (circa 469–399 BCE); the best known of the early Greek philosophers; creator of the "Socratic method" of teaching, also known as the dialectic method, in which truth and understanding were arrived at through a series of questions designed to challenge answers.

software operating and user programs controlling a computer system.

stackable credentials industry-recognized certificates, skills, and course attainments that build upon each other, leading to postsecondary education and/or entrance to a career.

stakeholder an individual or group that has a stake or strong interest in an enterprise or program, such as a local school; family members, taxpayers, and businesses all hold stakes in the quality, accountability, and overall educational impact of the local school.

standard a level of requirement; criterion; the expectation of what students should know and be able to do after completing a given area of instruction.

standard teaching license a license issued upon successful completion of an accredited teacher education program; additional requirements on both the national and state levels may include examinations, the submission of portfolios, and background checks; typically renewable every four to 10 years for teachers of agriculture.

Star Award top award for an FFA member who excels in supervised experience, FFA participation, and educational achievement.

State FFA Degree awarded by state FFA associations to students who have excelled in their FFA experience; the requirements for the State FFA Degree are set by each state association, provided those requirements are not inconsistent with the National FFA Constitution; there are also national requirements that a candidate for this degree must meet.

stereotype a conception or opinion resulting from the assignment of oversimplified characteristics to an entire group.

storage facility a secure location, such as a cabinet or locker, where materials and equipment can be inventoried, organized, and protected; different materials and equipment may require different storage facilities.

stream to interact with online audio and video content through an internet-connected device.

strengthening agriculture part of FFA program of activities that focuses on FFA chapter development, safety, agricultural advocacy, and agricultural literacy.

student advisement the process of offering assistance to students regarding their career and educational decisions or actions.

student-centered instruction a teaching method in which the processes of instruction are designed by the students and facilitated by the teacher; because the students create the processes themselves, they are more likely to be self-motivated to acquire the knowledge; the teacher, however, is ultimately responsible for determining what is best for the students.

student development the part of a program of activities that provides FFA members with opportunities to learn leadership and teamwork skills through participation in various training experiences.

student enrollment the total number of students within a school's agricultural education program; indicative of the impact of the agricultural education program within the school.

student teacher person in a preservice teacher education program who is placed in a cooperating school for an extended period of time under the supervision of a cooperating teacher; for much of that time the student teacher is the main instructor for the classroom, SAE, and FFA.

student-use material an instructional resource carefully written and otherwise prepared for use by students; examples include textbooks, activity manuals, computer-based modules, and record books.

summative evaluation evaluation occurring after instruction, often in the form of a grade; can provide feedback to determine the next level at which a student should be placed and what changes can be made the next time a topic is taught.

superintendent the one who has executive oversight and the authority to supervise and direct; the chief executive for the school system; provides for the preparation and administration of the school budget and the hiring of other school administrators, including school principals.

supervised agricultural experience the "student-led, instructor supervised, work-based learning experience that results in measurable outcomes within a predefined, agreed upon set of Agriculture, Food, and Natural Resources (AFNR) Technical Standards and Career Ready Practices aligned to a career plan of study" (National Council for Agricultural Education SAE guides, 2017, https://thecouncil.ffa.org/sae-resources/).

supervised agricultural experience program a series of individualized practical learning activities planned by the teacher, the student, the student's parents/guardians, and, if applicable, the employer; supervised by a qualified agriculture teacher; should develop competencies related to the interests and career goals of the individual student.

supervised study a method of instruction in which students are directly responsible for their own learning while under the direction of the teacher; can enhance student interest and develop problem-solving abilities.

supplementary projects (activities) specific projects or skills gained through experiential learning outside normal class time that add to the agricultural knowledge and competency of the student; these skills are not related to the major supervised agricultural experience and they are usually accomplished in less than a day.

supply a consumable material used to carry out instructional activities and/or work; supplies vary with the nature of the instruction and, depending on the field, may include potting soil, insecticide, metal, lumber, and oil, to name a few.

synchronous learning activities occurring live or in real time.

T

table of specifications a procedure used to chart test content, its relative emphasis, its educational level, and the types of questions used.

tablet computer a device with features of both a small computer and a smartphone, with an operation platform similar to that of a smartphone; apps can be used to achieve specific goals.

teach to impart knowledge or skill; to give instruction to.

teacher one who teaches, especially a person who is hired to teach.

teacher's manual a teacher-use material prepared to accompany a textbook and other student-use material; provides useful information, such as content summary and answers to questions in the student edition text, for the teacher to use in delivering instruction.

teacher-use material an instructional resource designed to aid the teacher in planning, delivering, and evaluating instruction; teachers' manuals, instructional resource guides, lesson plan libraries, and reference materials are examples of teacher-use materials.

teaching the art and science of directing the learning process; also a precept or doctrine.

teaching calendar a specialized teaching schedule that provides additional details on the time when units of instruction will be taught; the teaching calendar's increased specificity provides advantages over the basic teaching schedule, such as allowing a teacher to order supplies and materials, schedule field trips, and involve resource persons in a more timely and effective manner.

teaching license a certificate issued by a legally authorized state office documenting that an individual is qualified to teach and the conditions under which that person may teach; such conditions typically include subject area(s) and grade level(s).

teaching method the overall means, composed of a number of techniques and details, that a teacher uses to best facilitate student learning.

teaching schedule a plan showing the scope and sequence of content within an individual agricultural education course; typically includes such useful information as the broad units of instruction listed in the order in which they should be taught and the suggested number of class periods to spend on each unit.

technology-based development the use of curriculum databases, related matrixes, and technology-based access to assist in planning and developing the agricultural education program; improves the speed and ease of various processes, including standardization within a state and customization of curriculum to meet local needs.

tertiary system colleges and other institutions of learning after high school.

textbook a book systematically designed for use by students that deals with a specific subject, such as agriscience, horticulture, or landscaping; often contains certain characteristics, such as chapter or unit organization, motivational approaches, and suggestions for hands-on activities.

Thales an ancient Greek philosopher, from around 580 BCE, who found that all matter could be reduced to the quintessential element water, a finding that would later be debunked.

theorem a statement or proposition that can be demonstrated as a truth; often part of a larger theory.

theory of multiple intelligences the proposal by Howard Gardner (1983, 2008) that there are eight dimensions of intelligence: bodily-kinesthetic, interpersonal, intrapersonal, linguistic, logical-mathematical, musical, spatial, and naturalistic; puts forth the belief that everyone possesses intelligence in all eight dimensions, though most people excel in only a few.

tool a handheld device, such as a saw, that aids in performing manual or mechanical work.

toxic material a substance that is poisonous to humans, other animals, and plants; should be used carefully and stored in containers marked with a skull and crossbones.

traditional scheduling a class scheduling system that consists of six or seven one-instructional-hour class periods for five days per week for the entire school year; accordingly, students earn six or seven Carnegie units per school year.

training agreement a written statement, signed by the necessary parties, of the exact expectations, understandings, and arrangements of all parties involved in the supervised experience program.

training plan a written statement that documents the specific training activities in which the student is expected to participate; often based on the occupational competencies needed for the student to enter his or her tentative occupation.

training station the location on the job site where a student is placed as part of the supervised experience.

transfer goals what students should be able to do beyond the scope of the program

transition-to-teaching program a route to licensure for individuals who hold content degrees and have pursued careers in the content field; typically consists of intense education courses and an internship.

trauma-informed teaching requires teachers to anticipate unexpected behavior from students and respond to reduce the student's anxiety and stress; requires teachers to consider how trauma impacts student learning and behavior.

trimester scheduling an alternative class scheduling system that divides the school year into three 12-week trimesters; each class period may be 70 to 80 minutes in length, with students taking five classes each day; classes taken for two trimesters equal one Carnegie unit each.

True, Alfred Charles (1853–1929); director of the Office of Experiment Stations for the federal government and an early proponent of agricultural education in public schools; his work led to increased funding for agricultural education and to its continued growth in schools below the college level.

Turner, Jonathan Baldwin (1805–1899); a professor of classical literature at Illinois College, he brought the idea of agricultural colleges to American awareness; believed the federal government should pay for the establishment of universities to provide higher education in the arts and sciences.

U

unpaid placement supervised agricultural experience the type of supervised experience in which the student works in a job for experience only and is not compensated in any other manner for hours of labor.

V

vertical articulation articulation focused on preparing the student for higher grade levels; for example, students in grade 7 learn skills that will prepare them for grade 8, while students in grade 8 learn skills that will prepare them for grade 9.

virtual high school high school in which teachers deliver instruction primarily online and students are not physically located in one location.

Vocational Education Act of 1963 act that broadened agricultural education; revised the Smith-Hughes Act of 1917 by expanding agricultural education in the secondary schools to include a wide range of nonproduction agriculture, such as horticulture; opened the door for vocational training in other areas as well, such as business education.

voucher a system that allows parents to choose the school their children attend, whether public or private, and receive public funds through a voucher arrangement.

Vygotsky, Lev (1896–1934); a Russian psychologist who developed the sociocultural theory of cognitive development, a theory that emphasizes the role people play in a child's cognitive development; proposed that children develop through their interactions with adults; also proposed the "zone of proximal development."

W

webcam a camera that can capture live video and transmit it to the internet in live time.

work-based learning also known as supervised experience; a part of education that allows students to practice in a workplace what they have learned in the classroom or laboratory; a key component of agricultural education.

workplace specialist teaching license an alternative teaching license for an individual with substantial occupational experience but little or no advanced educational training; in many states, this license allows the teacher to teach only those courses that match his or her occupational specialization.

Z

zone of proximal development the level of proficiency in which students are unable to accomplish a task independently but can accomplish that same task with assistance.

Index

Page numbers in italics indicate boxes, figures, or tables.